Maintenance of Instruments & Systems

2nd Edition

Lawrence D. Goettsche, Editor

Maintenance of Instruments & Systems

2nd Edition

Lawrence D. Goettsche, Editor

Practical Guides
for Measurement and Control

ISA

Notice

The information presented in this publication is for the general education of the reader. Because neither the author nor the publisher have any control over the use of the information by the reader, both the author and the publisher disclaim any and all liability of any kind arising out of such use. The reader is expected to exercise sound professional judgment in using any of the information presented in a particular application.

Additionally, neither the author nor the publisher have investigated or considered the affect of any patents on the ability of the reader to use any of the information in a particular application. The reader is responsible for reviewing any possible patents that may affect any particular use of the information presented.

Any references to commercial products in the work are cited as examples only. Neither the author nor the publisher endorse any referenced commercial product. Any trademarks or tradenames referenced belong to the respective owner of the mark or name. Neither the author nor the publisher make any representation regarding the availability of any referenced commercial product at any time. The manufacturer's instructions on use of any commercial product must be followed at all times, even if in conflict with the information in this publication.

Printed in the United States of America.

ISBN-13: 978-1-64331-239-2

ISA
P.O. Box 12277
Research Triangle Park, NC 27709

Library of Congress Cataloging-in-Publication Data
Maintenance of instruments & systems / Lawrence D. Goettsche, editor.--
2nd ed.
p. cm.
Includes index.
ISBN 1-55617-879-4 (hardcover)

1. Engineering instruments--Maintenance and repair. 2. Systems engineering. I. Title:Maintenance of instruments and systems. II. Goettsche, L. D. (Lawrence D.)
TA165.M24 2004
681'.2'0288--dc22
2004019596

Table of Contents

About the Editor and Contributors

Lawrence (Larry) D. Goettsche, Editor of ISA's first (1995) *Maintenance of Instruments & Systems*, part of ISA's *Practical Guides Series*, has added considerable new information – and significantly updated other material in the 1995 publication – for this 2nd Edition.

Jim Tatera, owner of Tatera Associates, a process analytical systems consulting firm in Madison, Ind., authored Chapter 10, Maintaining Analytical Instruments. Mr. Tatera is a past director of ISA's Analysis Division; past chair of SP76, and technical advisor for IEC SC65D of the U.S. National Committee.

Diana Churchill Bouchard, a Scientist at the Pulp and Paper Research Institute of Canada who has lectured at McGill, Concordia University and Université du Quebec á Trois-Riviéres, contributed to Chapter 16, Software and Network Maintenance. **Robert Hasselbaum**, former Director, Services Marketing for Invensys Process Systems, Foxboro, Mass., contributed to the same chapter.

Mr. Goettsche has more than fifty years experience in instrumentation, measurement and automatic controls. Holder of a BS degree from New Hampshire College, he is owner of LG Process Automaton Services. During his career, he has been employed in the following capacities: Electrical and Control Designer for Kennecott Copper Environmental Engineering Projects Group, Electrical & Instrumentation Engineering at Roberts and Schaefer Company, Senior Systems Service Specialist at Johnson Yokogawa Corporation, Maintenance Engineer at Hercules Aerospace, Field Service Engineer at Leeds and Northrup Company, and Training Device Chief Petty Officer in the U.S. Navy.

Mr. Goettsche has authored and presented several articles on instrumentation, control system maintenance, fiber optics, and finding employment in the I&C field. He was the subject matter expert for the ISA Involve Interactive Training Program on *Interpreting Process Control Diagrams*. He is also a member of several ISA Standards Committees, including SP12.6, SP12.21, and SP5.1. Mr. Goettsche also was Editor of *The Maintainer*, newsletter for the ISA Maintenance Division, and *The Intermountain Transmitter*, the newsletter for the ISA Great Salt Lake Section. In Addition, he has held many positions and has been a member of the Executive Committee of the Great Salt Lake Section of ISA.

For the first (1995) *Maintenance of Instruments & Systems*, Mr. Goettsche was the sole author of the introductory chapter and chapters covering Fundamental Principals and Instrumentation and Control System Maintenance Personnel.

For this (2004) updated version, Mr. Goettsche also used much material authored by others for the 1995 book. Therefore, they also deserve credit.

Besides Mr. Goettsche, contributors to the 1995 *Maintenance of Instruments & Systems* included:

Lowell E. McCaw, president of ISA in 2004 who also served as one of Mr. Goettsche's technical reviewers for this book. For the 1995 book, Mr. McCaw – whose distinguished career includes considerable U.S. military service, professor emeritus at Monroe Community College and industrial instrument training consultant for Eastman Kodak – co-authored chapters on Maintenance Management and Tuning Details.

Dale E. Polmateer, a DuPont veteran and consultant/educator specializing in maintenance, instrumentation and controls, co-authored the chapter on Maintenance Management.

William Mostia, Jr., a longtime petrochemical industry engineer for Amoco. Previously, Dow Chemical and Texas Eastman, co-authored chapters on Maintenance Engineering and Configuration and Programming.

Keith D. Otto, co-founder and president of Otto Engineering Inc. of Neenah, Wis., and a pulp and paper industry veteran, co-authored the chapter on Maintenance Engineering.

Robert (Bob) E. Sherman, president of pulp and paper maintenance consulting firm USAFEMATIC Inc. who also served on Technical Advisory Boards of Dupont, Applied Automation Inc., and Precision Scientific Inc., co-authored chapters on Service/Contract Maintenance and Maintenance and Troubleshooting.

H. Pierce Rumph, an experienced electronics, instrumentation, process automation, and process optimization engineer, co-authored a chapter on Configuration and Programming.

Robert J. Rosenberg, then senior representative sales manager for Transmation Inc. with degrees in electrical engineering and business administration, co-authored a chapter on Calibration.

Robert Zielske, longtime chief instrument engineer for Georgia-Pacific Corp., Central Engineering, who also has held numerous ISA leadership positions, co-authored chapters on Calibration and Maintenance and Troubleshooting.

John Shaw, engineering consultant and owner of Process Control Solutions in Rochester, NY, whose more than 30-year career including working for Duke Power and Taylor Instrument Co., co-authored a chapter on Tuning Details.

Sheriff Ahmed, a Rosemount Inc. veteran who served as that company's Business Development Manager for China and held other business development, application support, and market development positions, co-authored a chapter on Safety.

1

Introduction

Overview

The *Maintenance* volume is key to the Practical Guides Series and certainly a key to the profitability of companies through ensuring that the control system is maintained so the plant can produce its products. This volume includes some history and speculates about future advances of instrumentation and control (I&C) system maintenance; it also covers some of the fundamental principles, vocabulary, symbolism, standards, and safety. It suggests the necessary basic knowledge required of I&C technicians and the interaction of maintenance in the retrofitting and start-up of control systems.

History of Instrumentation and Control Maintenance

From pneumatic instrumentation to computer-controlled systems — what a change! Is a seasoned instrument mechanic expected to troubleshoot a state-of-the-art computer-controlled system? Should a new instrument technician be expected to maintain pneumatic instrumentation? This volume documents experiences in the older types of systems as well as in the newer, state-of-the-art systems.

1930s

Distributed control is not new. In 1938, when *Chemical Processing* published its first issue, mechanisms for control were indeed distributed throughout the plant. Process control consisted of operator adjustments to hand valves that were based on direct readings of local gages. Control room instrumentation has taken some dramatic turns along the way — from large-scale pneumatic recorders to miniature analog electronic controllers to microprocessor-based digital systems.

Chemical and petroleum plants were among the first to use control systems for their processes. Pneumatic instrumentation became the leader in automatic control because of its safety. Pipe fitters were asked to perform maintenance on these early pneumatic instruments. In many cases, outmoded control room hardware is still operating effectively today — a tribute to the worldwide manufacturers of process control instrumentation.

In the late 1930s and early 1940s, operators relied on local instrument gages to monitor production processes. Control panels that did exist were located in the field near process sensing points. Typically, only a handful of indicators, recorders, and controllers were mounted on a local panel. Often, the process fluids were piped directly into control panels.

Where fill fluids were needed, mercury was commonly used. Control panels served as a convenient means for improving control coordination by allowing operators to adjust valves in response to visual instrument readings.

1940s

In the 1940s the use of pneumatic proportional controllers was increasing, so the early pipe fitters had to understand more of the theory of process and control. New words such as integral, derivative, sensors, and final control elements were added to their vocabularies.

By the late 1940s, a trend toward the concentration of controls in centralized locations had begun.

1950s

In the 1950s, operating unit control rooms were built to centralize operations and to accommodate operators assigned to monitor control boards on a full-time basis. With the growing number and complexity of the indicators, recorders, and controllers and the "need" to operate the plant remotely from these panels, the instrument mechanic was specialized to maintain the pneumatic control systems.

By the mid 1950s, electronic analog instrumentation had been formally introduced but did not win industry acceptance until the late 1950s and early 1960s. With the exception of chemical and petroleum plants, most new plants used electronic analog instrumentation because of the greater cost of tubing work between pneumatic transmitters and controllers and the expensive pneumatic auxiliaries, such as air compressors, filters, and dryers.

Increasing plant complexity necessitated increasing amounts of accurate, up-to-date operating information.

Now the instrument mechanic needed to know electronics and electricity in addition to pneumatics. Larger plants formed Electrical and Instrument (E&I), Instrument and Electronic (I&E), or Electrical and Control (E&C) groups; some formed an Instrument and Control (I&C) Group and had both instrument mechanics and instrument technicians. The knowledge required by I&C mechanics and technicians meant training was necessary, so vendors provided training on the equipment they sold.

1960s

Digital computers began to appear in control rooms in the 1960s. The computer's initial role was essentially that of a data logging device from which paper printouts could be obtained. However, the concept of direct digital control (DDC) gained notoriety in the 1960s.

1970s

By the mid 1970s, the drawbacks to DDC had become apparent. The central computer approach depended on the availability of a single large computer. Highly trained computer personnel were needed to maintain the computer hardware and to deal with the high-level software languages.

Single-loop analog control continued to flourish during the early 1970s. Thousands of electronic signal wires crisscrossed central control rooms, adding complexity to the pursuit of improved coordination. Recognizing multiple functions inherent in panel instruments, split architecture systems were introduced. Analog display stations were segregated from rack-mounted printed circuit cards in the quest for functional modularity.

I&C groups flourished, everyone was retrofitting and updating plants, and new plants provided more and more instrumentation requirements. Instrumentation vendors were training the instrument mechanics and electricians to maintain their equipment.

Standards for instrumentation were being developed, and manufacturers started listening to ISA when developing their new instruments.

A marriage between single-loop electronic analog control and pneumatic control developed because of the need for powerful control valve actuators.

The simplicity and accuracy of electronic controllers, recorders, and indicators made them the choice for instrument panels.

Current-to-pneumatic converters and pneumatic-to-current converters linked electronic instruments to pneumatic instruments and sensors and actuators. Chemical plants used pneumatic instruments in the hazardous areas along with signal wires to transmit the signals to central control rooms in safe areas.

Most plants built after the mid 1970s used electronic rather than pneumatic instrumentation. Pneumatic valves, however, are still used almost exclusively for throttling control and even on-off control. About the same time in this period Honeywell® and Yokogawa® introduced the first distributed digital control systems (DDCS), now called the distributed control system (DCS). Multiple minicomputers, geographically and functionally distributed, performed monitoring and control tasks that had been previously handled by the central DDC computer. Each microprocessor-based controller was shared by up to eight control loops. Serial bit communication over coaxial cable linked individual system devices.

As these distributed control systems became the standard for newer chemical and petroleum plants and the older single-loop pneumatic and electronic controllers were replaced, the I&C groups were trained on the new DCS. This was the first introduction of computers to the I&C technicians, and DCS manufacturers designed their systems to be configured and maintained by I&C groups — not highly trained computer personnel. As a technological breakthrough, the microprocessor accelerated advances in control system design. At the operator interface level, distributed control contributed to an unforeseen development. For the first time, CRT display consoles gained acceptance as the primary operator interface, and conventional single-loop analog stations were reduced to an emergency backup role at many early distributed control system installation sites. Long, floor-to-ceiling panelboards were replaced with low-profile CRT workstation consoles. Keyboards, CRTs and printers served as modern tools for seated control room operators.

By the end of the 1970s, control system innovations had advanced beyond industry's capacity to keep pace. Most plant sites contained an assortment of control technologies that spanned three decades. Instrumentation and control specialists (mechanics, technicians, and engineers) were commonplace in industry. Special I&C groups were established, as shown in the organizational chart of Figure 1-1.

1980s

DCS operator interfaces were refined in the 1980s (see Figure 1-2). Intelligent CRT stations utilized multiple-display formats to condense and organize extensive operating information. Hierarchical arrangements of plant-, area-, group-, and loop-level displays simplified on-screen database presentation. Real-time color graphics added further comprehensive overviews of unit operations.

Most microprocessor-based control systems had a vast array of alarms and diagnostics to help operators and maintenance personnel determine if there were any problems. Distributed control systems had many on-line and off-line diagnostics, including process and input alarms, reportable events, error messages, and hardware and software failure reporting.

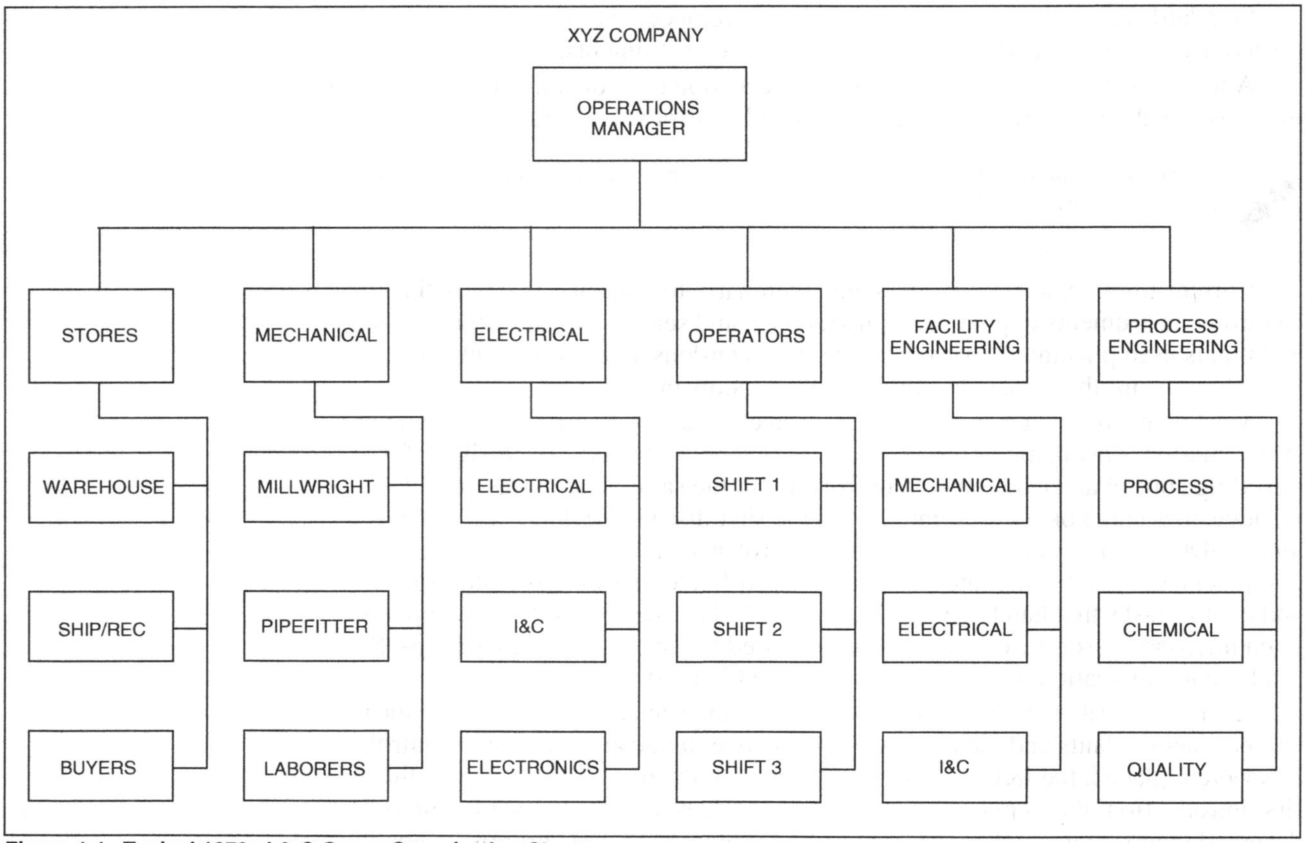

Figure 1-1. Typical 1970s I & C Group Organization Chart.

1990s

Trends for the 1990s were computer-integrated manufacturing (CIM) and management information systems (MIS). These interfaced the real-time devices (field devices at the machinery/process level) through distributed controllers to multiple-station coordination, then on to scheduling, production, and management information to the plant level for overall planning, execution, and control. Further development of artificial intelligence and expert systems gave advanced control new meaning.

With the introduction of computers and databases, maintenance management systems (MMS) helped maintenance and management personnel determine repair frequency and spare parts availability and made decisions on when to replace obsolete equipment.

Distributed control systems (DCS), programmable logic controllers (PLC), computer control systems (CCS), supervisory control and data acquisition (SCADA) and smart field devices were the norm. A digital signal was superimposed on the 4-20 mA signal for ranging and calibrating field devices. The International Organization for Standardization (ISO) Open Systems Interconnection (OSI) model and interconnection of devices made by different manufactures has opened systems architecture, replacing proprietary communications among devices.

2000s

Historically, factory floor maintenance methods and practices have been developed across a wide range of vertical industries, where the focus was to keep the assembly lines and processes running rather than preserving assets. Today, manufacturers are focused on the long-term benefits of factory floor support practices

Figure 1-2. Multiple-Display Distributed Control System.

that incorporate methods and procedures which ensure production lines are operational and preserve capital assets.

Skids and modular systems became the norm in the design of new plants. New gas electrical generating plants have been built from start to operational within a two year period. These plants are designed to be operated with a skeleton crew of 25 to 30 personnel, including operators, maintenance crew, and supervisors.

A crew of three operate and maintain in 12 hour shifts. Major overhaul periods are contracted to the system manufacturer, and contract maintenance is responsible for calibration. Knowledge of the complete plant, including operations and systems, are learned by all crews and supervision. Each crew member specializes in two or three systems.

A newer gas fired electrical generating plant organization chart is shown in Figure 1-3 which differentiates between maintenance and production. Because modern automation systems are installed, three units can be maintained and operated with 30 employees. Old coal-fired plants needed up to 200 people to operate them.

With the concept of skeleton crews to operate the plant, contractor type maintenance programs are becoming the norm. Many of the instrumentation tasks are completed by contract personnel. Work in the plant is becoming multi-disciplined.

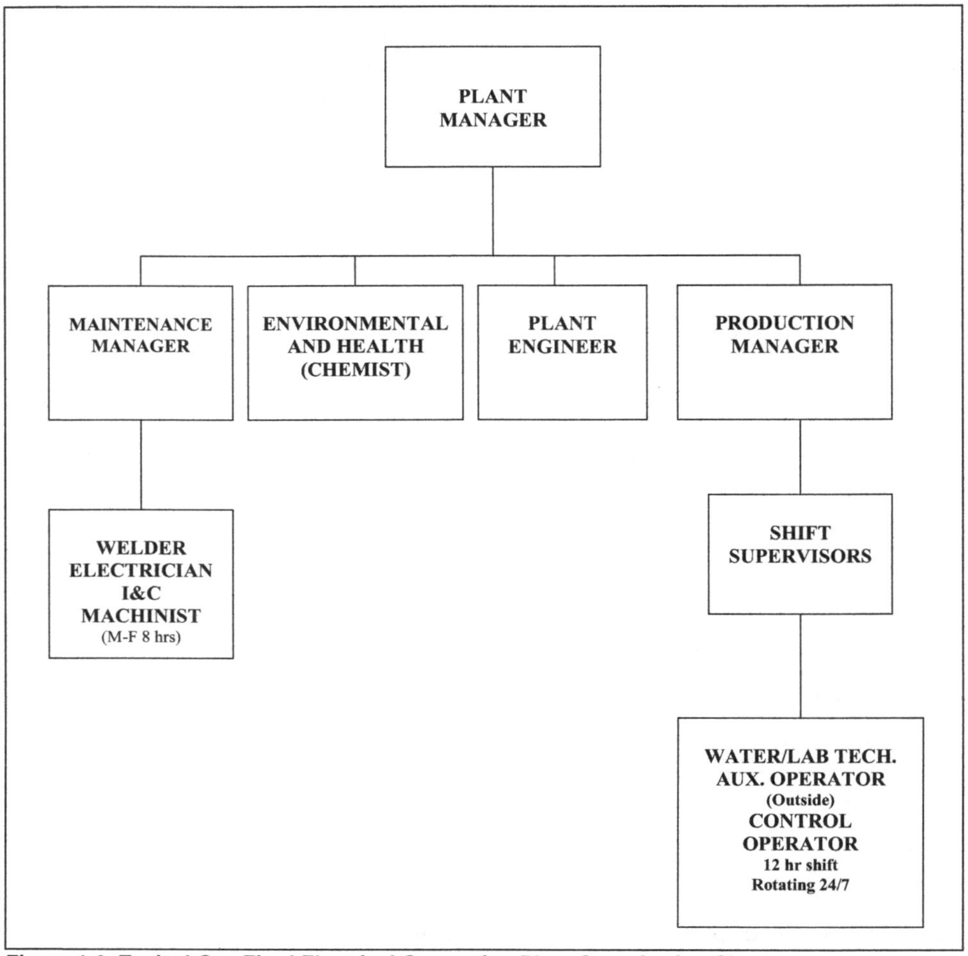

Figure 1-3. Typical Gas Fired Electrical Generating Plant Organization Chart.

Need for Instrumentation and Control Maintenance and Engineering

"Maintenance of instrumentation and process control systems from simple gages to complex distributed control systems is essential for the continuation of our industry." Statements such as this have been repeated thousands of times by company presidents, manufacturing directors, and production superintendents.

Maintenance personnel should be involved with new installations and upgrading older installations. They should ensure that the system is ergonomically easy to repair and well documented. Training should be done before a new system arrives so the maintenance department can help in installing and checking it out.

Equipment manufacturers provide engineering and start-up assistance. So the majority of the new opportunities to work in the I&C field is through original equipment manufacturers or service contract employees.

Because of the equipment's complexity, assistance is needed from the original equipment manufacturer. Configuration of control systems and instruments should be done by those very familiar with the system requirements and system/instrument capabilities.

Instrumentation tells us the process parameters in which we are operating. A simple gage tells the temperature or pressure; the more complex instrumentation tells much more about the process. Proper operation of all equipment is required to make a quality product and to do it safely.

The technological advances of the past few years and the trends for more technical and specialized equipment require better trained and educated maintenance personnel. The types of equipment in control systems cover many disciplines: mechanical, electrical, electronic, computer science, chemical, and environmental, among others.

The instrumentation and control field is more than electronics — it is a systems experience. It is necessary to know the physics of heat, light, noise, and mechanical advantage, as well as to have mechanical dexterity and aptitude, logical thought, computer literacy, process knowledge, and the ability to work with others in different disciplines.

Because of the many different knowledge factors, the individual crafts (electrician, mechanic, pipe fitter, etc.) have to work together, and finger pointing will sometimes occur. Electrical engineers, mechanical engineers, chemical engineers, and process engineers must understand each other and determine where their responsibilities start and stop.

The field has grown with the application of computers, artificial intelligence, self-tuning, computer-integrated manufacturing (CIM), and so on. Larger companies train pipe fitters to be instrument mechanics in pneumatic plants and electricians to be instrument technicians in electronic plants. Knowledge of the process is needed to design new systems; therefore, all engineering disciplines get involved with the instrumentation and control system. Those who were fortunate to get involved in early instrumentation and control systems have become the I&C maintenance personnel and the control systems engineers of today.

The complexity of control loops and systems requires specialists. The systems concept requires more varied knowledge and the overall concept of control rather than component troubleshooting and replacement.

When the control system doesn't work, the plant doesn't produce. The control system design can determine the profitability of a company. If it is maintainable and the mechanics, technicians, and engineers are trained, the production output of the plant will be high.

Corrective, preventive, and operational maintenance must be performed by qualified and experienced I&C maintenance personnel.

Because of the complexity of existing control systems that utilize many fields of expertise, several maintenance backgrounds are also required. This group is now required to maintain, troubleshoot, and calibrate pneumatic, electrical, electronic, and computerized instruments and systems. The systems approach, which looks at the whole picture to gain an understanding of the process, is the special attribute of I&C maintenance personnel.

When assistance is needed, I&C personnel must have someone to go to for help. In the past, maintenance supervisors had a broad knowledge of most of the equipment and could make decisions on how to repair, when to repair, and so on. A few years ago, many supervisors were instrument mechanics, but contemporary maintenance supervisors are managers who know very little about the operation and maintenance of the wide variety of instruments and control systems used today, since most have never been instrument mechanics or technicians. In fact, many of them know very little about pneumatics, electronics, or computers. Today, knowledge of the process, knowledge of the overall system, and knowledge of the expertise of their employees is far more important than knowledge of how to repair an individual instrument.

Who should the maintenance supervisors and managers go to for expert advice on the control system? Instrumentation and control system engineers or maintenance engineers with an I&C background. Instrumentation and control system engineers assist the mechanics and technicians and keep the supervisors and managers informed. They need to be a part of the design and start-up of the control systems.

Much money is being spent for training, fault tolerant systems, redundancy, and new techniques. One simple but essential area that may be neglected is the experience of the past and what that may teach about the present.

We learn from our past experiences. Being involved in the problems we encountered and the solutions that were found yesterday helps us make better decisions today. The learning technology that produces greater retention levels uses the most senses, such as hearing, seeing, and feeling. The applications of older systems should be used as the basis for designing newer and generally faster control systems. New problems are encountered in newer systems, but past application experience will help solve the new problems.

Don't neglect the knowledge and experience gained in the past.

Good maintenance saves money. With the equipment working properly, the process quality and production will be high. When equipment fails, production normally stops, and many production personnel cannot do their jobs. With good maintenance management, spare parts are available quickly to reduce the mean time to repair (MTTR). When the equipment is repaired properly, the mean time between failures (MTBF) is extended. The proper frequencies of preventive maintenance should provide less down time, and the down time that occurs can be scheduled. We can become pro-active instead of reactive.

2

Fundamental Principles

Overview

Maintenance of an instrument or system should ensure that it is operating within its normal operating parameters. Some describe maintenance as "fix what is broken." The preferred definition describes maintenance as activities that are intended to keep equipment in satisfactory working condition, including tests, measurements, replacements, adjustments, and repairs. This can involve preventive, predictive, and curative measures and requires technical knowledge of the system and a knowledge of past, present, and future states of the system as well as the skills to apply the knowledge. Some of this knowledge is very specific, while other knowledge is more generic but can be used to logically deduce more specific knowledge. The key is the acquisition, distribution, and utilization of the necessary knowledge and skills. Due to the many different types of instruments and systems available, basic technical knowledge and skills as well as specific knowledge and skills are necessary to maintain them. This chapter addresses some of this knowledge and skills.

Electronic Field Instrumentation

Electric transmission systems for in-plant use were developed as early as the mid-1930s. The two-wire, 4–20 mA current signal has since become the industry standard for in-plant transmission. The two-wire loop provides power for the transmitter circuitry, and current/feedback circuits within the transmitter force the current output on the loop to be proportional to the variable being measured. The operation, installation, calibration, and maintenance of a series of electronic field instruments—pressure, level, flow, temperature, and current-to-pressure converters will be covered.

Installation

Proper installation will ensure the instrument will meet the manufacturers specification. Always position the instrument in accordance with manufacturers specifications. Most liquid flowmeters must be placed where the pipe is full and not near any turbulent flow. Selecting a good ergonomic location of the instrument will make maintenance and troubleshooting easier.

Temperature specifications must be adhered to. Locating instruments in hazardous classified locations must be done in accordance with the National Electric Code.

Maintenance

Maintenance is any activity intended to keep equipment in satisfactory working condition, including test, measure, replace, adjusting and repair. Routine operations require keeping equipment functional.

Corrective maintenance includes locating causes of faulty performance, replacing defective components, adjusting service controls, or the like. It is a repair activity that is not normal to operating the equipment.

Preventive maintenance is intended to prevent faults from occurring during subsequent operation. Equipment maintenance is scheduled prior to failure, based on an estimate of the life of the equipment.

Predictive maintenance is a preventive maintenance program that anticipates failures that can be corrected before total failure. A variation from normal can indicate a system or equipment is approaching nonconformance. Vibration from normal can indicate a system or equipment is approaching nonconformance. Vibration, eccentricity, and noise monitoring are measurements that can predict failure. Also, an increase in diagnostic errors and retries can indicate a failure about to happen. Maintenance is performed when the need is indicated, based on regular or continuous monitoring of the equipment.

Unscheduled maintenance or non-scheduled maintenance must be done immediately when an emergency affects the productivity of the plant. This is maintenance to "fix what is broken" or provide a work–around so the operation that was started can be finished. Unscheduled maintenance is an urgent need for repair or upkeep that was unpredicted or not previously planned and must be added to or substituted for previously planned work.

Scheduled maintenance is planned. It includes preventive, predictive, corrective, modification, upgrading or retrofitting and calibration.

Information and its effective use are the keys to a good maintenance program. He who is not enlightened often blunders in the dark.

Why Maintain?

For most maintenance personnel the strongest motivation and the greatest satisfaction in their work arises from the job itself.

If one were in sales, keeping the customers satisfied would be an ever-present concern. However, most maintenance crews don't create a product or deal directly with customers, and they might feel that there is little reason to be taught good public relations. But maintenance does have customers to serve: the people within one's company.

In most manufacturing facilities, maintenance personnel rarely, if ever, come into direct contact with the company's customers. All employees, including management, are the maintenance department's customers, and they should be treated as such. It is important to maintain good customer relations no matter who the "customers" might be. The men and women in charge of different departments are, in effect, customers of maintenance. They must be pleased, and the work must be scheduled to be done largely at their convenience. Similarly, all the other employees in the building or plant are customers of maintenance; they are the "bosses." If one doesn't please these bosses, one may not have a job for long. Treat all "customers" with the respect and courtesy they deserve.

The function of maintenance is to keep equipment in a safe operating condition. It also must ensure that all instrumentation works when it is required to work. Emergency shutdown systems should be checked regularly as they must work every time. Improper maintenance could cause harm to personnel and the environment.

Safety standards seek to minimize the frequency and consequences of accidents related to toxic, flammable, or explosive chemicals in the U.S. processing industries. Procedures established for safety management are designed to protect employees of process plants by preventing or minimizing the consequences of chemical accidents that involve highly hazardous materials. Detailed written safety information must be compiled so that the employer and employees who operate the process can identify and understand the hazards associated with the process. This information must be communicated to all employees involved with the process. Safety information must include information on the hazards of the particular materials used in the process and detailed information on the process technologies and equipment used in the process.

Emergency shutdown systems must be understood, checked regularly, and *work when required.* The documentation must accurately represent the actual installation. Maintenance personnel must understand the control philosophy and must follow the specific maintenance procedures that are necessary to maintain equipment and system integrity.

An investigation of the various types of maintenance will emphasize its importance.

Corrective Maintenance

When a device, equipment, or a system fails to conform to the manufacturer's specification and the operation fails to meet its intended function, maintenance is required. All items must be repaired or restored to a serviceable condition following a failure or malfunction.

Corrective maintenance is an activity that is not normal in the operation of equipment and is expected to be performed by qualified personnel who are aware of the hazards involved. Such activities typically include locating causes of faulty performance, replacement of defective components, adjustment of service controls, or the like.

Maintenance personnel should have sufficient documentation to understand and repair the problem. After logical troubleshooting procedures have determined the faulty component, repair can be completed. Remember to check the operation and calibration after repair and document the repair activities and parts used.

Another name for corrective maintenance is remedial maintenance, which is defined as the maintenance performed following equipment failure, as required, on an unscheduled basis. Nonscheduled, unscheduled, or emergency maintenance is an urgent need for repair or upkeep that was unpredicted or not previously planned and must be added to or substituted for previously planned work — maintenance specifically intended to eliminate an existing fault.

Corrective maintenance is generally considered nonscheduled, unscheduled, or emergency maintenance if it affects the productivity of the plant. Normally, an emergency maintenance work order (see Figure 2-1) is written to document the parts and labor hours used and to be added to the history file. Corrective maintenance can be scheduled maintenance when the device or system does not affect the productivity of the plant.

Work-around should not be considered corrective maintenance, but it can eliminate emergency maintenance requirements. A work-around is action required to complete the process run, even though all equipment is not working satisfactorily. Part of the process may have to be run on manual.

A flow switch (a permissive) fails during a process run. Stopping the process when the failure occurs would cause product loss. In order to continue the run, the permissive input could be forced or the flow switch could be jumpered. When the run is complete, the problem must be corrected before another run is made.

Enclosure 2

MAINTENANCE WORK ORDER

____ Corrective ____ Preventive ____ Operational Priority ____ MWO # ____

Submitted by:	Department:	Phone Number:	Date:

For more information contact: Phone No.

Problem Loc: Bldg # ____ Room # ____ Unit ____ System ____ Loop # ____

Problem Description:

Planner Assigned:	Phone No.	Date Assigned:
Tech/s Assigned:	Shop/Phone	Date Assigned:

Corrective Action:

Instrument Data/Parts Required:

Manufacturer	Model	Serial Number	Part Number

Calibration Req: Yes No	Calibrated By:	Cal Rec #
Completed By:	Date:	OKd By:
Data Entry: By:	Date:	

Figure 2-1. Typical Maintenance Work Order.

Establishing a comprehensive maintenance plan is a way of formalizing and codifying the accumulation of all of the procedures and methods that are critical to maintaining the production systems. It represents a way of assessing the accumulated maintenance activities and capturing the best solutions. Every maintenance organization has some procedures and methods that qualify as a best practice. A maintenance plan that encompasses all production equipment and systems across all company plants and factories not only addresses the preservation of current assets, but provides a framework to organize best practice solutions and add others as they are recognized and validated.

Lowering the cost of business is a primary focus in manufacturing. Optimal production processes, lean factory operations, curtailed capital expenditures, and preservation of assets are manifestations of austere cost measures. Cost reductions have affected maintenance organizations with headcount reductions and scaled back training budgets. At the same time, maintenance staffs are under increasing pressure to keep equipment and automation systems up and running efficiently with a minimum amount of unscheduled downtime. Maintenance resources include people as well as specialized diagnostic equipment, software applications, communications, transportation, manuals, documentation, and access to supplier supported equipment.

Corrective emergency maintenance could generate a scheduled maintenance activity — for example, replacement of a whole unit or device to get the line or system working, followed by repair of the failed unit or device at a scheduled time. Care must be taken to ensure that the failed unit or device is either replaced or repaired for use at a later time. Some maintenance shops do not have a good program to ensure that boards and equipment taken out of service are either repaired and checked out immediately or discarded and replaced.

Corrective maintenance includes logical troubleshooting, which is a search for the cause of a malfunction or erroneous behavior, in order to remove the malfunction. Knowledge and experience are necessary to find the root problem. Qualified instrument technicians and instrument mechanics use logical troubleshooting concepts to ensure that the unit, device, or subassembly needs to be replaced and that the replacement corrects the problem.

Board or equipment swapping until the unit responds correctly is not considered acceptable troubleshooting practice.

Maintenance is not complete until the system has been repaired, reworked, or replaced and calibrated, tested, and, of course, logged into the equipment history files, as explained by the following paragraphs.

REPAIR

The word "repair" in a maintenance sense means to restore an item to serviceable condition following a failure or malfunction.

REWORK

Restoring an item to a condition that exactly conforms to original design specifications is called rework. This word is usually applied to corrective action taken when an item has failed an inspection but requires a relatively simple operation, such as replacing a part or component, to enable the item to pass an identical inspection.

REPLACEMENT

When an item fails or malfunctions, a decision must be made to repair or to replace it. If it is to be repaired, the proper maintenance work order and schedule must be submitted. If it is to be replaced, the proper material requisition must be submitted, and the malfunctioning item must be discarded if it cannot be repaired. The component or spare part may be replaced by a design equivalent replacement that is at least equivalent in performance to the item being replaced. It must also meet the requirements of form, function, and fit.

CALIBRATION

To ascertain outputs of a device that correspond to a series of values of the quantity that the device is to measure, receive, or transmit is called calibration. The output must be adjusted to bring it to the desired value within a specified tolerance, and the error must be ascertained by comparing the device output reading to a standard.

TESTING

Testing consists of checking out a system to determine the attributes or performance characteristics. The operators should verify that the malfunction has been removed and that no additional malfunctions exist.

RECORDING

The corrective action, parts used, and hours spent in repair must be recorded on the maintenance work order (see Figure 2-1).

INPUT

This information must be input into the maintenance management system (MMS) files. The service life of the equipment can be determined by the history files of the equipment in the MMS.

Service life is the length of time a mechanism or piece of equipment can be used before it becomes either unreliable or economically impractical to maintain in good working order.

Preventive Maintenance

Preventive maintenance (PM) consists of scheduled inspection and upkeep that is specifically intended to prevent faults from occurring during subsequent operation. Inept PM, however, can also cause problems.

Whenever any equipment is touched, it is exposed to potential damage. It is economically unwise to replace components prematurely. A good preventive maintenance program is the heart of effective maintenance, and inspection is the key to detecting the need for it.

A PM program requires an initial investment of time, parts, people, and money. Payoff comes months later. While there is little question that a good PM program will have a high return on investment, many people are reluctant to pay now if the return is not immediate, and some customers may perceive the PM activity only as, "That machine is broken again." PM supports a commitment of long-term life-cycle cost/total cost of ownership.

Most preventive maintenance is scheduled; that is, maintenance is carried out in accordance with an established plan. Scheduled maintenance or preventive maintenance may be scheduled by hours, uses, sequences, or the calendar. The use of performance intervals is itself a step toward basing PM on actual need instead of on a generality.

Inspection is a key to detecting the need for PM. It should be nondestructive so that it will not harm the equipment. The two main elements of fixed-interval PM are procedure and discipline. Procedure means that the correct tasks are done, the right lubricants are applied, and consumables are replaced at the best interval. Discipline requires that all the tasks be planned and controlled so that everything is done when it should be done.

Most organizations can establish a PM program based on the fact that certain equipment or systems can be taken out of operation for the time required to give them a good checkup. Sometimes this is not practical. Seasonal equipment, such as air conditioners and heaters, requires special maintenance care at the end of each season in order to clean and refurbish them when they are not being used so they will be ready for the next season. Failure patterns that have been developed show that equipment may fail within the first few weeks (infant mortality) and then will probably operate for quite some time before wearing out. Electronic components (once burned in) will last indefinitely until some outside force causes failure. A major cause of failures lies in the education and human reliability problems of the operators and maintenance personnel. Failures are frequently caused by persons who push buttons and turn on equipment before they read instructions.

PM records help determine the service life of equipment. Condition monitoring and prediction provide information on when problems are building to the failure point. If the scheduled interval indicates that the interval should be decreased, or if a failure occurs before the scheduled maintenance or preventive maintenance is performed, the maintenance supervisors and maintenance engineers should investigate. The investigation should determine a plan to ensure that the equipment will continue to operate until the equipment is replaced or retrofitted. Lead times of up to two years are frequently required to upgrade or retrofit a system. This information should be passed on to management personnel. If a malfunction is found during PM, plant policy will determine if corrective action should be accomplished as corrective maintenance or as a repair utilizing the Preventive Maintenance Work Order (PMWO).

Predictive Maintenance

Predictive maintenance is a preventive type of maintenance program that anticipates failures that can be corrected before total failure. Predictive maintenance can also determine that a failure is about to occur.

For example, vibration analysis determines the normal vibration that is acceptable; when the vibration starts to exceed this point, maintenance can be scheduled when the equipment is not being used in order to prevent a complete failure during operation. Noise, oil samples, additional heat, intermittent diagnostic errors, data highway retries, and unexplained glitches are other examples of warning signs for predictive maintenance. Most of the above can be measured, recorded, and annunciated.

Unscheduled Maintenance

Unscheduled or nonscheduled maintenance must be done immediately when an emergency affects the productivity of the plant. This is maintenance to "fix what is broken" or provide a work-around so that the operation that was started can be finished.

Unscheduled maintenance generally has to be acted upon quickly. Production pressures to hurry and correct the problem should not cause the maintenance personnel to shortcut safety procedures. Remember, maintenance is not complete until the system has been repaired, reworked, or replaced and calibrated, tested, and, of course, logged into the equipment history files.

Remedial Maintenance

Remedial maintenance (performed following an equipment failure) is generally unscheduled. Sometimes the item is repaired and restored to a serviceable condition, but not exactly to original design specifications, so that the operation can be completed.

Scheduled Maintenance

Scheduled maintenance is planned. It includes the following:

(1) Preventive maintenance, which is specifically intended to prevent faults from occurring during subsequent operation.

(2) Predictive maintenance, which uses special measurements to indicate equipment or a device is beginning to go out of specified limits and if not corrected may fail completely.

(3) Corrective maintenance, which can be scheduled when equipment is not being used.

(4) Modification, which incorporates design changes.

(5) Upgrading or retrofitting, which incorporates major changes to older systems.

(6) Calibration, which tests to certify measurements and makes adjustments accordingly.

Modifications, upgrading, and retrofitting may not be considered scheduled maintenance, but many of these functions are accomplished by maintenance personnel and are normally scheduled.

Operational Maintenance

Operational maintenance is activity, other than corrective, that is intended to be performed by the operator and is required in order for the equipment to serve its intended purpose. It includes the correcting of "zero" on a panel instrument, changing charts, adding ink, and the like. The level of operational maintenance should be well documented and agreed upon.

Troubleshooting

One of the goals in the establishment of a sound maintenance program was to have "no unscheduled down time." In order to meet this goal, a strong troubleshooting program must be established that gets to the root of a problem quickly and makes corrections that are lasting. Troubleshooting must be recognized as a critical skill and it must be continually developed and nurtured within the maintenance department.

Troubleshooting is defined as a method for finding a problem and correcting the problem. This even involves revisiting the problem area or equipment well after the problem has been resolved in order to be certain that the problem is going to remain solved. The following are some specific guidelines for the troubleshooter:

(1) The troubleshooter must use a clear and logical approach to finding the problem.

(2) The troubleshooter should work as quickly as possible to resolve the problem.

(3) The troubleshooter must work economically.

Logical analysis leads to some logical steps in the troubleshooting process:

(1) Verification

(2) Identification

(3) Repair

(4) Test

(5) Follow-up

Verification is the first step. It is necessary to be very sure that there really is trouble somewhere. This sounds so obvious, but often a trouble call comes in when there is no problem at all. Or a trouble call report points to the malfunction of a specific piece of instrumentation equipment when the equipment is fine but the process is at fault. A change in performance must be verified and the problem checked out before anything is done.

After assuring that something is wrong, it is necessary to *identify and locate* the cause of the trouble. This step calls for in-depth knowledge of the plant and thorough familiarity with the process. Generally, a process of isolation is used to track down the problem.

The next step in the process of troubleshooting is to *repair or fix* the problem. The problem should be corrected as quickly and as effectively as possible.

Once the problem has been corrected, it is necessary to *test* that the trouble has indeed been corrected. A thorough system check is made to be sure that everything is operating as it should. All interested personnel should be asked whether the problem has been corrected to their satisfaction.

The last step in the process is to *follow up*. This step might require some long-term study. Searching questions should be asked. What was the cause of the problem? Faulty engineering? Faulty installation? Operator error? Is some change in the process or equipment needed so that this trouble does not occur again? *Following up* also means revisiting the problem at some future dates just to be certain the problem is really solved.

Some other approaches to the troubleshooting process are the following:

(1) Equipment history

(2) Input/output (serial)

(3) Shotgun

EQUIPMENT HISTORY

Many times a problem is of the type that has occurred before. History repeats itself. The problem might be recognized from past experiences. Symptoms and similar problems, even with slightly different equipment, might well be recognized. The advantage of using this approach is that familiar problems can be solved quickly and with minimum down time. The disadvantage may be that the historical record could mislead the troubleshooter and, hence, valuable time could be wasted. Also, history generally does not apply to unfamiliar equipment: what may work for a previous generation of equipment may not work on the new version.

INPUT/OUTPUT (SERIAL)

This approach traces a signal through a loop or a piece of equipment. To use this approach one starts at one end of a loop and progresses through the loop from input to output, recording the signals and verifying as the progression goes on. Finally, the signal problem is located. The advantages of this approach are that (1) it works on all systems, and (2) it requires familiarity with the system. A disadvantage is that the approach can be very slow.

SHOTGUN APPROACH

This approach may also be called "pot luck" or intuition. There is no real rhyme or reason for the actions taken by the troubleshooter. Some mechanics know a process so well they can pinpoint problems almost before they occur. Skill and science are not really a part of the thought process here. However, it can work.

Troubleshooting charts are available for a variety of specific equipment, or they may be created by those who have knowledge of and experience with the equipment or system.

Calibration

Calibration is a test during which known measurement values are applied to a device and corresponding instrument readings are recorded under specified conditions. To ensure that the specifications are met, regular calibration cycles with records and reports are necessary. Most calibrations of instrumentation are performed by instrument technicians, and the maintenance department is responsible. Some organizations, however, have a separate calibration department. Chapter 13 gives details on calibrating instruments.

Since the definition of maintenance is "any activity intended to keep equipment in satisfactory working condition, including tests, measurements, replacements, adjustments, and repairs" and the accuracy of the instrument outputs affects the quality of the finished product, calibration is considered here as a part of maintenance.

The purpose of a calibration system is to provide for the prevention of tool or instrument inaccuracy through the prompt detection of deficiencies and the timely application of corrective action.

Instrument calibration requires accurate input signals and accurate measurement of the output, within the specifications of the instrument. In order to calibrate a temperature transmitter, a millivolt source representing a thermocouple or a decade box that represents an RTD must be accurate enough to provide an input within the specifications of the instrument. The output must be measured accurately enough to check the output specifications. Current or voltage is the normal output.

Input the value for the low end calibration, check the output, and record. Input the value for the high end calibration, check the output, and record. Check both points to determine if zero or span or both need adjusting. To determine the linearity of the instrument, input the value of several intermediate points and record. Repeatability should be checked by approaching several points repetitively from different sides of the point and ensuring the same results. If the measurements are within the instrument specifications, calibration adjustments should not be made.

The instrument calibration sheet will indicate when the instrument starts drifting out of calibration.

If the instrument is out of calibration, input the low end calibration point, measure the output, and adjust the zero calibration adjustment until the instrument reads the low end output value; then input the high end calibration point, measure the output, and adjust the span calibration adjustment until the instrument reads the high end output value. Repeat the low and high end calibration until there is no change in the adjustment. Then check a midpoint to determine if the linearity is proper. Record the final calibration data.

Every organization should prepare a written description of its calibration system. This description should cover the measuring of test equipment and standards, including the following:

(1) Establishment of realistic calibration intervals

(2) Listing of all measurement standards

(3) Establishment of environmental conditions for calibration

(4) Assurance of the use of calibration procedures for all equipment and standards

(5) Coordination of the calibration procedures for all equipment and standards

(6) Assurance that equipment is frequently checked by periodic system or cross-checks in order to detect damage, inoperative instruments, erratic readings, and other performance-degrading factors that cannot be anticipated or provided for by calibration intervals.

(7) Provision for timely and positive corrective action

(8) Establishment of decals, reject tags, and records for calibration labeling

(9) Maintenance of formal records to assure proper controls

After repairs or replacement of any component in a system, the calibration must be checked. In some instances, calibration that is traceable back to the National Institute of Standards and Technology (NIST) is required.

So, Why Maintain?

Good maintenance practices provide the long-term profitability of a company. Lowering overhead, ensuring equipment is within specifications to provide a quality product, increasing productivity by lengthening the mean time between failures (MTBF), and shortening the mean time to repair (MTTR) will make any company more competitive in the marketplace. Therefore, time must be scheduled for preventive maintenance, corrective maintenance (both scheduled and unscheduled), record keeping, and training. Parts control is necessary so the equipment is not down awaiting parts; thus, inventory must be controlled, as there is a cost of inventory (in dollars and space). Control systems engineers, instrument engineers, maintenance engineers, instrument supervisors, foremen, technicians, and mechanics must efficiently perform their functions to assure production stays on track (see Figure 2-2).

New systems or plants must be engineered for maintenance from the design stage through specifications and instrument selection, in start-up and checkout, in calibration, in setting up preventive maintenance programs, and in selection of spare parts. A few dollars saved on installation might cost many maintenance dollars annually in down time and upkeep after the project engineer's job is finished.

Maintenance vs. Troubleshooting

As has been previously stated, maintenance is any activity that is intended to keep equipment in satisfactory working condition, including tests, measurements, replacements, adjustments, and repairs. It could be calibration or cleaning; operational, predictive, preventive, remedial, or scheduled maintenance; testing, or tuning.

Place the instrumentation in easy reach for repair and calibration and provide adequate space to work around the equipment.

Troubleshooting is the process used to find the cause of a malfunction or an erroneous problem behavior in order to repair the malfunction. Troubleshooting can consist of measuring voltages, currents, resistance, pressures, temperatures, and so on, or it can be running on-line and off-line diagnostics on digital control

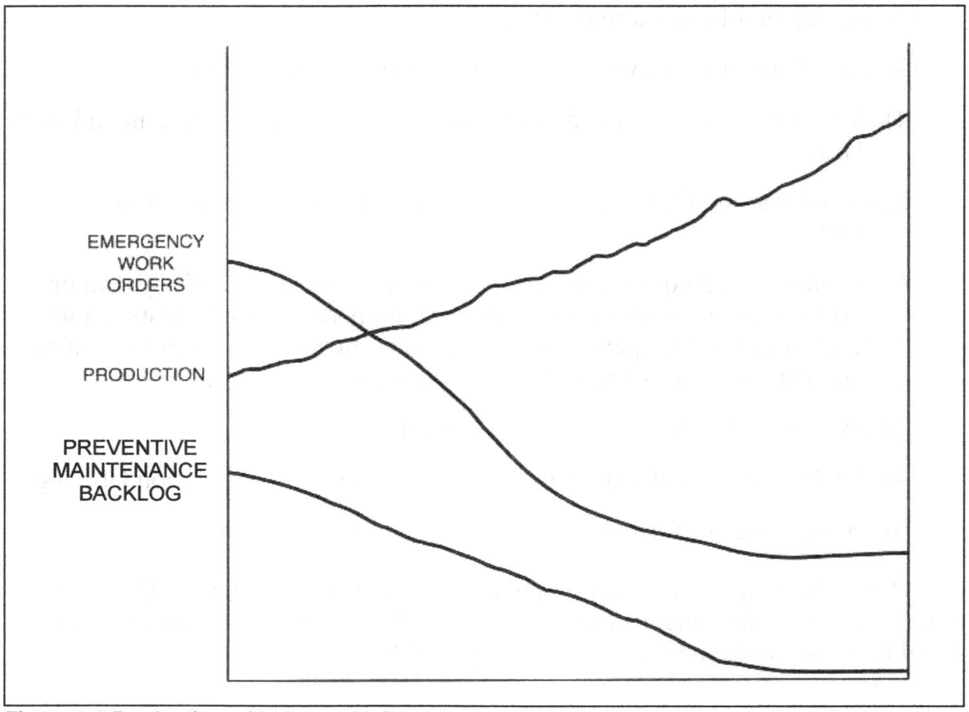

Figure 2-2.Production with Proper Maintenance.

systems. It can be visual checks for burnt or broken components, or it can be checking for blown fuses. After troubleshooting locates the malfunctioning component(s), it can be replaced. Eliminating the source of the trouble may also be considered part of troubleshooting, but it is usually called remedial maintenance or repair.

Calibration and Reasons to Calibrate

When one buys gasoline for the car, one expects to get exactly one gallon of gas when paying for one gallon. When one takes a child's temperature, one needs to be assured that the reading is correct. When the remote level indicator on a tank indicates that it is 90% full, can one be sure that the tank is not running over? Can the manufacturer be sure the process control system is providing the proper mix for the product? When a combustible mixture is being heated to just under the flash point, can one be sure that it won't ignite?

Continuing calibration assures that the equipment continually meets the specifications required at installation, and it should be checked regularly thereafter. Calibration after any maintenance ensures the equipment still conforms to the required calibration data. Customers require letters of conformance or certificates of calibration.

Calibration is a test during which known measurement values are applied to a device and a corresponding instrument reading is compared to a standard series of values the device was to measure, receive, or transmit. These known measurement values are usually obtained by using test equipment that has traceability to the National Institute of Standards and Technology (NIST), formerly the National Bureau of Standards (NBS). Data so obtained are used to determine the locations at which scale graduations are to be placed, to adjust the output to bring it to the desired value within a specified tolerance, and to ascertain the error by comparing the device output reading against a standard.

Chapter 13 gives particulars on procedures, standards, records, traceability, and calibration of specific types of equipment.

Vocabulary

Within every occupation, special meanings of words and phrases along with new words and special symbols are used to communicate among those individuals involved in the particular area of expertise. The word "instrument" may mean a musical instrument to a musician; it may mean an oscilloscope or a multimeter to an electronics technician; but to I&C personnel it is a device for measuring, transmitting, modifying, recording, or controlling the value of an observable attribute. It is a gage, controller, or recorder.

A "primary element" in chemistry or physics may mean a basic entity or element, but in the I&C world it is the device that first uses or transforms energy from a given medium or performs the initial conversion of measurement energy.

Within the I&C environment many words have more than one meaning. "To maintain" might mean "to keep equipment in satisfactory working condition," but it could also mean that an alarm remains on after removal of the signal until acknowledged.

New acronyms must be constantly learned, such as PID, P&ID, CRT, DCS, and so on, just as there are always new words and phrases, such as algorithm, process variable, and artificial intelligence.

The Automation, Systems, and Instrumentation Dictionary, definitions listed in ISA standards and recommended practices, and the definitions listed in Appendix A are helpful in understanding the vocabulary used by the instrumentation and control specialist.

Maintenance of an instrument or system should ensure that it is operating within its normal operating parameters. This can involve preventive, predictive, and other corrective measures. It also requires technical knowledge of the system and a knowledge of past, present, and future states of the system as well as the skills to apply the knowledge. Some of this knowledge is very specific, while other knowledge is more generic but can be used to logically deduce more specific knowledge. The key is the acquisition, distribution, and utilization of the necessary knowledge and skills. Due to the many different types of instruments and systems available, basic technical knowledge and skills as well as specific knowledge and skills are necessary to maintain them. This chapter addresses some of this knowledge and skills.

Troubleshooting

Information and its effective use are the keys to a good maintenance program. He who is not enlightened often blunders in the dark.

One of the most basic skills required to maintain an instrument or system is good troubleshooting. Good troubleshooting techniques can make the difference between timely, efficient repairs for one's department and a nightmare of long repair times with operations constantly on one's back to get things fixed.

Troubleshooting can be defined as the methods used to determine the cause of a system's failure to perform its appointed duties. The key word is "methods." Many different methods are used to troubleshoot problems, ranging from guessing to formalized troubleshooting procedures. All these methods work on occasion, and in some cases they may work for most problems. In many cases, troubleshooting skills are learned by "on the job" training. While, as with many skills, there is certain amount of "art" to troubleshooting and there are people who possess a talent for troubleshooting, troubleshooting is a trainable skill. Maintenance departments can enhance their skills by providing training in this area.

While there may be as many styles and methods of troubleshooting as people doing the troubleshooting, some basic techniques are the basis of all good troubleshooting methodologies. Some of these are discussed below.

Basic Troubleshooting Techniques

Logical Analysis

Troubleshooting is a systematic method of finding the cause of a problem in a system and correcting it. A good troubleshooting technique uses five general steps to safely, efficiently, and economically solve a problem.

VERIFY THE PROBLEM

The first step in troubleshooting is to verify that the problem exists. To determine whether to proceed with troubleshooting, efficiently gather information from every appropriate source. One valuable resource is the operator who monitors the system. The operator can provide insight into the system's performance. Another information source is historical data, such as maintenance records and trend charts. Consulting the system's control diagrams can offer an explanation of how the system should be operating. In addition, a loop diagram will indicate instrument power supplies, how instruments are connected, and other pertinent data about the intended operation of each loop.

An example of a system in which tank level is controlled appears in Figure 2-3. The level controller receives information from the level transmitter, then adjusts the control valve to maintain the tank level set point.

One troubleshooting scenario involves a level that has been dropping gradually, and yet the transmitter and indicators are accurate. Maintenance records show no previous problem with the system.

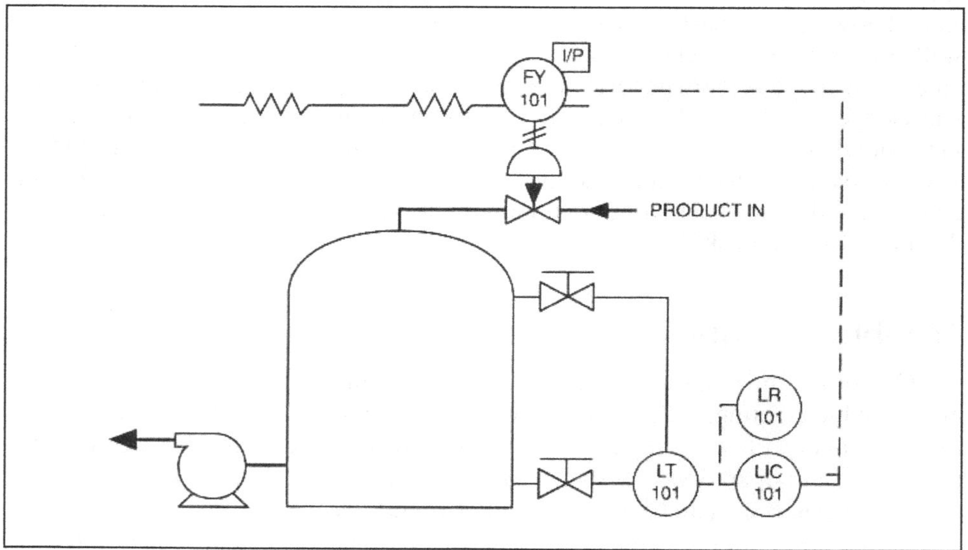

Figure 2-3. Logical Analysis Troubleshooting.

LOCATE AND IDENTIFY CAUSE

The second step in troubleshooting is to locate and identify the cause of the problem. In order to minimize the time that the process is upset, it is essential that this step be performed quickly. Begin by identifying a minimum number of possible causes based on the information gathered when verifying the problem. It saves time to perform the quickest checks of components first. One way to do this is to divide the system. Separate the portions that are operating correctly and concentrate only on the portion containing the problem.

In the system shown in Figure 2-3, all components except the control valve are operating properly. To identify why the control valve is not operating properly, the valve is placed in manual and stroked, and the I to P converter's output is checked. In this instance, a plugged filter in the air supply line to the I to P converter is the cause.

TAKE/RECOMMEND CORRECTIVE ACTION

After locating and verifying the problem, the next step is to take or recommend action to correct the problem. The choice of action will depend on the nature of the problem. In the level control system, replacing the plugged filter will likely remedy the problem. Corrective actions, whether taken or recommended, should always comply with facility procedures and guidelines.

VERIFY PROBLEM CORRECTED

Checking that the problem has been corrected is the fourth step of the troubleshooting procedure. Not only is this necessary to verify that the problem was solved, it also verifies that no other causes contributed to the problem. In the system where the plugged filter was replaced, stroking the valve again and verifying proper operation ensures the problem has been corrected.

Further troubleshooting is necessary if it is determined that the problem has not been corrected.

TAKE/RECOMMENDED FOLLOW-UP STEPS

The final step in troubleshooting is to take or recommend follow-up steps. There are two important elements of this step. The first is to recommend follow-up steps to prevent the problem from recurring. Replacing filters regularly is one measure that might be recommended for the level control system example. The other follow-up step is recording the problem and its correction to facility specifications.

Experience

This is the most simple method. It is simply that one has seen this problem before and knows what is wrong. This is an "on the job training" skill for the most part; however, good and accessible maintenance records can spread this skill around. Training given by experienced personnel can also spread this skill. A possible shortcoming of the method is that one may be treating a symptom rather than a root cause, i.e., one knows what is wrong but not why.

The Guess

This method is also known as "gut feel" or intuition. Sometimes looked down upon as not being logical or scientific, it works many times when other methods do not. While a guess can be just that—a guess—many times it is really a form of nonlinear thinking. The brain parallel processes the information known about the problem on a subconscious level. This can obviously be a very useful skill and, by exercising it, it can be developed.

Is/Is Not

This method approaches the problem by asking questions: "What is the problem?" and "What is not the problem?" It uses experience, training, and testing to isolate the problem into a more workable subset of the system to help isolate the cause of the problem. Many times this is done at an almost subconscious level, i.e., an instrument power light is on, therefore it IS NOT a power problem; the blown fuse indicator light is on, therefore it IS a problem related to a blown fuse.

Divide and Conquer

This method uses experience and training to divide the system into two pieces and check to see which section is performing correctly and which is not, then doing it again and again until the problem is solved. Sometimes the choice of location is one of convenience, i.e., where is it easiest to break into the system? When in doubt, break the system in half.

A 4-20 mA transmitter loop feeds an indicator at the control valve, a controller, and a recorder in the control room. The recorder and controller are not reading correctly. A check of the field indicator shows it is reading higher than the control room instruments. It is decided to break the loop at the transmitter utilizing the divide and conquer method. It is determined that the output of the transmitter is 16 mA when it is not connected to the other instruments. This is the same value that the field indicator is indicating but 8 mA more than the recorder and controller. It is then decided to break the loop at the field indicator. It is then determined that 16 mA are coming into the field indicator but only 8 mA appear to be leaving the field indicator. This apparent violation of the law of conservation of energy leads the technician to believe there is a ground loop. Replacing the field indicator solves the problem. Further examination of the bad field indicator shows that water had got into the indicator, and the resulting corrosion created a path to ground, thus creating a ground loop.

Remove and Conquer

In complex interconnected systems it is sometimes difficult to determine what box is causing the problem. A method to solve this kind of problem is to remove the boxes one at a time until the problem goes away. An alternate is to remove all the boxes and add them back one at a time to determine when the problem comes back.

The remove and conquer method was used in a system that had 16 digital DC motor drives that were talking to a VAX 3100™ computer via an RS-485/232 multidrop arrangement. The VAX was having regular but intermittent communication problems with the motor controllers. The communication link tested out OK, and it was suspected that one of the controllers was causing the problem by being an intermittent talker. To minimize the effect on operations, it was decided that each motor controller would be removed in turn from the communication link and run locally. This was done until the fourth one was removed and the problem went away. Replacement of the communication board on the controller solved the problem.

Walk-Through

This technique involves starting at one point in the signal path, testing that point, then going to another point downstream of the first point and testing, and so on. This is used if one does not know where to start the testing. Typically, it would start at the transmitter and go inward or at the controller and go outward.

Substitution

In this technique a known good component is substituted for a component in the system under test. This is commonly done in complex card-based systems. Sometimes a simulator is used to substitute for a device.

Vendor Troubleshooting Tables

These are tables provided by the vendor to troubleshoot their equipment. Generally, they are of limited use. They can, however, be supplemented by user-generated tables or trouble vs. solution tables.

Built-in Diagnostics

Most modern microprocessor-based equipment has built-in diagnostics. They don't always cover all the real-world problems one might encounter, but properly used they can be of great benefit. Some of this equipment is user-programmable. This gives users the opportunity to build in their own diagnostic routines. A little foresight in this area can greatly improve the maintainability of complex machinery. An example of this might be a PLC with additional diagnostic programming to provide troubleshooting information about the machine the PLC is controlling. Close-coupled rack-mounted DOS machines or state language machines can provide considerable flexibility in the programming for machine diagnostics.

The Second Head

Many times, in working on a tough problem, perspective of the problem can be lost. It is a case of not seeing the forest for the trees. At this time it is a good idea to bring in a second party to help out. This can be someone who is more knowledgeable or just another good troubleshooter. Many times this person does not actually solve the problem but rather serves as a catalyst by asking a few questions or discussing the problem, which often brings out a solution. The key is to apply appropriate resources to the problem but knowing when the point of diminishing returns has been reached and when more resources should be applied. Some people actually have a talent for being a "second head," i.e., asking the appropriate questions and having a different perspective, combined with good interpersonal skills. Once such people have been identified, they should be nurtured, as they are many times the key to the solutions of the most difficult problems.

Designed with Maintenance in Mind

Good troubleshooters are worth their weight in gold, but even average troubleshooters can do a good job if the systems they are troubleshooting are designed with testability and maintainability in mind. Too many times installations are designed for ease of installation and to meet operational requirements, with maintenance a distant third consideration. This happens many times because the design engineers have vested interests that are different from those of the maintenance department. The design engineer is generally evaluated on meeting budget and schedule, ease of installation, and assuring the installation meets operational requirements. The maintenance department is interested in keeping the equipment operating. A good working relationship with engineering and some joint accountability can go a long way in improving the life cycle of an installation.

Designed for Testability and Maintainability

In-house design projects should be designed with testability and maintainability in mind. The following are some of the concerns:

(1) Maintenance should be in the review chain of all engineering drawings before they are issued for construction.

(2) Adequate physical access for testing, removal, and replacement of components must be provided. No installation should require a technician with four-foot long tentacles with eyes in the end to work on the equipment.

(3) Adequate lighting for all areas should be provided where maintenance will be performed.

(4) Built-in test points should be provided, such as test diodes in current loops, test terminal blocks, test switches, etc.

(5) Maintenance-related indications should be provided, such as blown fuse indicators, status lights, meters, current indicators, and maintenance interfaces such as smart CRTs, LEDs, vacuum fluorescent displays, etc.

(6) The value of good tagging of instruments, wires, distribution panels, enclosures, fuses, terminals, etc., cannot be overestimated.

(7) Good maintenance documentation (loop drawings, motor schematics, one lines, ladder drawings, instrument spec sheets, all the vendor documentation, etc.) must be provided and kept up to date.

(8) Equipment and machinery maintenance records must be set upon installation and maintained throughout the equipment's life cycle.

(9) Good quality and appropriate test equipment must be provided. For example, looking for power transients with a multimeter is an impossible task, but it is easy with a power disturbance monitor.

(10) A good selection of spare parts must be available.

It is harder to enforce maintainability and testability requirements for equipment or systems designed and constructed outside the plant. The design and construction should be strictly monitored to ensure the equipment meets all the maintenance requirements. Maintainability and testability requirements should be part of the equipment design specifications.

Risks

Remember: Communication is always a two way-street!

The troubleshooting process starts from reports of ill-defined symptoms, which are then progressively refined until a source of the problem can be located and resolved. This process can be extremely challenging and potentially risky. For example, in nuclear power plants, the cost of both problems and errors can be extremely large. Consequently, managing and minimizing risks is an important aspect of troubleshooting. Some areas a troubleshooter must be concerned with are the following:

(1) Operational upsets that could incur economic cost or lead to safety problems.

(2) Plant shutdowns, safety system activation.

(3) Injection of a safety hazard as part of the troubleshooting process.

A smart CRT system was installed in a PLC installation that consisted of two PLCs with approximately 500 I/O. This CRT system was capable of graphics, alarming and printer outputs and talked to the PLCs through a serial communication link. In addition to displaying process information, the CRT had several maintenance displays. A maintenance technician was able to graphically view the status of al the PLC status registers. This provided the status of the PLC CPU, memory, local and remote I/O down to card level, communications status, register and bit values, etc. This, combined with the lights provided by the PLC, provided substantial maintenance knowledge concerning the system without ever having to go inside the PLCs. In fact, the design philosophy was that the maintenance displays should reduce the need to use a programming panel to determine the status of the PLCs by at least 95%. This was accomplished and was definitely a hit with the maintenance personnel.

(4) Bypassing safety systems as part of troubleshooting.

(5) Component or system damage.

(6) Errors introduced as a result of troubleshooting.

Communication

Because the systems that are being troubleshooted are connected to plant operations, communication with the plant operations personnel is required and good communication is many times critical to solving the problem. Sometimes it is also important to communicate with other maintenance and engineering personnel who might be involved.

The troubleshooter should communicate with the operations personnel to ensure that the system is safe to work on and in a state that minimizes the impact of the troubleshooting on the process.

Remember communication is an ongoing process not a one-shot deal!

Initial Steps

The initial steps in the troubleshooting process involve gathering information concerning what happened or did not happen versus what should have happened or not happened. This can involve a verbal or written description which includes location, symptoms, timing, magnitude, past history, damage, etc. A visual inspection is performed. Information concerning the system configuration and design (loop drawing, wiring drawings, motor schematics, system drawings, vendor prints, etc.) must also be reviewed. This information can then be used to determine the method(s) and the entry point that will be used to begin to troubleshoot the problem.

Safety

Troubleshooting must be done with safety in mind. Many times the systems that are being troubleshot were not designed with safety in mind for the maintenance personnel. A maintenance person should not, nor should they have to, perform an unsafe act to troubleshoot a piece of equipment. This should be a design requirement from day one. Maintenance procedures should stress safe practices and leaving the equipment after maintenance at least as safe as it was before the maintenance.

Poor maintenance safety practices lead to unsafe systems. Long-term practice of this as well as "poor boying" repairs leads to a maintenance environment where an accident is just waiting to happen. It could happen to you!

Some of the safety hazards typically encountered are electric shock, entanglement in equipment, burn hazards, sharp edges or points, head knockers, hazardous or toxic process materials, etc. Many of these hazards can be minimized by two techniques—guarding and identification. Guarding means placing a guard over the hazard. Where designs lack guarding, maintenance personnel should have them installed. Maintenance personnel should always re-install guards after they are done with them. Identification of hazards such a signs indicating high voltage or high temperatures can indicate to the maintenance person that a specific hazard is present.

It is good practice for operations to issue a "Ready to Work" permit and to tag appropriate equipment for any work that directly involves process materials or equipment. Necessary "Hot Work" permits should also be issued in regards to work in hazardous areas. Lockouts per OSHA (Occupational Safety and Health Administration) and NEC (National Electrical Code) requirements as needed should be performed on appropriate equipment prior to beginning work.

Before beginning to troubleshoot a problem, the troubleshooter should always get with the operational personnel to get the system in a state that is safe to work on and minimizes the impact of the troubleshooting on the process. The troubleshooter should never proceed unless he is absolutely sure that it is safe to proceed.

On-Line Troubleshooting

On-line troubleshooting (which involves interaction with the loop) always runs the risk of upsetting the process, which can lead to a safety hazard. Many times the troubleshooting of an instrument loop is done on-line. Steps must be taken to ensure that this can be done safely. Before starting the troubleshooting, the actions planned should be reviewed with the operator in charge of the loop. The operator should stand by to ensure that these actions are done safely. Any changes should be done in small steps. If possible they should be done while the loop is in manual. Care should always be taken to ensure that one is troubleshooting the right loop components.

An instrument technician was working on a temperature control loop and it was decided to put a run up box at the field connection of the thermocouple. However, when the technician went into the field, he disconnected the high temperature shutdown thermocouple instead of the control thermocouple and shut down the reactor. Having the operator identify the proper thermocouple could have prevented this.

Always double-check that you are working on the correct instrument, loop, circuit, and so on.

Any on-line modifications for maintenance purposes must be documented. This is particularly important for software changes or when instruments such as pressure, alarm switches, etc., are bypassed. On-line programming should be done only with great care. All software "forces" should be removed when the work is done or clearly identified that they are in effect. Physical bypasses should always be removed or clearly identified if left in place.

A section of a plant that had a complicated machinery interlock system was getting old and had not been well maintained. As problems arose involving process interlock switches over the years, many had been jumpered out when they became maintenance problems. This led to an interlock system that no one understood and that did not protect the machinery properly. Finally, a problem occurred in a small cooling pump that was not detected because the cooling water flow switch had been jumpered out. This caused major damage to one of the machines.

Failure to clear bypass jumpers sometimes occurs even with the best of intentions. The jumper is put in, the tech goes off shift, and another tech takes over and fails to remove the bypass jumper. A PLC contact is forced and the tech fails to un-force it when the work is done because it is not obvious that the contact is forced and so on. Maintenance procedures should be in place to identify and tag maintenance bypasses and to ensure that they are cleared. Tags on the equipment, special color wire for physical bypasses, and entries in operator and maintenance logs are some of the ways to minimize the problem.

Temporary bypass jumpers should be a color not normally used in your wiring so that bypass jumpers can be easily identified and removed after you have completed your troubleshooting. Tags attached to the jumpers can also be used.

If instruments are hot swapped, care should be taken to ensure that the replacement instrument has all the settings and calibration of the original instrument. Also, if the part that is hot swapped is part of a larger assembly, it should be remembered that configuration settings are typically scanned only on start-up and when reset and some systems may know if the card has been removed. Cycling of the power or resetting may be necessary to get the system to recognize the new card.

3

Diagrams, Symbols, and Specifications

Overview

Documentation comes in many forms, from general layout drawings to detailed schematics. It also includes parts lists, the theory of operation, specifications, purchase order data, preventive maintenance procedures, software configurations, training manuals, and any other helpful documentation. Process (piping) and instrumentation diagrams (P&ID), loop diagrams, and the equipment manufacturers' documentation will be the documents most used by the maintenance department. Some companies call the P&IDs "piping and instrumentation diagrams"; others call them "process and instrumentation" or other names, but the format is the same.

Good documentation means up-to-date information and as-built drawings. Improper revisions of the documentation can cause a bigger problem than a limited amount of good documentation. It is in maintenance's own interest to see that if any changes are made to the process or its controls, the respective P&ID and loop diagrams are revised immediately. Poor documentation can also be a safety hazard or a product loss. Maintenance is generally done during operation on most continuous processes and sometimes in the middle of a batch process. If improper documentation causes the wrong loop to be upset and the process goes into a critical stage, the product run could be lost or a hazardous malfunction could occur.

New to the maintenance department is the requirement to understand some software documentation. Some of this documentation is listed in a database and is not listed in manuals or on hard copy.

Diagnostics help maintenance personnel in locating the problem. In some cases diagnostics tells what card to change or what component on a card to change. The manufacturer's documentation is very helpful to one's understanding of the diagnostic codes and indicators used in larger distributed control systems.

Good documentation is the most beneficial asset in maintaining a facility.

Process (Piping) & Instrumentation Diagram

The process and instrumentation or piping and instrumentation diagram (P&ID) is probably the document most used by instrumentation and control system maintenance personnel. It gives an overall view of the entire process loop of the facility. The P&ID uses the symbols listed in ISA–5.1–1984–(R1992), Instrumentation Symbols and Identification and ISA–5.3–1983, Graphic Symbols for Distributed Control/Shared Display Instrumentation, Logic, and Computer Systems and shows the interconnection of process equipment and the instrumentation used to control the process. In some cases the analog controls are too complex to show on P&IDs; in this case, separate analog functional drawings are created, using ISA or SAMA symbols. (The Scientific Apparatus Makers Association has withdrawn and no longer supports its standards. They are available for historical purposes and general information only.) The process equipment is shown by symbols such as bubbles (circles or balloons) and others to denote valves, actuators, flowmeters, and so on.

Each instrument is represented on diagrams by a symbol, along with tag numbers and instruments or functions identified by alphanumeric codes. See Table 3-1.

The instrument loop number may include coded information such as plant area designation. It is also possible to set aside specific series of numbers to designate special functions; for instance, the series 900 tag numbers to 999 could be used for loops whose primary function is safety-related. Therefore, the type of symbol associated with its tag number gives maintenance personnel an idea of the function of the instrument, its loop identification, and its location (see Figure 3-1).

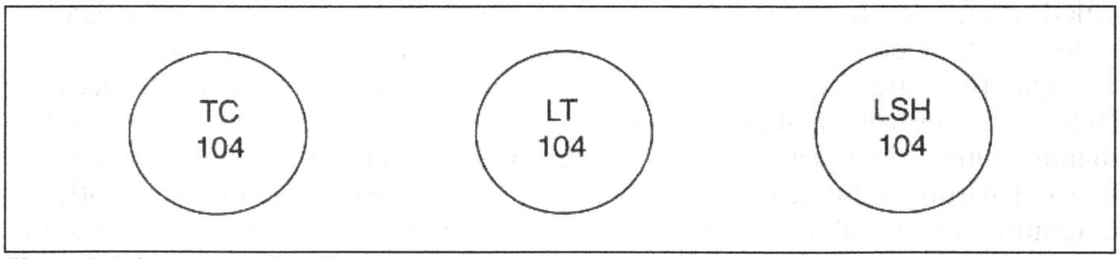

Figure 3-1. Instrument Tag Number.

Instrument Loop Diagrams

The process (piping) and instrument diagram applies to the whole process loop, while the loop diagram gives further information on the control loop of an individual parameter. An instrument loop diagram must contain the information needed to understand the operation of the loop. It also must show all connections to facilitate start-up and maintenance of the instruments. The loop diagram must show the components and accessories of the instrument loop, highlighting special safety and other requirements (see Figure 3-2).

Instrument loop diagrams are used for installation and troubleshooting.

USES

Some uses of an instrument loop diagram are listed in the chronology of project development:

(1) Engineering—Illustrate control philosophy and confirm the completeness of submitted data; an extension of P&ID, which shows the components and accessories of the instrument loop, connections between devices, and identification of component action; the specification of instrument hardware items and a means of communicating requirements to vendors.

(2) Construction—Initial construction of panels, field installation of instrumentation, instrument connections, instrument loop checkout, inspection, and documentation.

(3) Commissioning and Start-up—Prestart-up checking and calibration; as an extension of the P&ID for commissioning and start-up personnel, they highlight safety and other requirements; as an extension of the P&ID, they can also be used as training tools.

Table 3-1. Identification Letters (Source: ISA–5.1–1984 (R1992))

	FIRST-LETTER		SUCCEEDING-LETTERS		
	MEASURED OR INITIATING VARIABLE	MODIFIER	READOUT OR PASSIVE FUNCTION	OUTPUT FUNCTION	MODIFIER
A	Analysis		Alarm		
B	Burner, Combustion		User's Choice	User's Choice	User's Choice
C	User's Choice			Control	
D	User's Choice	Differential			
E	Voltage		Sensor (Primary Element)		
F	Flow Rate	Ratio (Fraction)			
G	User's Choice		Glass, Viewing Device		
H	Hand				High
I	Current (Electrical)		Indicate		
J	Power	Scan			
K	Time, Time Schedule	Time Rate of Change		Control Station	
L	Level		Light		Low
M	User's Choice	Momentary			Middle, Intermediate
N	User's Choice		User's Choice	User's Choice	User's Choice
O	User's Choice		Orifice, Restriction		
P	Pressure, Vacuum		Point (Test) Connection		
Q	Quantity	Integrate, Totalize			
R	Radiation		Record		
S	Speed, Frequency	Safety		Switch	
T	Temperature			Transmit	
U	Multivariable		Multifunction	Multifunction	Multifunction
V	Vibration, Mechanical Analysis			Valve, Damper, Louver	
W	Weight, Force		Well		
X	Unclassified	X Axis	Unclassified	Unclassified	Unclassified
Y	Event, State or Presence	Y Axis		Relay, Compute, Convert	
Z	Position, Dimension	Z Axis		Driver, Actuator, Unclassified Final Control Element	

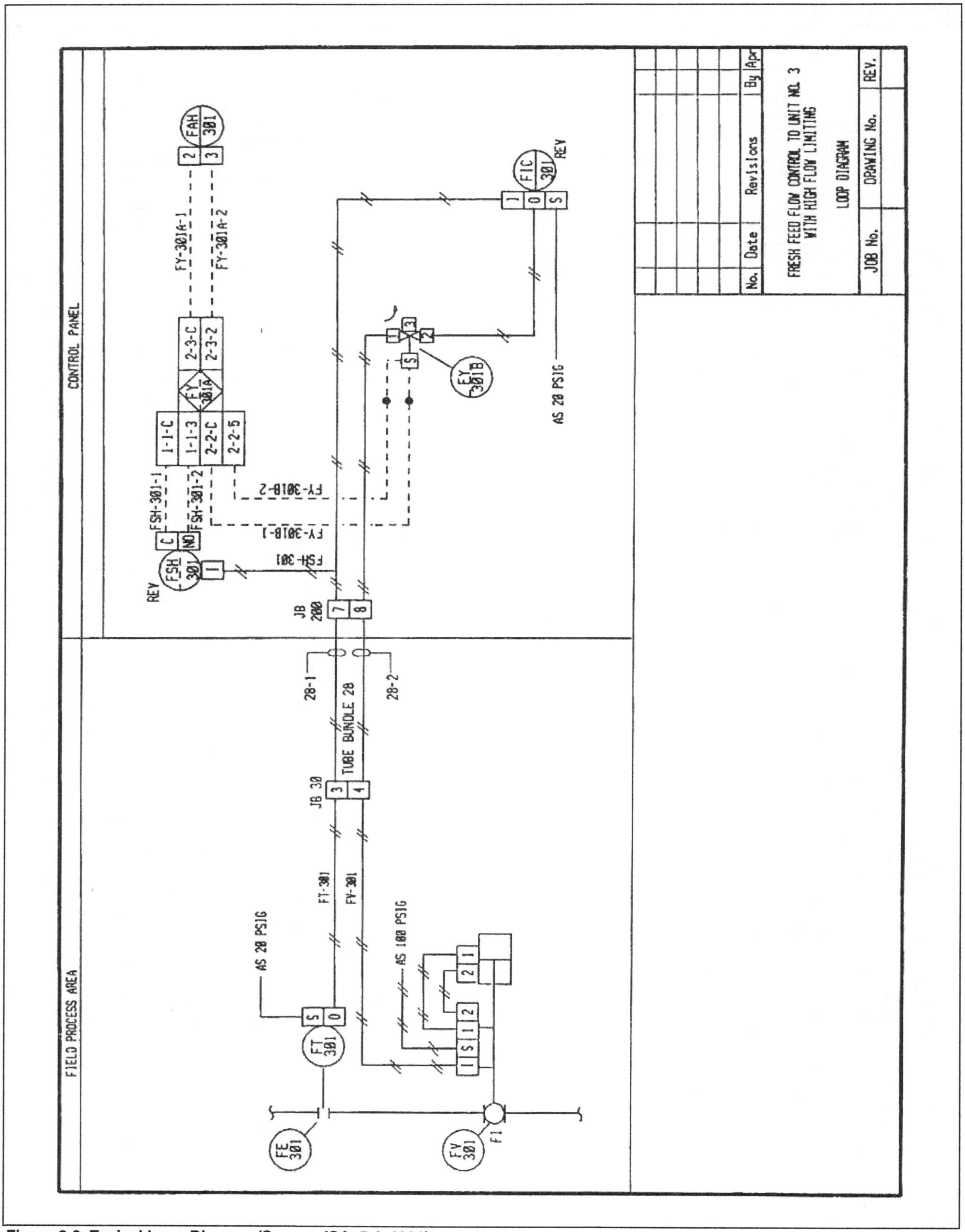

Figure 3-2. Typical Loop Diagram (Source: ISA–5.4–1991).

(4) Operation—Communications medium between operations, maintenance, and engineering personnel; operations training device.

(5) Maintenance—Troubleshooting, routine calibration, preventive and corrective maintenance tool.

(6) Modification—Rearrangement, reconstruction, enhancement.

CONTENT

The instrument loop diagram is a composite representation of all associated instrument loop information. It contains all associated electrical and piping connections and should contain all the information needed to accommodate the intended uses. Instrument loop diagrams use the same symbols and identification used in process (piping) and instrument diagrams. All tagging or numbering must agree with the P&ID.

To accomplish minimum needs, a loop diagram shall contain at least the following information:

(1) Identification of the loop and loop components shown on the P&IDs. Other principal components of the loop to be shown and identified under ISA–5.1, Instrumentation Symbols and Identification.

(2) Word description of loop functions within the title. If not adequate, use a supplemental note. Identify any special features or functions of shutdown and safety circuits.

(3) Indication of the interrelation to other instrumentation loops, including overrides, interlocks, cascaded set points, shutdowns, and safety circuits.

(4) All point-to-point interconnections with identifying numbers or colors of electrical cable, conductors, pneumatic multitubes, and individual pneumatic and hydraulic tubing. This identification of interconnections includes junction boxes, terminals, bulkheads, ports, and grounding connections.

(5) General location of devices such as field, panel, auxiliary equipment racks, termination cabinet, cable spreading room, I/O cabinet, etc.

(6) Energy sources of devices, such as electrical power, air supply, and hydraulic fluid supply. Identify voltage, pressure, and other applicable requirements. For electrical sources, it is desirable to identify circuit or disconnect numbers.

(7) Process lines and equipment sufficient to describe the process side of the loop and provide clarity of control action. Include what is being measured and what is being controlled.

(8) Actions or fail-safe positions (electronic, pneumatic, or both) of control devices such as controllers, switches, control valves, solenoid valves, and transmitters. These are to be identified in accordance with ISA–5.1, Instrumentation Symbols and Identification.

It is desirable that loop diagrams contain the following additional information (see Figure 3-3):

(1) Process equipment, lines, and their identification numbers, source, designation, or flow direction.

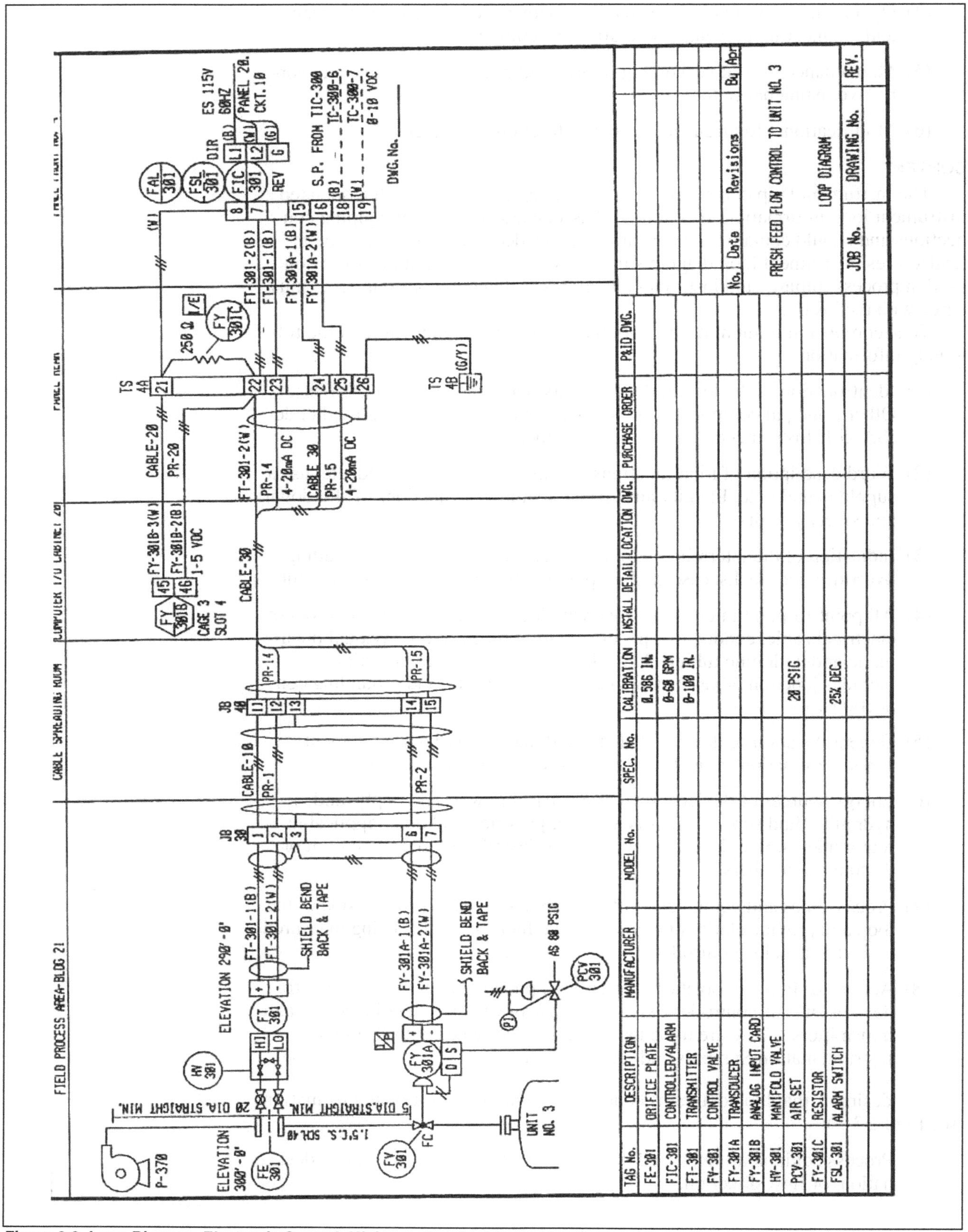

Figure 3-3. Loop Diagram, Electronic Control, Minimum Required Items plus Optional Items (Source: ISA–5.4–1991).

(2) Reference to supplementary records and drawings, such as installation details, P&IDs, location drawings, wiring diagrams or drawings, and instrument specifications.

(3) Specific locations of each device, such as elevation, area, panel subdivision, rack or cabinet number and location, I/O location, etc.

(4) Cross reference between loops that share a common discrete component, such as multiple recorder, dual indicator, etc.

(5) References to equipment descriptions, manufacturers, model numbers, hardware types, specifications, or data sheets, purchase order numbers, etc.

(6) Signal ranges and calibration information, including set point values for switches and alarm and shutdown devices.

(7) Software reference numbers, such as I/O addresses, control block types and names, network interfaces, point names, etc.

(8) Engraving or legend information that helps identify the instrument or accessory.

(9) Accessories, tagged or otherwise identified, such as regulators, filters, purge meters, manifold valves, root valves, etc.

(10) References to manufacturer's documentation such as schematics, connection details, operating instructions, etc.

(11) Color code identification for conductors or tubes that use numbers for differentiation.

The location of the instrument on the instrument loop drawing represents its location in the facility. Generally, the left side of the drawing is near the process and the right side is the operator's window in the control room. The center sections of the loop diagram are other connection points between the sensor and the control room.

FORMAT

A loop diagram should be a small drawing that is easy to handle, preferably 11 in. × 17 in. A loop diagram should normally contain only one loop. Avoid showing a loop on multiple pages or sheets, where practical. Judgment should be used to accommodate the individual situations in which loops that share common components can be adequately and completely communicated on a single diagram. Overcrowding should be avoided and space provided for future additions and loop data.

SYMBOLS

The symbols used for P&IDs are suitable for loop diagrams but are expanded to include connection points, power sources (electrical, air, hydraulic), and instrument actions and ranges to clarify certain connection and operation details required on loop diagrams.

The general terminal or bulkhead symbol identifies a junction box or termination assembly and each connection used. Location of the terminal or bulkhead is indicated by the section of the loop diagram in which it is drawn. The junction box number is designated by the number located above the symbol. Instrument terminals and/or ports show the connection to the instrument. It is suggested that identifying numbers or letters be the manufacturer's designation.

The balloon symbol and identification are the same as shown on the P&ID.

ENERGY SUPPLIES

Instrument system energy supply identifications are shown on loop diagrams. The identification also indicates the appropriate voltage or pressure.

INSTRUMENT ACTION

Identification of instrument action shows the direction of the instrument signal by appropriate letters placed close to the instrument bubble.

Direct-acting controllers, controllers in which the value of the output signal increases as the value of the input (measured variable) increases, are represented by the letters DIR. An alternate way to represent a direct-acting instrument is an arrow pointing up (↑).

Reverse-acting controllers, controllers in which the value of the output signal decreases as the value of the input (measured variable) increases, are represented by the letters REV. An alternate way to represent a reverse-acting instrument is an arrow pointing down (↓).

SET POINT VALUES

Set point and calibration values are optional and may be indicated on loop diagrams. Set point values are generally indicated only on devices that are adjusted only by maintenance personnel, such as temperature or pressure switches and field-mounted discrete alarm units. The symbol to indicate the set point is a diamond adjacent to the instrument symbol, with the set point written in the diamond. If several devices require set point information, a table should be included to the left of the title block to include set point data. Set point values should be put on one document only.

CALIBRATION INFORMATION

Calibration information may be indicated by a rectangle adjacent to the instrument symbol with the data written inside the rectangle. It is suggested that, if this optional information is shown on the loop diagram, a table to the left of the title block should include this data (see Figure 3-3).

ALARMS

General alarms and instrument system alarms are shown on loop diagrams. A hardwired device or a separate alarm unit will be shown as a distinct device. If the alarm is included as part of another device, such as a controller, recorder, or shared display and/or shared control, the alarm letter designator may be placed on the input side or the output side of the device symbol as indicated in Figure 3-3. Multiple alarm capability is provided in most systems. Alarms on measured variables shall included the variable identifier; alarms on controller output shall use the undefined variable identifier.

Block Diagrams

A block diagram shows the graphical representation of the hardware in a system. The primary purpose of a block diagram is to indicate the paths along which information or control flows between various parts of a system.

A block diagram may show a block only for each major piece or card in a system or device, or it may show the functions or hardware in more detail. The interconnections between the blocks are shown, but the circuits within the blocks are generally not shown.

Function Blocks

Function blocks, with the function identification in the blocks, are used in configurable devices such as controllers and signal conditioners. The function designations associated with controllers, computing devices, converters, and relays may be used individually or in combination. The use of a box avoids confusion by setting off the symbol from other markings on a diagram and permits the function to be used as a stand-alone block on conceptual designs.

Table 3-1 lists common function block symbols. It is expected that functions associated with the use of succeeding-letter Y (relay, compute, or convert) will be defined outside a bubble on a diagram when further definition is considered necessary.

INSTRUMENT INDEX

An instrument index is a listing of all instruments in a system, usually in tag number order. Tag number order could be by function or loop identification. The use of an instrument index avoids duplicated tags.

Logic Diagrams

Logic diagrams represent a logic design and sometimes the hardware implementation or a logic operation or set of operations. Boolean logic is illustrated by Figure 3-4. Logic diagrams aid in troubleshooting interlocks and start-up/shutdown systems (see Figure 3-5 for ISA symbols). Interlock and start-up and shutdown programs use logic instructions and details that should be shown on logic diagrams to aid in troubleshooting newer control systems. Logic diagrams can be discrete, binary, or Boolean. Frequently, problems in control systems are caused by an external switch or a permissive. On older PLCs the digital part has been documented using relay ladder (relay contacts and coils) or Boolean logic diagrams (ANDs and ORs). To document the analog part, fill-in-the-blank forms are used.

Engineers, configuration integrators, and maintenance technicians need a prescribed way to design, illustrate, and document control logic used in the performance of their profession. In order that every one concerned can work with the same format, several types of logical illustrations have evolved over the years. Some forms are more descriptive for batch logic, while other forms are possibly better for continuous process logic, and still others possibly better for motor logic.

The intent of this discussion is not to prove one method superior to the others but simply to describe the various methods used and to provide a working knowledge of how to use each of these methods.

Some basic forms of logic are described below. An example of a very simple timed and temperature controlled batch cooker will be used to illustrate the topic, using the forms of logic displays discussed in this section on logic. If you have used and understand one of the forms of logic illustrated, you should be able to interpret and understand the other forms of illustrated logic.

In all cases, the operator begins the cooking operation by pushing the "start" button or entering a "start" command. The heating media is controlled by a solenoid valve. Once started, the cook proceeds until the temperature has exceeded 150 degrees F for ten minutes. The operator may stop the cook at any time by pushing the "stop" bottom or entering a "stop" command. The operation is over-simplified in order to emphasize the method of illustrating the logic methods and displays and not the logic itself. See Figures 3-6 through 3-11.

Logic diagrams are used for interlocking and sequencing. The logic diagrams can be documented using a matrix, Boolean expressions, or true/false flow charts.

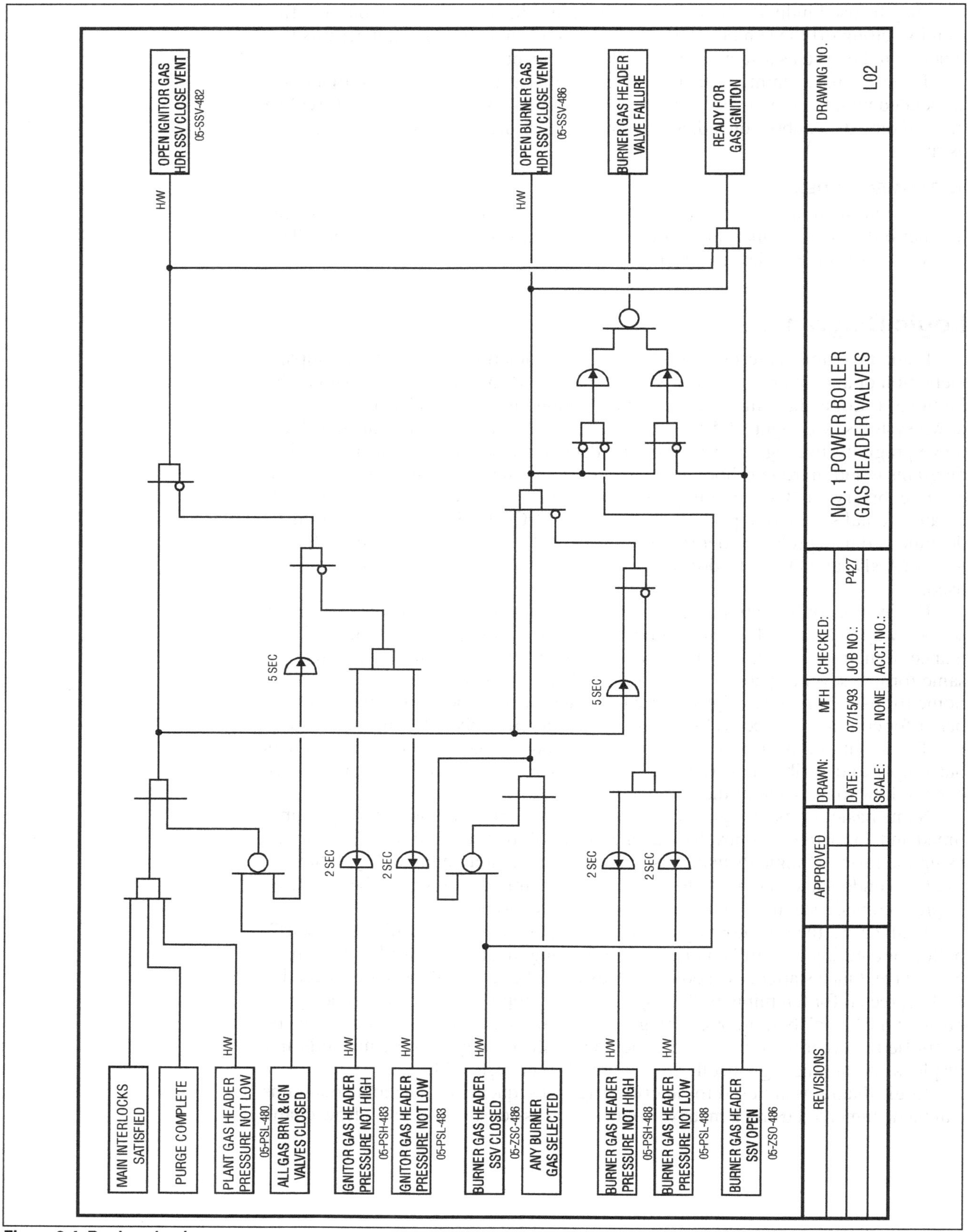

Figure 3-4. Boolean Logic.
Courtesy of Orion CEM, Inc.)

FUNCTION	SYMBOL	DEFINITION	EXAMPLE
4.1 INPUT	Statement of Input Alternatively: Statement of Input Initiating instrument or device number, if known	An input to the logic sequence	The start position of a hand switch *HS-1*, is actuated to provide an input to start a conveyor. Alternative diagrams: a) HS-1 Start Conveyor Manually b) HS 1 Start Conveyor Manually
4.2 OUTPUT	Statement of Output Alternatively: Statement of Output Operated instrument or device number, if known	An output from the logic sequence.	An output from the logic sequence commands valve *HV-2* to open. Alternative diagrams: a) Open Valve HV-2 b) Open Valve HV 2
4.3 AND BASIC	A B — A — D C	**Logic output *D* exists if and only if all logic inputs *A*, *B*, and *C* exist.**	Operate pump if suction tank level is high and discharge valve is open. Tank Level High Valve Open — A — Operate Pump
4.4 OR BASIC	A B — OR — D C	**Logic output *D* exists if and only if one or more of logic inputs *A*, *B*, and *C* exist.**	Stop compressor if cooling water pressure is low or bearing temperature is high. Water Pressure Low Bearing Temperature High — OR — Stop Compressor

Figure 3-5. ISA Symbols for Diagramming Binary Logic (Source: ISA–5.2–1976 (R1992)).

FUNCTION	SYMBOL	DEFINITION	EXAMPLE
4.5 QUALIFIED OR	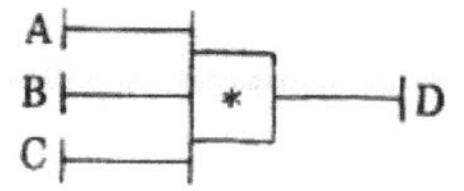 *Internal details represent numerical quantities (see "Definition").	Logic output *D* exists if and only if a specified number of logic inputs *A*, *B*, and *C* exist. Mathematical symbols, including the following, shall be used, as appropriate, in specifying the number: a. = equal to b. ≠ not equal to c. < less than d. > greater than e. ≮ not less than f. ≯ not greater than g. ≤ less than or equal to [equivalent to *f*] h. ≥ greater than or equal to [equivalent to *e*]	a) Operate mixer if two, and only two, bins are in service. 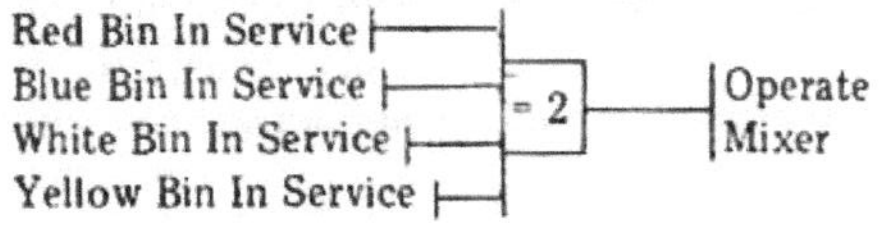 b) Stop reaction if at least two safety devices call for stop. 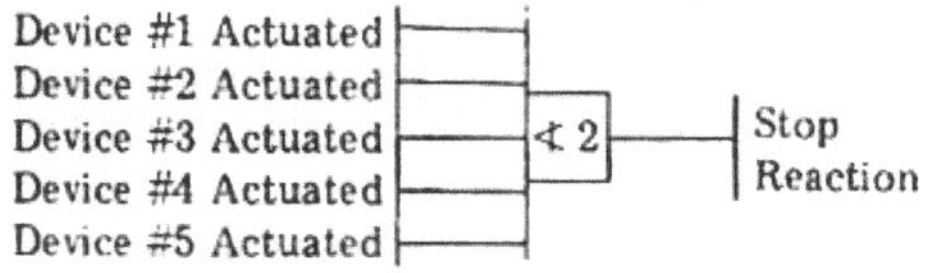 c) Operate materials feeder if at least one and no more than two mills are in service. 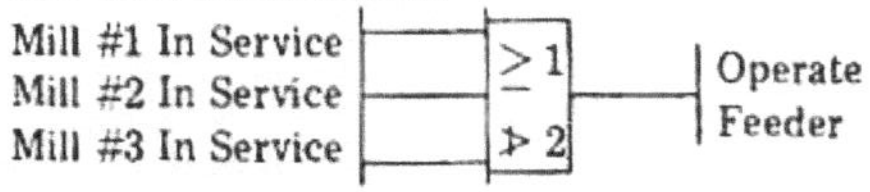
4.6 NOT	BASIC 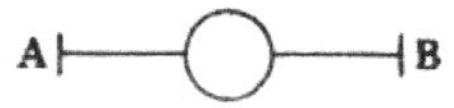 **The *NOT* symbol may be drawn tangent to an adjacent logic symbol.**	**Logic output *B* exists if and only if logic input *A* does not exist.**	Shut off fuel gas if burners no. 1 and no. 2 are not on. 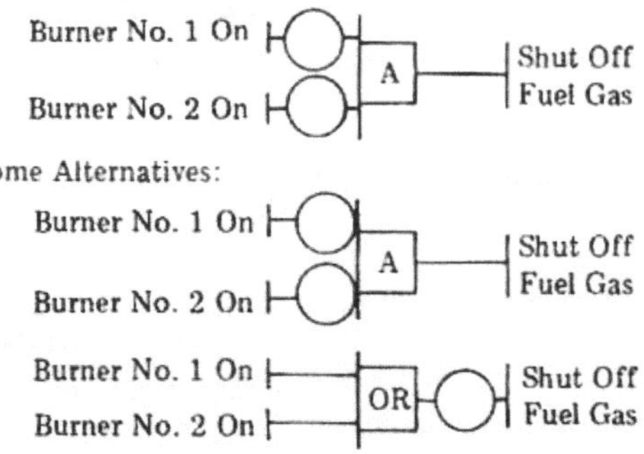

Figure 3-5. ISA Symbols for Diagramming Binary Logic (Source: ISA–5.2–1976 (R1992)) cont'd.

FUNCTION	SYMBOL	DEFINITION	EXAMPLE
4.7 MEMORY (Flip-Flop)	BASIC 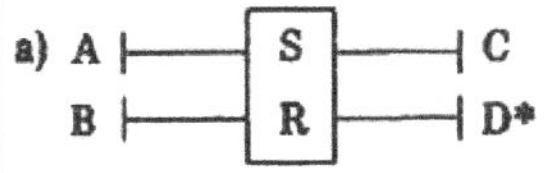 *Output *D* shall not be shown if it is not used.	**S represents *set memory* and R represents *reset memory*.** **Logic output *C* exists as soon as logic input *A* exists. *C* continues to exist, regardless of the subsequent state of *A*, until the memory is reset, i.e., terminated by logic input *B* existing. *C* remains terminated regardless of the subsequent state of *B*, until *A* causes the memory to be set.** **Logic output *D*, if used, exists when *C* does not exist, and *D* does not exist when *C* exists.**	If tank pressure becomes high, vent tank and continue venting, regardless of pressure, until venting is stopped by manual actuation of hand switch, HS-1, provided that the pressure is not high. If the venting is stopped, a compressor may be started. 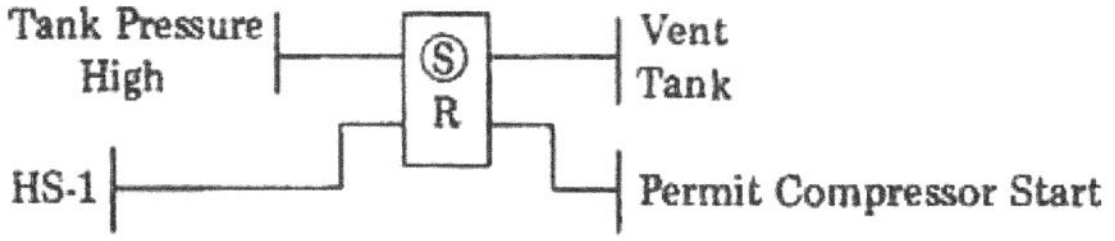
		Input-Override Option If inputs *A* and *B* exist simultaneously, and if it is desired to have *A* override *B*, then *S* should be encircled, i.e., Ⓢ ; if *B* is to override *A*, then *R* should be encircled, i.e., Ⓡ *Loss-Of-Power-Supply Option* The unmodified letter *S* denotes that *no consideration* has been given to the action of the memory on loss of the logic power supply. See paragraphs 4.7 b, c, and d, below, and 3.8.	
	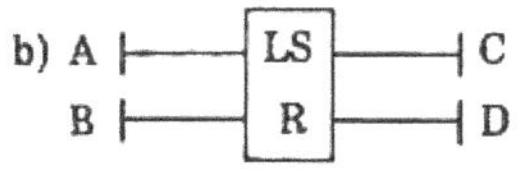 (See Appendix C)	Similar to definition of symbol (a) except that the memory shall be *lost* in the event of loss of the logic power supply.	If feed begins to flow, the cooler shall operate until the feed tank is empty. In the event of loss of the logic power supply, the cooler shall not operate 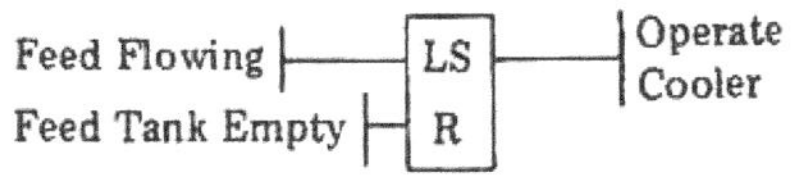
(cont'd)	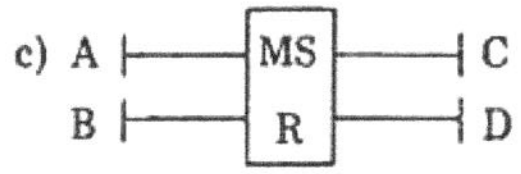	Similar to definition of symbol (a) except that the memory shall be *maintained* in the event of loss of the logic power supply.	If standby pump operation is initiated, the pump shall operate, even on loss of the logic power supply, until the process sequence is terminated. The pump shall operate if start and stop commands exist simultaneously. 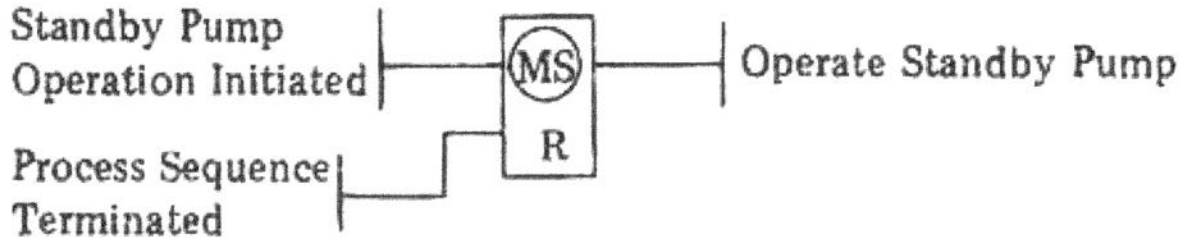

Figure 3-5. ISA Symbols for Diagramming Binary Logic (Source: ISA–5.2–1976 (R1992)) cont'd.

FUNCTION	SYMBOL	DEFINITION	EXAMPLE
4.7 (cont'd)	d) 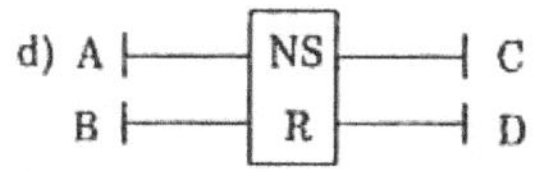	Similar to definition of symbol (a) except that *after consideration* it is deemed *not significant*, so far as the process is concerned, whether the memory is maintained or lost in the event of loss of power supply.	If reservoir level is low, operate fill pump until either level is high or water quality is unsatisfactory. It is not significant to the process what happens to the pump on loss of the logic supply. If start and stop commands are simultaneous, the pump shall stop. 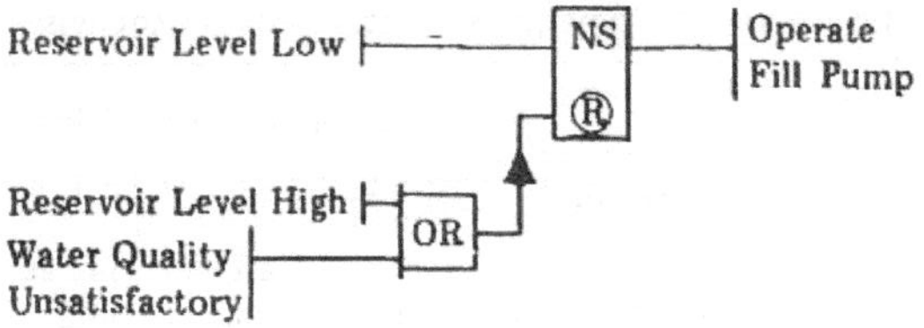
4.8 TIME ELEMENT	a) 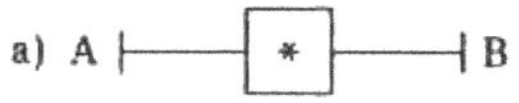 *For functional details, see the following (also see Section 3.9):	Logic output *B* exists with a time relationship to logic input *A* as specified.	
BASIC	b) 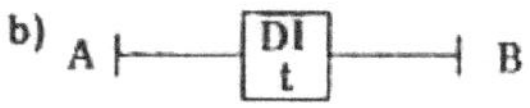 (Delay Initiation of output)	The continuous existence of logic input *A* for time *t* causes logic output *B* to exist when *t* expires. *B* terminates when *A* terminates.	If reactor temperature exceeds a high limit continuously for 10 seconds, block catalyst flow. Resume flow when temperature does not exceed the limit. 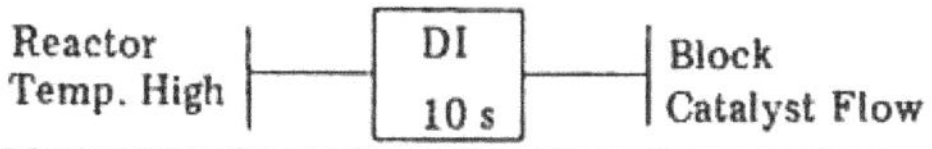
BASIC	c) 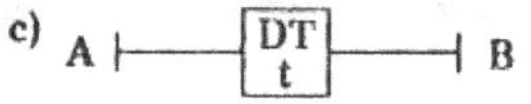 (Delay Termination of output)	The existence of logic input *A* causes logic output *B* to exist immediately. *B* terminates when *A* has terminated and has not again existed for time *t*.	If system pressure falls below a low limit, operate compressor at once. Stop the compressor when pressure is not low continuously for one minute. 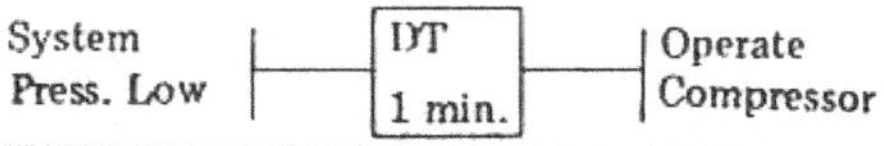
BASIC	d) 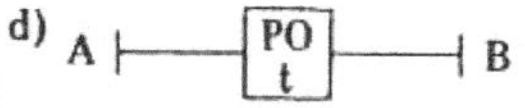 (Pulse Output)	The existence of logic input *A*, regardless of its subsequent state, causes logic output *B* to exist immediately. *B* exists for time *t* and then terminates.	If vessel purge fails for any period of time, operate evacuation pump for 3 minutes and then stop the pump. 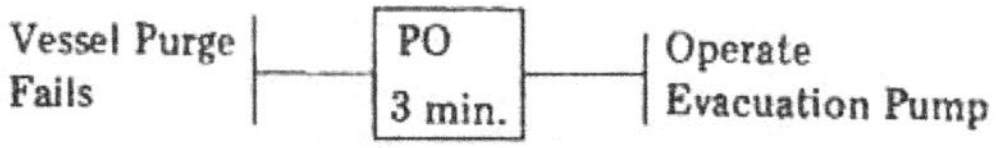

Figure 3-5. ISA Symbols for Diagramming Binary Logic (Source: ISA–5.2–1976 (R1992)) cont'd.

LADDER LOGIC

Probably the oldest and most used is "ladder logic." Ladder logic received its name from the appearance of the document, which takes the shape of a ladder with the logic displayed as contacts in each rung of the ladder. The left side of the ladder rail is the power source side of the logic, while the right side of the ladder rail is the neutral or return side of the power source. Power is transferred from the power rail on the left side of the drawing to each successive contact or function on its right.

Comparing a printout of the "real-time" logic with what is on file may be helpful in determining whether the logic has been changed recently and might be causing the problem.

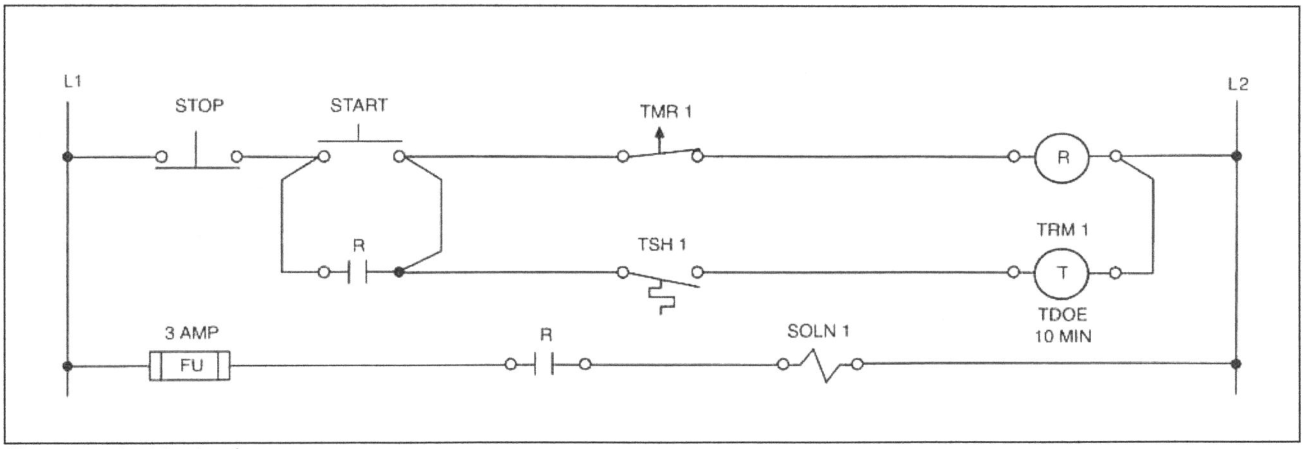

Figure 3-6. Ladder Logic.

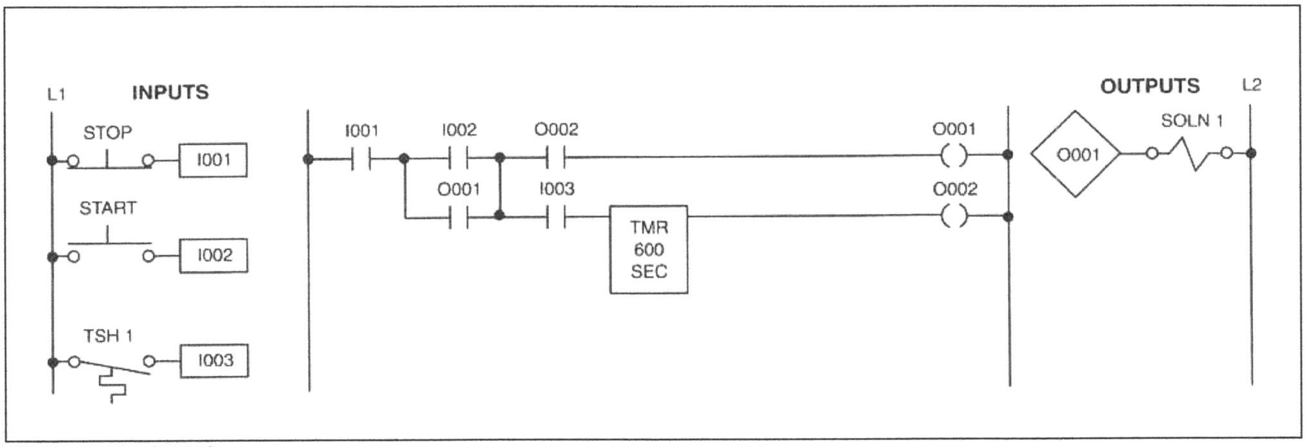

Figure 3-7. PLC Logic.

PLC LOGIC

This type of drawing originated in the days of relays and electromechanical timers, etc., but the same format is used today to document PLCs that use software logic. The power flow is the same in a PLC; however, the logic is scanned from top to bottom at a fast rate, and any contacts shown above a given ladder rung will not function until the PLC has completed one scan. This slight but important difference needs to be remembered when dealing with a PLC versus a relay network.

PLC ladder logic can be displayed in "real time," on a programmer or CRT screen, so that the technician can follow the power flow through the logic to determine what is preventing the logic from working. Therefore, it is important to use the programmer when troubleshooting, because it will be the logic that is actually being used and not the logic on an old document that has since been changed.

Sometimes software jumpers are installed to temporarily resolve a problem and are then forgotten. By reviewing the original logic and comparing it to the "real-time" logic, these forgotten jumpers are easily found.

Until recently, most automated process plants used two separate subsystems: a strictly digital programmable logic controller (PLC) to handle the digital on/off devices and a shared display distributed control system (DCS) to handle the analog displays and control loops.

SAMA LOGIC

SAMA logic is used mostly for continuous process control definition but can also describe batch processes, although the Scientific Apparatus Makers Association no longer supports the standards upon which this logic is based. This logic includes both the symbols and the diagramming format used to represent the measurement, control, and computational aspects of a control loop or loops. The discussion below will pertain only to the logic for analog systems.

Several major instrument companies began using SAMA Logic Diagrams (see Figure 3-8) when they went to split-architecture hardware, separating the control unit from the display unit. SAMA diagrams provided a structured and reproducible way to work with and display the various functions of split-architecture control.

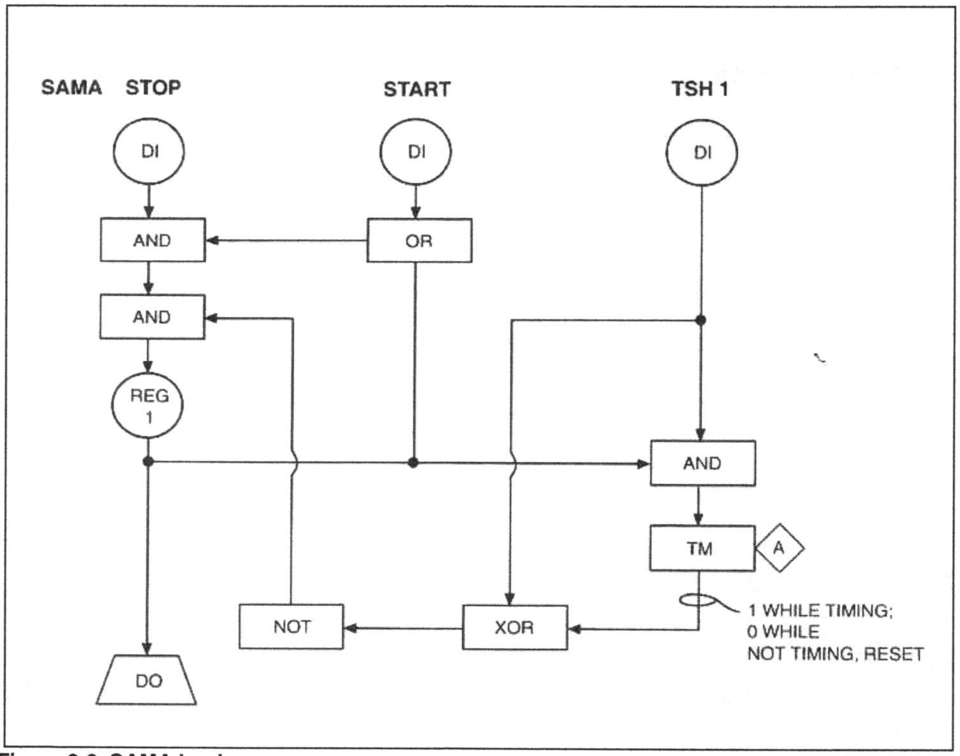

Figure 3-8. SAMA Logic.

When distributed control systems emerged, some manufacturers again went to the SAMA form of functional diagrams. Since all of the logic resided in the computing portion of the DCS, they had to find a structured way to design and display this logic. SAMA diagrams have a specific format for all the basic functions that can be grouped together to produce either simple or complex diagrams that describe the process control and logic.

SAMA drawings are usually generated during the design stage of a project and, in rough form, are used to review and alter the control and logic, as necessary, until agreement is reached. In many cases, the SAMA diagram is generated only after agreement is reached. In other cases, the SAMA diagram is generated after the P&IDs are in their final stages. Due to the complexity of today's instrumentation, it is impossible to show all the logic on the P&IDs or loop drawings. SAMA drawings are an essential part of the engineering stage of a project and many times are even more important during equipment and software checkout and start-up. SAMA symbols are used on drawings by several boiler control vendors.

GATE LOGIC

Ladder logic can be duplicated using GATE logic or vice/versa. Two contacts, in parallel, are shown as a rounded arrow symbol that represents an "or" command. Two contacts, in parallel, are shown as a bullet nosed symbol that represents an "and" command. The outputs of some commands are the inputs of other commands or is the solution of the logic just performed. The drawing looks somewhat similar to a ladder diagram, in that the logic flows from left to right and from top to bottom, at least in most cases. The drawings are somewhat easier to develop and understand because one symbol can perform several ladder functions. If the physical device chosen to produce this logic is programmed in GATE, the programming is simplified. If the final device chosen to produce the logic is a PLC, the GATE logic must still be converted to relay ladder logic.

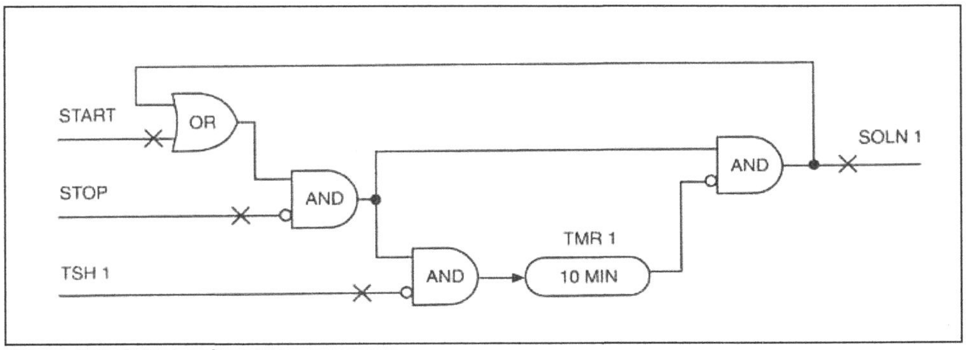

Figure 3-9. Gate Logic.

TIME SEQUENCE LOGIC

Time sequence logic is used to define an automatic system that relies heavily on events happening at specific times and/or specific lengths of time. Events may have to overlap and the time sequence logic format can represent and display the overlapping events much better than other logic formats. Time sequence logic formatting was and is used when mechanical timing devices are used, such as rotating drum, pin-set timers and sequencers.

FLOW CHART LOGIC

Flow charting is used mostly as a design tool to show detailed functions as they apply to a computer program or logic program. It is used to describe sequential events that interact and rely on certain checks to start, stop or interrupt a continuous program.

Flow charting describes and illustrates the flow of information, inputs, outputs, checks and interrupts, etc. that might be controlled by a computer. They were used in the early days of computers to illustrate what the computer does or how it handles the tasks of batch and continuous processing.

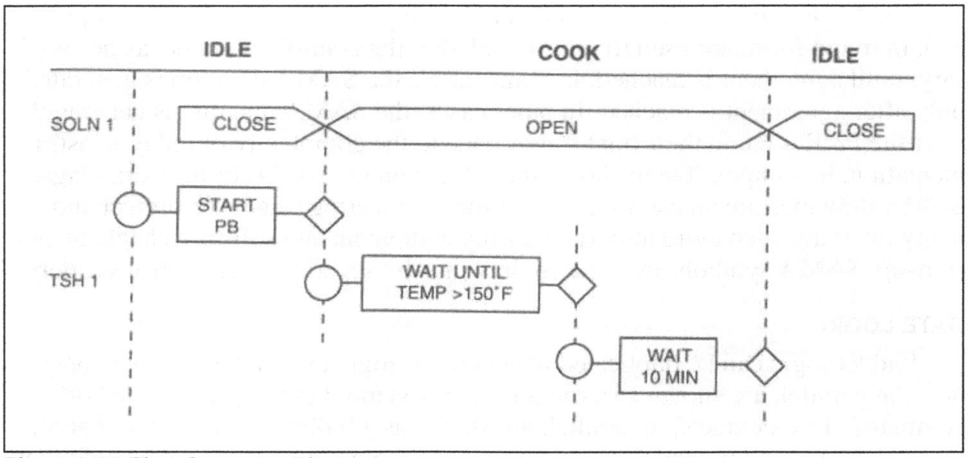

Figure 3-10. Time Sequence Diagram.

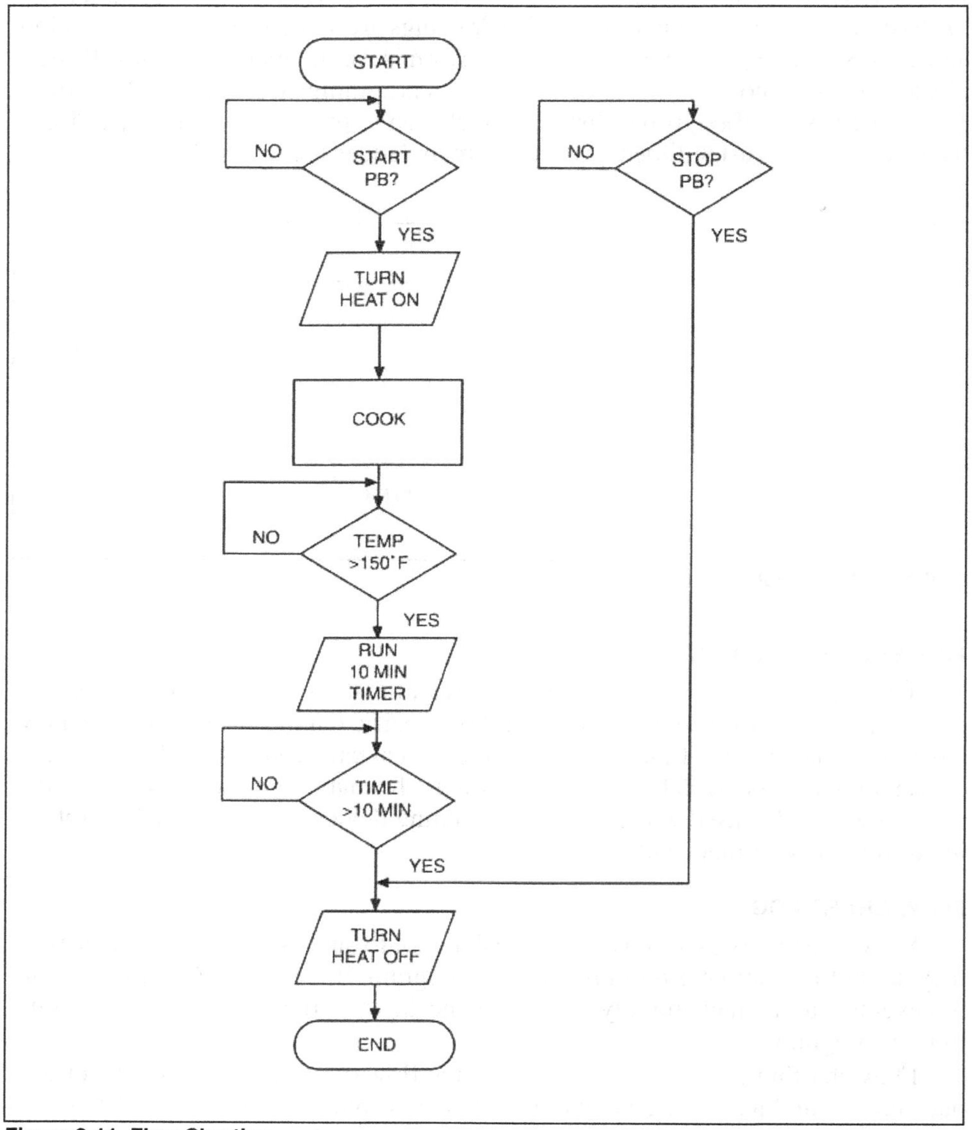

Figure 3-11. Flow Charting.

BASIC PROGRAMMING LANGUAGE LOGIC

With the onset of personal computers (PCs) being used for control or as an additional window to the process, the BASIC software program began to be used extensively.

The present generation of DCSs and PLCs have integrated both the regulatory and sequence control functions, and with further enhancements in the batch capabilities of these systems to handle multi-recipe, multi-product plants, a combination of ladder logic, SAMA logic, GATE logic, flow charts, time sequence diagrams, block diagrams, and basic computer languages and written process operating procedures must be used to document the functional requirement of the software.

Most process plants consist of a combination of batch and continuous control functions:

(1) Analog (continuous) regulatory control loops (e.g., level control loop of a distillation column).

(2) Digital (discrete) control devices (e.g., control of motors, on/off valves, etc.)

(3) Interlock logic (this function usually defines the relative status of two or more discrete devices to ensure the safe operation of the plant).

(4) Sequence control logic (the basis for batch process control. Actions are initiated in response to trigger events such as time, special operating conditions, modification of product recipes, etc. Sequence may be fixed or variable, and they may depend on the status of the process or on external factors such as market demand for different product grades).

Highway Drawings

The highway drawings of a DCS system show the highway, any local area networks, screens, and consoles. They show what is on the highway, what consoles are tied to common console electronics, and what the printers, computers, redundant controllers, etc., are tied to. It is a must on a large system. See Figure 3-12.

Schematics

Instrument schematics are generally supplied by the manufacturer of the instrument. A schematic is defined as a generalized diagram or scheme. In electronics, the schematic shows all the interconnections, both power and signal, and is used to troubleshoot the circuit.

Schematics are used to aid in board-level troubleshooting. The maintenance manuals supplied with the device or control system have schematics of all circuit wiring. Each component generally has an identification, such as R3 for a resistor or C14 for a capacitor. These identifications will cross reference to the material list by the part number of the component. The values of some components are also listed on the schematic.

Older devices used hard-wiring and individual components. It was common practice to troubleshoot and replace an individual component. Then came printed circuit boards with soldered-in components, and again it was common practice to troubleshoot and replace an individual component. Schematics were needed to troubleshoot and repair these circuits. With the newer microprocessor-type equipment and devices using high-speed circuits, more sophisticated types of test equipment are required.

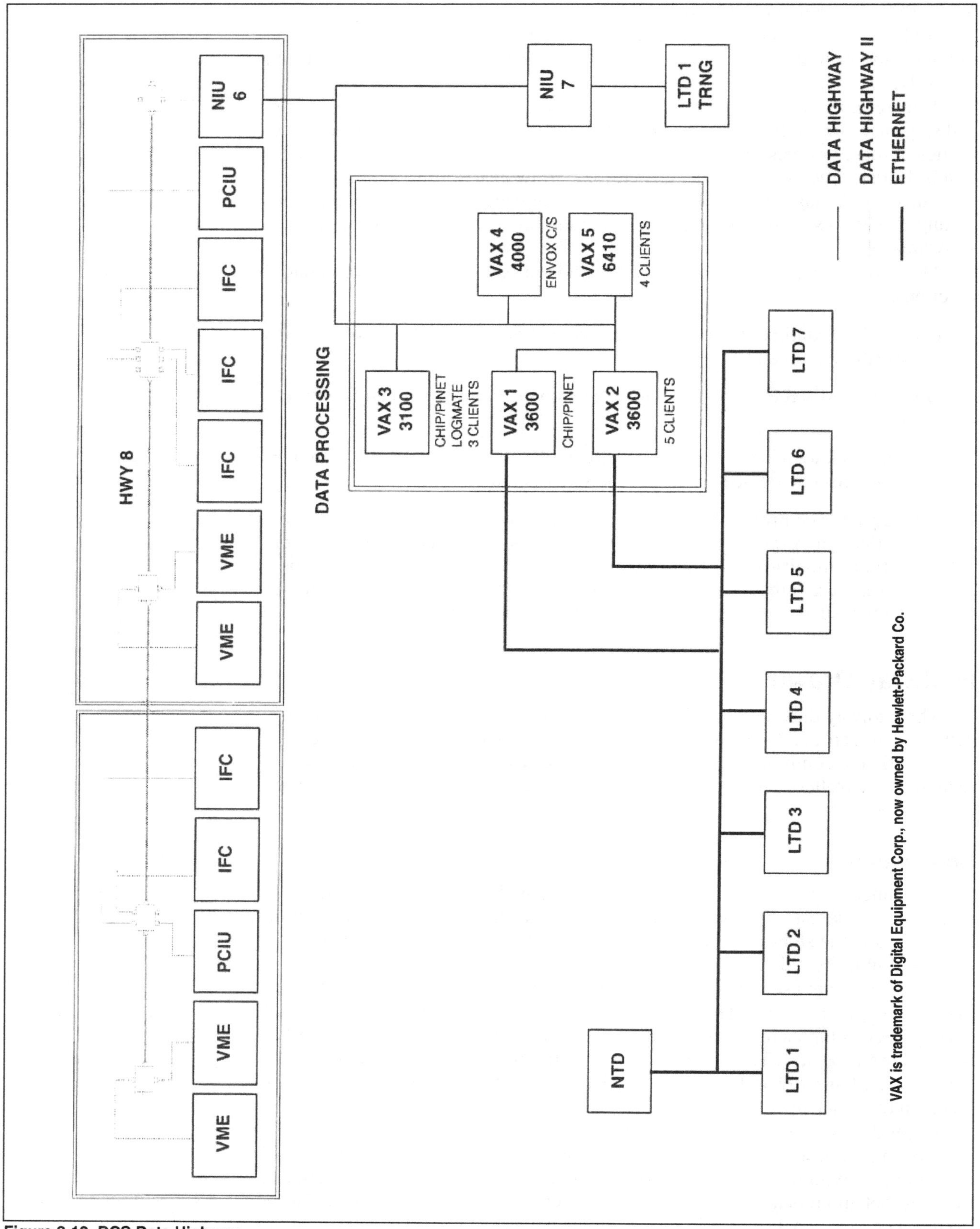

Figure 3-12. DCS Data Highway.
(Courtesy of Fisher-Rosemount)

The use of schematics on mulitlayered printed circuit boards is not normally necessary for maintenance and troubleshooting (unless there are plug-in replaceable components); instead, the circuit board is replaced.

Replacement of individual components on multilayered printed circuit boards is not suggested.

Specifications

Instrument Specification

Each instrument is built to meet certain specifications such as those for operating, performance, and construction. A specification is a detailed and exact statement of particulars, especially a statement prescribing dimensions, construction, tolerances, bills of materials, features, and operating conditions. Maintenance personnel use these documents to ensure that after repair or rework the instrument meets the performance and operating specifications that the instrument was designed to meet.

After repair or rework, ensure that the instrument still meets required performance and operating specifications.

Calibration is generally required after an instrument is repaired or reworked. The calibration record and reports should be maintained in the equipment history files. In some instances, certification and calibration traceability are required.

The "as-installed" instruments should be described in detail on the relevant ISA–20–1981, Specification Forms for Process Measurement and Control Instruments, Primary Elements and Control Valves specification sheets. Or, if the user prefers, ISA–TR20.00.01–2001, Specification Forms for Process Measurement and Control Instruments, Part 1: General Considerations, should be a primary part of plant documentation.

Specifications for instruments can be found in manufacturers' product data bulletins, instruction manuals, installation and service instructions, and other manufacturers' literature. These should be available to maintenance personnel.

PERFORMANCE SPECIFICATIONS

(1) Accuracy—The degree of conformity of an indicated value to a recognized accepted standard value, or ideal value; the maximum positive and negative deviation observed in testing a device under specified conditions and by a specified procedure; generally indicated as a percent of input/output span or a unit of measurement. Examples are:

- Factory calibrated to ± 0.25% of input span for all ranges;
- ±1°F for base-metal thermocouple ranges;
- Output deviation ± 0.008 V;
- Error % of span ± 0.2; and
- Input accuracy ± 0.25% of span ± 1°C.

(2) Stability—Resistance to changes in accuracy due to environmental changes; accuracy due to changes in temperature, relative humidity, and/or altitude.

(3) Immunity—Common or normal mode rejection and radio frequency interference.

(4) Repeatability—Closeness of agreement among a number of successive measurements of the output for the same value of the input under the

same operating conditions, approaching from the same direction, for full range traverses.

(5) Resolution—The least interval between two adjacent discrete details that can be distinguished one from another. Generally determined in decimals, such as 1 degree or 0.1 degree, percent resolution such as 0.01% or 12 bits.

(6) Linearity—Maximum deviation of the calibration curve from a straight line so positioned as to coincide with the calibration curve at the lower range value and to minimize the maximum deviation, deviation specified as ± percentage, ± unit of measure, or ± LSB.

OPERATING SPECIFICATIONS

(1) Power requirements—Line (supply) voltage limits, both voltage and frequency; generally listed as 120 V ± 10%, 50/60 Hz, less than 5 VA or 120 V AC (95-135 V), 60 Hz, 40 VA average.

(2) Line voltage effect—Effect of line voltage changes.

(3) Load effect—Effect of load from 0 to maximum load resistance; generally listed as percentage of span (± 0.01 % of span).

(4) Inputs/outputs specifications—Includes types of inputs and outputs (analog/digital), load voltage and current, contact ratings, isolation, and so on.

(5) Ambient temperature effect—Effects of ambient temperature changes in accuracy of instrument.

(6) Scan rate—A single sweep of PC applications program operation. The scan operates the program logic based on I/O status, and then updates output and input status, sampling in a predetermined manner each of a number of variables intermittently. The time required for this is called the scan time.

(7) Shock and vibration—Amount of shock and vibration the instrument can withstand and continue to conform to other specifications.

(8) Adjustments—Changing proportional band, reset, cycle time, zero offset, span (includes ranges of adjustments), and so on.

CONSTRUCTION SPECIFICATIONS

(1) Electrical classification—For hazardous location installations in Class, Division, and Group areas, such as Class I, Div 2, Groups A, B, C, and D areas.

(2) Housing—NEMA classification such as NEMA 1 – General purpose indoor, or NEMA 4 – Watertight and dust-tight, indoor and outdoor. See Figure 3-13.

(3) Dimensions—For example, for panel cutout: faceplate dimensions, length, width, height, etc.

(4) Weight—Average weight.

(5) Terminal connections—Type, size, and location.

(6) Mounting—Mounting specifications such as panel, DIN rail, etc.

Maintenance engineers and technicians must be familiar with the instrument specifications when replacing, repairing, or calibrating instruments. Accuracy and safety standards should not be compromised when replacing older instruments with newer instruments. However, most newer instruments have tighter specifications than the older instruments they replace.

TYPE NO.	DEGREE OF PROTECTION
1	INDOOR USE
2	INDOOR USE, LIMITED AMOUNTS OF FALLING WATER
3R	OUTDOOR USE, UNDAMAGED BY THE FORMATION OF ICE ON THE ENCLOSURES.
3	SAME AS 3R PLUS WINDBLOWN DUST
3S	SAME AS 3R PLUS WINDBLOWN DUST, EXTERNAL MECHANISMS REMAIN OPERABLE WHILE ICE LADEN.
4	OUTDOOR USE, SPLASHING WATER, WINDBLOWN DUST, HOSE-DIRECTED WATER, UNDAMANGED BY THE FORMATION OF ICE ON THE ENCLOSURE.
4X	SAME AS 4 PLUS RESISTS CORROSION
6	SAME AS 3R PLUS ENTRY OF WATER DURING TEMPORARY SUBMERSION AT A LIMITED DEPTH.
6P	SAME AS 3R PLUS ENTRY OF WATER DURING PROLONGED SUBMERSION AT A LIMITED DEPTH.
11	INDOOR USE, CORROSIVE LIQUIDS AND GASES — BY OIL IMMERSION OF ENCLOSURE.
12, 12K	INDOOR USE, DUST, DRIPPING NONCORROSIVE LIQUIDS.
13	INDOOR USE, DUST, SPRAYING WATER, OIL, AND NONCORROSIVE COOLANTS.

Figure 3-13. NEMA Enclosure Classifications.

System Specifications

FUNCTIONAL SPECIFICATIONS

A key document generated in the planning phase of any project is the functional specification. It is used not only for planning, bidding, and checkout of new systems but also should be available to maintenance personnel. This reference indicates what the system is designed to do and to what accuracy. An operator or process engineer might ask that tighter limits than called for be set in the overall system specifications. The functional specification is a document that tells exactly what the system should do, what will be supplied to the system, and what is expected to come out of it. The functional specifications should be used for start-up and checkout to ensure that the system purchased conforms to the functional specifications. If the system does not conform to the system goals, responsibilities should be determined during checkout; otherwise, the maintenance department will be asked to make the changes.

After a system or project specification book is completed and approved, it is used for review and even checkout. After start-up is complete, an engineering manual should be provided; this is a more detailed, as-built document that has more information than the system specification. This usually isn't furnished until six months after the start-up and includes detailed programming information, etc.

Maintenance engineering should be a part of any project team, from preparing the functional specifications and selecting vendors and equipment through start-up and checkout and throughout the life of the equipment.

Maintenance personnel such as instrument and control system mechanics and technicians should also be involved in start-up and checkout. They will gain experience in the system and should also be providing the training necessary to maintain the system.

OPERATIONAL PROCEDURES

Operational procedures are generally provided by the process or manufacturing engineering departments. They should also be available to maintenance personnel during system checkout. It is suggested that operating personnel be available during checkout, preventive maintenance, and corrective maintenance. They, the operating personnel, are more familiar with the process and instrumentation and control system idiosyncrasies.

Instrument Symbols

Symbols are used on drawings to indicate devices, types of communication lines, connection points, vessels, functions, and so on. The ISA–5 series of standards includes some of these symbols and figures. These standards illustrate the symbols that are intended to depict instrumentation on diagrams and drawings. The term "instrument," which is a device used directly or indirectly to measure and/or control a variable, includes primary elements, final control elements, computing devices, and electrical devices such as annunciators, switches, and push buttons. The term does not apply to parts (e.g., a receiver bellows or a resistor) that are internal components of an instrument.

The circular symbol, used to denote and identify the purpose of an instrument or function, is called the bubble or balloon. Squares, diamonds, and other shapes indicate instruments or functions such as shared display/control, computer functions, or programmable logic control (see Figure 3-14).

Discrete Instrument

A circle by itself is used to indicate a discrete or distinct instrument, that is, an individual instrument that performs one function such as a primary element or a single-loop temperature controller.

Shared Display

The bubble or balloon may include a box around the circle, which indicates shared display and/or shared control. A shared display is an operator interface device (usually a video screen) used to display process control information from a number of sources (time-shared) at the command of the operator. The signals and/or data (i.e., alphanumeric and/or graphic) reside in a database from which selective accessibility for display is at the command of the user. The video screen may be called a visual display unit (VDU), operator's station (OpSta), monitor, or color graphics terminal. In almost all cases there will be a solid horizontal line or a pair of solid horizontal lines that designate this device (display) as accessible to the operator.

Several device faceplates can be represented on the video screen. They might indicate controllers, digital or analog indicators, control stations, or switches. Graphic representations of the process are also used on the shared display, along with displays for individual loops, a group of loops, or an overview of a process or plant. Alarm annunciation is also indicated on shared display screens as well as real-time and historical trending.

The computer function symbol, a hexagon, is also a shared display that utilizes a computer. The Technical Association of the Pulp and Paper Industry (TAPPI) and the pulp and paper industry uses the hexagon with a horizontal line as an accessible computer function.

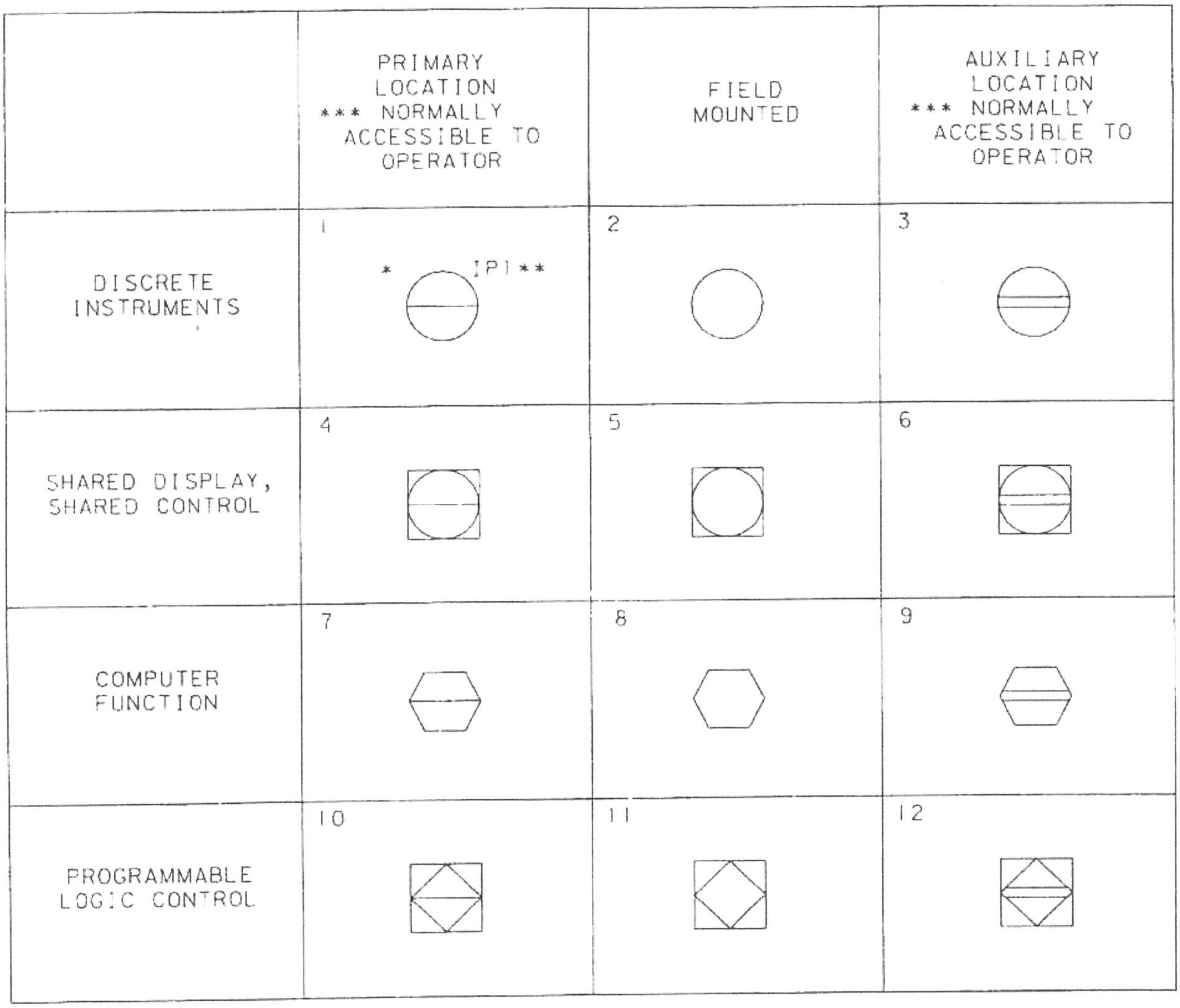

* Symbol size may vary according to the user's needs and the type of document. A suggested square and circle size for large diagrams is shown above. Consistency is recommended.

** Abbreviations of the user's choice such as IP1 (Instrument Panel #1), IC2 (Instrument Console #2), CC3 (Computer Console #3), etc., may be used when it is necessary to specify instrument or function location.

*** Normally inaccessible or behind-the-panel devices or functions may be depicted by using the same symbols but with dashed horizontal bars, i.e.

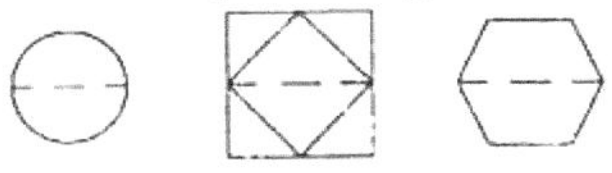

Figure 3-14. General Instrument or Function Symbols (Source: ISA–5.1–1984 (R1992)).

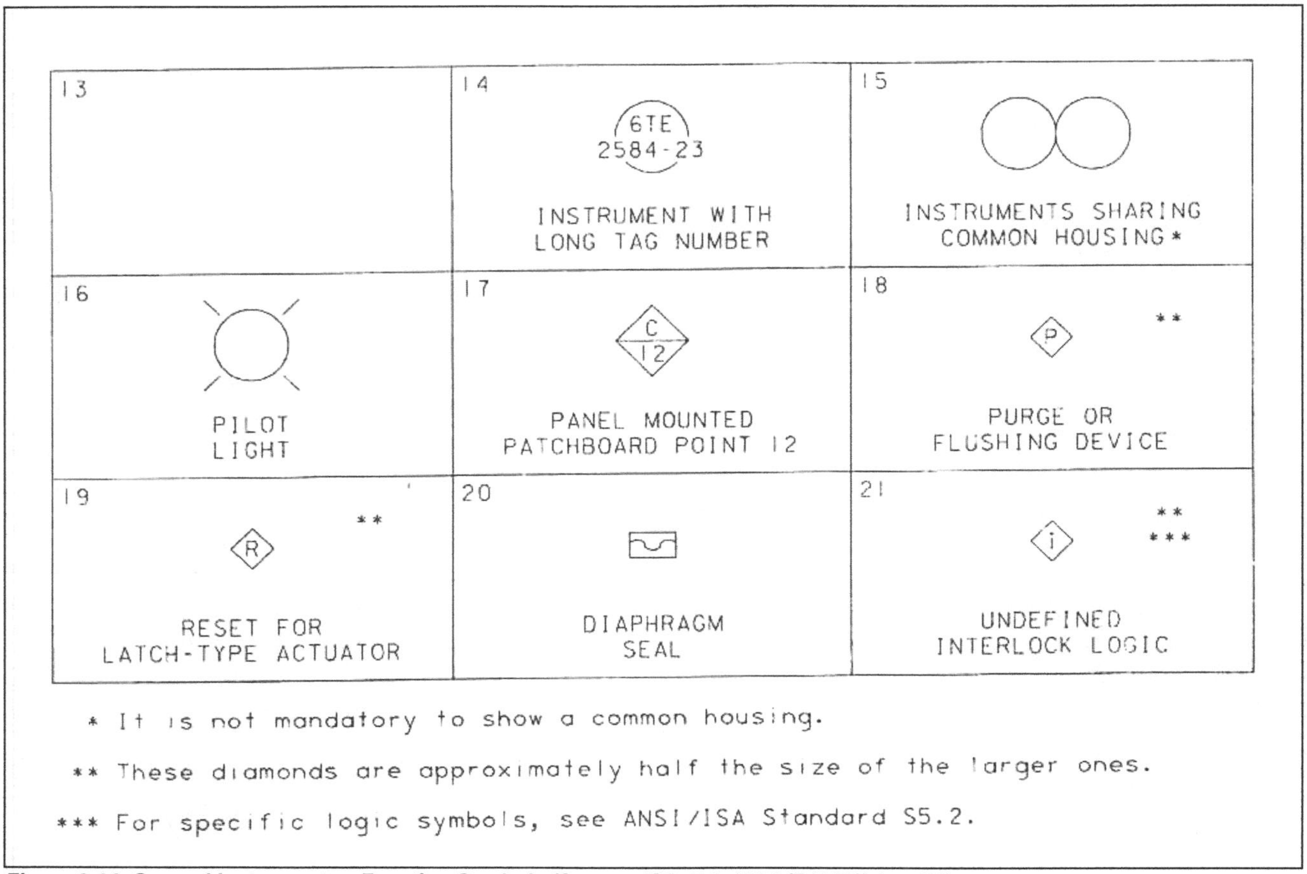

Figure 3-14. General Instrument or Function Symbols (Source: ISA–5.1–1984 (R1992)) (cont'd).

Trending information is valuable for tuning the loops and for going back in time to find the cause of a problem.

Shared Control

Shared control, which permits a number of process variables to be controlled by a single device, utilizes a shared controller that contains a plurality of preprogrammed algorithms that are usually accessible, retrievable, configurable, assignable, and connectable and allows user-defined control strategies or functions to be implemented. The symbol for shared control is the same as for the shared display (a circle surrounded by a square). The shared control instrument might be a part of the shared display instrument, but, in some cases, field-mounted processors will not have an operator's display and are not accessible to the operator. Some manufacturers provide a means for attaching a portable calibrator or display unit with keyboard to assist maintenance in troubleshooting the shared controller without communication with the main control room. This portable display is not considered accessible to the operator.

Distributed control systems (DCS) utilize shared controllers placed throughout the plant on a data highway to transfer set point and other information from the shared display in the main control room and to transfer process data to the shared display from the shared controller.

A distributed control system (DCS) is one that, while being functionally integrated, consists of subsystems that may be physically separated and remotely located form one another. Control, logic, calculations, and measurement functions are normally performed in the remote processor units close to the process. The control and logic continue whether or not the data highway is connected. The data highway provides a path to monitor the process and to request to the remote processors that changes be made.

Multiloop controllers may be considered shared control and/or shared display. Multiloop controllers may be panel mounted near the process.

Computer Functions

Computer functions are indicated by a six-sided polygon (see Figure 3-14). Computer symbols should be used where systems include components identified as "computers," as distinct from integral processors, which drive the various functions of a distributed control system. The computer component may be integrated with the system via data line, or it may be a stand-alone computer.

A computer function symbol also could indicate a higher-level computer, generally considered a mainframe type, that provides remote setpoints to distributed control systems and setpoint tracking for manual analog backup. Mainframe computers are the link between gathering information about the process from the process control systems and the management information system (MIS). They provide information so that (1) management can make decisions or (2) the computer-aided manufacturing system can determine whether changes should be made in the process. A mainframe is also used for gathering data for the MIS; it can also be part of a computer-integrated manufacturing (CIM) system.

When the mainframe computer is not used for control, a tag number does not have to be used in the symbol; the symbol for "not accessible to the operator" is used (a bare polygon). An I or and O (or both) indicates input or output (or both). It is advisable to identify the computer function with a tag number.

Direct digital control (DDC) usually uses a mainframe computer and is indicated by the computer function symbol. All acquisition, linearization, logic, and control functions are accomplished in the central processing unit of the computer. Tag numbers are generally used when the mainframe computer is used for control.

There are several disadvantages of this type of control. One is that if the computer malfunctions, all loops are inoperative. But the biggest reason that DDC is not used very often in process control is the time and cost of programming or making changes to the computer program.

The computer function symbol is also used with a personal computer-based control system. Tag numbers are used when the PC is used for control. This symbol could be used when a standard PC is used as an aid or a window to a distributed control system and is not performing acquisition or control functions. If the PC is used only for data acquisition, the letter O is placed in the polygon on the P&ID or loop diagram. If the PC is an additional operator's console or a distributed control system, the shared display symbol could be used.

Programmable Logic Controller

A programmable logic controller (PLC) may be indicated by a diamond within a square box (see Figure 3-14), or it may be indicated as a shared display/shared control symbol. The programmable logic controller symbol is normally used with a packaged PLC or stand-alone digital logic controls that may be part of the distributed control system (DCS). If both systems are supplied by the same manufacturer, the DCS symbols (shared display/control) are normally used.

General Logic or Sequential Control

A diamond that is not within a square is the general logic or sequential control symbol for undefined complex interconnections. An example is a diamond with the letter I inside to designate an undefined logic function.

Location Symbols

Symbols indicate the location of the instrument and whether it is accessible to the operator.

A symbol with no horizontal line indicates that the instrument is field-mounted near the process and not on an auxiliary panel. Some example of field-mounted instruments are thermocouples, pressure transmitters, flowmeters, pH electrodes, and flow or pressure switches.

A single horizontal line indicates the instrument is panel-mounted on a primary panel and is accessible to the operator. Some examples are single-loop controllers, recorders, indicators, manual stations, and switches. "Accessible" is a term applied to a device that can be used or be seen by an operator for the purpose of performing control actions, e.g., setpoint changes, auto-manual transfer, or on-off action.

A single broken or dashed horizontal line indicates than the instrument is located behind the panel and is normally inaccessible to the operator. Some examples are square-root extractors, current-to-pneumatic converters, alarm units, and so on.

Two horizontal solid lines indicate that the instrument is located on the front of a panel in an auxiliary location (not on the main control panel). Some examples are backup manual stations, analytical transmitters and controllers, local alarms, and bypass switches.

Two broken or dashed lines may indicate instruments located behind the auxiliary panel. These instruments are not normally accessible to the operator.

Valve, Damper, Actuator, Regulator, and Primary Element Symbols

Symbols are standard for control valves, dampers, actuators, regulators, and primary elements, as are graphic symbols for tanks, vessels, columns, and so on. Figure 3-15 shows some of the valve body symbols; Figure 3-16 shows some actuator symbols. The ISA–5 series of standards show additional symbols.

The general symbols for valves and primary elements are often used on P&IDs, but a more descriptive symbol might be used on loop diagrams. Instrument maintenance personnel must recognize these symbols in order to locate instruments and know their functions. Engineers and maintenance personnel need to know the operations of the instruments and the correlations of the instruments to the plant process in order to properly maintain the systems.

Instrument Symbols

Instrument line symbols are used to determine the type of energy that is transmitted from one device to another. Process piping lines are bold, and all instrument line symbols are fine in relation to process piping lines. Direction-of-flow arrows may be used on process piping lines.

Instrument Supply

The instrument supply line symbol is used to indicate the instrument supply or connections to the process; it is a fine, solid line. Abbreviations are suggested to denote the type of power supply, and the value (level) of the supply should be indicated, such as ES 120 V AC (see Figure 3-17). Supply lines are not normally shown on the P&ID but are usually shown on instrument loop diagrams.

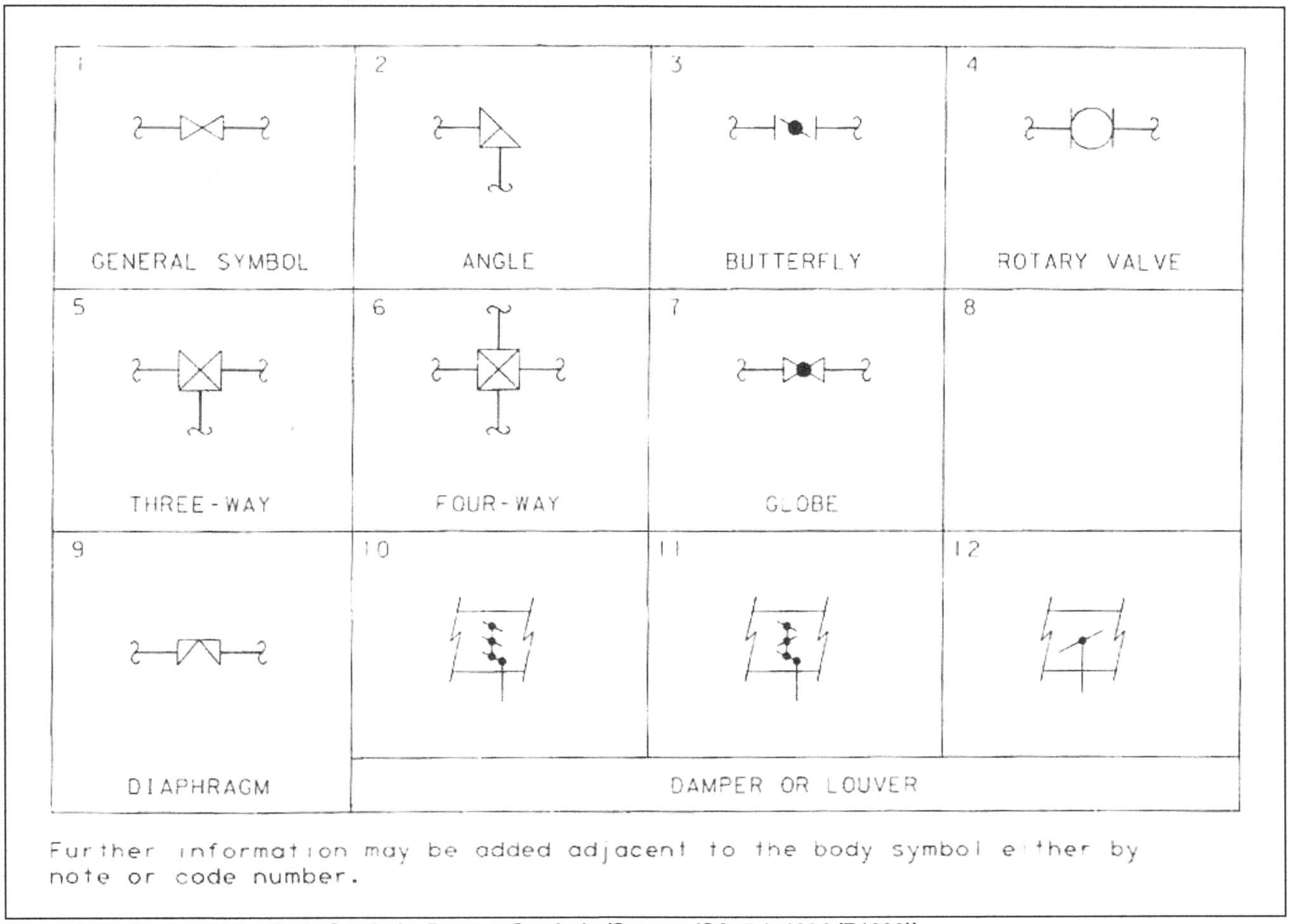

Figure 3-15. Control Valve Body Symbols, Damper Symbols (Source: ISA–5.1–1984 (R1992)).

There are three types of air lines. Generally, AS (air supply) is used, but, in order to differentiate between plant air lines and instrument air lines, the letters IA and PA are used. Instrument air lines should meet ANSI/ISA–7.0.01–1996–Quality Standard for Instrument Air. Instrument air must be clean and dry.

Connection to a Process

A fine, solid line is also used to indicate connection between the process piping lines and the device (see Figure 3-18).

Signal Lines

Signal lines lie between the primary element and other components in the instrument control loop. Figure 3-19 shows line symbols for the medium used to transmit data.

PNEUMATIC SIGNAL LINE

The pneumatic signal line symbol represents pneumatic tubing and is shown by two short diagonal lines on the fine, solid line about every one half inch. This indicates that an air (pneumatic) signal, usually 3–15 psig, is used as the energy medium. The level of the pneumatic signal may be written near the signal lines on loop diagrams.

The pneumatic signal symbol applies to a signal than uses any gas as the energy medium. If a gas other than air is used, the gas may be identified by a note on

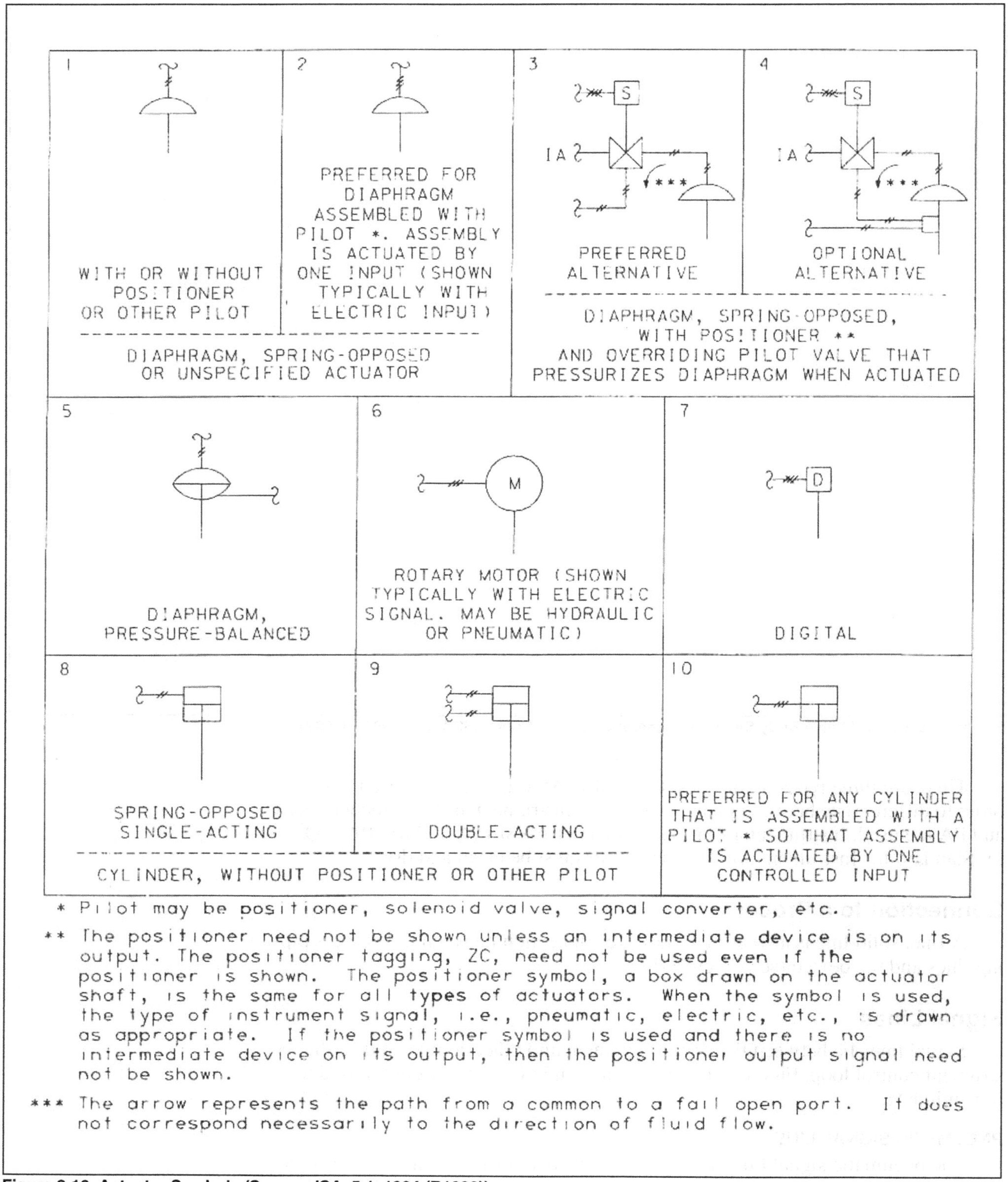

Figure 3-16. Actuator Symbols (Source: ISA–5.1–1984 (R1992)).

the signal symbol or otherwise. This pneumatic signal line is used for pneumatic connections between instruments but not for energy supply lines. Energy supply lines are always fine, solid lines.

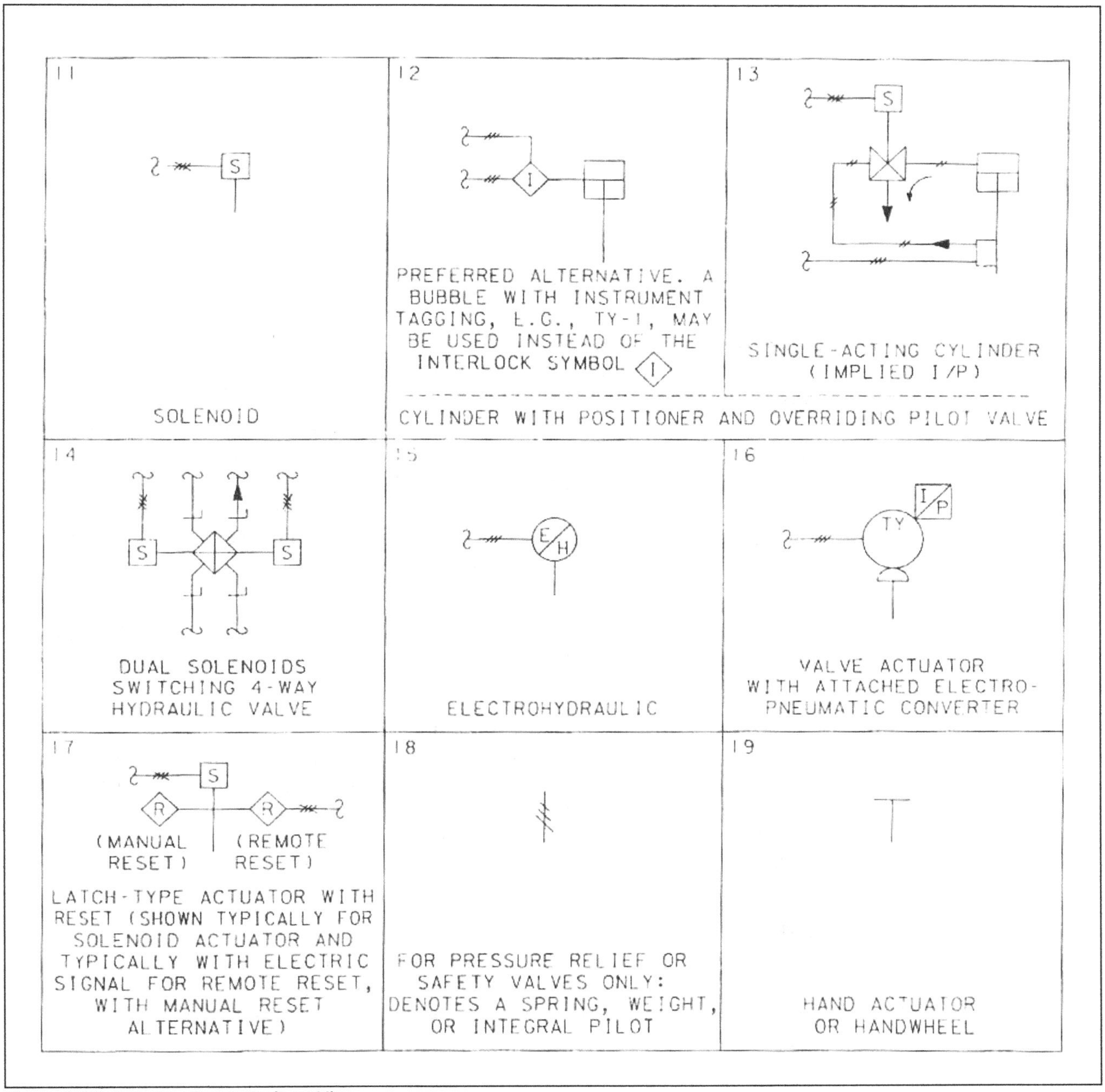

Figure 3-16. Actuator Symbols (cont'd).

AS	AIR SUPPLY	HS	HYDROGEN SUPPLY
IA	INSTRUMENT AIR	NS	NITROGEN SUPPLY
PA	PLANT AIR	SS	STEAM SUPPLY
ES	ELECTRIC SUPPLY	WS	WATER SUPPLY
GS	GAS SUPPLY		

Figure 3-17. Instrument Supply Line Symbols (Source: ISA–5.1–1984 (R1992)).

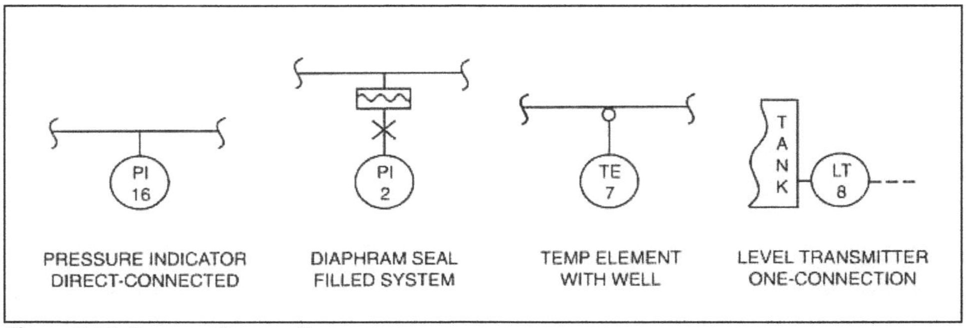

Figure 3-18. Process Connection.

ELECTRICAL SIGNAL LINE

The electrical signal is represented by dashed lines or by three short diagonal lines every one half inch on a fine, solid line. "Or" means this can be the user's choice, but consistency is recommended. These symbols represent signals that connect instruments or devices and not power supplies. If the power supply is 120 V AC, it is indicated by a fine, solid line as stated above.

The electrical signals may be current or voltage (AC or DC, analog or digital). Binary signals may be indicated with a back slash. The level of the signal may be written near the signal lines on instrument loop diagrams.

HYDRAULIC SIGNAL LINE

Hydraulic signal lines are represented by a capital letter L on the fine, solid line. Hydraulic lines use hydraulic fluids as the medium. Hydraulic lines are generally used when high pressures are required.

ELECTROMAGNETIC OR SONIC SIGNAL LINE

Electromagnetic phenomena include heat, radio waves, nuclear radiation, and light. "Sonic" pertains to audible sounds and ultrasonic sounds or acoustics. The symbol is a sine wave, and when guided it will also have a fine, solid line. "Guided" is a term that means the medium is confined within an enclosed area such as a wave guide, fiber optic cable, or other. "Not guided" means the medium travels through the atmosphere.

INTERNAL SYSTEM LINK

With the introduction of computers and microprocessors, it became necessary to indicate software or data link communications. A small circle breaks the solid line about every one half inch. This symbol is used to show communications within a microprocessor-based controller.

MECHANICAL LINK

The mechanical link symbol is a series of solid circles, each surrounded by another circle, that break the solid line. It is used when the transmission medium is mechanical, such as gears or a shaft.

CAPILLARY TUBE SYMBOL

The capillary tube symbol is a fine, solid line with an X every one half inch. This symbol is generally used only from the primary element to the first device it is connected to. Some examples are filled temperature bulbs and pressure transmitters in which the diaphragm is separated from the electronics.

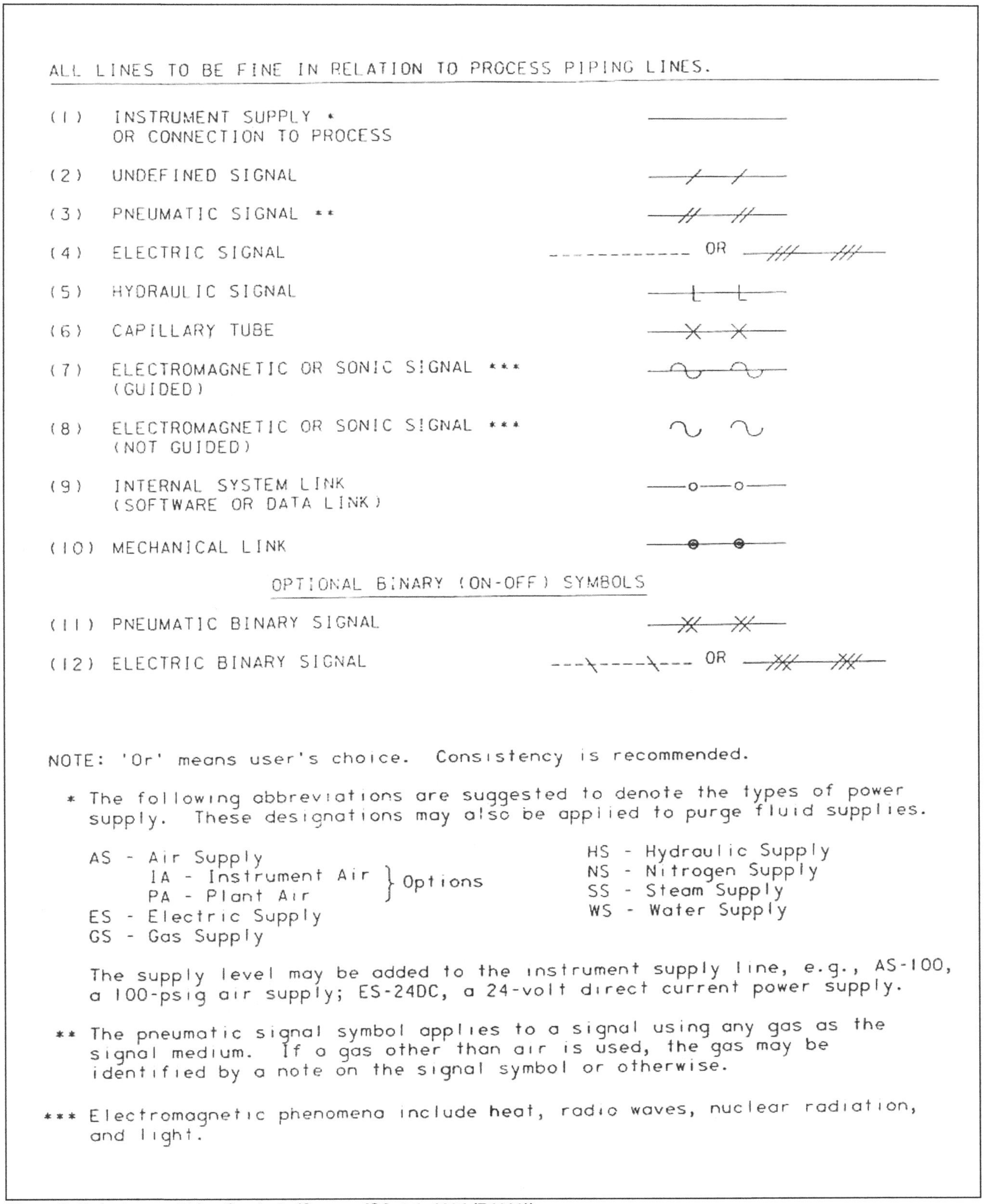

Figure 3-19. Instrument Line Symbols (Source: ISA–5.1–1984 (R1992)).

UNDEFINED SIGNAL LINE

A solid line with one short diagonal line every one half inch is the symbol for an undefined signal line. When this symbol is used, a statement or note should be put on the drawing to indicate the actual medium.

BINARY SIGNAL LINES

Optional binary (on-off) symbols may indicate with a back slash on the normal signal line. An electrical binary signal may be back slashed on a dashed line. This symbol is used when it is advisable to differentiate between the normal 4–20 mA signal lines and binary lines.

Internal System Function Block Symbols

The internal system function block function designations may be used to designate the computer function for computation and signal conditioning. The use of a box around the appropriate symbol avoids confusion by setting off the symbol from other markings on a diagram and permits the function to be used as a stand-alone block on conceptual designs (see Figure 3-20). The function designations associated with controllers, computing devices, converters, and relays may be used individually or in combination.

Additional Symbols Used on Drawings and Diagrams

All of the symbols mentioned above are used on process and instrumentation diagrams (P&ID) and instrument loop diagrams. Some of the symbols are used on other process drawings, and, of course, some additional symbols are used on other drawings, diagrams, and schematics. the general terminal or bulkhead symbol is used on instrument loop diagrams to identify a junction box or termination assembly and to identify each connection used (see Figure 3-21).

Instrument terminals and/or ports show the connection to the instrument (see Figure 3-22). It is suggested that identifying numbers or letters be the manufacturer's designation. The balloon symbol and identification are the same as shown on the P&ID.

Instrument system energy supply identification is shown on instrument loop diagrams rather than on P&IDs. The identification includes all energy supply connections and also indicates the appropriate voltage or pressure, as shown in Figure 3-23.

Additional information that may be found on instrument loop diagrams is included later in this chapter.

Tag Numbers

Instrument symbols usually contain a tag number, which is the unique identification number of the instrument (see Figure 3-1). The tag number indicates the function of the instrument and its loop number (see Figure 3-24).

Function

The definition of "function" is the "purpose of" or "an action performed by" a device. The functional identification of an instrument or its functional equivalent consists of letters from Table 3-1 and includes a first letter, which designates the measured or initiating variable (i.e., T indicates temperature, P indicates pressure, F indicates flow, DP indicates differential pressure, etc.) and one or more succeeding letters, which identify the functions performed.

The sequence of identification letters begins with a first letter selected according to Table 3-1. Readout or passive functional letters follow in any order, and output functional letters follow these in any sequence, except that output letter C

THE FUNCTION DESIGNATIONS ASSOCIATED WITH CONTROLLERS, COMPUTING DEVICES, CONVERTERS AND RELAYS MAY BE USED INDIVIDUALLY OR IN COMBINATION (ALSO, SEE TABLE 1, NOTE 14.). THE USE OF A BOX AVOIDS CONFUSION BY SETTING OFF THE SYMBOL FROM OTHER MARKINGS ON A DIAGRAM AND PERMITS THE FUNCTION TO BE USED AS A STAND-ALONE BLOCK ON CONCEPTUAL DESIGNS.

NO	FUNCTION	SYMBOL	MATH EQUATION	GRAPHIC REPRESENTATION	DEFINITION
1	SUMMING	Σ	$M = X_1 + X_2 + \ldots + X_n$	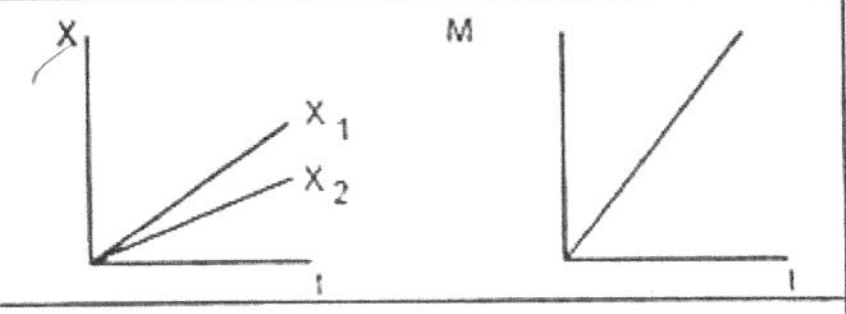	THE OUTPUT EQUALS THE ALGEBRAIC SUM OF THE INPUTS. (THE INPUTS MAY BE LABELED WITH POSITIVE OR NEGATIVE SIGNS).
2	AVERAGING	$\Sigma/_n$	$M = \frac{X_1 + X_2 + \ldots + X_n}{n}$	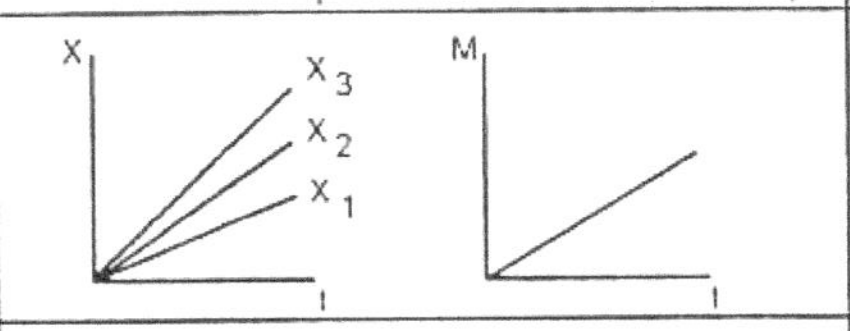	THE OUTPUT EQUALS THE ALGEBRAIC SUM OF THE INPUTS DIVIDED BY THE NUMBER OF INPUTS
3	DIFFERENCE	Δ	$M = X_1 - X_2$	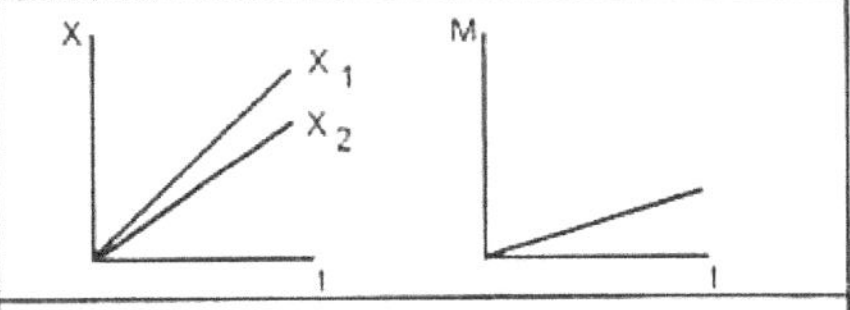	THE OUTPUT EQUALS THE ALGEBRAIC DIFFERENCE OF THE TWO INPUTS.
4	PROPORTIONAL	K 1:1 2:1	$M = KX$	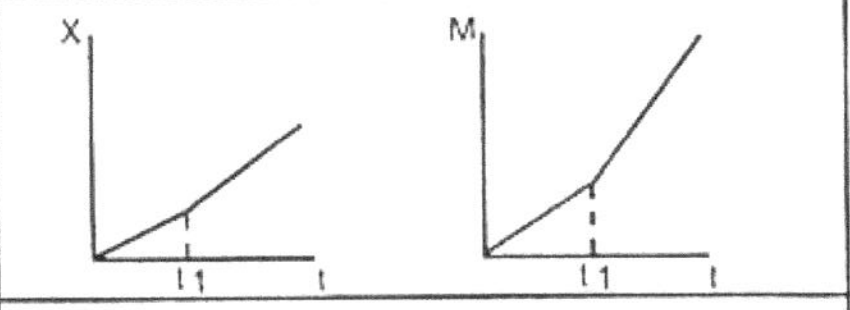	THE OUTPUT IS DIRECTLY PROPORTIONAL TO THE INPUT. IN THE CASE OF A VOLUME BOOSTER, "K" MAY BE REPLACED BY 1:1 FOR INTEGER GAINS, 2:1, 3:1, ETC., MAY BE SUBSTITUTED FOR K.
5	INTEGRAL	$\int$	$M = \frac{1}{T_I} \int X dt$	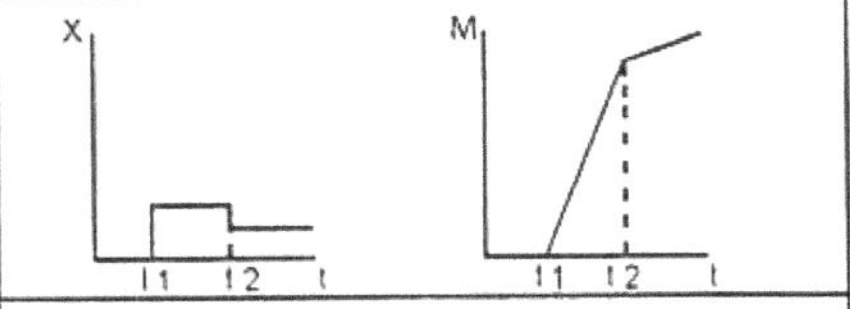	THE OUTPUT VARIES IN ACCORDANCE WITH BOTH MAGNITUDE AND DURATION OF THE INPUT. THE OUTPUT IS PROPORTIONAL TO THE TIME INTEGRAL OF THE INPUT.
6	DERIVATIVE	$^d/_{dt}$	$M = T_D \frac{dX}{dt}$	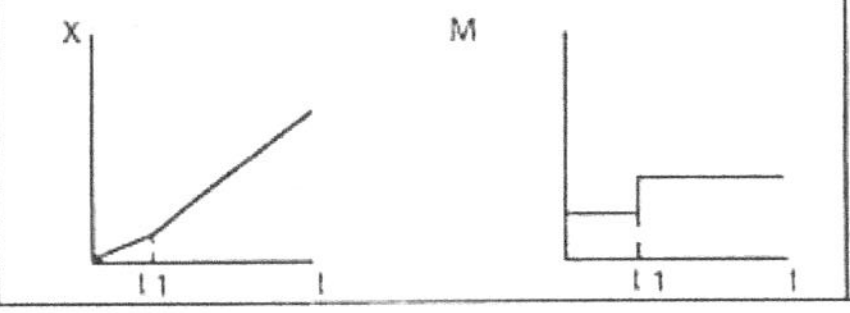	THE OUTPUT IS PROPORTIONAL TO THE RATE OF CHANGE (DERIVATIVE) OF THE INPUT.

Figure 3-20. Function Block Symbols (Source: ISA–5.1–1984 (R1992)).

NO	FUNCTION	SYMBOL	MATH EQUATION	GRAPHIC REPRESENTATION	DEFINITION
7	MULTIPLYING	X	$M = X_1 X_2$	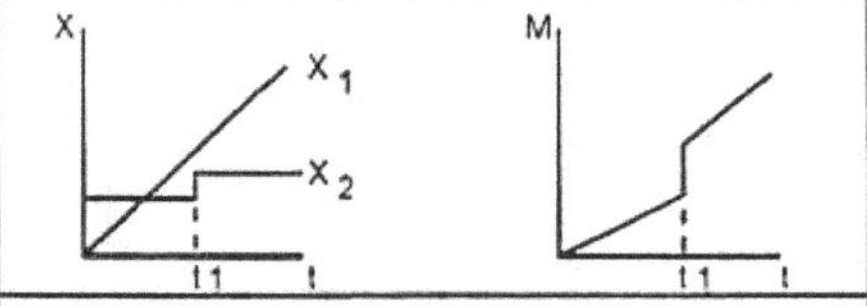	THE OUTPUT EQUALS THE PRODUCT OF THE TWO INPUTS.
8	DIVIDING	÷	$M = \frac{X_1}{X_2}$	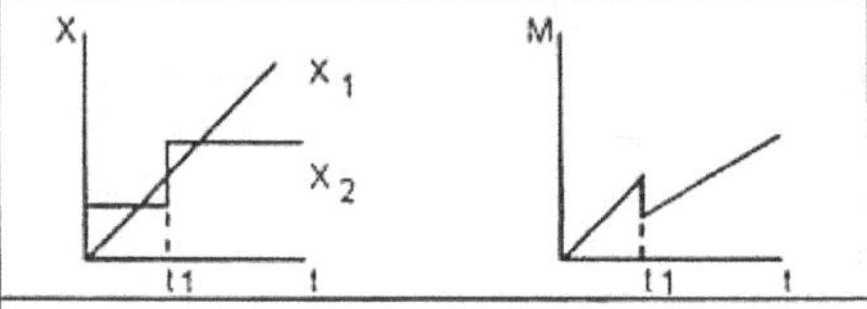	THE OUTPUT EQUALS THE QUOTIENT OF THE TWO INPUTS.
9	ROOT EXTRACTION	$\sqrt[n]{\ }$	$M = \sqrt[n]{X}$	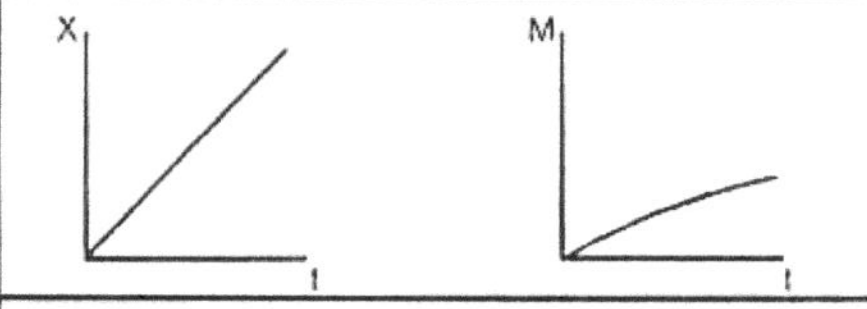	THE OUTPUT EQUALS THE ROOT (I.E., CUBE ROOT, FOURTH ROOT, 3/2 ROOT, ETC.) OF THE INPUT. IF n IS OMITTED, A SQUARE ROOT IS ASSUMED.
10	EXPONENTIAL	x^n	$M = X^n$	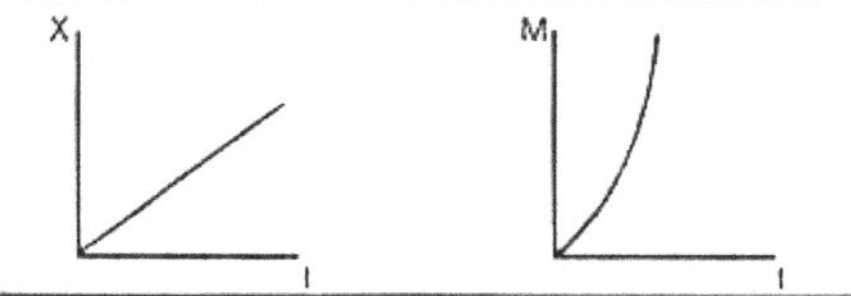	THE OUTPUT EQUALS THE INPUT RAISED TO A POWER (I.E., SECOND, THIRD, FOURTH, ETC.).
11	NONLINEAR OR UNSPECIFIED FUNCTION	f (X)	$M = f(X)$	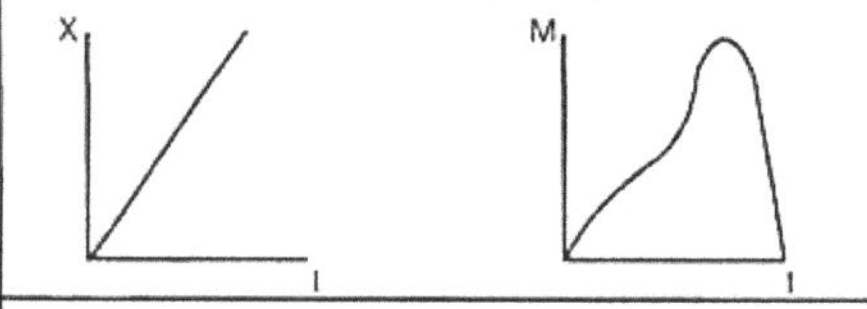	THE OUTPUT EQUALS SOME NONLINEAR OR UNSPECIFIED FUNCTION OF THE INPUT.
12	TIME FUNCTION	f (t)	$M = Xf(t)$ $M = f(t)$	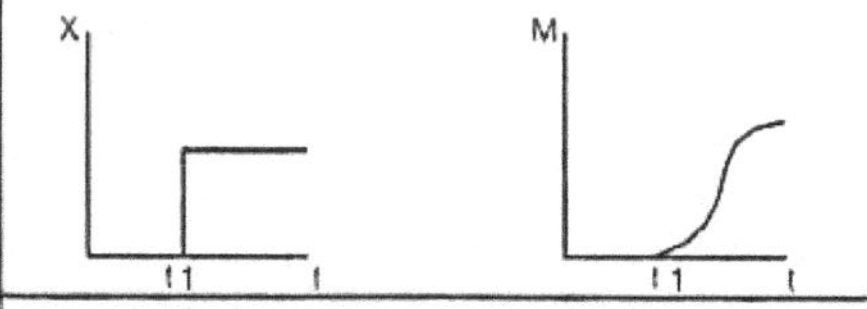	THE OUTPUT EQUALS THE INPUT TIMES SOME FUNCTION OF TIME OR EQUALS SOME FUNCTION OF TIME ALONE.
13	HIGH SELECTING	>	$M = \begin{cases} X_1 \text{ FOR } X_1 \geq X_2 \\ X_1 \text{ FOR } X_1 \leq X_2 \end{cases}$	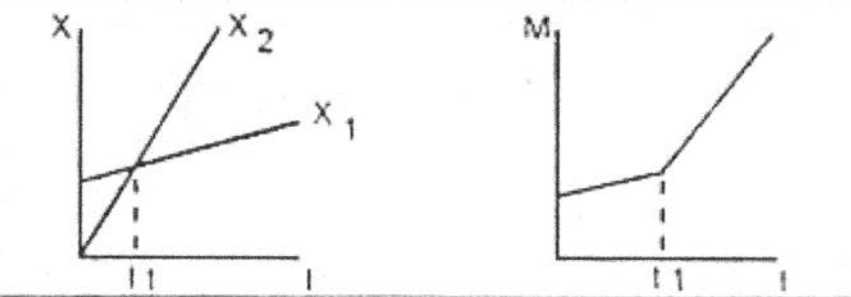	THE OUTPUT IS EQUAL TO THE GREATER OF THE INPUTS.

Figure 3-20. Function Block Symbols (Source: ISA–5.1–1984 (R1992)) cont'd.

NO	FUNCTION	SYMBOL	MATH EQUATION	GRAPHIC REPRESENTATION	DEFINITION
14	LOW SELECTING	<	$M = \begin{cases} X_1 \text{ FOR } X_1 \leq X_2 \\ X_1 \text{ FOR } X_1 \geq X_2 \end{cases}$	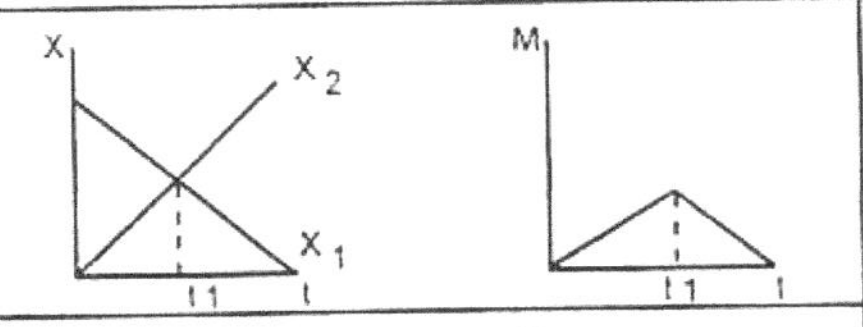	THE OUTPUT IS EQUAL TO THE LESSER OF THE INPUTS.
15	HIGH LIMITING	≯	$M = \begin{cases} X \text{ FOR } X \leq H \\ H \text{ FOR } X \geq H \end{cases}$	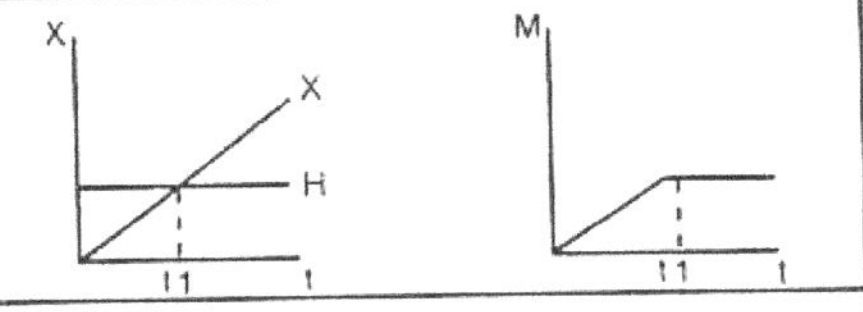	THE OUTPUT EQUALS THE INPUT OR THE HIGH LIMIT VALUE WHICHEVER IS LOWER.
16	LOW LIMITING	≮	$M = \begin{cases} X \text{ FOR } X \geq L \\ L \text{ FOR } X \leq L \end{cases}$	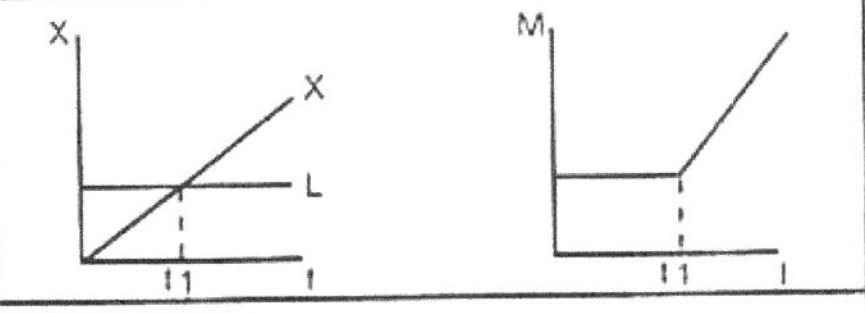	THE OUTPUT EQUALS THE INPUT OR THE LOW LIMIT VALUE WHICHEVER IS HIGHER.
17	REVERSE PROPORTIONAL	-K	$M = -KX$	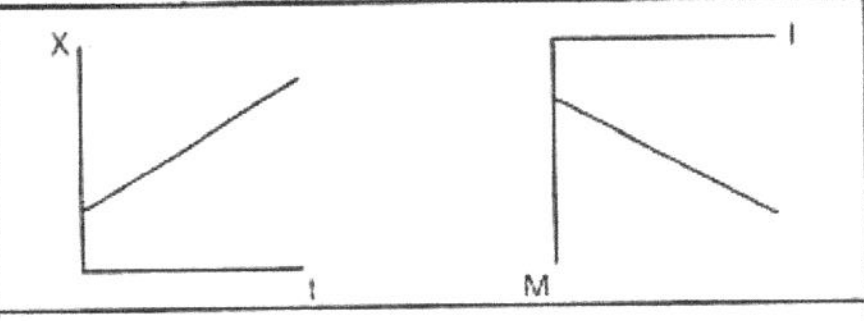	THE OUTPUT IS REVERSELY PROPORTIONAL TO THE INPUT.
18	VELOCITY LIMITER	V̸	$\frac{dM}{dt} = \frac{dX}{dt} \begin{cases} \frac{dX}{dt} \leq H \text{ AND} \\ M = X \end{cases}$ $\frac{dM}{dt} = H \begin{cases} \frac{dX}{dt} \geq H \text{ OR} \\ M \neq X \end{cases}$	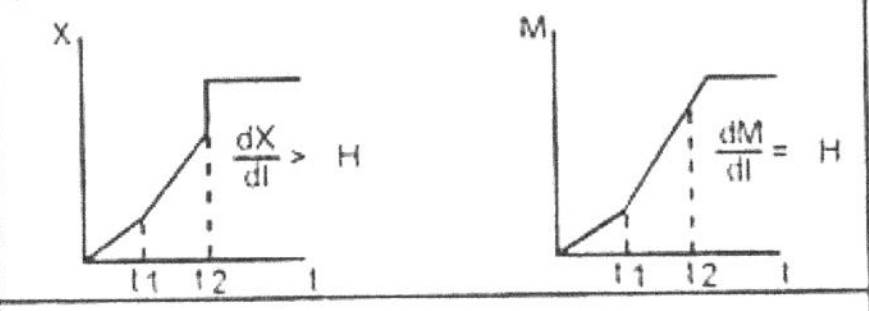	THE OUTPUT EQUALS THE INPUT AS LONG AS THE RATE OF CHANGE OF THE INPUT DOES NOT EXCEED A LIMIT VALUE. THE OUTPUT WILL CHANGE AT THE RATE ESTABLISHED BY THIS LIMIT UNTIL THE OUTPUT AGAIN EQUALS THE INPUT.
19	BIAS	+ − ±	$M = X \mp b$	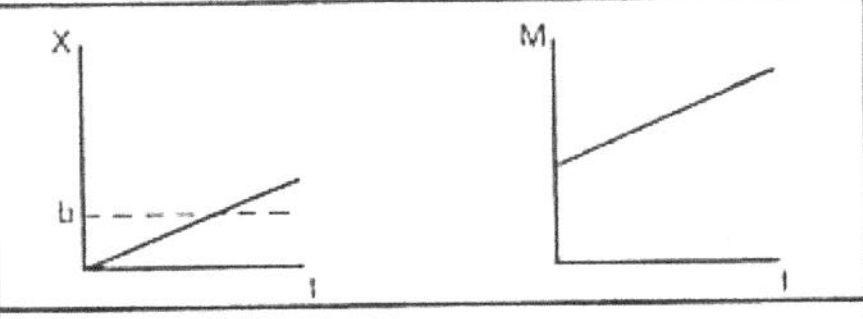	THE OUTPUT EQUALS THE INPUT PLUS (OR MINUS) SOME ARBITRARY VALUE (BIAS).
20	CONVERT	*/*	OUTPUT = f (INPUT)	NONE	THE FORM OF THE OUTPUT SIGNAL IS DIFFERENT FROM THAT OF THE INPUT. E - VOLTAGE I - CURRENT P - PNEUMATIC A - ANALOG B - BINARY H - HYDRAULIC O - ELECTROMAGNETIC, SONIC R - RESISTANCE (ELECT.) D - DIGITAL

Figure 3-20. Function Block Symbols (Source: ISA–5.1–1984 (R1992)) cont'd.

NO	FUNCTION	SYMBOL	MATH EQUATION	GRAPHIC REPRESENTATION	DEFINITION
21	SIGNAL MONITOR	** H	STATE 1 $X \leq H$ STATE 2 (ENERGIZED OR ALARM STATE) $X > H$	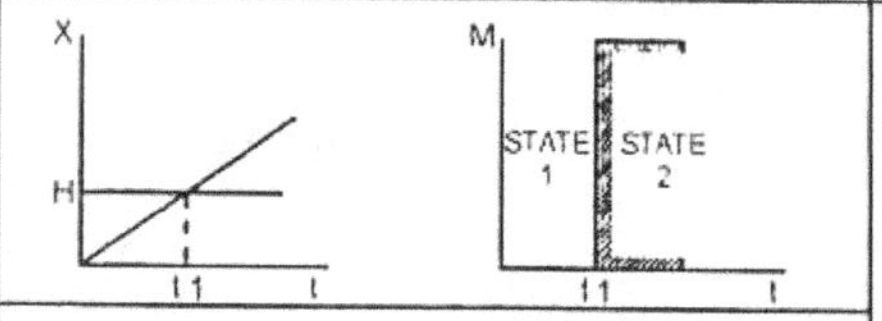	THE OUTPUT HAS DISCRETE STATES WHICH ARE DEPENDENT ON THE VALUE OR THE INPUT. WHEN THE INPUT EXCEEDS (OR BECOMES LESS THAN) AN ARBITARY LIMIT VALUE THE OUTPUT CHANGES STATE.
		** L	STATE 1 (ENERGIZED OR ALARM STATE) $X < L$ STATE 2 $X \geq L$	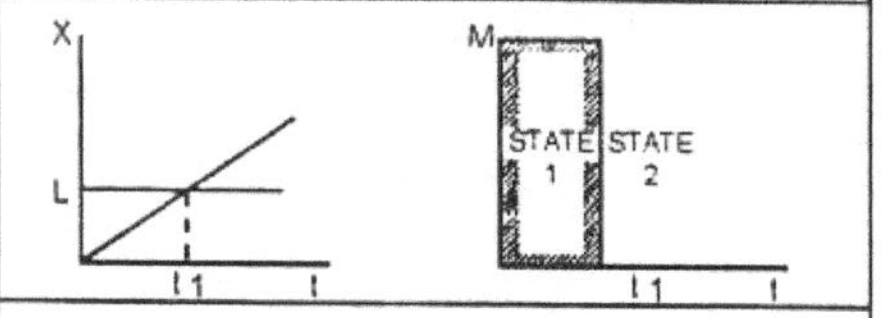	
		** HL	STATE 1 (FIRST OUTPUT M ENERGIZED OR ALARM STATE) $X < L$ STATE 2 (BOTH OUTPUTS INACTIVE OR DE-ENERGIZED) $L \leq X \leq H$ STATE 3 (SECOND OUTPUT M_2 ENERGIZED OR ALARM STATE) $X > H$	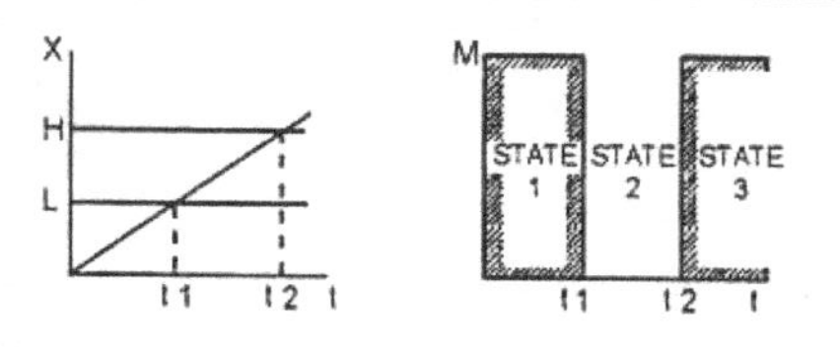	

THE VARIABLES USED IN THE TABLE ARE:

- b - ANALOG BIAS VALUE.
- $\frac{d}{dt}$ - DERIVATIVE WITH RESPECT TO TIME.
- H - AN ARBITARY ANALOG HIGH LIMIT VALUE.
- $\frac{1}{T_1}$ - INTEGRATING RATE.
- L - AN ARBITARY ANALOG LOW LIMIT VALUE.
- M - ANALOG OUTPUT VARIABLE.
- n - NUMBER OF ANALOG INPUTS OR VALUE OF EXPONENT.
- t - TIME
- T_D - DERIVATIVE TIME.
- X - ANALOG INPUT VARIABLE.
- $X_1, X_2, X_3, \ldots, X_n$ - ANALOG INPUT VARIABLE (1 TO N IN NUMBER).
- * - TABLE 1 LETTER DESIGNATORS.

NOTE: THE SQUARE MAY BE USED AS A FLAG

I-O	ON-OFF
REV	REVERSE ACTION

THIS TABLE HAS BEEN MODIFIED FROM SAMA PMC 22-11-1981 WITH PERMISSION OF THE COPYRIGHT HOLDER, SCIENTIFIC APPARATUS MAKERS ASSOCIATION. COPIES OF PMC 22-11-1981 ARE AVAILABLE FROM SAMA, 1101 16TH STREET NW, WASHINGTON, D.C., 20036.

Figure 3-20. Function Block Symbols (Source: ISA–5.1–1984 (R1992)) cont'd.

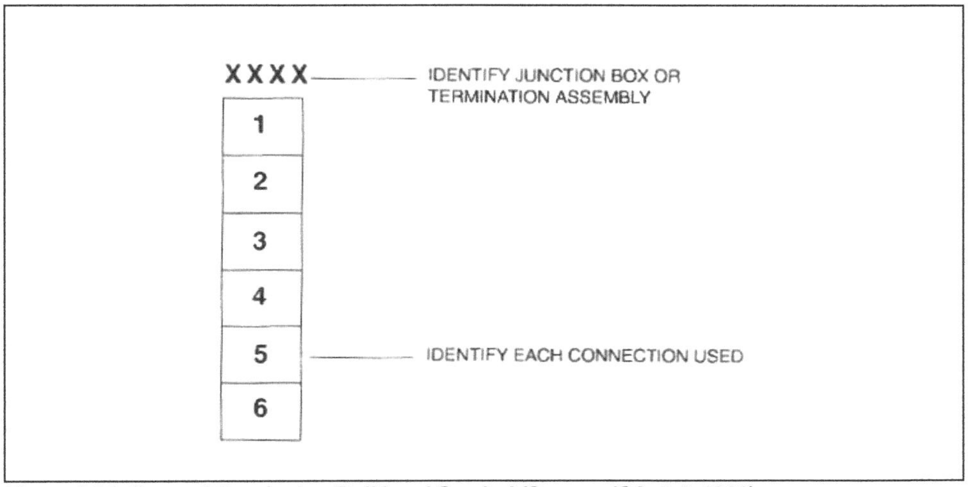

Figure 3-21. General Terminal or Bulkhead Symbol (Source: ISA–5.4–1991).

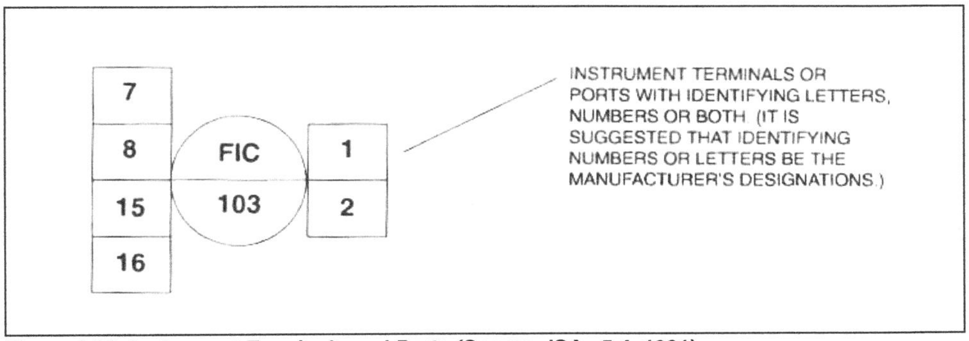

Figure 3-22. Instrument Terminals and Ports (Source: ISA.–5.4–1991)

(control) precedes output letter V (valve) (e.g., PCV, a self-actuated control valve). However, modifying letters, if used, are interposed so that they are placed immediately following the letters they modify.

The functional identification of an instrument is made according to the function and not according to the construction.

The functional identification of a differential pressure recorder used for flow measurement is FR; a pressure indicator and a pressure-actuated switch connected to the output of a pneumatic level transmitter are identified by LI and LS, respectively. The first letter of the functional identification is selected according to the measured or initiating variable, not according to the manipulated variable. Thus, a control valve that varies flow according to the dictates of a level controller is an LV (level valve), not an FV (flow valve).

Any first letter, if used in combination with the modifying letters D (differential), F (ratio), M (momentary), K (time rate of change), Q (integrate or totalize), or any combination of these, is intended to represent a new and separate measured variable, and the combination is treated as first letter entity.

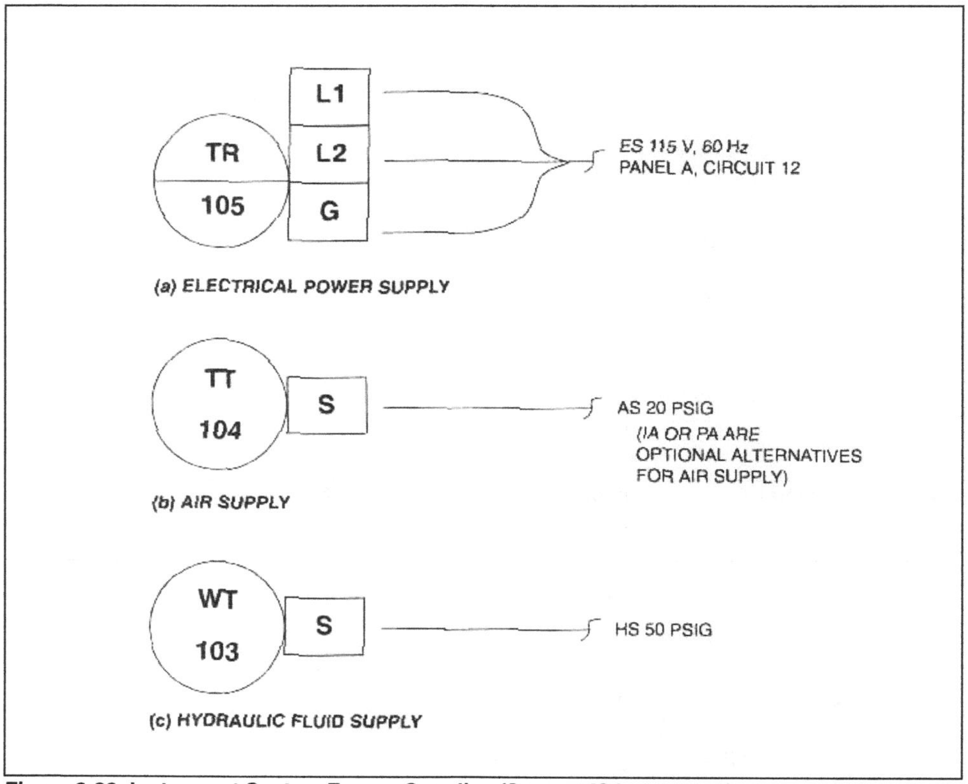

Figure 3-23. Instrument System Energy Supplies (Source: ISA–5.4–1991).

TYPICAL TAG NUMBER

TIC 103	— INSTRUMENT IDENTIFICATION OR TAG NUMBER
T 103	— LOOP IDENTIFICATION
103	— LOOP NUMBER
TIC	— FUNCTIONAL IDENTIFICATION
T	— FIRST-LETTER
IC	— SUCCEEDING-LETTERS

EXPANDED TAG NUMBER

10-PAH-5A	— TAG NUMBER
10	— OPTIONAL PREFIX
A	— OPTIONAL SUFFIX

NOTE: HYPHENS ARE OPTIONAL AS SEPARATORS.

Figure 3-24. Tag Numbers.

Remember, the first letter is being modified. Use the modifier after the first letter, not the modifier after succeeding letters.

The succeeding letters of the functional identification designate one or more readout or passive functions and/or output functions. A modifying letter may be used, if required, in addition to one or more other succeeding letters.

Expanded Tag Number

If a given loop has more than one instrument with the same functional identification, a suffix may be appended to the loop number, e.g, FV-2A, FV-2B, FV-2C, etc., or TE-25-1, TE-25-2. The suffixes may be applied according to the following guidelines:

(1) An uppercase suffix letter should be used, i.e., A, B, C, etc.

(2) For an instrument such as a multipoint temperature recorder that prints number for point identification, the primary elements may be numbered TE-25-1, TE-25-2, TE-25-3, etc., corresponding to the point identifications number.

(3) Further subdivisions of a loop may be designated by serially alternating suffix letters and numbers.

4

Maintenance Personnel

Overview

What should instrumentation and control system personnel know to do their jobs? What training is needed to work in the I&C field? What type of experience is needed? Should they have a mechanical, electronic, or computer background? What skills does the job require? These are questions that must be considered when pursuing a job in instrumentation, measurement, control, and manufacturing systems.

Training skilled support is key to best practices. One issue receiving a considerable amount of concern by production operation and maintenance managers is improving and maintaining the skill set needed by the support staff to maintain equipment and systems. Manufacturers have established training programs to keep their support staff current on emerging technologies. However, there is a growing concern that as the complexity of automation systems steadily increases, the availability of qualified technicians will diminish.

As maintenance staffs are trained to support more complex automation systems, their enhanced skill set becomes highly marketable, and they leave due to more marketable skills. Training support staff is essential as companies upgrade automation systems and equipment, but the very training that is required makes them vulnerable to staff attrition.

Companies recognize that the complexity of plant floor systems will increase due to emerging technologies and more sophisticated automation. There is concern that obtaining and retaining technicians with the necessary skill sets to support this new technology could become an issue.

Operations and maintenance managers are aware of the problem of retaining their in-house trained technicians, but there is currently no way to prevent them from leaving due to the demand for their specific skill set.

Companies will continue to provide the majority of in-house production automation systems support. In terms of asset management, some companies consider maintenance and support of their operations as a core competency. This implies more internal training and limiting outsourcing maintenance to very complex systems and equipment. A best practice approach would be to foster a lifetime career attitude through a continuous training program.

Multi-Disciplined

Operations and maintenance managers are finding that it is becoming necessary to train their staff to be multi-disciplined with regard to the equipment and technologies. Where a maintenance technician was once an expert in a single system or piece of equipment, he or she must now be able to service and support a much more diversified range. This is accomplished by internal training where technicians are required to teach specific skills to other staff members, and similarly must learn skills of their team members.

Continuous Training

The majority of operations managers plan to sustain or even decrease current staffing levels based on pressures to reduce cost. At the same time, there is no reduction in equipment and production systems to be supported. They will meet these added demands by "stretching" the support staff, with current support staff required to maintain equipment and systems beyond the present scope. Maintenance managers realize the need for continuous training to meet the maintenance needs of production operations. However, the need for increased training budgets conflicts with the overall operations budgets being slashed across most manufacturing operations.

Training of Maintenance Workers

What is meant by "training programs" for maintenance employees? These are the instructional programs that will enable employees to acquire the knowledge and skills necessary to perform planned, preventive, and predictive maintenance tasks.

Some experts say one way to measure the value of training is to compare it with the often ruinous cost of down time. Trained workers can keep down time to a minimum.

Training programs are important because they help the employee better accomplish his or her job with quality and efficiency. Training should not be thought of as remedial learning to correct some deficiency in a worker's skills, nor should it be considered a process that ends when the worker reaches a certain predetermined level of knowledge or skill. The most productive, satisfied workers tend to be those who continually grow with their jobs, developing new talents over time, and training can be a part of that continual growth. Managers can create an atmosphere in which employees will want to learn and grow.

The companies that recognize the need for ongoing training stand a better chance of surviving and prospering. Those that don't run the risk of technological obsolescence, decreased productivity, and increased costs. Training of maintenance employees should represent a significant part of the maintenance department's activity and should be included in planning for plant or equipment upgrades.

Having the newest, best control hardware available isn't worth much without well-trained people to operate and maintain it.

The continuous training of maintenance workers is a necessity in a quality maintenance organization. In fact, a new worker's training needs should be assessed when the worker is invited to join the maintenance department. During the hiring process, management may administer aptitude tests to determine whether the candidate has the capability to do the job he or she is seeking, psychological tests to determine a candidate's temperament, and interviews to gage the candidate's level of commitment to the job. Obviously, the candidate who is hired will perform fairly well on all these tests, but the tests should reveal areas in which training will benefit the new employee.

Ideally, after all tests and interviews are completed, an individual training program (ITP) should be developed for the new employee. Here, an attempt is made to match the skills of the employee against specific job assignments. Job task analysis is a significant help in the identification of skills needed to perform a specific job. Job assignments and position descriptions will change with time, and the constant updating of the ITP is essential to the career growth of each maintenance employee. It is suggested that this information be reviewed with the employee every year.

There is some confusion between *training* and *education* as it relates to the training function. The word "education" generally refers to formal education that leads to a degree. Some companies and some jobs may have formal education requirements in order for a candidate to be hired. Also, a company may support a worker's education after hiring, to help the worker progress in his or her job.

ISA offers certification programs for control systems technicians (CCST) and maintenance mechanics (CIMM). For details, go to www.isa.org/certify.

Training, contrasted with a formal education curriculum, can be seen as very job- or task-related. A training course might be designed to a specific skill—for example, to train the employee to calibrate a magnetic flowmeter. There might be some theory in the course, but the emphasis would be on "hands-on" calibration of a specific flowmeter. Upon completion of the training, the employee would demonstrate to an instructor or supervisor that he or she could successfully calibrate the flowmeter.

To summarize, training is job- or task-specific. Education may include some training skills but is not meant to help an employee accomplish specific tasks. Also, education may not eliminate the employee's need for task-specific training.

Objectives

Planning and clearly stating department objectives for the training program is extremely important.

A maintenance department should have written objectives that define and detail the tasks for which the department is held responsible. These department-wide objectives can drive the development of training objectives for each individual in the department. These would include identification of the skills needed to do the routine tasks in a specific job at the present time. But in planning training objectives, employees and their supervisors should ask questions such as: What sort and what level of cross-training (training Employee A in Employee B's job duties) are needed so that the department can perform essential functions even if a key employee is absent? Are employees ready to deal with emergencies or other nonroutine situations? Is refresher training needed to reinforce emergency procedures or other skills that are not used routinely? What skills are going to be needed in the future when some new process comes on stream? Individuals in the department should also have objectives for career growth. What skills are needed to grow in the job?

Resources

Many resources for education and training are available to maintenance employees. There is no "best" training resource; employees or supervisors should choose the one best suited for the training objectives or the individual employee's learning style.

Some companies try to adopt generic training programs that are quickly set in motion, easily administered, and rarely adjusted for different company or individual needs. This is not the ideal approach. A better approach is the individual training plan method in which the employee and the supervisor map out a training plan that may include a number of resources, always keeping in mind the objectives of the training.

What specific resources are available? These can be divided into a few categories:

(1) Institutional classroom instruction: local community college courses, private training centers

(2) In-plant training course instruction: vendor training, company-designed courses, video instruction

(3) Individualized training: programmed instruction, computer-based instruction

INSTITUTIONAL CLASSROOM INSTRUCTION

Formal instruction might include the following:

(1) *Community college* courses and programs directed towards branches of technology such as electronics, computer engineering technology, or instrumentation and automatic process control.

(2) *Private training center* courses in more specific subjects such as statistical process control, HVAC applications, microelectronics, or instrument calibration. ISA, the publisher of this book, operates a training center with courses that range in length from one day to a week or more.

IN-PLANT TRAINING COURSES

Companies often arrange classroom training for employees at the work site.

Vendor-taught courses are part of the technical support many major equipment manufacturers provide to their customers. Vendor courses usually pertain to the operation and maintenance of that vendor's equipment but can include basic instrumentation courses as well. These courses vary in length; they may be as short as three days or as long as thirty days.

Some software vendors offer courses in the use of their software products. Types of software that may be supported by training include maintenance management systems and computer-integrated manufacturing programs.

Company-taught classroom training in instrumentation and control is common. Large companies may have training departments that teach operations and procedures required by the company. Company-taught courses might include computer operations, OSHA requirements, plant safety, and new equipment and systems operator's training. These are subjects that are important to many companies, but an in-house training program can make the course specific to the needs of the company. Sometimes this training is supplemented with vendor or contract-taught courses that are used as part of company apprenticeship programs.

Large corporations can organize training interventions when sudden changes in policy or procedure require many or all employees to be retrained. An example was asbestos awareness training. Sudden legal concerns about a subject in which nobody had a background meant that almost everyone in some plants needed training.

Training consultants offer customized training programs that are tailored to the needs of the employees and can be very specific. Training organizations also come to plants bringing "canned" courses to be offered at the plant site. Again, these activities are generally for group participation and all employees receive the same experience. Typically, training consultants work under short-term contracts to teach individual courses. Courses in specific requirements (the National Electrical Code, for example) and quality programs are often taught by training contractors who specialize in such areas.

INDIVIDUALIZED TRAINING

This is increasingly popular with maintenance departments. Here, an employee sets goals and objectives based upon his or her own specific skill needs. Individual training also allows flexibility in scheduling.

Programmed instruction involves having the student follow lessons from text materials. Usually, sets of exercises that the student works periodically test the student's comprehension and reinforce the lessons. This approach allows the student to work at his or her own pace.

Interactive or computer-assisted instruction has largely replaced the older text-based approach. A student works at a computer terminal and follows a menu of instruction. In interactive video programs, an employee sees a video image of a specific instrument and identifies component parts with a mouse or a touchscreen. The software can give multiple-choice tests and record the student's score automatically, right from a computer terminal. It has been proven that this multimedia approach to individualized training, engaging the senses of sight, hearing, and touch, is much more effective in ensuring students' retention of knowledge and skills.

Methods

The methods used for training maintenance workers are varied and diversified, but the two main categories are the traditional lecture/demonstration and practical "hands-on," in addition to those discussed previously.

In the *lecture/demonstration method* information is presented in a typical classroom setting. Students listen and then respond to the information presented. This method is fast, allows a high density of material in a short period of time, but is relatively inefficient in terms of new information or skills learned and retained. If this method is compared to that of the interactive video, a much higher retention rate can be seen in the latter.

Hands-on training may be conducted in a variety of settings. The science laboratory may have been the first attempt at hands-on instruction. It must be stated that to develop a hands-on classroom/laboratory is a very expensive undertaking. In addition to equipment costs, the ratio of students to instructors must be low, allowing lots of individual help from the instructor. Some skills are best taught in the laboratory or plant process center. Although the cost of hands-on training is high, it is still the preferred method for skills training. In apprentice training the laboratory exercises are generally followed up by field training with more senior mechanics or technicians.

Training Schedules

What is the best way to schedule training? This is a much-debated question. Some union contracts demand a certain number of hours of training each month. A more practical approach is to schedule training as needed, with the amount of time not fixed but flexible. The trainee should take enough time to master the new skill to the appropriate level of competence.

Certainly not all training can be delivered this way. Some must be much more structured. For example, training might be needed on a new piece of process equipment, with this training to be supplied by the equipment vendor. The vendor decides on a specific day for two sessions of four hours each to be delivered on the plant site. The maintenance department proceeds to schedule two groups in four-hour blocks. Training costs are paid to the vendor in accordance with the contracted prices for the training.

This is an example of training that is expedient and easy to schedule, but is it cost effective in terms of what is learned by each of the maintenance workers who attended the training? An alternate approach to meeting the training needs would be to select three lead mechanics to meet with the vendor. In this approach, after orientation to and discussion of the new piece of equipment, the lead mechanics work on the unit in the plant and perform the required maintenance under the supervision of the vendor. Each of the three mechanics performs the operation to the satisfaction of the vendor and the customer. At a later time each of the three leaders forms groups of three mechanics to perform the skill training required. This activity is rotated until all the mechanics are able to perform the required skill to the satisfaction of the customer. This might be a more effective way to schedule training and meet the needs of the customer and to motivate the employees.

Training Evaluation

Evaluating a training course after it has been given is time-consuming and can be tedious, but it is very important. The lack of evaluation of a training experience can be frustrating to management and to trainees. Frequently, the numbers of hours employees spend in training are noted and checked off, but that should not be the extent of the evaluation. It is important to determine how well the objectives of the training have been achieved. Otherwise, no one may know whether an important skill has been learned correctly until the skill has to be applied, and by then it may be too late.

An evaluation mechanism should be an important part of every training activity.

One simple test is to have immediate supervisors review skill competence by observing a demonstration by the employees who have just received training. Cumulative training evaluation reports can be helpful to management at annual evaluations of employees.

Instructors also should be made accountable in the evaluation process. Employees should have an opportunity to make criticisms or suggestions regarding the quality of the course and the skill of the instructor, and to give this feedback on a confidential basis if they prefer.

Training Costs

Training can be the best investment the maintenance department makes. If proper training leads to maximum safety and efficiency and minimum down time, training costs are a real bargain.

In an industrial climate where technological obsolescence is an ever-present threat in a plant, a good rule of thumb for training budgeting purposes is the expenditure of about $1000.00 per year per mechanic.

For formal education leading to a degree, many companies offer tuition assistance programs whereby a worker is reimbursed for part or all tuition costs. Remember, however, that educational programs are generally not job/task specific.

Multicraft/Multiskilled, Multi-Disciplined

The definition of multicraft as stated in *The Automation, Systems, and Instrumentation Dictionary* is "maintenance personnel who are proficient in more than one craft such as instrument technician, or electrician and instrument mechanic; responsible to maintain a variety of equipment used in control systems." The definition of multiskilled is "maintenance personnel who are skilled in more than one craft." A systems mechanic could be required to cross craft lines if responsible for maintaining a system that requires electrical and mechanical knowledge and skills.

In years past, plants used specific crafts such as electricians, pipe fitters, etc. They usually learned their trade through an apprentice program and were perhaps licensed as journeymen. Most of the apprentice programs were tested and controlled by unions. Strong unions ensured that the electrical work would be done only by journeyman electricians.

The first electronics technicians learned their skills in the military and at technical and community colleges. No apprenticeship programs were available, and the unions did not become involved.

The knowledge factors and skills required to be an instrument technician overlap several crafts. Where does the electrician's work stop and the electronics technician begin? Or is there a dividing point at all? For example, installing and maintaining motors is the electrician's job. But what if a variable frequency drive were added to control the motor? These have transistors and microprocessors. Is the electrician stepping on the electronics technician's toes by working on something with microprocessors in it? Then what happens when the speed of the motor is controlled with a process controller such as a DCS system?

Multicraft, multiskilled system mechanics, system technicians, and instrument mechanics are titles used in some organizations to allow crossing craft lines. Area responsibilities have been a driving force to institute cross-training. Another driving force for smaller companies is the savings in manpower required to be competitive. The choice has been to be non-union and hire those who can do the jobs required or have several craft unions renegotiated with one workers' union. Workers' unions have allowed the multicraft and multiskilled concepts as long as a rate schedule is approved, generally with additional pay.

Older pneumatic plants at one time used pipe fitters and electricians as two separate shops to maintain their plants. When these older plants were upgraded and retrofitted, they installed distributed control systems and, in some cases, new electronic field instruments. In the interim they may have converted from pneumatic to electrical and electrical to pneumatic to use their pneumatic sensors and output devices. Whose responsibility is an I/P or P/I transmitter—a pipe fitter's or an electrician's? Nowadays, the answer is neither—the responsibility falls with a maintenance worker with some of the skills of both pipe fitters and electricians.

There are many examples of systems in which overlapping crafts require system technicians to have a variety of different skill sets. The repairer of a robot may have the title "system technician." Pneumatics, hydraulics, electricity, electronics, and computer expertise are required to maintain and operate a robot. The system technician could be considered multiskilled or multicraft.

Power systems use high-pressure pneumatics and hydraulic systems; control systems are used as an input to these high-pressure systems. Very large valves and mixers are sometimes operated pneumatically or hydraulically. The position or speed may be determined by an error signal between the set point and the process variable. One system could use *all* energy sources, and to maintain the entire system, instrumentation and control system technicians must have both mechanical and electronic skills.

The trend toward multicraft and multiskilled personnel raises some questions: Are maintenance personnel becoming jacks of all trades and masters of none? Or are they becoming more knowledgeable and more proficient in their jobs? Do their jobs (or future jobs) require more varied and/or detailed knowledge? Is there a need to specialize?

Can one be a master of all skills? It is necessary to be a master of more skills and knowledge than in the past but it isn't possible for everyone to fix everything; therefore, some specialization is necessary. This issue has been discussed earlier in this chapter.

However, it should be stressed that instrument maintenance personnel today use skills that in the past belonged to several different job crafts.

Knowledge Factors

General Knowledge Factors

Here are some of the general subjects any prospective employee needs knowledge of in order to work in the field of instrumentation and control systems:

(1) Proficiency in mathematics is necessary to understand and perform almost any technical function. The depth of mathematics skills will determine how far one can advance, such as from technician to technologist and then to engineer.

(2) Physical sciences knowledge is necessary to understand electricity, electronics, and manufacturing processes. Physics and mechanics are the basis for the more advanced physical sciences. In a wastewater treatment facility, for example, a knowledge of basic water chemistry, an understanding of pressures and the differences in absolute and gage pressure, and the effects of differences in cylinder size in hydraulic systems are necessary. In order to be proficient in pneumatics one must have a fairly good understanding of basic mathematics and physics. In the analytical systems, the higher mathematics and physical sciences are necessary. Pressure, level, flow, and temperature principles utilize physical science knowledge. Knowledge and understanding of the physical sciences makes for a good instrument technician. The terminology must be understood, and one must be able to read and interpret drawings such as a P&ID.

(3) Electricity and electronics knowledge is required, especially for the newer electronic and microprocessor-based controllers. Understanding both analog and digital electronics, which includes fundamental computer theory, is necessary to maintain the newer instruments and control systems. Understanding the proper use of test equipment is also a requirement to maintain instruments and control systems, from the basic single-loop controllers, pneumatic or electronic, to the larger, more advanced computer-controlled systems. Electrical and electronic maintenance and repair techniques as well as pneumatic maintenance techniques are part of the knowledge required in the instrumentation and control systems fields.

Basic knowledge of process measurement and instrumentation concepts and terminology is the part of the I&C technician, technologist, or engineer that sets them apart from the basic electronics technician.

(4) Knowledge of the different types of primary elements is another basic factor for I&C personnel. The primary measuring element is the component of a measuring or sensing device that is in direct contact with the substance whose attributes are being measured. This is the beginning of the control loop. It is the first system element that responds quantitatively to the measured variable and performs the initial measurement operation. A primary element performs the initial conversion of measurement energy.

(5) Knowledge of signal processing or conditioning equipment is necessary. This type of equipment converts signals from one form to another, such as current-to-pressure or analog-to-digital. Conditioning is often necessary to convert a signal to usable energy that is compatible with devices that perform mathematical functions such as square root, sum, integration, and counting. (An instrument that changes a sensor's output to a standard signal is properly designated a transmitter, not a converter. For example, a temperature element (TE) typically may connect to a transmitter (TT), not to a converter (TY).) A converter is also referred to as a transducer; however, "transducer" is a completely general term, and its use specifically for signal conversion is not recommended. Indicators and recorders can be considered signal processing/conditioning equipment.

(6) Process control instrumentation knowledge includes basic automatic control theory, and an understanding of feedback control loops is a must. The terminology of gain, proportional band, integral action, manual bias, etc., must be understood in both analog and digital control circuits as well as in both pneumatic and electronic systems. Knowledge of final control elements such as positioners and actuators is necessary to know how the control valve affects the process loop.

(7) In addition to the technical knowledge requirements, following correct formal procedures in performing and documenting maintenance tasks is necessary to properly maintain instrumentation and control systems. National or industry-wide codes, standards, and recommended practices must be understood as well as plant procedures.

Special Knowledge Factors

Maintenance personnel often develop specialized knowledge according to the type of plant in which they work. For instance, a petroleum refinery will probably have a distributed control system and will certainly have process analyzers. Thus, knowledge of DCS, distillation columns, and, of course, relevant safety standards and codes is a prime requisite for working in a refinery.

In a wastewater treatment facility, one needs to know some chemistry such as pH, specific ion, and conductivity measurement. Controlling pumps and valves that use programmable logic controllers and distributed control systems is a special knowledge that one might need in this industry.

An instrument technician in a steel mill needs to know the basic temperature fundamentals; some metallurgical instrumentation knowledge is also necessary.

Nuclear plant instrumentation and control technicians are tested in the knowledge and skills outlined in ANSI/ISA-67.14.01–2000, Qualifications and Certification of Instrumentation and Control Technicians in Nuclear Facilities. Table 4-1 lists special knowledge factors for the certification of Control Systems Technicians and Instrumentation and Control Technicians in nuclear facilities.

Those who work in the electrical utilities must have knowledge of turbines, water treatment, DCS systems, boiler controls, automatic start-up and shutdown systems, and material delivery systems. Special knowledge of rotating machinery, cooling towers, heat exchangers, and evaporators is helpful. Temperature, pressure, flow, and vibration analysis are necessary basic knowledge factors.

More examples could be given of the special knowledge needed in various industries. Technological advances and trends in industry practice also have an impact on the types of knowledge maintenance personnel may need. Not all plants are equipped with the latest computer control hardware, for example, or fiber optic sensing devices, or robots. But new devices and procedures are becoming more widely used all the time, and they present new challenges and new opportunities for the maintenance technician.

Programmable logic controllers, distributed control systems, and other computer-based systems are being used in more and more plants and are often applied in process control system loops. The knowledge of the particular brand of PLC or DCS used will be necessary in order to maintain the plants that utilize such microprocessor-based hardware.

Some maintenance tasks involve the software that these devices require in order to operate. Configuration and programming of PLCs and DCSs are often performed by engineering firms or process engineers. In some plants the maintenance personnel cannot change configuration because the software documentation is controlled by a software control configuration board. However, technicians and technologists are learning to configure DCS and PLC systems.

Table 4-1. Typical (but not necessarily inclusive) Knowledge List for Qualification and Certification of Instrumentation and Control Technicians in Nuclear Power Plants
(From ANSI/ISA-67.14.01–2000.)

Knowledge

(1) Mathematics
- Arithmetic Operations
- Ratio and Proportions
- Algebra
- Geometry
- Trigonometry
- Fundamental Calculus

(2) Physical Sciences
- Physics
- Mechanics
- Heat Transfer Fundamentals
- Nuclear Physics
- Health Physics
- Chemistry Fundamentals
- Pneumatics/Hydraulics
- Basic Water Chemistry

The following topics will include Theory of Operation and Failure Mechanisms and Modes

(3) Electricity and Electronics
- Basic Electricity
- Basic Electronics
- Analog Electronics
- Digital Electronics
- Fundamental Computer Theory
- Electrical/Electronic Measurement and Test Equipment
- Electrical/Electronic Maintenance and Repair Techniques

(4) Process Measurement and Instrumentation
- Basic Concepts and Terminology
- Temperature Measurement Fundamentals and Applications
- Fluid Thermometry
    - Thermocouple
    - Bimetallic Devices
    - Resistance Temperature Detectors
    - Thermistors
- Pressure Measurement Fundamentals and Applications
    - Bourdon Tube Devices
    - Diaphragms
    - Bellows
    - Capsules
    - Strain Gages
- Level Measurement Fundamentals and Applications
    - Float Systems
    - Displacer Systems
    - Differential Pressure Devices
    - Radiation Level Systems
    - Sonic Level Systems
- Flow Measurement Fundamentals and Applications
    - Primary Elements
    - Differential Pressure Devices
    - Magnetic Flow Measurement
    - Turbine Flowmeters
    - Rotameters
- Miscellaneous Transducers
    - Fire Detection Transducers
    - Security System Transducers
    - Meteorologic Transducers
    - Environmental Transducers
    - Seismic Transducers

- Analytical Transducers
  - In-line and Laboratory pH Meters
  - O_2 Analyzers
  - H_2 Analyzers
  - Conductivity/Salinity Cells

(5) Signal Processing/Conditioning Equipment
- Transducers
  - Pneumatic/Electronic
  - Force Balance
  - Motion Balance
  - Moment Balance
- Signal Conditioners
  - Integrator/Ratemeters, Function Generators, and Square Rooters
- Signal Converters
  - P/I & I/P & E/P & E/I & AD/DA, etc.
- Summers
- Amplifiers
- Bistables
- Control Relays/Solenoids
- Indicators
  - Visual
  - Audible, etc.
  - Annunciators
- Recorders

(6) Process Control Instrumentation
- Automatic Control Theory
- Controllers
  - Pneumatic
  - Electronic
- Analog Control Systems
- Digital Control Systems
- Final Control Elements
  - Positioners
  - Actuators

(7) Radiation Detection Instrumentation
- Portable Radiac
- Area Radiation Monitor
- Portal Radiation Monitor
- Process Radiation Monitor
- Reactor/Nuclear Monitoring Systems

(8) Nuclear Power Plant Systems
- Reactor Theory/Construction
- Primary/Secondary Systems and Flowpaths
- Auxiliary Systems
- Turbine-Generator and Auxiliary Systems
- Emergency/Safeguard Systems
- Reactor Control and Protection Systems
- Major Electrical Distribution

(9) Basic Plant Operation
- Plant Layout
- Plant Procedures
- Company Policies

(10) Advanced Plant Instrumentation
- Computer Systems
- Multivariable Control Systems
- Multiplexed Control and Information Systems
- Microprocessor/Minicomputer-Based Systems
- Programmable Logic Controllers
- Distributed Control Systems
- Supervisory Control Systems
- Data Acquisition Systems

Ideally, technicians, technologists, and engineers should all have the knowledge to determine whether a maintenance problem is a software or a hardware problem, to reconfigure basic programs, and to call up and troubleshoot logic on both DCS and PLC applications. Also, instrument technicians as well as technologists and control system engineers should know how to configure single-loop microprocessor controllers and recorders.

Configuration and tuning of process loops is a special work element in which the additional knowledge of tuning parameters and configuration of a particular vendor's systems is very helpful in maintaining and troubleshooting process control loops. Knowledge of the process dynamics is required to properly tune process loops.

Smart devices such as self-tuning controllers and fuzzy logic controllers are part of the latest generation of digital devices. They "think" by responding to new process situations according to past experience or rules of pattern recognition. Maintaining and configuring these controllers requires special knowledge.

Industrial robots have been used in the manufacturing industries for years. Robots are now being used in repetitive or dangerous process applications. Special knowledge of robot elements or programming is helpful in plants in which robots are used. Engineers are now designing robots for use in process applications.

Communication systems such as local area networks (twisted pair, coaxial, or fiber optic), or long-range communication such as telemetry (phone lines or microwave) have also been included in data acquisition and control and, therefore, are a part of the I&C arena.

A piece of equipment in widespread use in the I&C field is the personal computer (PC). In some cases they are hardened (they have better cooling and filtering and are sometimes more dusttight) for use on the factory floor. Most people now use the personal computer in an office environment. Basic computer knowledge is necessary to operate most PCs. Most programs are now menu-driven or Windows-driven for ease of use. The computer is used in maintenance management systems and computer-integrated manufacturing systems. If the job requires I&C personnel to maintain the PCs used in the process, special knowledge is required.

Knowledge of power and grounding considerations is becoming necessary when using the newer microprocessor equipment, which must have clean power. Some knowledge of uninterruptible power supplies (UPS) is also helpful. Proper grounding and shielding are important considerations in instrumentation and control systems.

Heating, ventilating, and air conditioning (HVAC) systems have become a special work element of the I&C groups. Temperature control has always been a part of I&C work, but the concern was process temperature control, such as furnaces. With heating, cooling, and humidity control now being a factor in the requirements of many processes, I&C personnel must maintain these systems.

Knowledge of codes, procedures, standards, and recommended practices is required, especially in hazardous areas. One needs to know how to work on an explosionproof device in hazardous areas. Also, knowledge of intrinsic safety wiring practices is necessary when working on intrinsic circuits.

Fiber optic sensors and communication media are starting to be used in the I&C industries. These measurement and communication means will become part of the I&C world.

Table 4-2 lists some special knowledge requirements, many of which have been discussed in this section. Of course, no one can become an expert in all of these areas, but personnel in most plants or shops will find they need knowledge in some of them.

Table 4-2. Special Knowledge Areas

Emergency Shutdown Systems
Programmable Logic Controllers
Distributed Control Systems
Communication Systems
Manufacturing Control Systems
Heating, Ventilating, and Air Conditioning Systems
Configuration/Programming
Process Control Applications
Codes, Standards, and Recommended Practices
Nuclear Power Plant Systems
Basic Plant Operations
Advanced Plant Instrumentation
Pneumatic Control Systems
Analytical Systems
Metallurgy
Boilers/Turbines
Distillation
Wastewater Treatment
Fiber Optics
Maintenance Management Systems
Computer-Integrated Manufacturing
Batch Control Systems
Industrial Robotics

Skills

Skill is defined as proficiency, ability, or dexterity. It is an art, trade, or technique, particularly one requiring use of the hands or body. In this chapter we are making a distinction between knowledge and **skill**—the ability to apply knowledge to perform practical tasks. However, developing a skill, like developing knowledge, may require specialized ability or training.

Basic Skills

Some basic skills required in maintaining and repairing instrumentation, measurement, and control systems are use of one's hands and body, use of words, numbers, and senses. Analytical thinking, intuition, and originality also help in diagnosing problems in I&C systems.

Mechanical dexterity is important in order to use test and measurement equipment properly and to align, calibrate, test, troubleshoot, adjust, and repair equipment and systems. Some basic pneumatic skills are necessary even in electronic control systems because pneumatics are used in many final control elements. Mechanical skills are required in most I&C job requirements. Electrical skills such as wiring and soldering are a large part of the job since most equipment must be powered by electricity. In plants that do not separate electricians from instrument and control technicians, the latter will install and repair electrical circuits.

Among the mental skills that are considered basic are communication skills, both oral and written. Not only is it necessary to know the symbols and terminology used in instrumentation and control, but one should also have the skill to read and interpret drawings such as process instrumentation drawings, schematics, logic diagrams, and flow sheets, as well as written documents such as procedures and standards.

The habit of working safely and being aware of safety issues is a crucial skill that must be foremost in the mind at all times. Due to the interaction of control circuits, working on one circuit may affect others, so the maintenance worker must know and understand the complete application and take care to consider all the possible effects of his or her work.

Special Skills

Computer skills will help when configuring or programming process systems. Some computer literacy is required to troubleshoot computer-based equipment.

The instrumentation and control systems field is so large that one cannot have thorough knowledge and effective hands-on skills in all aspects of it. All I&C personnel eventually concentrate on certain areas of instrumentation and control, becoming specialists. The specialization is often dictated by circumstances of one's job or career; it is natural to become skilled at the type of work one performs most often.

Becoming skilled at the processes and equipment used in a specific industry is a common way to specialize. The specialization might be in electrical power plant instrumentation, chemical processing, paper and pulp, food, wastewater treatment, or steel mill processes. Table 4-3 offers an example of specialization by industry; it lists skills required for certification of instrument and control technicians in nuclear power plants.

There are types of instrument systems used in various industries that can represent a specialty. Boiler and furnace controls, distillation instrumentation and controls, batch reactors, or heating, ventilating, and air conditioning (HVAC) systems are all specialized work areas in which special skills are required. Another specialization might be medical or biomedical equipment or manufacturing control systems. Analytical equipment is a specialty that requires special knowledge and skills. Expert skill in computer control systems, their configuration, programming, and application in particular processes is another potential specialty.

It is possible to become a specialist in a particular type of equipment—even one particular manufacturer's equipment, where one might become a field service representative for the manufacturer. One might specialize in programmable logic controllers (PLCs) in general or the Allen-Bradley PLC-5™ in particular.

The development of leadership skills is required for those who hope to supervise maintenance groups. Like any special skill set, leadership skills are acquired through work experience and training.

Table 4-3. Excerpt from ANSI/ISA-67.14.01–2000. Qualification and Certification of Instrumentation and Control Technicians in Nuclear Power Plants

Skills

(1) Proper use of the following technical information:
- Maintenance/Instrumentation/Administrative Procedures
- Technical Manuals
- Wiring and Termination Drawings
- Design Specification Sheets
- Solid-State Schematics
- Power Distribution Drawings
- Assembly Drawings
- Unit Conversion Tables
- Functional Control Drawings
- Integrated Circuit Schematics
- Electromechanical Drawings
- P&IDs
- Electronic Control Drawings
- Piping and Elementary Drawings
- Relay Logic Drawings
- Complex Logic Diagrams
- Computer Flow Charts
- Instrument Loop Diagrams

(2) Proper use of the following measurement and test equipment:
- Miscellaneous Hand Tools

- Digital VOM
- VOM
- Electrometer
- Function Generators
- Differential (Null) VM
- Power Supplies
- Calibrators, Pneumatic/Electronic
- Calibrated Resistance/Decade Boxes
- Oscilloscope/Oscillograph
- Calculator
- Calibration Standards
- Various Gages
- XY Plotters/Recorders
- Wheatstone Bridge
- N_2 Pressure Regulator Calibration Unit
- Strip Chart Recorder/Visicorder
- Millivolt Potentiometer
- Kelvin Bridge
- Vacuum Testers
- Deadweight Tester/Comparator
- Capacitance Bridge
- Transistor Tester
- Noise/Vibration Analyzers
- Megohm Bridge
- Oscilloscope Photography/Camera
- Tubing Bender
- Frequency Spectrum Analyzer
- Pulse Counter
- Pico Amp Source
- Pico Ammeter
- High Speed Event Recorders
- Meter Shunts
- Inductance Bridge
- Stroboscope

(3) Align, Calibrate, Test, Troubleshoot, Adjust, and Repair
- Typical Analog and Digital Circuitry
- Process Measurement Sensing Elements and Transducers
- Signal Processing/Conditioning Equipment
- Process Control Loops
- Complex Measurement and Control Systems
- Nuclear Radiation Instrumentation Systems
- Reactor Plant Control/Instrumentation Systems
- Advanced Plant Instrumentation
- Proper Valving Technique

(4) Proper Use of the Following Plant Administrative Documents or Systems
- I & C Shop Practices/Procedures/Forms
- Information/Library/Records Management Systems
- Equipment Utilization Forms
- Maintenance Orders/Work Orders
- Bypass Control/Jumper/Wire Removal Forms
- Plant Tagging System
- FSAR and Technical Specifications
- Radiation Work Permits/Forms
- Channel Check Procedures/Forms
- Functional Test Procedures/Forms
- Calibration Procedures/Forms
- Alignment Procedures/Forms
- Receipt/Inspection Procedures/Forms
- Document Control Systems/Forms
- Surveillance Test Procedures/Forms
- Confined Space Entry Techniques/Forms
- Nonconformance Items
- Review, Approval, and Control of Procedures/Changes
- Training and Certification Procedures/Forms

- Temporary Procedure Changes/Forms
- Facility/Equipment Change or Modification
- Procedures/Forms

(5) Proper Plant Safety Practices
- First Aid/CPR
- Electrical Safety Techniques
- Handling of Toxic Materials
- Handling of Radioactive Sources/Contaminated Materials
- Use of Safety Equipment
- Use of High Pressure/Temperature/Pneumatic/Hydraulic and Fluid Systems
- (PSM), Process Safety Management. Procedures and Documentation

Job Titles and Descriptions

It is difficult to define job titles in a general way. From one company to another the same title may apply to completely different job duties. Two terms that are fairly universal are *engineer* and *technician*. But there are different levels or types of engineers and technicians in many companies (e.g., Senior Technician, Maintenance Engineer, etc.) that are hard to generalize about. And there are many other titles, such as Specialist, Craftsman, or Technologist, which may have very different job descriptions in different organizations.

Technician

The Automation, Systems, and Instrumentation Dictionary defines a technician as "an expert in a technical process or a person whose occupation requires training in a specific technical process."

The following is one system of defining technician categories and describing the duties within each category. This system loosely follows the four-level classification system used by the National Institute for Certification in Engineering Technology (NICET).

Technician I: A person who performs entry-level tasks on basic devices, working under direct supervision. NICET refers to a Level I technician as a Technician Trainee.

Technician II: A person who performs the tasks of a Technician I and intermediate tasks such as instrument troubleshooting, calibration, and repair, working under general supervision; a person who can provide technical assistance or instruction or both during on-the-job training to a Technician I. NICET refers to a Level II technician as an Associate Engineering Technician.

Technician III: A person who performs the tasks of a Technician II as well as independent work with little or no supervision on jobs covered by written plans, specs, or instructions; a person who can provide technical assistance or instruction or both during on-the-job training to a Technician I or II. NICET refers to a Level III technician as an Engineering Technician.

Technician IV: A person who performs the tasks of a Technician III as well as independent work including delegated responsibilities and duties for which engineering precedent exists; who can provide supervision to other technicians; who can provide technical assistance or instruction or both during on-the-job training to a Technician I, II, or III. NICET refers to a Level IV technician as a Senior Engineering Technician.

Instrument Maintenance Technician: A person who assembles, installs, maintains, troubleshoots, and repairs various components of measurement and control systems. A high school diploma or the equivalent is required, but with today's technology changing so rapidly, vocational/technical school or community college training is strongly recommended. Most employers will also provide on-the-job training.

Instrument Technician: Usually requires certification or graduation from a technical college program. The technician works with theoretical and analytical problems, helping engineers find ways to improve the performance of an instrument or system, as well as helping mechanics troubleshoot system components.

Instrument Engineering Technician: An engineering technician is one who, in support of engineers or scientists, can carry out in a responsible manner either proven techniques, known to those who are technically expert in a particular technology (instrumentation and control systems), or those techniques especially prescribed by engineers. Performance as an Engineering Technician requires the application of principles, methods, and techniques appropriate to a field of technology (instrumentation and control systems), combined with practical knowledge of the construction, application, properties, operation, and limitations of engineering systems, processes, structures, machinery, devices or materials, and, as required, related manual crafts, instrumental, mathematical, or graphic skills. Under professional direction, an engineering technician analyzes and solves technological problems, prepares formal reports on experiments, tests, and other projects, or carries out functions such as drafting, surveying, designing, technical sales, advising consumers, technical writing, teaching, or training.

The education of an engineering technician places great emphasis on mathematics and applied physics with intensive laboratory work in which the technician develops practical knowledge of engineering theory and methods. They differ from engineers by reason of their more specialized technical background and skills.

Mechanic

Instrument Mechanic: An instrument mechanic installs, calibrates, inspects, tests, and repairs instruments and control system devices. Mechanics must be able to work well with their hands, be willing to improve their skills through training, and be interested in learning new technologies. Many instrument mechanics started in the pneumatic instrumentation area, but technology is changing to electronic and computer equipment. However, mechanical instrumentation such as flowmeters, control valves and actuators will still provide a need for the mechanic.

Technologist

Instrument Engineering Technologist: The engineering technologist is qualified to practice engineering technology by reason of having the knowledge and the ability to apply the well-established mathematical, physical science, and engineering principles and methods of technological problem-solving that are acquired by engineering technology education and engineering technology experience. The engineering technologist will usually have earned a baccalaureate degree in engineering technology or gained considerable technical experience on the job.

The technologist is a member of the engineering team, which will normally include technicians and engineers and, for special projects, may include scientists, craftsmen, and other specialists. The configuration of technical personnel possessing complementary capabilities that facilitate the engineering process is, by necessity, peculiar to each situation. The technologist is expected to have a thorough knowledge of the equipment, applications, and established state-of-the-art design and problem-solving methods in a particular field (instrumentation and control systems).

Field Service

Instrument Field Service Representative: The instrument field service representative fulfills customer service requirements on company-produced instruments and equipment, including (a) analyzing and correcting instrument operating problems, (b) contacting a supervisor or field engineer for assistance when necessary, (c) promoting scheduled maintenance agreements and sales of replacement or spare parts, (d) training new service personnel, (e) maintaining tools and equipment in good operating order, (f) reporting time, activities, and expenses in accordance with company requirements, and (g) keeping up-to-date on products and product applications.

Field Service Representatives also provides NIST calibration on their equipment and also other third-party equipment. They take part in upgrading or installing new instruments and control systems.

Engineers

Field Service Engineer: Installing distributed control systems and programmable logic controllers in the field is overseen by the field service engineer. The engineer is generally from the company that provided the control system and understands the process application and also the equipment theory, operation, and configuration and programming. Light engineering, reconfiguration and equipment changes can be installed or directed by the field service engineer, who also provides field engineering services at installation and start-up sites, ensuring specification fulfillment and operating ability. The field service engineer also defines any potential problems that may arise, including (a) managing installation and initial servicing, (b) maintaining technical abilities and awareness of new methodologies, (c) checking out operation of panels and instruments, (d) providing instruction in the maintenance and repair of equipment to company service personnel, and (e) consulting on service problems in the field.

Maintenance Engineer: The maintenance engineer assists maintenance supervisors, technicians, and mechanics in maintaining the facilities/systems by providing in-depth engineering assistance on high maintenance equipment. He or she assists design and project engineers in plant start-up, retrofit, and modifications to ensure that the design is engineered for maintenance and ensures standardization of equipment and systems. The maintenance engineer determines service life of equipment by cost to repair and frequency of repair as well as the frequency of preventive maintenance. He or she ensures that the equipment continues to conform to the technical specifications, is kept in calibration, and operational maintenance is being performed.

The maintenance engineer (a) works with management to ensure customer/production/maintenance cooperation by providing systems and equipment so that the end item meets all specifications and production schedules, (b) ensures that proper documentation is provided and kept up to date, (c) suggests and provides training when required to ensure that the maintenance personnel have the knowledge to maintain the equipment or system, and (d) suggests when contract maintenance is needed instead of using in-house maintenance personnel.

Instrument Engineer: The instrument engineer applies standard engineering standards and practices to the specification, sizing, and functional design of instrumentation hardware or control systems, which involves a clear understanding of the manufacturing or scientific process to be controlled. The instrument engineer also serves as the key person on the instrumentation design and operation team, often supervising and reviewing the team's efforts.

Under supervision, the instrument engineer participates in the design and planning of control and instrumentation systems as required by project assignment, including (a) collecting background information, (b) preparing drawings and calculations, (c) designing or modifying systems, (d) assisting in selection and procurement of equipment, and (e) ensuring compliance with applicable standards and codes.

Control Systems Engineer: The control systems engineer designs, assembles, builds, and operates instrumentation and control systems, and applies standard engineering standards and practices to the specifications, sizing, and functional design of instrumentation hardware and control systems. A clear understanding of the manufacturing or scientific process to be controlled is necessary. The control engineer serves as the key person on the instrumentation design and operation team, often supervising and reviewing the teams's efforts.

Project Engineer: The project engineer is generally assigned to a specific project from the very beginning through start-up and checkout and is involved with engineering activities associated with designing and constructing a manufacturing or processing facility as well as engineering activities related to a specific objective, such as solving a problem or developing a product.

Credentialing

Credentialing is an umbrella term that encompasses post-secondary program accreditation, engineering registration, and professional certification. Credentialing is a formal means of recognizing individuals who meet certain criteria of proficiency in their career field. Many professions, both technical and nontechnical, have associated credentialing programs. Generally, qualification criteria for such programs include specific requirements of education, job experience, and examination.

Earning a credential is a sign of one's commitment to competence and integrity in job performance and to thorough, up-to-date knowledge in one's field of expertise. A credential is a way to demonstrate achievement when applying for a new job, a promotion, or a salary increase. For many specialized technical fields, credentialing programs add to the stature of those fields as viable career paths and important areas of knowledge. And for the general public, credentialing programs help assure that industrial operations are safe and that industrial products are of a high quality.

ISA – The Instrumentation, Systems, and Automation Society is a leader in developing and offering certification programs for manufacturing practitioners. ISA offers the Certified Industrial Maintenance Mechanic (CIMM) program, the Certified Automation Professional (CAP) program, and the Certified Control Systems Technician (CCST) program. ISA also supports the development of the Control Systems Engineer (CSE) P.E. license. Visit ISA's website at www.isa.org/certify to get details on these programs.

The National Institute for Certification in Engineering Technology (NICET) is an examining body for engineering technicians and technologists. NICET offers certification in Industrial Instrumentation Engineering Technology. There are four levels of NICET certification, with eligibility requirements as follows:

Level I:	Less than 18 months work experience; no written test.
Level II:	Two years relevant work experience; written test.
Level III:	Five years relevant work experience; written test.
Level IV:	Ten years work experience plus significant involvement in one major project; written test.

Candidates for Levels II, III, and IV must pass an exam in order to receive certification. Table 4-4 lists general and special work elements for NICET certification.

For more information on P.E. registration, the Certified Specialist program, or NICET certification—including eligibility requirements, application procedures, and exam dates and locations—contact ISA at the address in the front of this book.

Table 4-4. NICET Work Elements
(Excerpt from NICET Industrial Instrumentation Engineering Technology, 02300)

LEVEL I — GENERAL WORK ELEMENTS

- Communication Skills
- Basic Individual Safety
- Industrial First Aid
- Basic Mathematics
- Basic Metric Units and Conversions
- Physical Science Fundamentals
- Nomenclature
- Pneumatic, Hydraulic, Electric, and Schematic Diagrams
- Fundamentals of Pneumatics
- Basic Electrical and Electronic Quantities and Components
- AC and DC Circuits
- Basic Tools and Test Equipment
- Basic Drafting

LEVEL I — SPECIAL WORK ELEMENTS

- Manufacturers' Product Documents
- Pneumatic Instruments — Test Equipment, Calibration, Repair
- Kirchoff's Laws
- Operational Amplifier and Transistor Circuits
- Electronic Instruments — Test Equipment, Calibration, Repair

LEVEL II — GENERAL WORK ELEMENTS

Core Elements

- Pressure Measurement Principles
- Level Measurement Principles
- Flow Measurement Principles
- Temperature Measurement Principles
- Control Valves and Final Control Devices
- Pneumatic Controllers, Operation
- Electronic Controllers, Operation

Noncore Elements

- Intermediate Mathematics
- Basic Digital Circuits
- Primary Pressure Sensors and Transducers
- Primary Level Sensors and Transducers
- Primary Flow Sensors and Transducers
- Primary Temperature Sensors and Transducers
- Control Valve Actuators and Positioners

LEVEL II — SPECIAL WORK ELEMENTS

- Analytical Instrumentation
- Pneumatic Signal Transmission
- Electronic Signal Transmission
- Fiber Optics
- Signal Transducers
- Microprocessors
- Microprocessor Troubleshooting
- Process Instrumentation Diagrams and Symbols
- Single-Loop Feedback Control
- Programmable Logic Controllers
- PLC Numbering Systems and Codes
- PLC Logic Concepts
- PLC Hardware Components
- PLC Installation, Start-up, and Maintenance
- Internal Control Documentation

- Pressure Transmitters and Calibration
- Level Transmitters and Calibration
- Flow Transmitters and Calibration
- Temperature Transmitters and Calibration
- Control Valve Body Repair
- Actuator and Positioner Repair
- Pneumatic Controllers, Hardware and Repair
- Electronic Controllers, Hardware and Repair

LEVEL III — GENERAL WORK ELEMENTS

- Business Communications
- Technical Libraries
- Troubleshooting and Problem Analysis
- Equipment
- System Operating Procedures
- Safety Regulations
- First Aid Responsibility
- Multivariable Control Loops
- Process Diagrams
- Control Systems Troubleshooting Principles — Pneumatic
- Isolating Problems — Troubleshooting Pneumatic Control Loops
- Control Systems Troubleshooting Principles — Electronic
- Isolating Problems — Troubleshooting Electronic Control Loops
- Process Dynamics
- Control Loop Tuning Concepts
- Closed-Loop Tuning Methods
- Trial and Error Tuning
- Open-Loop Tuning
- Integral Tuning Method
- Fluids

LEVEL III — SPECIAL WORK ELEMENTS

- Final Control Element Selection and Sizing
- Control Valve Selection
- Control Valve Sizing
- Flow Measurement — Advanced Techniques
- Flow Measurement Hardware
- Installation of Primary Devices
- Differential Pressure Flowmeter Primary Element Sizing
- Process Analyzers
- Advanced Digital Circuits for Measurement and Control
- Tuning Multivariable Loops
- Smart Transmitters
- Adaptive Tuning
- Single-Loop/Stand-Alone Digital Controllers
- Self-Tuning Controllers

LEVEL IV — GENERAL WORK ELEMENTS

- Programmable Logic Controllers
- PLC Software Components
- PLC Programming and Documentation
- PLC Data Highways
- Computer Control Systems — General
- Data Highways and Multiplexers
- Analog Hybrid Computers
- Analog Computer Maintenance
- Digital Computer Program Languages and Algorithms
- Direct Digital Control
- Supervisory Control
- Distributed Systems, Shared Controllers
- Digital Systems - Installation, Start-up, and Maintenance
- Digital Control System Testing
- Project Planning
- Crew Scheduling
- Test and Maintenance Procedures
- Computer Diagnostics
- Personal Computers

- Maintenance and Interruption of Service
- Maintenance Records
- Cost Savings Maintenance Methods
- Inventories

LEVEL IV — SPECIAL WORK ELEMENTS

- Process Control Applications
- Application — Rotating Machinery
- Application — Cooling Towers, Chillers, HVAC
- Application — Heat Exchangers and Condensers
- Application — Evaporators
- Application — Boiler Controls, Furnaces
- Application — Distillation
- Application — Batch Reactors
- pH, ORP, Specific Ion Control
- Industrial Robots
- Structure Elements of an Industrial Robot
- Industrial Robot Programming
- Application — Robots
- Leadership Skills

Certification

ANSI/ISA-98.00.01-2002 *Qualification and Certification of Control System Technicians* identifies the recommended criteria for certification of control system technicians. These criteria address qualifications based on education, experience, training, and job performance. The purpose of this standard is to provide industry with the basis for certifying the qualification of control systems technicians who work on facility equipment important for safe and efficient operations, in order to help reduce the possibility that unqualified personnel could perform improper maintenance on such equipment. It is intended solely as a recommendation for functional organization, and offers a structured basis for certification of control system technicians in each specific facility where they may be used.

The standard also recognizes that each facility or company needs a certification program, procedure, and/or plan to cross-reference the relationship between the facility job descriptions/categories and the three (3) functional levels. The key to this cross-reference is the existence of a good representative job description of each type of technician at a facility.

This standard can be used independently in circumstances where no equivalent, formally accredited program exists, although formal accreditation by a recognized and reliable accreditor is strongly recommended. It is not intended, however, to set forth qualifications, which differ according to the individual job descriptions in each organization. For example, the "Typical knowledge and skills lists" included as Tables 4-1, 4-2 and 4-3 are meant to be a guideline only.

Technician Categories

An alternative to NICET's definition of technician categories is the following:

- *Technician I* – shall be a person who performs calibration, loop checking, troubleshooting, and start-up with appropriate direction on devices that perform basic functions such as temperature, pressure, flow, level and analytical measurements. Shall be a person who can perform Technician II level tasks under the direction of a Technician II. Shall be a person who can provide technical assistance or instruction, or both, during on-the-job training to a Technician I.

- *Technician II* – Shall be a person who can perform the tasks of a Technician I. The technician shall also be capable of performing maintenance, repair, or organizing tasks without direction. Shall be a person who can perform Technician III level tasks under the direction of a Technician III. Shall be a person who can provide technical assistance or instruction, or both, during on-the-job training to a Technician I and a Technician II.

- *Technician III* – Shall be a person who can perform the tasks of a Technician I and Technician II, and can perform complex troubleshooting, calibration, and repair without direction on instrumentation, control systems, or equipment. Shall be a person who can perform Technician III level tasks under the direction of a Technician III. Shall be a person who can perform technical assistance or instruction, or both, during on-the-job training to a Technician I and a Technician II. Shall be a person who can perform project organization tasks.

Minimum Experience Requirements

In addition to the knowledge and skills items, there shall be a minimum experience requirements for each technician category. Individual programs shall specify the type and duration of experience required for participation in that program.

Certification of Qualifications

Each technician shall be tested on at least a representative sample of the knowledge and skills items for the job description of the category for which the technician is to be certified.

5

Maintenance Management and Engineering

Overview

Maintenance management has progressed over the years from a rather "hit-or-miss" attempt at keeping the plant going through a "trial-and-error" approach to complicated computer-assisted management of many complex maintenance operations. Record keeping and inventory control are just a couple of the necessities that have gone "high tech" in the maintenance organization today. Maintenance training has also taken on a new look. Periodic job/task analyses allow matching individual skills to the required tasks of maintenance personnel.

Training for upward progression in one's job has become vital to a dynamic organization. Many opportunities for effective use of maintenance workers can be analyzed from a review of a typical maintenance worker's day.

A recent "Factory Automation Maintenance Survey," conducted by Dick Slansky, Senior Analyst at ARC Advisory Group, showed industrial plants typically spend 75% of their maintenance budget in-house. The remaining 25% is done under contract.

Yes, maintenance management is big business and very important to the nation as it views its competitive position in world markets.

Maintainability and maintenance go with product reliability, quality assurance, and availability. Keeping the plant or process going at top performance will help assure products are of top quality and are competitive in world markets.

An objective of a sound maintenance management program must be customer satisfaction, making sure customers have the equipment and systems needed to perform the production function in the most efficient manner possible within the limits of the equipment and systems under their control (see Figure 5-1).

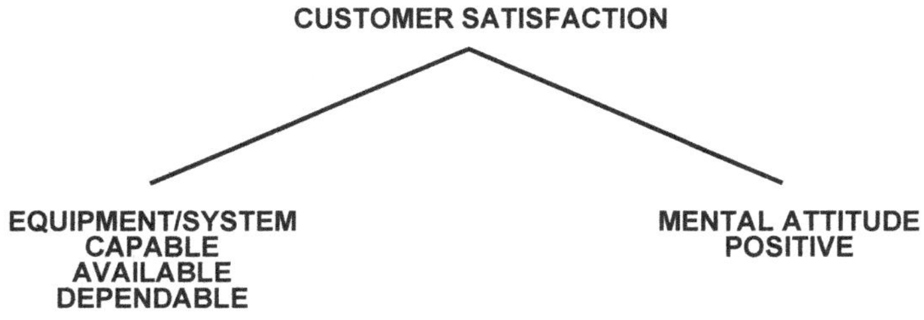

Figure 5-1. The Objective of Maintenance

The Need for Maintenance Management

The bottom line is profit. Sound maintenance practices, driven by a well organized maintenance management program, will contribute to profitability. Petrochemical producers spend around $13 billion dollars annually on equipment maintenance. Being able to control these expenditures is a necessity. Good maintenance and good service to the customer are not accidental. Planning is the key, and it is required. A machine is designed and built once but must be serviced many times. Engineering creates the design, the design is constructed, the plant organization accepts the design and construction, and then the maintenance organization comes into the loop because it is expected to keep the system going at peak performance. Some maintenance organizations come into the loop early in the design phase, which is a good practice and pays dividends in the long run. Other maintenance organizations do not come into the loop until the plant is built and operating, which results in very slow acclimatization to the plant's operations.

Excellent reliability built into instrumentation today has shortened maintenance time, compared to even a decade ago. Meanwhile, the skill set needed to work with today's computerized automation equipment has changed significantly. Maintenance personnel have needed to undergo considerable training to keep abreast of the new and emerging instrumentation and control technologies.

Maintenance Philosophy

In order to arrive at an optimum organization for instrumentation and control system maintenance, an understanding of maintenance philosophy is required by both the operating and maintenance personnel.

A key to a successful maintenance program is assuring its philosophy is written and communicated through the maintenance and operating organizations. Writing the philosophy requires going beyond platitudes and generalities. Rather, confront "head-on" and resolve some very basic issues.

Some typical issues are:

(1) How much is appearance worth?

(2) How many outages, and of what duration, are acceptable?

(3) What are the priorities?

(4) Is centralized or decentralized maintenance functionality needed?

(5) What are the shift coverage requirements?

(6) What work should be done by contractors and what should be done in-house?

As a part of preparing the maintenance philosophy, the responsibilities as they are distributed within the organization must be determined and analyzed.

The Operating Group

The operating group has the ultimate responsibility for any maintenance program. The operating group (the proprietor) owns all the equipment and has the ultimate responsibility for the proper operation of the equipment and operating maintenance (operational maintenance). This group approves PM (preventive or predictive maintenance) and plans, schedules, and establishes the maximum acceptable allowable down time (frequency and time duration). The bottom line for the proprietor is to determine if and when all maintenance will take place to maintain equipment and to meet customer standards and satisfaction.

Maintenance (Shops and Support) Department

Maintenance department personnel have responsibility for planning and controlling maintenance programs. They inspect, lubricate, and repair equipment. Their mission is to provide quality maintenance service and activities through efficient and effective systems and technologies that enable production facilities to manufacture at the lowest cost. They keep equipment histories, especially with regard to standards and traceability. They keep track of spare parts inventories, spare parts needs, and extra machinery lists. They prepare maintenance procedures and training courses. They suggest and manage training of operating personnel on equipment so a high proficiency of operation is possible.

The maintenance organization responds to the needs of the operating organization within the guidelines established by that organization (down time and frequency) with the type of maintenance required to deliver the forecast. Maintenance group personnel determine the "how" and the "who" for the maintenance function.

Operations and Maintenance Cooperative Efforts

Some maintenance functions require the operating group to cooperate with the maintenance department to determine the equipment to be included in any maintenance program. The group must justify new equipment to be purchased based upon equipment history, costs, and level of quality to match customers' needs. This cooperative group must determine what needs to be done and where maintenance will be performed.

Some organizations must keep statistics and cost data on the maintenance of equipment. This data may be a part of the history file kept by the maintenance department, or it may be a separate function kept by a support group (such as the accounting department). Information and statistics are kept for both the operating group and the maintenance department. Purchase costs, frequency of repair, repair costs, and other statistical data become a part of the crunch cost data. This type of information allows for continuous improvement of equipment and procedures.

Maintenance Management Organization

There are maintenance organizations and then there are other maintenance organizations. Is there a specific formula for the right organization? Perhaps not.

Department members should have the skills and stature to earn the respect of those with whom they work. Trades people need to be able to demonstrate that they have the skills to meet the demands of the plant.

A good rule of thumb for determining the size of the maintenance department is an operating dollar value equal to 2% of the replacement value of the facility. A high tech rule of thumb is one person full time for each one to two million dollars of automatic control systems investment.

Many companies have adopted a team approach to maintenance management in which various department members are assigned to specific teams (shown in the star organization chart of Figure 5-2 as function points). Each function point becomes a team with a specific mission. Team members represent specific duties or areas such as operations, personnel, planning, workplace, and coordination.

Regular meeting times are established, and team members share concerns at these meetings. This is thought to be a more effective way to solve problems and provide a more efficient department with much more involvement in the decision-making process. Each department member has input in managing the department.

Department Supervisor

The position of department supervisor requires a very dynamic and versatile individual. First, this individual must be extremely knowledgeable of the plant and plant processes. Ten to twenty years experience is a requirement for first- or second-line maintenance supervision. Plant knowledge generally comes from years of experience in a variety of different assignments. Background should be engineering or technology with good people skills, organizational abilities, etc. In the past it was generally thought that a skilled trades person with good people skills could advance to this position. Modern thinking favors the engineering base because of the rapidly advancing technologies and complexity of the modern plant and processes. An instrumentation and control background seems to prepare a person well for this position.

The department supervisor has traditionally held a position of leadership within the department, but traditional roles are changing. In addition to the leadership function, the role of resource person has been added to the position. A resource person within an organization may be called upon from the bottom up as well as from the top down and is the spokesperson for the department. He or she is also easily identified with the department.

Figure 5-2. Team Functions.

Basic Requirements for a Maintenance Department

It is very important that all equipment be properly identified and critical instrumentation be backed up by up-to-date history files.

Planned Maintenance Program

A planned maintenance program should include the following important component programs: predictive maintenance, preventive maintenance, repair, corrective maintenance, and protective programs. A total program requires using each, based on data from the history file, cost reports, and critical applications. The following are definitions of important components of a planned maintenance program:

Planned maintenance — Each equipment piece is identified and its maintenance defined.

Repairs maintenance — Equipment fails and the required maintenance is performed.

Preventive maintenance — Equipment maintenance is scheduled prior to failure on an estimate of the life of the equipment.

Predictive maintenance — Maintenance is performed when the need is indicated based on regular or continuous monitoring of the equipment.

Corrective maintenance — Maintenance is reduced or eliminated by analyzing prior maintenance work and making changes in design, installation materials or procedures.

Critical equipment — Equipment failure that adversely affects safety, environment, system availability, yield, quality or costs.

Noncritical equipment — Equipment failure that can be identified and repaired within a reasonable period of time so adverse effects, if any, are within acceptable limits.

Equipment component information is kept in the history file of the component. This information is generally computerized and is readily available to all maintenance employees. A sample life cycle chart should be included as a reference (see Figure 5-3).

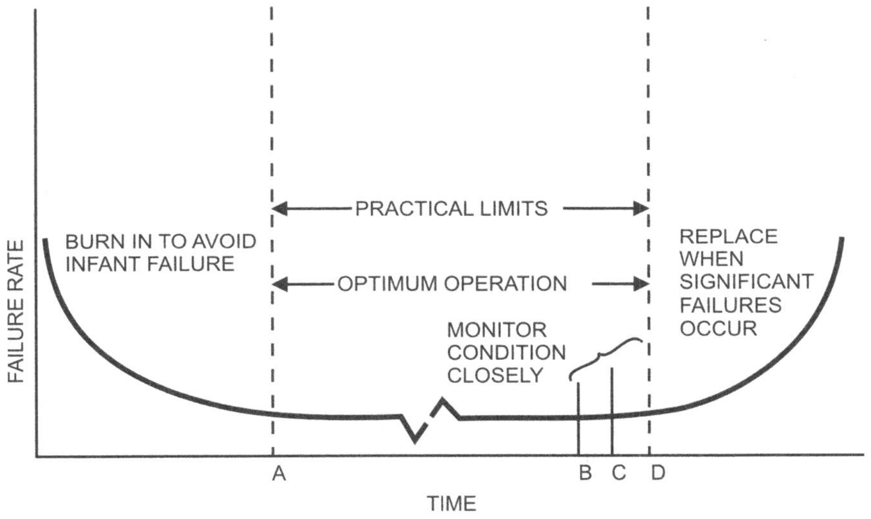

Figure 5-3. Sample Life Cycle Chart.

PREDICTIVE MAINTENANCE

Maintenance scheduled prior to failure and based on an estimate of the life of the equipment is called predictive maintenance. Predictive maintenance is used on equipment that has been classified as critical, for which unscheduled down time would be detrimental to plant operations. Depending on the critical nature of the equipment and process, the frequency of predictive maintenance may vary. In some plants this type of maintenance is performed at regularly scheduled shutdown times. Some examples of predictive maintenance programs are: vibration analysis on mixers, air compressors, generators, pumps, and motors; oil analysis on vacuum pumps, air conditioning compressors, and diesel engines; thermography on outside electric lines, main interior switch gear, and roofs. Any change in performance is indicated by monitoring equipment.

PREVENTIVE MAINTENANCE

Maintenance scheduled prior to failure, based on an estimate of the life of the equipment or from the maintenance department history and repair files, is called preventive maintenance. Usually the manufacturer can and will provide failure rate data to help plan this type of maintenance program. Initially, the equipment history records will indicate the true frequency for shutdown to maintain a "no unscheduled down time" concept.

CORRECTIVE OR REMEDIAL MAINTENANCE

When the equipment fails, required maintenance is performed to get the equipment back in operation. The nature of repair, time, and cost are recorded. This type of maintenance is acceptable but is not complete until the entire problem has been solved.

Planning and Scheduling

Planning and scheduling form the "genesis" of an effective maintenance management organization. With planning, each job is analyzed, resources carefully chosen, and the process designed to accomplish the task. Scheduling allows for customer interaction to achieve desired goals. Parts, labor, and time tables are established and agreed to by the organization so work can be done safely, at the right time, and in the most efficient manner (see Figure 5-4).

Formal planning and scheduling provide a work order system structured for better cost control and job progress and which provides report data on equipment and the system itself.

The advantages of planning and scheduling include the following:

(1) For production—Reduces cost of maintenance while improving service; provides data on maintenance performance; provides an orderly process by which work is accomplished.

(2) For maintenance—Establishes goals for the work force; eliminates delays due to waiting for materials and equipment, other skills, tools, etc.; applies specialized knowledge to planning/scheduling of shutdowns.

(3) For management—Permits accurate forecasting of labor and materials; enables management to level out peak work loads; provides a full planned day's work for each person and increases productivity.

(4) For all technical staff—Provides clear instruction; fewer obstacles to accomplish tasks; and fewer interruptions and delays.

Work Order System

Although a solid preventive maintenance program can dramatically increase equipment operational availability, it cannot prevent all equipment breakdowns. Therefore, a work order system is a necessary part of the maintenance program.

This section examines a work order system and presents sample work orders. The work order system or cycle begins with the work request. The work request contains an identification number; a description; a labor estimate; important dates; and originator information. The work order is generated from the information that appears in the work request. The work order is the most important document in the work order system.

The function of the work order system is to maximize effectiveness.

A sample of a work order is shown in Figure 5-5. In addition to data taken from the work request, the work order must show the approval signature of the person approving the order.

A typical daily schedule sheet is shown in Figure 5-6. A typical weekly schedule is shown in Figure 5-7. Scheduling of work is done on a priority basis. Occasionally, it is necessary to generate an emergency work order (see Figure 5-8).

JOB: CHANGE FEEDWATER PUMP SEAL

STEPS:

1. LOCK SWITCH OFF
2. LOCK VALVES OFF
3. OBTAIN LIFT
4. OBTAIN PALLET
5. DISCONNECT PIPING
6. DISCONNECT MOTOR
7. MOVE ASSEMBLY TO PALLET
8. MOVE UNIT TO SHOP
9. OBTAIN SEAL
10. CHANGE SEAL
11. MOVE UNIT TO SITE
12. REINSTALL ASSEMBLY
13. RECONNECT PIPING
14. RECONNECT MOTOR
15. UNLOCK VALVES
16. UNLOCK MOTOR
17. RETURN LIFT
18. RETURN PALLET

(a) MAINTENANCE JOB PLAN

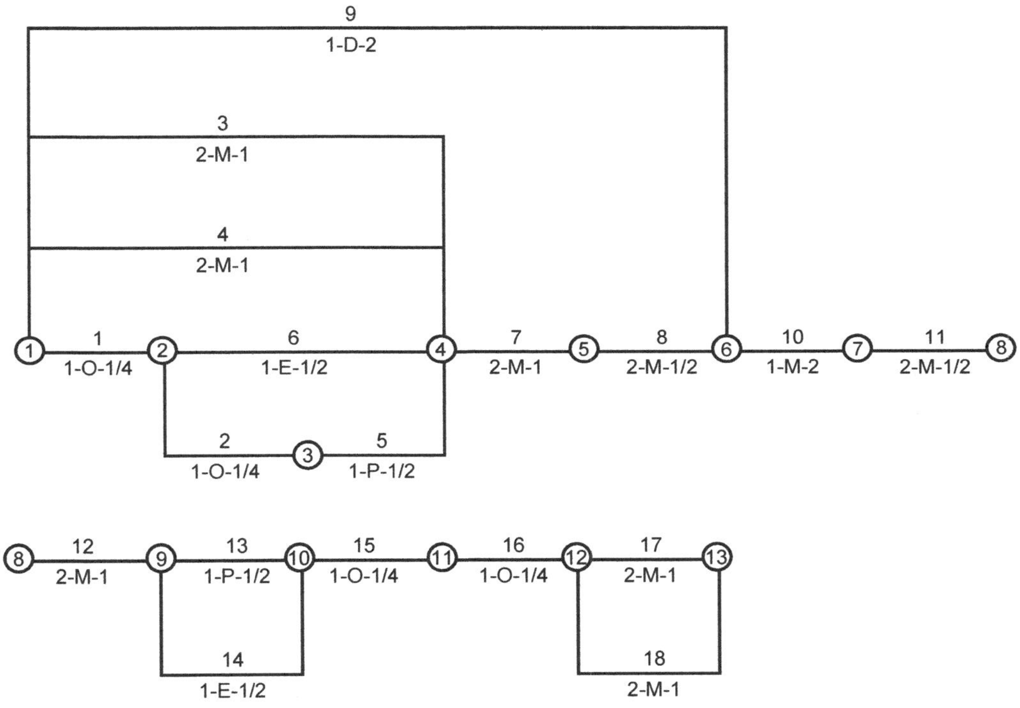

(b) ARROW DIAGRAM, PUMP REPAIR

JOB NO.	TIME	JOB NO.	TIME
1–2	1/4 HR.	6–7	2 HRS.
2–3	1/4	7–8	1/2
2–4	1/2	8–9	1
3–4	1/2	9–10	1/2
4–5	1	10–11	1/4
5–6	1/2	11–12	1/2
1–6	2	12–13	1

CRITICAL PATH
1 – 4 – 5 – 6 – 7 – 8 – 9 – 10 – 11 – 12 – 13

(c) PUMP REPAIR

Figure 5-4. Planning and Scheduling.

PLANT ENGINEERING WORK ORDER

RETURNED TO AREA COORDINATOR	LOCATION ______	TYPE WORK CODES
REASONS:	CHARGE No. ☐ (Area Letter) ☐ ☐ (Equipment Number)	☐ 1. Repairs
INSUFFICIENT DETAILS ☐		☐ 2. Alterations - Improvements
PROJECT REQUIRED ☐	PROJECT No. ______	☐ 3. Aid to Operations
		☐ 4. Preventive Maintenance

TITLE

DETAILED DESCRIPTION OF WORK (To by typed)

CRAFT	ESTIMATED LABOR NO. MEN	S.D. HRS.	REG. HRS.	TOTAL HRS.
Carpenters				
Painters				
Pipe Fitters				
Control				
Electricians				
Machinists				
Yard				
Total				

Shutdown Required	Yes	No
Safety Priority	Yes	No
Equipment Modified	Yes	No

If yes, Safety Acceptance required

______ Area Coordinator

Material STORES ☐ ORDERED ☐ Req. No. Del.

Issued by ______ Area Coordinator Approval ______ Date ______ Date Received ______

Completed ______ (Mechanic/Foreman Signature) Date ______ Actual Hours ______

Figure 5-5. Sample Plant Engineering Work Order.

Approximately 90% to 95% of available manpower is scheduled, leaving 5% to 10% for emergencies.

Computerized maintenance management programs will assist in generating sample work order formats. Totally integrated maintenance management (TIMM) is such a program.

MTTF, MTTR, and Availability

Equipment availability is an essential part of the maintenance management system. The following are some important terms and definitions:

(1) MTTF — Mean time to failure, the average time/distance/events a product delivers between breakdowns. The average or mean-time between initial operation and the first occurrence of a failure or malfunction. The number of measurements between failures on many pieces of identical equipment, approaching infinity.

(2) MTBF — Mean time between failures, the limit of the ratio of operating of equipment to the number of observed failures as the number of failures approaches infinity. The total operating time divided by the quantity (n+1), where n is the number of failures during the time considered.

DAILY SCHEDULE SHEET

ENGINEERING DEPARTMENT

Technician ______________ Craft ______________

______ Shift ______ No. Men ______________

Date	W. O. No.	Title of Job								Distr. Est.	Act

Figure 5-6. Typical Schedule Sheet.

(3) MTTR — Mean time to repair, the average time it takes to repair a failed item.

(4) Availability — The probability a system or equipment will, when used under specified conditions, operate satisfactorily and effectively. Also, the percentage of time or number of occurrences for which a product will operate properly when called upon. An example of availability is shown in Figure 5-9.

MECHANICAL AREA WORK SCHEDULE

Corp. ______ Elec. ______
Mach. ______ Paint. ______
Plumb. ______ Gen. Main. ______
C. Shop ______

Area Shutdown ______

Period ______ Through ______

Req. No.	Charge No.	Description	ESTIMATED HOURS Carp.	Mach.	Plumb.	Elec.	Paint	G. Main.	Y. Men	C. Shop	Remarks

Figure 5-7. Sample Mechanical Area Work Schedule.

EMERGENCY REPAIR ORDER

Filled In By Area

Area ______________ Shift ______ Date ______________

Equipment Name ______________

Description of Problem ______________

____

____

Work Requested By (Name) ______________

Filled In By Mechanic

Repair Action Taken ______________

____

____

____

Parts Used ______________

____

Shop ______________ Cost Code ______ Hours ______

Original to Mechanic
Copy to Area Coordinator

Signature(s) ______________

Figure 5-8. Typical Emergency Repair Order.

Training Maintenance Workers

Training programs for maintenance workers should prepare them to carry out the tasks assigned by their company and to satisfy the needs of their customers. The management function of such programs involves choosing the correct training methods and materials, the correct subjects for training, instructors, and a place for this training. It also involves detailed scheduling so there is ample time for sometimes complex instruction as well as for performance of the job at hand.

Without an ongoing training program, any company eventually may face technical obsolescence, decreased productivity, and increased costs.

Many avenues are open, such as individual training programs, classroom training on site or at a training institute, or on-the-job training, to name but a few. Apprentice programs are again becoming a popular method. It should always be remembered that most people learn more quickly and retain more of what they learn when they are involved in a hands-on manner.

Some method for evaluating the progress of trainees must be chosen. No amount of training is beneficial if its results cannot be measured and applied.

Training is covered in some detail in another section of this book. Without thoughtful management, this important concern can be a useless and costly exercise. With concerned management, it can be one of the most productive areas in the life of a company.

$$\text{AVAILABILITY} = \frac{\text{TOTAL TIME} - \text{LOST TIME}}{\text{TOTAL TIME}}$$

24 HR X 365 DAY OPERATION = 8760 HOURS

$$A = \frac{8760 - TL}{TT}$$

$$\text{AVAILABILITY} = \frac{\text{MTTF}}{\text{MTTF} + \text{MTTR}}$$

(a) DEFINITION

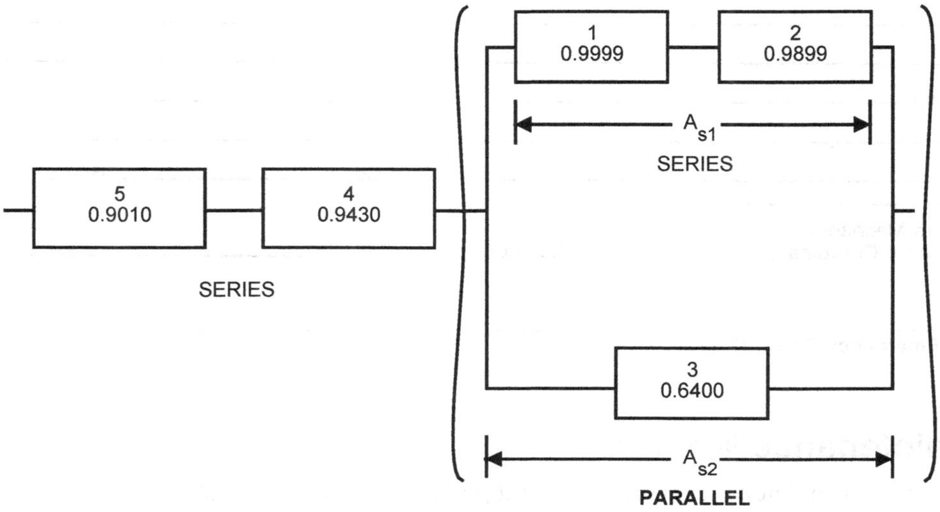

(b) SERIES/PARALLEL AVAILABILITY

SERIES:

A_{s1} = (0.9999) (0.9899) = 0.9898

PARALLEL:

A_{s2} = 1 – (1 – 0.9898) (1 – 0.6400) = 1 – (0.0102) (0.36) = 0.9963

TOTAL SYSTEM:

A_{s3} = (0.9963) (0.9430) (0.9010) = 0.8465

WHERE:

A_1 = 0.9999
= 0.9899
= 0.6400
= 0.9430
= 0.9010

(c)

Figure 5-9. Availability.

Preparing Functional Specifications

Overview

The key document generated in the planning phase is the "Functional Specification." This document tells exactly what the system should do, what will be supplied to the system, and what is expected to come out of it. The functional specification should be written as generically as possible to allow for selection of the optimum supplier.

Select a Project Team

Before preparing a functional specification, a project team should be selected to define the project. This team should have representation from the following groups; Facilities Engineering, Project Engineering, Manufacturing Engineering, Process Engineering, Maintenance Engineering, Operations Department and Safety and Loss Prevention Departments.

Project Team Meetings

The first meetings should review the project definitions, requirements and company philosophies. Then determine the responsibilities of each member. In the next meetings you should prepare a logic network, defining user needs, functional requirements, and expansion requirements during all phases of design, insure installation conforms to codes and standards.

Next, start preparing process and engineering flow sheets, instrument drawings, and input/output tabulations. Define the man-machine interface.

Do not design a minimum system and feel you can add to it later.

Determine musts, should, and likes

Clear distinctions should be made between functional necessities and desirable operations.

First list all the musts; next the shoulds, then the wishes or likes. Your original functional requirements should include all of the above. The priorities might change if very little or no cost is involved in the less important priorities.

The control system costs are generally a very small part of the overall project and can be a benefit in production/manufacturing quality and costs. It is easier to not cut back on control system cost, rather than adding later.

There will be changes to the process before the new system goes on line. The system must be flexible so additions and changes can be made without hardware obsolescence or project startup being delayed.

Supplier Qualifications

The vendor shall state how its company is committed to the control and information systems business. The system/product shall be flexible and modular in design utilizing state-of-the-art microprocessor technology. The system/product shall have a program for upgrading the system to keep up with technology and extending the longevity of the system.

Determine the suppliers willingness to take system responsibility and also the vendor's track record. Try to determine vendors long-term survival and growth and their service ability.

Codes and Standards

Insure the equipment specified is to be designed in accordance with national standards, as listed below:

American National Standards Institute (ANSI)
American Society for TEsting and Materials (ASTM)
The Instrumentation, Systems, and Automation Society (ISA)
Insulated Cable Engineers Association (ICEA)
National Electrical Manufacturers Association (NEMA)
American Society of Mechanical Engineers (ASME)
National Fire Protection Association (NFPA)
Underwriters' Laboratories (UL)
Factory Mutual (FM)

Responding to Functional Specifications

Each and every line item must be responded to by the words: comply, exception, non-compliance, or noted. Include documentation to support compliance.

Exceptions must detail the difference between the context of the line item and the vendor's system. How will the exception effect the system goal, functional requirements, owner's strategies, utilities, or any other aspect which could effect the process, maintenance, operation, etc.

Non-compliance is cause for rejection if it is a must or does not meet codes and standards. Vendors may state an alternative.

When appropriate, note that you understand the context of the line item but it cannot be answered with comply, non-compliance or exception, (an example would be a statement that owner will supply equipment, service, etc.)

Owner-Vendor Meeting

This document, the functional specification, will be used for the owner-vendor meeting.

Computerized Maintenance Management Systems

Why CMMS?

In the 1980s and 1990s the single most important piece of hardware that was accepted by industry and revolutionized some areas of industry was the computer. Maintenance management has been strongly affected by the computer, and its arrival on the scene brought computerized maintenance management systems (CMMS) (see Figure 5-10).

To be competitive, industry must find ways to reduce maintenance costs. Manual systems can no longer meet today's need for the up-to-date and accurate information necessary for effective maintenance and materials management. Computerized information systems can provide data on costs, labor use, equipment activity or inactivity, down time, inventory control, and actual vs. planned performance and scheduling.

Objectives for a CMMS Installation

Some objectives for installation of CMMS in a plant might be:

(1) to provide management with timely and accurate information that will assist management to plan, schedule, budget, staff, direct, and control plant operation maintenance;

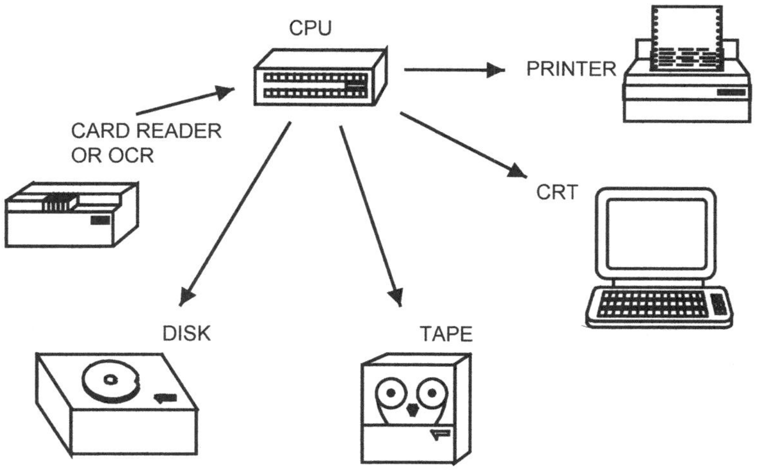

(a) BASIC HARDWARE CONFIGURATION

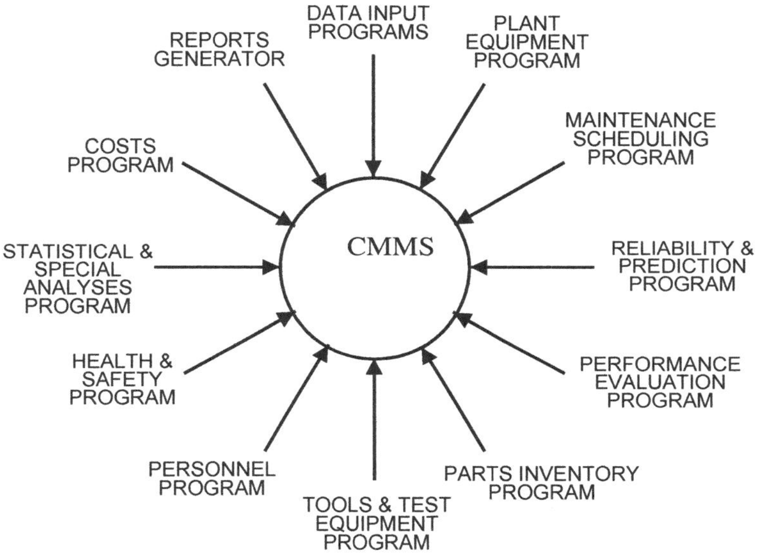

(b) COMPUTERIZED MAINTENANCE MANAGEMENT SYSTEM, BASIC SOFTWARE STRUCTURE

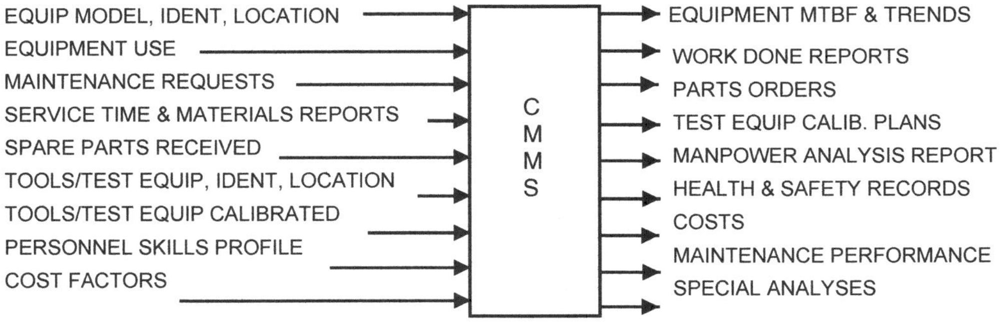

(c) INPUTS/OUTPUTS

Figure 5-10. CMMS.

(2) to reduce maintenance costs by having more and better quality information;

(3) to provide maintenance workers with tools that will allow them to perform better and more productively on the job; and

(4) to help the plant to produce a better quality product because of a better running plant.

CMMS Functions

EQUIPMENT AND MAINTENANCE

Some equipment and maintenance functions available with CMM systems are as follows:

(1) Provides descriptions of maintainable assets

(2) Provides records of where parts are used

(3) Provides a track on repairable spares, such as motors and gear boxes

(4) Provides an estimate of tasks

(5) Automatically triggers preventive and predictive tasks

(6) Reschedules preventive maintenance

(7) Schedules work orders

(8) Provides benefits from "opportunity scheduling"

(9) Schedules shutdowns

(10) Documents repair history

(11) Collects manpower and material costs

MATERIALS MANAGEMENT

Some materials management functions available with CMM systems are as follows:

(1) Identifies parts

(2) Reserves inventory

(3) Records inventory

(4) Allows for lower inventory levels

(5) Conducts inventory analysis

(6) Coordinates maintenance tasks with availability of materials

(7) Automatically records complete parts inventory

(8) Consolidates purchase orders

(9) Facilitates automatic costing

(10) Institutes automatic purchasing

(11) Identifies multiple suppliers

(12) Processes different types of purchase orders

(13) Expedites procurement

(14) Evaluates supplier performance

FINANCIAL MANAGEMENT

Some financial management functions available with CMM systems are as follows:

Accounting Functions Related to Purchase of Goods and Services

(1) Invoice/purchase order matching

(2) Invoice adjustments

(3) Taxes, freight, duties, and accrual distribution

(4) Check requisitioning

(5) Check reconciliation

(6) Cash requirements forecasting

(7) Currency adjustments

General Accounting Functions

(1) On-line transaction processing

(2) Journal voucher entries

(3) Access to general ledger

(4) Auditing through a comprehensive, detailed audit trail

(5) Financial reporting

(6) Interfacing to other systems

SPECIAL FUNCTIONS

In addition to the functions mentioned above, some special functions are also available when a CMM system is purchased for maintenance management. First is a possible approach to more closely monitor project management with the system. Project managers have access through the system to stores, purchasing, inventory control, and accounting functions. Database development is also possible with the system. Items such as computer-assisted instruction (CAI) can be available with the computer, among others. Also, CMMS is an interactive process with various steps being repeated as new knowledge is gained.

Benefits of CMMS

Table 5-1 shows some benefits from a computerized maintenance management system.

CMMS provides an important tool in linking the training function to the scheduling function. This can be done by providing a "match" of employee skills and certifications with the job tasks required to keep the plant running smoothly. Prior to CMMS, maintenance departments scheduled from a "man hours available" bank. The problem has been that the job-required skills were not available for the specific task scheduled. Typically, this left hours unused in a given work week. It can be seen that a goal would be to keep percentages as high as possible. In plants that adopt a computerized maintenance scheduling system, these percentages can be improved. By keeping accurate training records in the computer, it is easy to match skills available with skills required to accomplish specific tasks. The match is done by the computer.

CMMS will also provide a communication link back to the department that requested the work so that they might be able to better plan their production based upon the job being completed. The bottom line is better, more efficient scheduling with higher percentages shown on the scheduled vs. work completed charts.

Table 5-1. Benefits of CMMS.

Reports	Benefits
1. Scheduled preventive maintenance	1. Anticipate hardware problems, identify worn out or faulty equipment
2. Equipment reliability (predictive maintenance)	2. Improved maintenance • Materials and people • Productivity
3. Maintenance performance	3. Improved planning and control of plant maintenance, finances; $ variances and productivity
4. Parts inventory control and automatic re-ordering	4. Reduced plant down time due to parts availability; efficient use of $
5. Personnel skills inventory, planning, and utilization	5. Better manpower planning, career development, and skills utilization
6. Tools and test equipment	6. Improved use of capital for new/replacement equipment; calibration
7. Costs	7. Performance to budget cost of ownership measurement
8. Health and safety	8. Meet OSHA and EPA requirements; safety
9. Exceptions and specials	9. Identification of critical items; special analyses

Accurate records are kept of employees' training activities with an accurate listing of the skills each employee has acquired. This information has come about as a result of a task analysis and training provided to update skills needed to accomplish specific tasks. This information is documented in a database that is readily accessible for scheduling through CMMS.

A specific task (job) comes into the shop as a work order. Maintenance managers go to the database and search for the skills needed to accomplish the specific job. The search shows that employee "A" has the necessary skills to complete the job. Then the manager checks the availability of employee "A" for the estimated time that the job will require. Again a match is made for the time available and the time required. Then the employee and the job are scheduled.

Another variation of the above could occur when a specific job requires that a certified employee be matched to a specific job. OSHA is requiring more careful matching of skills with specific job requirements. The data on specific employees who have certifications in specific skills can be readily documented. Again, this information can be matched against job requirements.

A variation on the same theme could be the establishment of a match of priority assignments with specific employees. If a top priority job or individual cannot be matched, the next best can be identified and scheduled for the job based on availability.

Expert Systems

The increasing sophistication of manufacturing and processing has increased pressure upon maintenance delivery systems. Maintenance systems must match the high levels of integration and performance criteria of the equipment itself. To meet this demand, more companies are considering the use of expert systems maintenance programs. A list of the typical maintenance problems that expert systems can help solve are as follows:

(1) Lack of concern about equipment maintenance and service prior to production start-up.

(2) Limited training time for personnel on new equipment, processes, or methods.

(3) Too few persons in an organization with enough experience and knowledge in the maintenance of particular processes or equipment to be considered experts.

(4) Inability to easily distribute troubleshooting knowledge among service people with varying skills and abilities.

(5) Knowledge and experience pool shrinking as older, highly trained experts leave.

(6) Poor maintenance on production equipment that allows premature equipment failure leading to customer complaints about quality and delivery.

(7) Inaccurate diagnostic techniques or inadequate training causes technicians to resort to parts swapping, resulting in unnecessarily high levels of good parts being returned for repair or discarded.

There are different approaches to building an expert system for CMMS. Using programming language such as PROLOG® or LISP®, an engineer can interview the experts and extract the knowledge, which is, in turn, translated into a program. The second approach is to use a rule-based artificial intelligence shell. By filling in the shell, an expert system is built, which will then create a system of production rules that sequence the diagnostic process. The third approach is to use a model-based system. This is the most recent development in commercial expert technology for diagnostic and other applications.

Computer maintenance management systems will help management be more effective even while reducing manpower. Where the human factor can be replaced by a more reliable, variable, and quality-oriented computerized system, it behooves management to consider the advantages of CMMS. Computers, given the right direction and information by humans, can provide timely, consistent, accurate analysis, and valuable assistance for improved maintenance management.

Office/Shop Layout

In order to maintain a facility, the maintenance personnel and the maintenance supervisors should have a place from which to work. This space could be the centralized maintenance area for all types of maintenance, or it can be a special shop for just the I&C groups. The differences in centralized, decentralized, and specialty shops will be covered later. When maintenance personnel are given an area of their own they feel part of the organization. The size of the shops are determined by the number of personnel and the diversity of the work that is required from the maintenance personnel.

Office Area

The office area usually houses the maintenance supervisors and support personnel. Sometimes the control systems engineers have their offices in or near the I&C maintenance shop offices. Reference material such as standards, drawings, and equipment manuals are generally in or near the office area.

A private office should be available for the managers, supervisors, and foreman so they have a place to talk individually to other maintenance personnel and other supervisors. The maintenance personnel should know that the managers and supervisors are a part of their team and that the doors are generally open to them through the proper chain of command. Depending on the size of the organization and the number of different shops required, the first line supervisor, the foreman, or the leadman may have their offices closer to the shop area.

Near the supervisor's office is generally an area for the support personnel. This space is generally more open and may use modular furniture, but each of the support personnel should have a specific work area.

Technical Library

The reference material required to maintain a large plant should be available to those who should have access. Reference material, which includes equipment manuals and drawings such as P&ID's, loop drawings, flow diagrams, and schematics, should be readily available. The master or as-built drawings are kept here. Vendors' manuals and literature should be available for every instrument or system that is to be maintained. There must be control to ensure the reference material is up to date and available to the maintenance personnel. Larger plants may have someone dedicated to the document library, but smaller plants may have this task assigned as a part of one individual's job function. This area is sometimes called document control or the reference library.

It is very important that this reference material is always kept up to date. When the manufacturer or engineering requests a change in any system, these changes must be documented. When the maintenance personnel replaces an instrument in the process, that information must be documented.

General training material such as videos and training manuals should be available to all maintenance personnel. Configuration and/or calibration data are kept in the reference area.

A copy machine should be in the library so that a page or a drawing can be copied for use in the field without taking the master drawing or manual. In some plants, the planners provide a layout or drawing with the maintenance work order to expedite the repair of the instrument or system.

A computer that is part of the maintenance management system should also be in the library. This system should be the database for all the drawings and parts locations.

Appropriate readers should be available for all drawings, parts lists, and manuals. In addition, tables must be available to lay out the drawings as well as to provide a place to study manuals and other materials. Proper lighting is also a must.

Training Area

A dedicated classroom is needed to provide a quiet learning atmosphere. A continuous training program should be available to all maintenance personnel. In addition to a classroom, a private study area could be provided to use interactive instructional materials.

Shop Area

Most plants provide a workbench area for each maintenance mechanic or technician. This bench generally has electrical power and air, if required. The bench area is where tools and personnel reference material may be kept.

In addition to the private workbench area, benches are set up to maintain a particular type of instrument such as a pneumatic bench with air, valves, and gages provided. Bench setups for repair and calibration of specific instruments should be available. An area for electronic repair utilizing clean power, several DC power supplies, and grounding for static electricity should be provided. A maintenance and calibration area is necessary for different types of equipment such as oxygen analyzers, liquid analysis equipment, flowmeters, temperature baths, etc.

If a plant has a distributed control system, a hot backup unit should be in this area to check out cards of the system. This hot backup unit may also be utilized for training by putting a monitor in the classroom area. The area can also be used for staging new systems.

To feel a part of the organization, each person should have an area that can be considered his or her own work area. The maintenance personnel workbench is the "office" of mechanics and technicians.

An area should be set aside for drill presses, grinders, lathes, saws, tube benders, threaders, welders, and parts cleaning. This shop should be outfitted with the equipment required to maintain the plant, as some plants separate the electrical/electronic/instrumentation shops from the mechanical shops. Large doors and overhead cranes should be provided, if required.

Decontamination Area

In some hazardous plants, equipment must be decontaminated before the maintenance personnel can troubleshoot. Decontamination is also required for equipment being sent off-plant for repair. The decontamination area should be a separate room with safety controls determined by the hazard. Decontamination ovens, wash tanks, evacuators, and so on should be in this area.

Parts and Equipment Warehouse

The spare parts and equipment warehouse should be located near the maintenance shop area. This area is generally not a part of the maintenance department but is frequented by the maintenance personnel on a regular basis, so it should be readily accessible. The maintenance management system (MMS) should list the parts and equipment that are available in stores.

Tools and Test Equipment Area

Special tools and test equipment may need a controlled checkout procedure. It is not necessary for each technician or mechanic to have special tools and test equipment that may be used only once or twice a year. Most of these special tools and test equipment are very expensive, so a tool and test equipment booth should be provided. Larger plants may have someone dedicated to check out this equipment, while smaller plants will have procedures available to use the special tools and test equipment.

Locker and Lounge Area

Because workers need a place to change and store their clothes, each worker should have a personal locker. Restrooms and showers should be available near the locker area. Areas must be provided for both male and female workers.

A lounge area should be provided with kitchen facilities, vending machines, tables, and chairs. Lunch and breaks may be taken in this area.

Staging Area

In a large plant a modernization project is always going on. An area should be set aside in or near the shop to stage the new system. Here, the system for hot backups is available. If a system is available for training, this is a good place for the remote controllers. An operator's station should be provided in the classroom or a classroom should be located near the staging area.

Centralized/Decentralized Shops

The size (number of maintenance personnel and physical area) of the plant and the maintenance organization concept will determine whether the plant has more than one maintenance office.

A centralized shop could be one large maintenance shop with all crafts centrally located in one building or area. The centralized shop may have different foremen for the different crafts or maintenance teams with different supervisors or leads. The centralized shop may be the central point for the team concept instead of separate shops for each team. The concept utilizes teams to maintain a system or physical area. A team may consist of a mechanical person, an electrician, an instrument mechanic, and a control system technologist.

The decentralized shop concept is generally used in larger plants. The mechanical crafts may have a separate shop from the electrical crafts and the instrumentation technicians. Another concept of decentralized shops is area responsibility, utilizing teams to maintain a system or physical area. Team shops might be near the system for which the teams are responsible.

Centralized Shops

The centralized maintenance shop is generally cost effective, as less duplication of equipment and less supervision is needed. Figure 5-11 shows a suggested layout of a centralized shop. The centralized shop concept seems to promote harmony between the maintenance personnel because they are closer together. When shops are split, finger pointing and jealousy may exist. Generally, small groups do form either by craft lines or by teams, but since they are together at the start of the shift, during lunch, during training, and sometimes in the shop, they become friends and feel part of the organization.

When the area concept is used with the centralized shop, a small area near the process is generally set aside for maintenance personnel. This is their "hangout" near the process and a place to work from. If needed, they can go back to the main maintenance shop to do the heavier or specialized work that can only be done in the larger, better equipped shop.

When using the area concept with the centralized or decentralized shop concept, it is important to work with the equipment operators. They should be involved when repairs are being made to operate the controls. Operators should do some operational maintenance, such as changing paper and filters, cleaning and oiling, on-line diagnostic checks, etc. In so doing, the operators may call the maintenance personnel only when help is needed, not every time they *think* they need help.

In larger plants, service trucks are provided to carry some of the basic parts and equipment necessary to maintain the area. Such trucks can be used by personnel who are assigned jobs out of the central maintenance shop, who must go out to the work area to perform the tasks and return to the shop for bench-type repair functions.

Decentralized Shops

Decentralized maintenance shops are common in very large plants. The concept is to put the maintenance shop near the work area. These shops could be team shops with the crafts needed teamed together or they could be decentralized along craft lines.

The biggest disadvantage is the duplication of equipment needed. Each shop needs a physical area to work in, a lounge or locker room, office space, tools, and

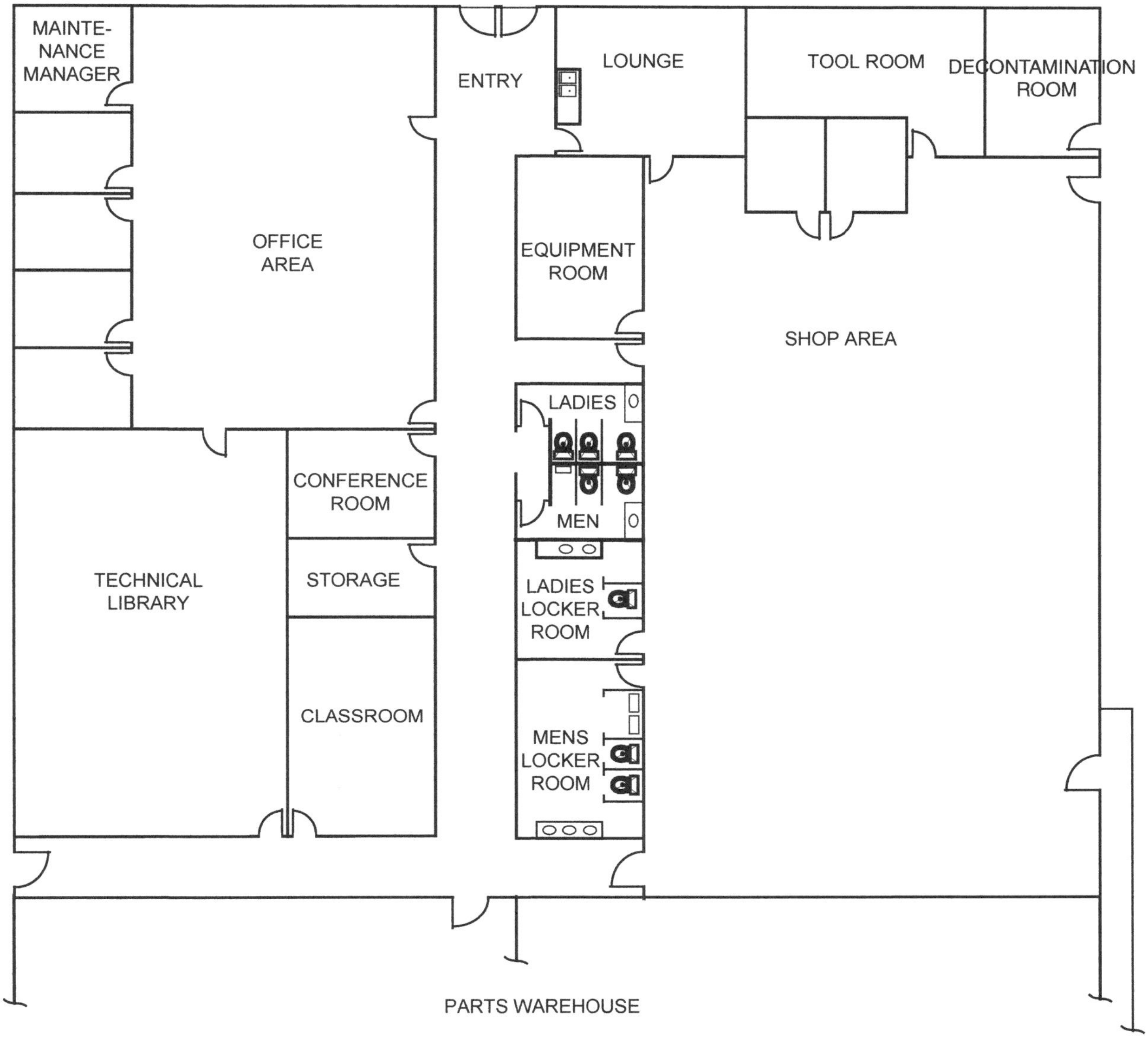

Figure 5-11. Centralized Shop Layout.

test equipment. Advantages are the closeness to responsibilities and working closely with the operators of the equipment.

Union plants seem to decentralize maintenance shops along craft lines. When different union locals are represented, each local or craft may have a different shop. When nonunion technicians start maintaining the instrumentation and control systems, another dividing line appears.

Shop Vans

Shop trucks or vans are utilized in larger plants where the transportation is needed. They are also utilized by contract maintenance personnel. The vans are stocked with the equipment needed for maintenance. Contract personnel generally have responsibility for more than one plant, so they may be in the plant only one or two days a month.

6

Pressure and Flow Instruments

Overview

Pressure instruments are used to determine the pressure in a vessel and sometimes the level of a vessel. It is necessary to measure pressure for safety reasons, such as over-pressuring a vessel or controlling the pressure of a vessel. Pressure instruments can also be used to measure flow, such as a differential pressure transmitter across an orifice. Pressure instruments can also measure vacuum. The pressures may be calibrated in gage or absolute pressure.

Flow instruments may be the differential pressure instrument across an orifice or one of the more sophisticated mass flowmeters, magnetic flowmeters, turbine flowmeters, vortex meters, or many other types.

Also explained in this chapter are level transmitters. Some level transmitters use the pressure method; others use the capacitance and electromechanical methods.

Pressure Transmitters

Theory of Operation

Pressure is defined as force per unit area. If a weight of 100 pounds acts on an area of one square inch, the pressure would be stated as 100 pounds/1 square inch or 100 pounds per square inch (psi). However, this definition of psi by itself is meaningless because reference conditions are not defined.

Pressure can have one of two reference points: atmospheric pressure or absolute zero pressure. A pressure measurement using atmospheric pressure as the reference point is known as gage pressure (psig). The pressure measurement referencing absolute zero pressure is known as absolute pressure (psia).

Recall that atmospheric pressure at sea level is 14.7 psi. Gage pressure uses this atmospheric pressure as a reference and calls it 0 psig. Absolute pressure, as its name would suggest, uses absolute zero as a reference; therefore, –14.7 psig equals 0 psia. Figure 6-1 shows the relationship between gage and absolute pressure measurements.

Of course, there are other units of measure besides pounds per square inch. U.S. process engineers still talk about pressure above atmospheric in terms of psi, but at lower pressures (below 10 psig) they frequently use inches of water (in. H_2O) instead. And (to add confusion), below atmospheric pressure, millimeters of mercury (mm Hg) are commonly used. European engineers generally use kilo-

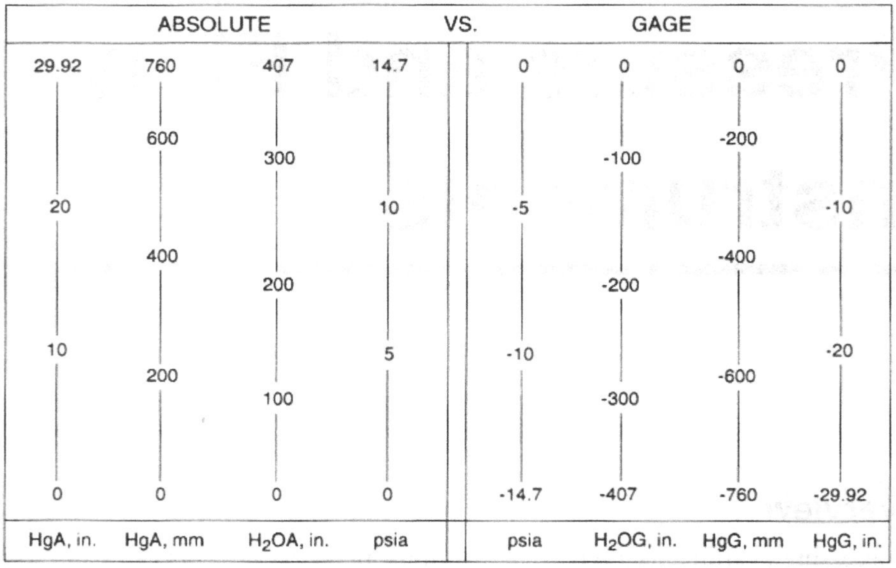

Figure 6-1. Absolute to Gage Conversions.

grams per square centimeter or millimeters of water for lower pressures and bars or millibars for the higher-pressure measurements. Figure 6-2 shows conversions between commonly used units.

TO OBTAIN	Atmospheres	psi	mm Hg, torr	in. Hg	bar	millibar
	MULTIPLY BY					
Atmospheres	1	14.7	760	29.92	1.013	1013
psi	0.0680	1	0.0193	2.036	0.069	68.95
mm Hg/torr	0.00132	0.0193	1	0.0394	0.00133	1.333
in. Hg	0.0334	0.4911	25.4	1	0.03386	33.86
bar	0.987	14.5	750.1	29.53	1	1000
millibar	0.000987	0.0145	0.750	0.0295	0.001	1

Figure 6-2. Pressure Unit Conversion Table.

Today's electronic pressure transmitters predominantly use one of two methods for sensing pressure: capacitance and strain gage. A capacitance pressure-sensing element has a flexible metal membrane placed between two fixed capacitor plates. Pressure inputs move the membrane, changing the capacitance relationship between it and the fixed plates. The differential capacitance signal is amplified, conditioned, and converted into a 4–20 mA signal. Figure 6-3 depicts a cross-sectional view of a capacitance sensor.

Strain gage measurements are based on the principle that the resistance of an electrical conductor is a function of its cross-sectional area. A change in the area means a change in resistance. To utilize this principle, a conductive material is bonded to a thin piece of silicon by one of several means. Pressure inputs flex this

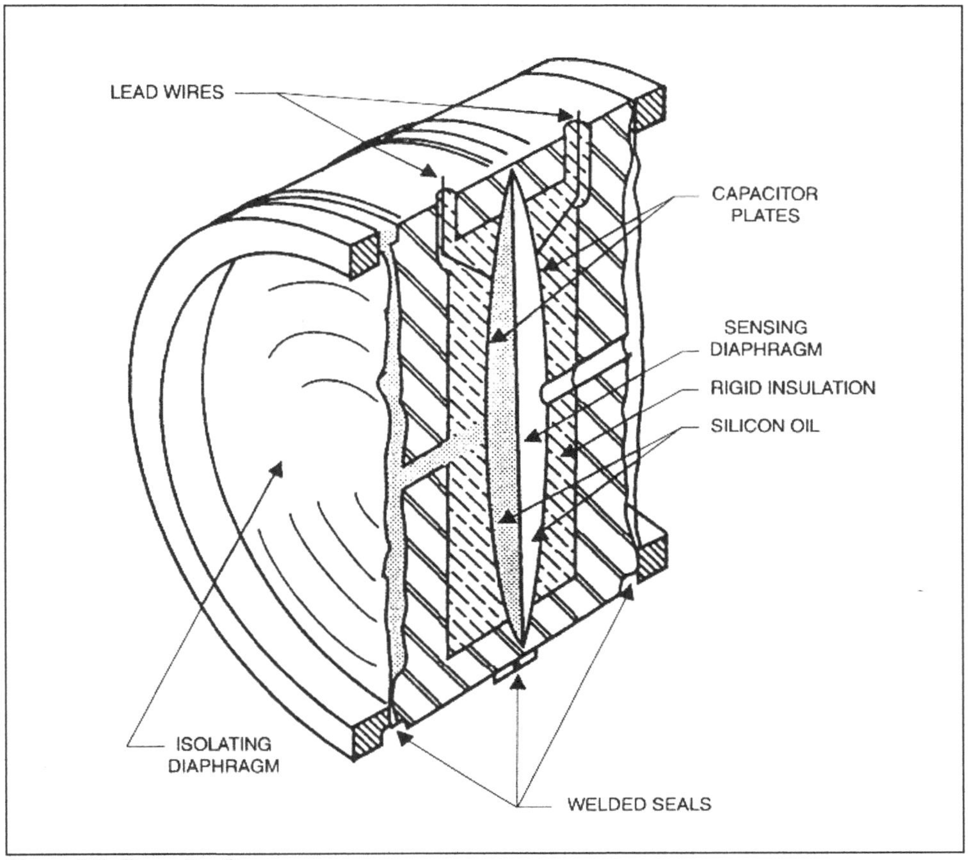

Figure 6-3. Capacitor Sensor.

element and the change in resistance is measured. This signal is amplified, conditioned, and converted into a 4–20 mA signal. Figure 6-4 shows a top view of a strain gage sensor.

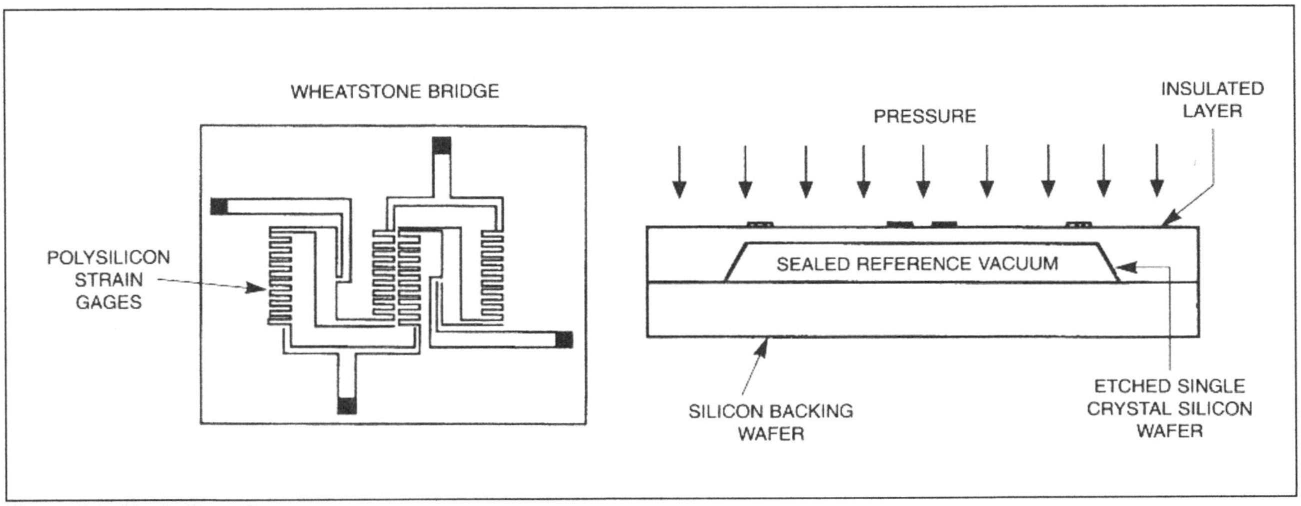

Figure 6-4. Strain Gage Sensor.

Installation

The most important step in the installation of an electronic pressure transmitter takes place before piping it up to the process or making the electrical connections. The key to obtaining a good, reliable measurement is selecting a transmitter with specifications that fit the application.

These specifications fall into two categories: functional and performance. Functional specifications define the transmitter's operating limits and performance specifications describe how effectively the transmitter will operate within those limits. Functional specifications include such things as temperature and pressure limits, hazardous area compatibility, power supply requirements, and elevation/suppression capability. Performance specifications include accuracy, stability, temperature and static pressure effects, vibration and power supply effects, and mounting position effect.

Several considerations for choosing a pressure transmitter are as follows:

(1) The transmitter should be the proper type for the measurement required—differential, gage, or absolute.

(2) Transmitter features should also match the nature of the application. For instance, remote seals or sanitary connections should be readily available for dairy or food installations.

(3) The wetted parts, such as O-rings, flanges, and diaphragms, must be compatible with the process.

(4) In the case of sour gas measurement, NACE MR0175/ISO 15156 requirements must be met.

(5) Nonwetted materials, such as housings and bolts, must be able to resist environmental conditions.

(6) The transmitter should be capable of operating in environments with up to 100% relative humidity and should be designed and packaged to prevent damage or failure caused by moisture.

(7) Lightning protection should be considered when mounting the transmitter in an open area where lightning strikes are possible.

(8) The transmitter must be able to contain the process pressure and have overpressure protection.

(9) The output signal must be compatible with the readout device being used and able to operate on the available power. Standard analog 4–20 mA loops typically operate on 24 V DC.

Once the correct transmitter has been specified, the actual manner in which the transmitter is installed has much to do with the quality of the measurement. Proper mounting procedures should never be compromised and should take the following tips into consideration.

Installation Tip 1: Transmitter Location

Proper transmitter location with respect to the process pipe is important. Consider the following to determine the best transmitter location:

- Keep corrosive or hot process material away from the transmitter.
- Avoid sediment deposits in the impulse piping.
- Keep liquid head balanced on both sides of the transmitter.

- Keep impulse piping as short as possible.
- Avoid ambient temperature gradients and fluctuations.

Installation Tip 2: Process Tap Connections

Process tap connections are located differently depending upon the application. Gas, liquid, and steam each require a different mounting position. Process connection taps for gas measurement should be either on the top or side of the piping. If on the side, the impulse piping should be pitched down toward the piping to allow any condensate to drain. A general rule is to slope piping at least one inch per foot (eight centimeters per meter) down toward the process connections for gas service.

Process connections for liquids should also be on the side but should pitch up from the transmitter at least one inch per foot toward the process connection to allow any gas that might collect in the transmitter to bubble upward.

When dealing with steam applications it is important to protect the transmitter from possible over-temperature conditions. The best way to do this is to take the impulse piping out the side of the process piping into the outlet of a tee. The transmitter is mounted below the connection on the run of the tee. The impulse pipe is filled with water and the other tee opening is plugged. This ensures hot steam is kept out of the transmitter. All three of these configurations are shown in Figure 6-5.

Installation Tip 3: Impulse Piping

Impulse piping between the process and the transmitter must transfer the pressure seen at the process taps to the transmitter. The following conditions could cause errors in pressure transfer:

- Leaks
- Friction loss—particularly if purging is used
- Trapped gas in a liquid line
- Liquid in a gas line
- Temperature-induced density variation between legs

Doing the following can keep errors to a minimum:

- Keep impulse piping as short as possible and free from bends, turns, and high and low points. When using impulse piping to cool the process, the general rule is a 100°F drop in temperature per foot of piping.
- Use wet legs on steam service and other high-temperature applications. Ensure that both piping legs are filled to the same level when using sealing fluid.
- Keep both impulse legs at the same temperature.
- For liquid service, avoid high points that can trap gas. Slope piping at least one inch per foot up toward the process connection to further ensure that no gas will be present.
- For gas service, avoid low points where condensate can accumulate. Slope piping at least one inch per foot down toward the process connection to ensure no liquid accumulation.
- Vent gas from transmitters in liquid service and liquid from transmitters in gas service.
- Use impulse piping of sufficient diameter to avoid friction effects.

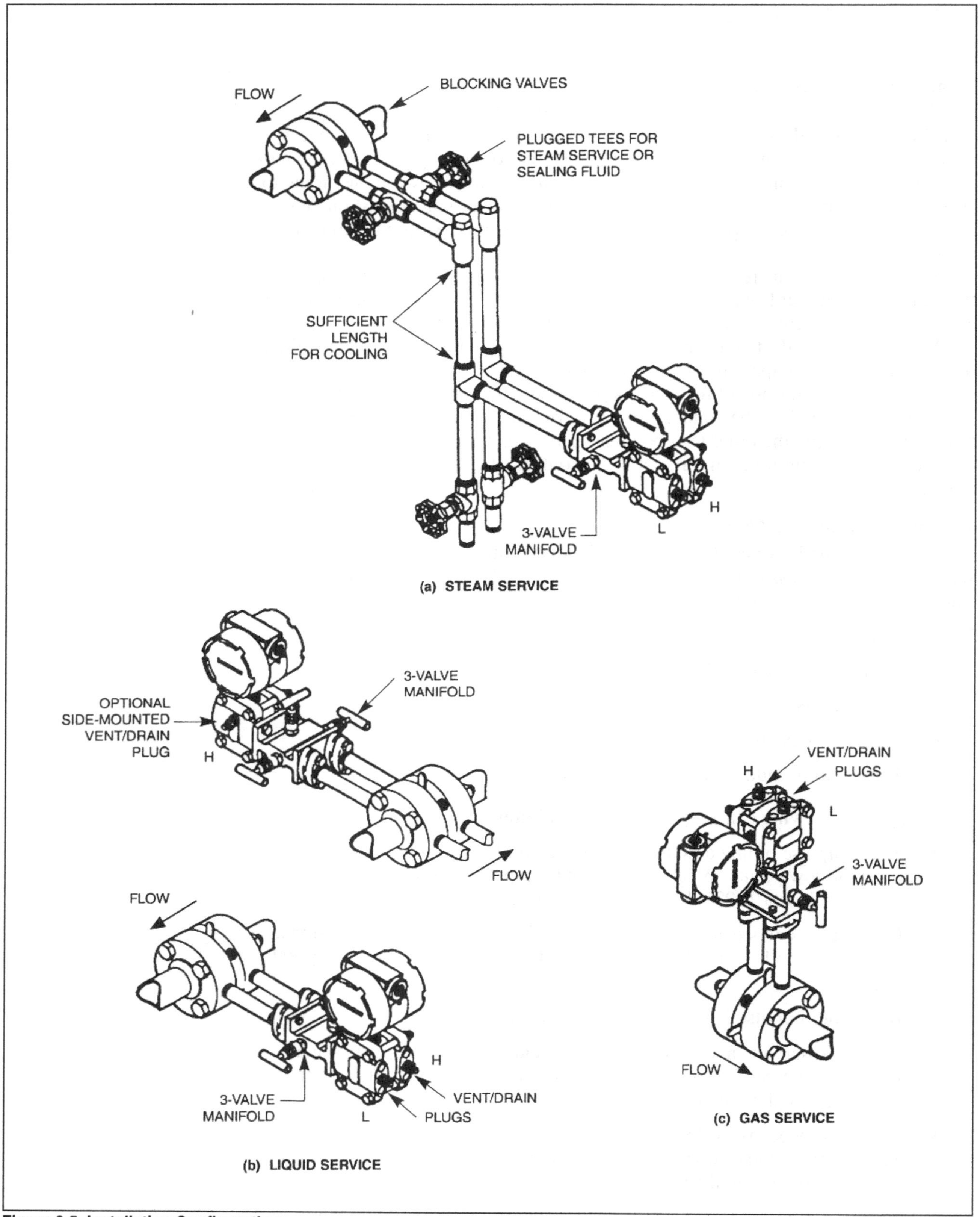

Figure 6-5. Installation Configurations.

- Avoid purging through the transmitter. Make the purge connection close to the process taps and purge through equal lengths of the same size pipe.

Instillation Tip 4: Block Valves

Valves and three-valve manifolds installed in the impulse piping should be designed for use during commissioning, calibration, and removal for transmitter repair. Port openings should be close as possible in size to the inner diameter of the impulse piping.

Instillation Tip 5: Environmental Considerations

Hazardous locations require equipment to have either an explosionproof or intrinsic safety rating. Conduit seals may or may not be required depending upon local code or insurance requirements.

Process and ambient temperature limits should be observed. Failure to do so could result in poor performance and possibly premature equipment failure. To protect the transmitter from process heat, use one foot of impulse piping for every 100°F the process temperature is above the transmitter upper temperature limit. In the case of the other extreme, heating boxes or heat tracing should be used in very cold atmospheres. Wet legs in cold environments should contain antifreeze.

In areas where electrical storms are prevalent, some form of transient protection is advised.

Instillation Tip 6: Support Mounting

Transmitters must be adequately supported. Pipe or wall mounting should be used. With the advent of small, light-weight transmitters, brackets may be dispensed with in some cases.

Installation Tip 7: Electrical Wiring

The manufacturer's guidelines for wiring should be followed. Wiring should be twisted-pair encased in conduit or shielded to prevent emi/rfi interference. Transmitter circuits should be properly grounded and multiple grounds should be avoided. Installation must be in accordance with all applicable national and local codes. The question of whether or not conduit seals are required is best settled by either the local electrical inspector or insurance underwriter.

Calibration

Calibration is the procedure by which the output of a transmitter is adjusted to properly represent a known pressure input. It is the most frequently performed maintenance operation on pressure transmitters and plays a major role in determining how well the transmitter will operate. A pressure input is used to provide zero and span adjustments to the transmitter. When complete, the transmitter should be spanned for operation in a given application. Therefore, it is important that the calibration procedure be done properly. Using the following hints will ensure that the multi-step process of calibration goes smoothly.

Calibration Tip 1: Procedures

Read and understand the calibration procedure as written in the manufacturer's instruction manual. Following the instructions carefully is the first step in ensuring a proper calibration.

Calibration Tip 2: Equipment

Use the proper calibration equipment. The pressure source and readout device must be of greater accuracy than the instrument being calibrated. As a general rule, the pressure source and readout device should be at least four times more accurate than the device being calibrated. High-accuracy measurements cannot be obtained when calibration is performed with low-accuracy equipment.

In addition, test equipment should be checked against a standard on a regular basis to ensure that accuracy is maintained.

Calibration Tip 3: Leaks

Eliminate all leaks in the calibration system. Leaks are a big source of calibration error.

Calibration Tip 4: Liquids

Drain all liquids from the transmitter and impulse piping before calibrating. Remember, on steam lines the condensate legs must be refilled after calibration.

Calibration Tip 5: Linearity Adjustments

Linearity adjustments should only be made at one point. All other points are used only to check the adjustment.

Calibration Tip 6: Damping

Damping should be set after the transmitter is placed in service.

Calibration Tip 7: Three-Valve Manifold

Use of a three-valve manifold adds some complexity to the process. When zeroing a transmitter with a three-valve manifold, the following sequence should be used. First, place the manifold in bypass and then close the low-side valve. Adjust the zero and then open the low-side valve before closing the bypass. Failure to do so could cause a zero shift.

On steam service, it is necessary to close the low-side block valve before starting the zeroing procedure. If this is not done, the low-side wet leg level will be reduced by an amount equal to the differential.

Calibration Tip 8: Improving Performance

Several other things can be done to further improve performance. All require some special equipment or preparation.

- Calibrate at temperature. Transmitter performance is affected by changes in temperature. To minimize these temperature effects, calibration at the expected ambient temperature is recommended. If temperatures are expected to fluctuate, it is good practice to calibrate between the extremes. Seasonal recalibration is another good practice, especially where there is a wide variance between summer and winter temperatures.
- Calibrate at static line pressure. As with temperature, transmitter performance is affected by changes in static pressure. These effects can be reduced by calibrating at line pressure. If this is not practical, rezeroing after operating pressure has been reached can make a substantial improvement in transmitter performance.
- Rezero after installation. Transmitters should be rezeroed after installation. This will eliminate any errors induced by mounting position effects or piping stresses generated during installation.

Maintenance

A transmitter that has been properly selected for its application and is properly installed, calibrated, and maintained should have reduced need for repair. A preventive maintenance program should include the following aspects:

- Calibration checks should be performed to determine what, if any, drift has taken place and if recalibration is necessary.
- Process connections should be checked for plugging or leakage.

- The transmitter should be examined for signs of leakage, corrosion, or damage.
- Part of any good maintenance program involves keeping good records of which checks have been performed and when.

Troubleshooting

In the case of a suspected transmitter malfunction, the procedures listed in Table 6-1 can assist in isolating the problem and its source. The information is separated according to three primary malfunction symptoms. Under each symptom, checking procedures are listed for conditions that could be causing the malfunction. To troubleshoot, select the symptom most closely resembling the unit's malfunction and use the information accordingly. Always deal with the most likely and easiest-to-check conditions first.

Isolate a failed transmitter from its pressure source as soon as possible. Pressure that may be present could cause serious injury to personnel if the transmitter is disassembled or ruptures under pressure.

Pressure and Differential Pressure

These transmitters are usually used for clean fluids only, unless combined with a bubble tube or purge system that keeps the transmitter body clean and free of solids or plugging media. When troubleshooting these types of transmitters look for the following problems or follow the suggestions discussed below.

Before performing any service on a working loop, inform the operator and have the loop placed in manual control. Check for any alarms that may shut down the process if the level signal is varied as part of the calibration or checking.

(1) Check the transmitter range and density against the actual conditions. Many problems are the result of changing the product being measured and not changing the range of the transmitter to compensate for a different density or specific gravity of the product.

(2) If the transmitter does not respond to level changes, and if the product is harmless and under low pressure, open the drain or vent on the body of the transmitter to purge it of entrapped air or any material that may be causing the problem. If the product is hazardous, perform these operations only according to the rules that govern hazardous materials.

(3) Check all purges and process tubing for leaks, plugging, or a closed isolation valve. Check that purge pressures are at least three pounds above the process backpressure, which includes any padding or tank pressure.

(4) Check the transmitter location to the bottom level of the tank or vessel, and if the transmitter is located below the bottom or empty line of the tank, adjust the transmitter output to read zero when the tank is empty.

(5) When checking calibration of a level transmitter, always set the 100% level at the overflow of the tank or chest. If the 100% level is set at some point below the overflow, for an assumed safety margin, the operator will never be certain of the level readings.

(6) If the operator knows that 100% is not full or overflowing, the tendency is to let the level exceed 100% at times, and this can cause overflow conditions.

(7) If a differential pressure transmitter is used for the level measurement of a pressurized tank, a dry or wet leg must be used to compensate for the pressurization of the tank. If a dry leg is used, it must be kept dry. Extend the dry leg above the upper tank tap, using ½-to 5/8-inch tubing or 1-inch pipe, and slope the connecting tubing at least 1 inch per foot to the tap

Table 6-1. Troubleshooting Guide for Transmitters

Symptom	Potential Source	Corrective Action
High output	Primary element	Check for restrictions at primary element.
	Impulse piping	Check for leaks or blockage.
		Check that blocking valves are fully open.
		Check for entrapped gas in liquid lines and for liquid in dry lines.
		Check that density of fluid in impulse lines is unchanged.
		Check for sediment in transmitter process flanges.
	Transmitter electronics failure	Determine faulty circuit board by trying spare boards.
		Replace faulty circuit board.
	Power supply	Check output of power supply.
Erratic output	Loop wiring	Check for adequate voltage to the transmitter.
		Check for intermittent shorts, open circuits, and multiple grounds.
	Process fluid pulsation	Adjust electronic damping pot.
	Impulse piping	Check for entrapped gas in liquid lines and for liquid in dry lines.
	Transmitter electronics connections	Check for intermittent shorts or open circuits.
		Make sure post connectors are clean and check the sensor connections
	Transmitter electronics failure	Isolate faulty circuit board by trying spare boards.
		Replace faulty circuit board.
Low or no output	Primary element	Check installation and condition of element.
		Note any changes in process fluid properties that may affect output.
	Loop wiring	Check for adequate voltage to transmitter.
		Check for shorts and multiple grounds.
		Check for polarity of connections.
		Check loop impedance.
	Impulse piping	Check that pressure connection is correct.
		Check for leaks or blockage.
		Check for entrapped gas in liquid lines.
		Check for sediment in transmitter process flange.
		Check that blocking valves are fully open and that bypass valves are tightly closed.

Table 6-1. Troubleshooting Guide for Transmitters (cont'd)

Symptom	Potential Source	Corrective Action
		Check that density of fluid in impulse lines is unchanged.
	Transmitter electronics connections	Check to see that calibration adjustments are in control range.
		Check for shorts in sensor leads.
		Make sure post connectors are clean and check the sensor connections.
	Transmitter electronics failure	Isolate faulty circuit board by trying spare boards.
		Replace faulty circuit board.

connection. This will keep any condensation, spray, or slight overfill from filling the dry leg. Provide a trap on the dry leg, below the transmitter, to catch any liquid, and drain the trap periodically. If the trap fills too frequently, the dry leg can be modified by using a 1:1 diaphragm sensor that isolates the dry leg from the process.

If a wet leg is used, the level of fluid in the wet leg must be kept constant, or the zero of the level transmitter will change. For this reason it is wise to supply a capacity or pot at the top of the wet leg, that helps to maintain a constant level, even if a small amount of liquid is lost. The volume of an inch of 1-inch pipe or ½-inch tubing is quite small. The pot can provide more than 100 times the volume of the same section of pipe or tubing. Also check the filling fluid density (if different from the tank product) and suppress the transmitter accordingly. Also check that the wet leg is filled to the top connection. For suppression and elevation definitions, refer to Chapter 13.

(8) When checking a differential pressure transmitter used on an open tank or vessel, confirm that the low side of the transmitter is vented to the atmosphere.

Diaphragm Transmitters

Diaphragm transmitters are used when a viscous fluid or one with entrained solids can cause plugging of a standard transmitter. The diaphragm should be chosen to be compatible with the process and can be flush or extended to allow the face of the diaphragm to actually be flush with the process even when the flange may be a few inches away from the process.

When venting any transmitter in the southern part of the U.S. or anywhere there are nesting bugs, provide a bug screen on the vent to prevent them from nesting in the vent, plugging it, and causing a pressure shift.

Diaphragm transmitters use about the same troubleshooting and installation techniques as standard transmitters. There are, however, a few differences.

On stock chests, large slurry tanks or any vessel that can hold many days supply of product, a means of servicing or removing the diaphragm level transmitter from the vessel, without first draining it, is a necessity. A special sliding gate valve is manufactured that bolts to the transmitter and also to the vessel flange. The valve is designed to be close coupled. Do not extend any flange mounting more than a very short distance from the vessel flange, as any extended piping will only fill with the product and possibly plug the opening. If the piping exceeds more than 6 inches from the vessel flange, it may require a manual purge connection to periodically purge the opening. Sometimes, a continuous purge is needed. Purge the sensor before removing the transmitter, as this will tend to keep the guides in the gate valve clean and reduce the force needed to open or close the valve.

A drifting or nonlinear transmitter output indicates a dented or leaking diaphragm. When replacing a diaphragm level transmitter, closely check the flange size, diaphragm and body material, and range, so that a duplicate can be selected. If the unit was used on vacuum service, check for any special options required for this type of service.

Volumetric Diaphragm Transmitters

The only difference between a diaphragm transmitter and a volumetric diaphragm transmitter is that the latter has the diaphragms attached to the body of the transmitter using volumetric filled capillary. This type of transmitter is used almost exclusively on viscous and corrosive fluids or fluids with solids, where the measurement is in inches of water or in the low pressure range. The capillary-connected, low-side diaphragm allows it to be placed at the top of the tank without fabricating a sometimes complex wet or dry leg or purges.

When troubleshooting capillary systems, look for capillary leakage or damaged diaphragms. A slow response to a level change indicates a loss of some filling fluid. No response indicates a dead system or a malfunction of the transmitter mechanism.

If both capillaries are not the same length, temperature effects can be large. If the capillary lengths are the same but only one side is in the direct sun, compensation can also be compromised. Move the coiled high side capillary to receive the same amount of sun as the low side to help improve temperature compensation.

Any time a capillary system is bench-calibrated and reinstalled on the process, an elevation or suppression adjustment must be made to obtain a true zero reading. Check the vendor's instructions or specifications for limits to these adjustments.

There are limits to how far that transmitter can be mounted above the lower sensor before the weight of the filling medium actually separates the fluid in the capillary and causes the equivalent of bubbles in the transmitting medium. If the capillary lengths are long and the transmitter looses signal, check the vendor's instructions for possible limits that the body of the transmitter cannot be above the lower sensor.

If readings are erratic, check the temperature limits of the diaphragm and also the capillary fill. These limits may be exceeded on a hot process or hot ambient area.

When moving or mounting capillary systems, never kink the capillary. Check the specifications for the minimum bending radius allowed. Coil any excess capillary, but remember to keep both capillaries as close to the same temperatures as possible.

Differential Pressure Technology

Theory of Operation

Differential pressure or head-type flowmeters represent one of the most commonly used flowmeter technologies. Differential pressure is the basis of flow because without it there can be no flow. For example, picture a thin section of fluid in a pipe. If an equal pressure were applied to both sides of the section ($P_1 = P_2$), the fluid would just sit still. However, if one pressure is greater than the other ($P_1 > P_2$), flow would take place from the high-pressure region to the low-pressure region. The greater the differential pressure, the higher the rate of flow. Given this information, it is easy to see that flow rates are proportional to differential pressure. Mathematically, flow is expressed as a square root function of the differential pressure, as shown in the following equation:

$$Flow = Constant \times Differential\ Pressure^{(1/2)}$$

Some type of primary element is used in the pipeline to restrict flow and create the differential pressure. This pressure is measured by a differential pressure transmitter and a flow rate is indicated.

Installation

General installation practices that apply to pressure transmitters should also be applied to differential pressure transmitters measuring flow. In addition, some special considerations are listed below.

Installation Tip 1: Straight Pipe Runs

A general rule is that there should be four pipe diameters of straight pipe upstream of the orifice plate (or other primary element) and two pipe diameters of straight pipe downstream.

Installation Tip 2: Piping

The most popular type of orifice plate is concentric. It is important to mount this type of orifice so that the bore is exactly centered within the pipe. This is more critical in small diameter pipes, but should be adhered to as much as possible in any installation. Other possible types of orifice plates include eccentric, segmental, and quadrant edge. Each has special installation practices that should be adhered to.

The plate should also be kept clean and free from accumulations on either the upstream or downstream side. Failure to do this will affect the profile of the flow and product measurement errors.

Installation Tip 3: Pressure Tap Location

There are several different types of pressure taps. Flange taps are bored into the flanges that hold the orifice plate. They are generally located one inch from the plate both upstream and downstream. This is the most common tap location for pipe sizes 2 inches and larger.

Pipe taps are another type of pressure tap. They should be located 2-1/2 pipeline diameters upstream and 8 diameters downstream of the restriction.

Finally, D-D/2 taps should be located one diameter upstream and ½ diameter downstream from the orifice plate.

Calibration

Calibration procedures given for pressure transmitters also apply to differential pressure transmitters measuring flow. It should be noted that the orifice plate, or any other head-producing primary element, is an inferential flowmeter. Therefore, it is important that the physical condition of the orifice plate and the bore be checked prior to calibration. Any errors due to the primary element will be seen in the measurement from the differential pressure transmitter.

Maintenance

Maintenance tips given for pressure transmitters also apply to differential pressure transmitters. In addition, the maintenance of the primary element is also critical to the accuracy of the measurement.

Maintenance Tip 1: Sharp-edged Orifice Plates

Most orifice plate applications utilize sharp-edged orifice plates. If the sharp edge becomes dulled, measurement drift will result. The flowmeter measurement will be lower than the flow in the pipe as less differential pressure is developed across the larger orifice plate bore. The plate should be checked periodically to ensure that neither wear nor corrosion are a problem.

Maintenance Tip 2: Removal

Differential pressure transmitters should be removed from service with care. The transmitter could be damaged or have a calibration shift if full pipe pressure is applied to one side of the transmitter. To ensure against this, first open the bypass valve. This will expose both sides of the transmitter to the same pressure. Then other valves can be opened without damaging the transmitter.

Maintenance Tip 3: Placing in Service

Differential pressure transmitters also should be placed in service with some care for the same reasons. To do this, first open the bypass valve with the flowmeter system not in service. This exposes both sides of the transmitter to equal pressure. Then the valves to the taps are opened, allowing fluid to flow through the impulse piping. Finally, the bypass valve is closed and the transmitter is in service.

Troubleshooting

Troubleshooting tips given for pressure transmitters may be used for differential pressure flow transmitters as well.

The Differential Pressure Flowmeters are the most applied flowmeter. This type of device has a flow restriction in the line that causes a differential pressure or "head" to be developed between the two measurement locations. Types of restrictions are:

- orifice
- Venturi
- nozzle
- Pitot tube
- Pitot-static tube
- elbow
- wedge

Orifice

The most commonly applied orifice is a thin, concentric, and flat metal plate with an opening in the center (Figure 6-6), installed perpendicular to the flowing stream in a circular conduit or pipe. In most applications, a sharp edged hole is bored in the center of the orifice plate. As the flowing fluid passes through the orifice, the restriction causes an increase in velocity. As the fluid leaves the orifice, its velocity decreases and its pressure increases. There is always some permanent pressure loss due to friction.

In all cases, the sensing line or device must be kept full of fluid during flow measurement, or an error will result. All flow sensing devices assume the line is full and do not account for air bubbles, air pockets, or a partially full line.

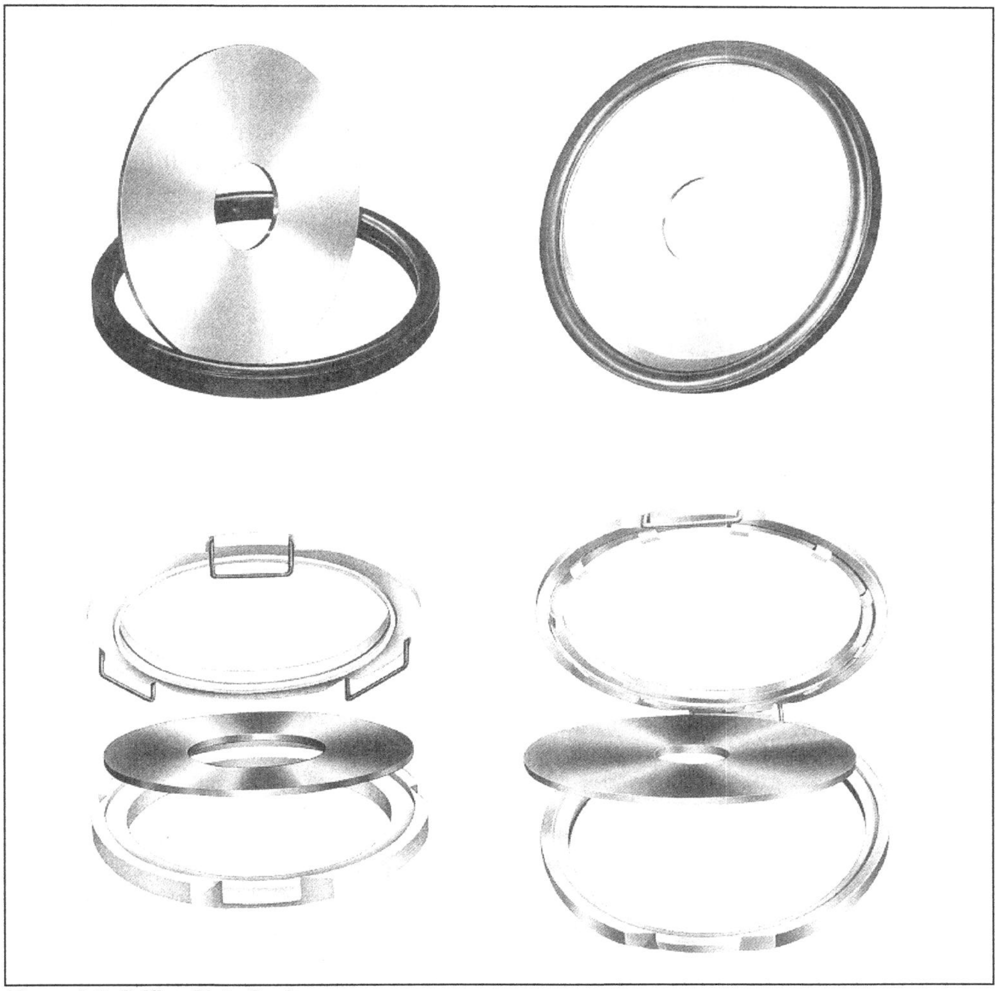

Figure 6-6. Orifice Plate Seals.
(Courtesy of Daniel Industries, Inc.)

ORIFICE PLATES

Most pressure drop flow devices have somewhat limited turndown ratios. An orifice plate has roughly only a 3-to-1 turndown; a magnetic flowmeter has roughly a 30-to-1 turndown. Be aware of this rangeability when selecting a flowmeter. For orifice devices, try to place the operating flow rate at 70% of full scale or above to allow a reasonable accuracy at the lower flow rates.

If there are complaints about an orifice plate installation, use the following checklist:

(1) Nonlinear flow devices, (the orifice plate is the most common) must be linearized when used in a controlled loop. Always know the calibration curve of the flow device being used and be sure that all parts of the loop are matched or linearized. If the flow is not accurate, check to see that it was not linearized twice, once in the transmitter and once in the readout software.

(2) Use recommended installation practices when installing a dP transmitter on a flow loop. Process tubing and the physical mounting of the transmitter are critical to a stable, accurate flow measurement. See calibration comments in Chapter 13.

(3) Check that there is sufficient straight pipe run upstream and downstream of the orifice. Use ASME charts and graphs (see the Bibliography).

(4) Is the orifice plate centered and installed with the sharp side facing upstream? Orifice data is usually stamped on the sharp-edge side of the orifice handle or paddle, but if the data is missing or it is felt that the orifice itself may be worn or damaged, remove the orifice and examine it thoroughly. Check for the correct dimensions as defined in the orifice plate data sheet. Is the edge still sharp without any serious nicks or dents?

(5) Is the bore of the piping in the orifice run smooth and of the correct diameter? See pipe schedule tables for IDs.

(6) Is there any trash or buildup in the line in front of the plate?

(7) Are the "drop legs" to the transmitter at the same starting elevation? Are they full of liquid or condensate (in the case for steam flow)? (Notify the operator before performing any test on a working transmitter.)

(8) Check the orifice plate data sheet for range, density, bore size, units, etc., to confirm that all the data is in agreement with the transmitter calibration.

(9) Check that the orifice run is full when flow is being measured. Check for proper mounting of the transmitter, above or below the orifice, etc.

An orifice plate reading did not agree with a supplier's gas flow turbine meter after the turbine meter had been replaced with a new unit, but did agree reasonably close before the replacement. The orifice plate measurement was thought to be correct since it agreed with the turbine meter reading for quite some time. The turbine meter had been replaced because the bearings were worn. When a new turbine meter caused a lower reading that just happened to be about the same as the orifice reading. After transmitter calibration checks, the two readings still did not agree. A study of the orifice calculation sheet revealed that the density factor for natural gas had been changed slightly and was now incorrect for the type of gas being used. Using the correct density for the gas and recalibrating the dP transmitter corrected the problem. This sounds like a simple mistake, but the error accounted for about one million dollars per year additional cost to the plant, once it was proven that the gas company's turbine meter was now correct and the plant's orifice meter agreed with the turbine meter.

Venturi

A venturi is a restriction with a relatively long passage with smooth entry and exit (Figure 6-7). It produces less permanent pressure loss than a similar sized orifice, but is more expensive.

Nozzle

Flow nozzles have a smooth entry and a sharp exit (Figure 6-8). For the same differential pressure, the permanent pressure loss of a nozzle is of the same order as that of an orifice, but it can handle dirty and abrasive fluids better than the orifice. Nozzles are primarily used in steam service because of rigidity.

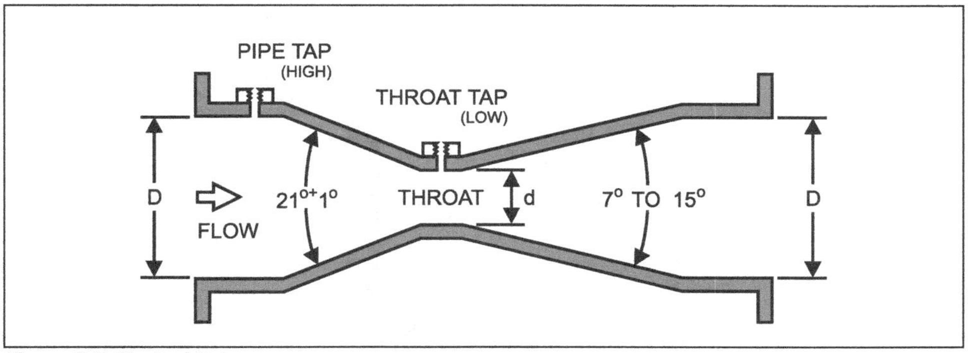

Figure 6-7. Venturi Tube.

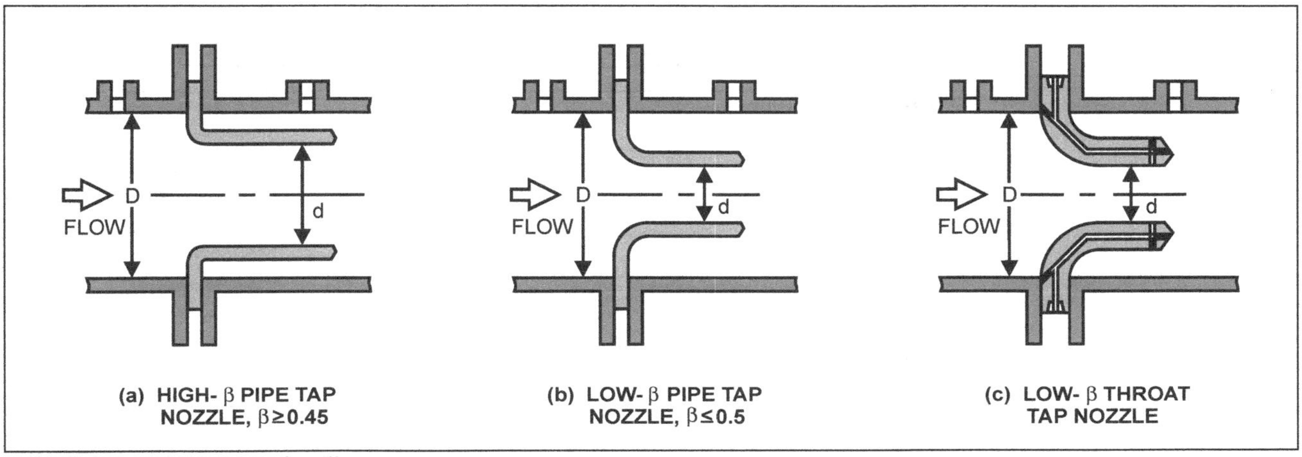

Figure 6-8. ASME Long Radius Flow Nozzle.

Pitot Tube

A Pitot tube is a point velocity measuring device (Figure 6-9). It has an impact port; as fluid hits the port, its velocity is reduced to zero and kinetic energy (velocity) is converted to potential energy (pressure head). The pressure at the impact port is the sum of the static pressure and the velocity head. The Pitot tube has negligible permanent pressure drop in the line, but the impact port must be located in the pipe where the measured velocity is equal to the average velocity of the flowing fluid through the cross section.

Pitot-Static Tube

The Pitot-static tube (Figure 6-10) has an integral construction of a Pitot tube and an annular tube with static pressure ports. The differential pressure between the two ports is the velocity head. The impact port must be located to measure the average fluid velocity.

Elbow Meter

This flowmeter is a pipe elbow in which a pressure differential is created by centrifugal force between the inside diameter and the outside walls of the pipe elbow. This flowmeter does not introduce any additional pressure loss in the system other than that already caused by the elbow, but its pressure differential is very low.

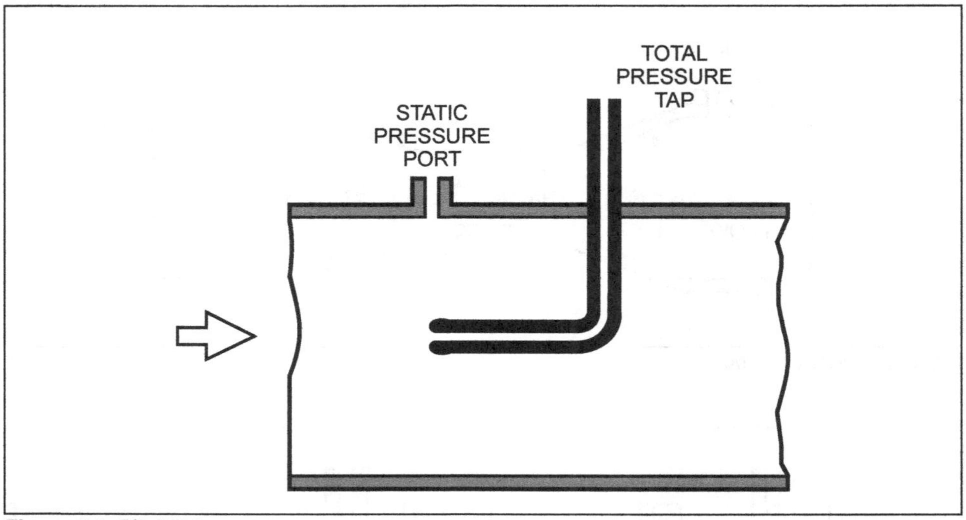

Figure 6-9. Pitot Tube.

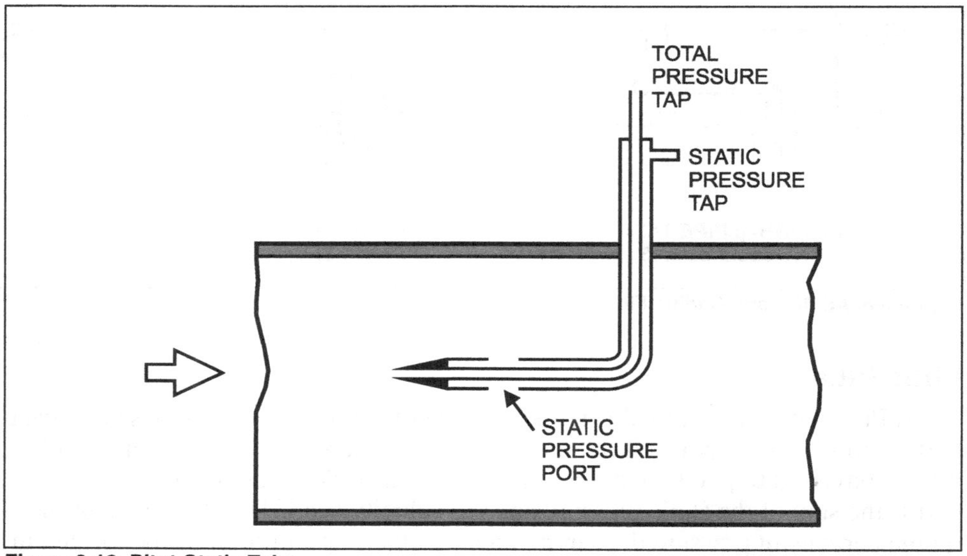

Figure 6-10. Pitot-Static Tube.

Wedge Meter

This flowmeter consists of a wedge-shaped element placed in perpendicular to the flow at the top of the conduit (Figure 6-11). The bottom of the conduit is unrestricted, making the wedge meter particularly useful in slurry measurement.

Level Transmitters

Level Sensors

Level may be measured by a wide variety of level devices. Probably the most common principle used today is still the measurement of product backpressure. If a clean liquid level is measured the backpressure measurement is usually made by a standard pressure or differential pressure transmitter. It is important to know if the product being measured is under pressure or at atmospheric pressure. It is also

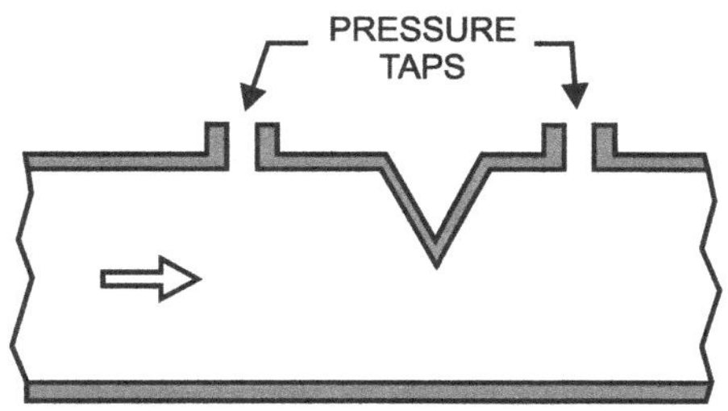

Figure 6-11. Wedge Meter.

important to know the density of the product being measured. If the product density is changed, the level transmitter must be recalibrated.

When product viscosity is high or the product tends to plug or coat, a diaphragm level transmitter, a purge system, a capacitance probe, or ultrasonic level sensor or other types of sensors must be used to reduce maintenance or operating problems. Each type of level device has its good and bad features. An attempt will be made to cover most of the more common sensors and discuss ways to troubleshoot and maintain them.

There are quite a few types of electronic level sensors. A rough listing includes the following:

(1) Capacitance

(2) Radio frequency

(3) Electromechanical

(4) Float/displacer

(5) Load cells

(6) Microwave

(7) Nuclear

(8) Optical/laser

(9) Resistance

(10) Sonic/ultrasonic

Level can usually be checked visually, in open tanks, and inferentially, in closed tanks. Before removing a level device for calibration, try to check its reading at both a high level and a low level. Establish the true zero and make every effort to establish the 100% level indication to be the actual tank or vessel overflow point. Check dry and/or wet legs, if used, and drain or fill as required for a true zero and full indication. Check the density of the filling fluid and all calculations based on the fluid and length of the wet leg.

Theory of Operation

Pressure transmitters can also be used to make level measurements. The pressure at the bottom of a liquid-filled tank is directly proportional to the level in the tank. Therefore, the tank level can be determined simply by measuring the pressure.

However, there is one additional consideration to be made. All liquids do not have the same weight per unit volume. Water weighs 62.4 pounds per cubic foot, but gasoline only weighs 41.2 pounds per cubic foot. Therefore, the pressure at the bottom of a tank of gasoline would be 41.2/62.4 = 66% that of an equivalent tank of water. The ratio of the weight of the volume of a liquid to that of an equivalent volume of water is called specific gravity. In order to determine a liquid level, it is essential that the *specific gravity* of a liquid be known. Level is then determined by multiplying the liquid height above the tap by the specific gravity of the liquid, as shown in the following formula:

$$Level = Height \times Specific\ gravity$$

The measurement is independent of volume or vessel shape and is generally expressed in inches of water (in. H_2O).

Installation

Standard installation practices for pressure transmitters given earlier for pressure transmitters also apply to level transmitters. In addition, liquid level installations require some special attention. It is necessary to first determine if the tank level to be measured is in an open or vented tank or a pressurized tank. Other considerations include whether or not diaphragm seals are involved and if vacuum conditions are present. Each of these variables affect the installation.

Open tanks are the simplest. The pressure transmitter should be mounted at or below the tank connection and measures the pressure of the liquid above. The pressure is sensed by the process flange and transmitted to the high-pressure side of the sensing element. The low-pressure side is vented to the atmosphere. The calibration is simply determined as a function of the desired level and the specific gravity of the process material. Figure 6-12 shows an example of open tank level measurement.

Pressurized tanks require reference connections. A reference connection is a connection from the transmitter to the top of the tank. Its purpose is to balance out the pressure above the liquid so that a true level measurement can be made. If there were no reference connection, the transmitter would indicate the level plus the tank pressure.

The low-side transmitter piping that provides the reference connection will remain empty if gas above the liquid does not condense. This is called a dry leg condition. In this case range determination calculations are the same as those described for bottom-mounted transmitters in open vessels shown in Figure 6-12.

However, condensation of the gas above the liquid can cause the low side of the transmitter piping to slowly fill with liquid. Instead of allowing this to happen, the pipe is purposely filled with a convenient reference fluid (either the process liquid or some compatible material) to eliminate this potential error. This is called a wet leg condition.

In these cases the reference fluid will exert a head pressure on the low side of the transmitter. To account for this, the level must be determined and the negative effect of the reference leg must also be subtracted. The resulting calculations will result in a negative calibration (–120 to –15 in. H_2O). such calibrations are said to have an elevated zero. Figure 6-13 shows an example of a wet leg calibration. Under some conditions it is necessary to utilize diaphragm seals on transmitters that measure level. Sanitary applications, applications where clogging is a potential problem, and applications with noncontamination requirements are examples of applications where seals are used. In addition, vacuum applications always require diaphragm seals. Without them, wet legs could change level as the vacuum changes.

Vacuum applications also require another special installation consideration. Transmitters must always be mounted at or below the bottom process connection. Failure to do so could result in diaphragm failure.

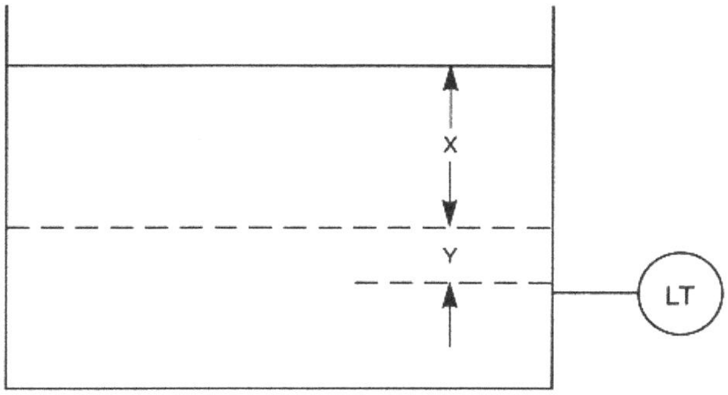

Let X equal the vertical distance between the minimum and maximum measurable levels (500 in.).
Let Y equal the vertical distance between the transmitter datum line and the minimum measurable level (100 in.).
Let **SG** equal the specific gravity of the fluid (0.9).
Let h equal the maximum head pressure to be measured in inches of water.
Let e equal head pressure produced by Y expressed in inches of water.
Let **Range** equal e to $e + h$

Then h = (X)(**SG**)
= 500 0.9
= 450 in. H_2O
e = (Y)(**SG**)
= 100 0.9
= 90 in. H_2O
Range = 90 to 540 in. H_2O

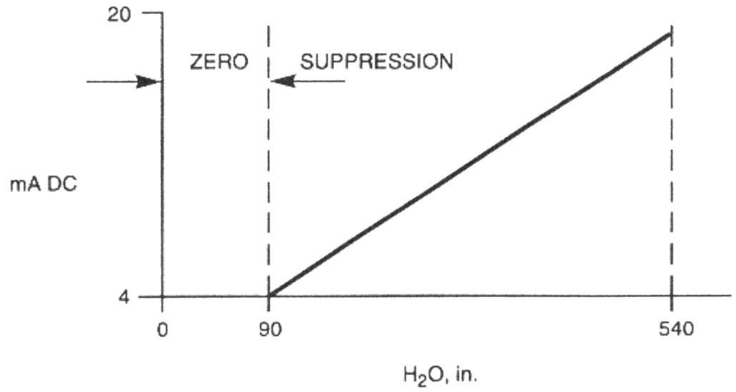

Figure 6-12. Open Tank Level Measurement.

Calibration

Calibration practices applicable to pressure transmitters in general, also apply to level transmitters. It should be noted that transmitters used in making level measurements frequently are calibrated at spans other than zero-based. For instance, they may be calibrated from –100 to –20 in. H_2O or 50 to 100 in. H_2O. Such calibrations are said to have elevated or suppressed zeroes. Elevation and suppression capabilities of a transmitter are expressed in percent and are determined by dividing the lower calibrated value by the span and multiplying by 100.

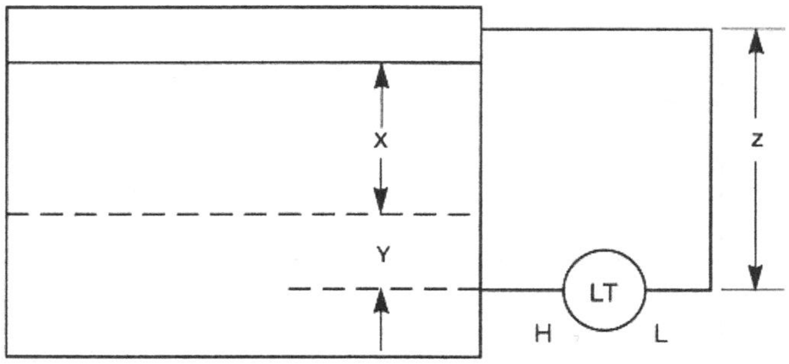

Let *X* equal the vertical distance between the minimum and maximum measurable levels (500 in.).
Let *Y* equal the vertical distance between the transmitter datum line and the minimum measurable level (50 in.).
Let *z* equal the vertical distance between the top of the liquid in the wet leg and the transmitter datum line (600 in.).
Let $\mathbf{SG_1}$ equal the specific gravity of the fluid (1.0).
Let $\mathbf{SG_2}$ equal the specific gravity of the fluid in the wet leg (1.1).
Let *h* equal the maximum head pressure to be measured in inches of water.
Let *e* equal head pressure produced by *Y* expressed in inches of water.
Let *s* equal head pressure produced by *z* expressed in inches of water.
Let **Range** equal *e s* to *h* +*e* − *s*.
Then $h = (X)(\mathbf{SG_1})$
$= 500\ 1.0$
$= 500$ in. H_2O
$e = (Y)(\mathbf{SG_1})$
$= 50\ 1.0$
$= 50$ in. H_2O
$s = (z)(\mathbf{SG_2})$
$= 500\ 1.1$
$= 550$ in. H_2O
Range = *e s* to *h* +*e* − *s*
= 50 550 to 500 + 50 660
= 610 to 110 in. H_2O

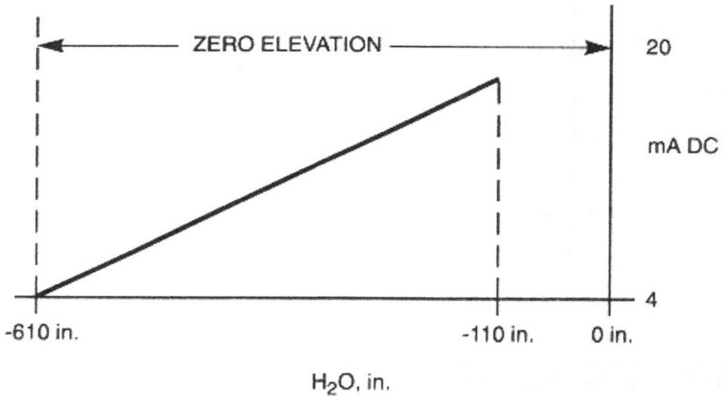

Figure 6-13. Wet Leg Example.

Maintenance and Troubleshooting

There are no special considerations for the maintenance and troubleshooting of a level transmitter. The procedures and tips given for pressure transmitters should be followed.

Flow Transmitters

Flow is measured by a number of different methods. This section will cover flow measurement made using the following types of flowmeters:

Differential Pressure
Magnetic
Mass Flow
Turbine
Open Channel
Vortex
Positive Displacement
Target
Thermal Mass
Tracer Dilution
Ultrasonic
Variable Area
Insertion

Fluid and Gas Properties

You should be familiar with the properties of liquids and gases when selecting, installing, calibrating and maintaining flowmeters. Some of the properties that effect accuracy and lifetime of flowmeters are:

Temperature
Pressure
Density
Specific Gravity
Viscosity
Liquid State
Gaseous State
Vapor Pressure
Boiling Point
Electrical Conductivity
Sonic Conductivity
Specific Heat
Velocity
Flashing/Cavitating

Flow Sensors

Flow can be measured with direct or inferential devices. In most cases, flow is measured by the pressure drop developed across a flow device, such as an orifice plate, venture, or any specific restriction in a process flow line. Newer flow devices work on various physical or magnetic principles. The ISA Practical Guides Series volume, *Flow Measurement, 2nd Edition*, provides a complete description of all the various flow devices available today. Although flow is measured in many different ways, the following checklist will assist you in troubleshooting flow metering problems.

Selection

Selecting and using the proper flowmeter for your application will give you a more accurate reading of your process. The wrong flowmeter for your application will make controllability of the process harder, if not impossible.

Installation

Flowmeter installation must be performed correctly to achieve the stated performance of the flowmeter. The effects of improper installation vary from reduction in accuracy to a flowmeter that does not operate at all.

Piping Considerations

It is preferred to install flowmeters in vertical piping with flow upwards. Some flowmeters require special orientation such is the case of the magnetic flowmeter, which requires that electrodes be mounted in the horizontal plane. The proper physical installation of a flowmeter in the piping is critical to achieving accurate measurement.

Most flowmeters require the pipe installed upstream and downstream of the flowmeter to be straight, be the same diameter as the flowmeter, have smooth inside surface, not be out of round, and contain no pipe fittings to develop a uniform velocity profile with minimum distortion and/or swirl upstream of the flowmeter. Flow conditioners may be used upstream to minimize upstream straight run requirements.

Pressure taps should be located upstream of the flowmeter. The tap must be flush with the inside of the pipe and all burrs removed to ensure it will not disturb the velocity profile. Liquid and gas and steam pressure taps may be located off the side of the pipe.

Temperature taps should be located downstream of the flowmeter so the well can be in the flow stream for accuracy, while not disturbing the velocity profile upstream of the flowmeter.

Liquid differential pressure transmitters should be located below the pipe with the impulse tubing constantly sloping upward towards the pipe so gas bubbles do not collect in the impulse tubing and cause a bouncy signal.

Gas differential pressure transmitters should be located above the pipe with the impulse tubing sloping constantly downwards toward the pipe so any condensate will drain into the pipe.

Electrical Considerations

GROUNDING

Grounding of piping and flowmeter body is often overlooked or performed improperly. Most flowmeters require the meter body to be grounded to the upstream and downstream piping to keep the flowmeter from being at different electrical potential than the pipe, and to prevent static buildup in the flowmeter form the flow liquids.

POWER

Most flowmeters are powered by 24 V DC, 110 V AC, or 220 V AC. Power should be in a conduit system separate from the signal conduit system. 24 V DC powered transmitters do not require separate power conduit.

SIGNAL WIRING

Signal wiring should be shielded twisted pairs with separate shields for each signal. The shield should be continuous from the flowmeter to the sensing device and grounded at one end only, typically at the end closest to the sensing device. The end of the shield that is not grounded should be insulated so it does not come in contact with the other conductors or flowmeters body.

LIGHTING

The flowmeter and flow transmitter should be located where the technician will have sufficient light while servicing the instrument, including at night.

Clearance and Accessibility

The area around the flowmeter and flow transmitter should be accessible and clear of uninsulated hot or cold pipes that would impair the ability of the technician to work safely.

Maintenance

Most properly applied and installed flowmeters require very little maintenance and can operate for extended periods of time with few problems. A few, however, may require some routine service. Maintenance problems and frequency of routine maintenance vary with the process fluid, type of flowmeter, and nature of the upset conditions.

Some flowmeters lose their accuracy, as they get dirty from the process fluid. Rotating flowmeters such as turbine meters have bearings that may need to be lubricated or changed periodically. Magnetic and thermal flowmeters may require periodic cleaning of the sensors. Vortex shedders may build up a coat on wetted parts and require periodic cleaning.

Troubleshooting

After a new startup, or a process modification, many problems may occur.

CHECK THE FLOWMETER

- First visually inspect the flowmeter to see if it was installed properly.
- Check the specification of the flowmeter to see if it is the proper flowmeter for the application.
- Check all electrical connections.
- Check the pressure connections.
- Check for entrapped gas in liquid lines.
- Check for sediment in the lines.
- Check that blocking valves are fully open and the bypass valves are tightly closed.
- Check for vibration of flowmeters that are susceptible to vibration.
- Ensure that there is flow.

CHECK THE TRANSMITTER

- Visually inspect the transmitter installation.
- Check the specification of the transmitter to see if it is the proper range for the process.
- Check all electrical connections.
- Check the range of the signal (zero, span, linearity).
- Simulate an input to the transmitter, if possible, and check the transmitter output.

- Check the zero by turning off the flow or bypassing and blocking the flowmeter.

CHECK THE CONTROL SYSTEM

- Is the control system properly ranged?
- Does the control system properly linearize the signal? (Linear, Square Root, etc.)
- Check wiring to the control system.

After prolonged use, or changes in weather conditions, failures and inaccurate readings require troubleshooting. Some of the problems may be suggested above. However, process fluids and gases may cause buildup inside the pipes, insulate the electrodes, round the orifice, or fail electrically or electronically.

CHECK THE TRANSMITTER

- Simulate an input and check the control monitor, if possible.
- Determine if the problem is in the flowmeter, transmitter, or control.

CHECK THE FLOWMETER

- The transmitter and control system should be checked first because it is more difficult to check the flowmeter.
- Check at the pressure taps using another flow measuring device.
- Check the electrical power.
- In most cases, if the problem is in the flowmeter, it must be removed from the process to be cleaned, repaired, or replaced.
- Check the calibration of the flow tube.
- Check the orifice.

AFTER REPAIR

- Check for leaks.
- Calibrate the loop.
- Re-insulate, if required.
- Document your findings and repair.

Calibration

Calibration can be performed in any location. A specific calibration can range from a simple dimensional check to extensive measurement of multiple variables. More detailed on calibrating flowmeters are located in the Calibration Chapter. In general, flow measurement devices are calibrated by three methods: (1) a "wet" calibration using actual fluid flow; (2) a dry calibration using flow simulation by electronic or mechanical means; or (3) a measurement check of physical dimensions and use of empirical tables relating flow rate to these dimensions.

Magnetic Flowmeters

Magnetic flowmeters (magmeters) are designed to measure the flow of electrically conductive liquids in a close pipe. With the obstructionless design, there are no moving parts to wear and no pressure drop other than that offered by the section of pipe of equal length and inside diameter. Because the flow path is relatively unobstructed, magnetic flowmeters are well suited to use with high-viscosity fluids and process liquids that contain solids. They can be selected for compatibility with most corrosive of chemicals as well as meet sanitary requirements for food applications.

Magnetic flowmeters do not create pressure drop. Velocity is measured directly, so variations in density do not affect their accuracy. Magnetic flowmeters are free of Reynolds number constraints; they may be used for measuring the velocity of liquids with any flow profile. The output signal is linear and directly proportional to the flow velocity. Magnetic flowmeters can measure flow of conductive liquids in both forward and reverse directions. However, flow in the forward directions develops a voltage that is opposite in phase or polarity from the flow in the reverse direction.

Theory of Operation

This section discusses the fundamental operating principles of magnetic flowmeters, the function of the flow tube, and operation of modern pulsed field coil design. Although Faraday's law explains the theoretical principles of magnetic flow measurement, practical application of these principles is more complicated.

The operating principle of the magnetic flow tube is based on Faraday's law of electromagnetic induction, which states that a voltage is induced in a conductor moving through a magnetic field:

$$E = kBDV$$

The magnitude of the induced voltage E is directly proportional to the velocity of the conductor V, conductor width D, and the strength of the magnetic field B.

Figure 6-14 illustrates the relationship between physical components of the magnetic flowmeter and Faraday's law. Field coils placed on opposite sides of the pipe generate a magnetic field. As the conductive process fluid moves through the fields with average velocity V, between electrodes, a voltage is extracted. An insulating liner prevents the signal from shorting to the pipe wall. The only variable in this application of Faraday's Law is conductive liquid velocity V, because field strength is a controlled constant and electrode spacing is fixed. Therefore, output voltage E is directly proportional to liquid velocity, resulting in the inherently linear output of the flowmeter system.

Electrical Design

The pulsed DC design represents an advanced application of Faraday's Law. Since ideal environments (free form extraneous voltage and noise) seldom exist, this design primarily avoids conflict rather than compensates for it.

Unwanted noise has many sources. These include:

(1) electrochemical voltage resulting from electrolytic reaction between the metal electrode and the ion-conducting process fluid;

(2) chemical reactions within the process fluid itself;

(3) inductive coupling of the magnetic field to the internal electrode wiring and the process fluid (commonly called quadrature voltage with AC systems);

(4) capacitive coupling of the coil voltage or outside power systems to the electrode circuit;

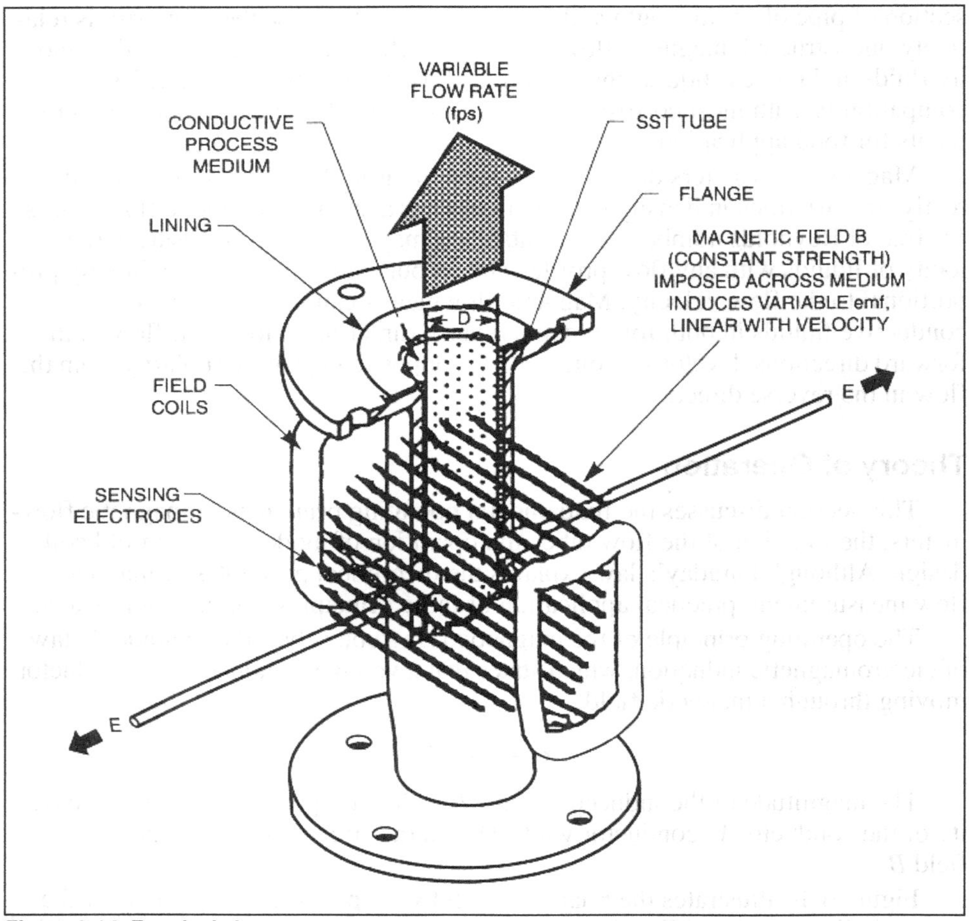

Figure 6-14. Faraday's Law.

(5) transmission losses or phase shifts resulting from fluid impedance and transmission cable capacitance; and

(6) stray voltage or current loops within the process fluid.

Pulsed DC systems are immune to electrochemical noise, and they have several advantages such as automatic zeroing, low power consumption, and simplified installation. Pulsed DC systems power the coils with controlled amplitude, low-frequency square wave. The flow signal is a matching square wave with an amplitude that is proportional to the velocity of the liquid conductor.

Although stray noise may be picked up with induced flow voltage at the electrodes, it is easily separated by the following procedure:

(1) Measure voltage with the magnetic field high and note the amount.

(2) Measure voltage with magnetic field low and note the amount.

(3) Subtract the measured voltage in step 2 from that in step 1. This difference is the actual flow voltage.

Installation

The importance of proper magmeter installation cannot be overemphasized.

OPERATION PLANNING

Operation planning is necessary for successful installation. Process fluid conductivity should be 5 microsiemens/cm (5 microhos/cm) or greater.

Flowtube site and process fluid characteristics are also important considerations. Proper flowtube orientation, sizing, and bypass piping minimize the effects of corrosive, fatty, and abrasive process fluids. Careful installation reduces start-up delays, facilities maintenance, and ensures optimum performance.

Process conditions and loop elements must be considered during site selection to ensure an accurate, noise-free output signal. Avoid placement near pumps (especially positive displacement pumps) or flow variations may appear in the output signal. Also, keep slurries as homogenous as possible to prevent damage to the flowtube electrodes. Homogeneity can often be achieved by locating the flowtube five straight pipe diameters downstream of loop pumps, with additives upstream of the pumps. Finally, design the process loop conditions to ensure that the flowtube always remains full, with no entrapped air.

UPSTREAM PIPING

Magnetic flowmeters are less sensitive to liquid velocity profile variation than most other flowmeters. However, to ensure specification accuracy over widely varying process conditions, install the flowtube with a minimum of five straight pipe diameters upstream and two straight pipe diameters downstream from the electrode plane. This will allow disturbances to settle out prior to measurement. This procedure is usually adequate to allow for disturbances created by elbows, valves, and reducers.

FLOWTUBE ORIENTATION

A flowtube may be installed at any angle between horizontal and vertical without impeding specification accuracy. However, some orientations are better than others. Orientation concerns include keeping the flowtube full, minimizing the effect of entrapped gas, and liner wear. Vertical installations with upward flow are preferred.

SIZING

Correct flowtube sizing is a very important consideration, because sizing affects flow velocity. It is often necessary to install a magmeter that is larger or smaller than the adjacent piping to ensure the fluid is in the range of 1 to 30 ft/s.

Many magmeters are built with a flow direction arrow to indicate the proper direction of flow through the meter.

The preferred orientation for a magmeter is vertical with flow upward through the meter or sloping upward. The most common way is horizontal, but a full pipe is required and the electrodes must be on the sides of the flowtube.

Magnetic flowmeters are more forgiving of piping considerations. Elbows should be located a minimum of three pipe diameters upstream. All that is required on the downstream side is sufficient backpressure to keep the magmeter full of liquid. Two diameters of downstream should be acceptable. Control valves should be placed downstream of the magmeter.

Electrical Connections and Grounding

Ensure that the proper voltages (120 V AC, 220 V AC, or 24 V DC) are provided to the magmeter. Magmeter grounding is really a combination of standard grounding procedures and of bonding the meter body to the process liquid. The most important is bonding, which is ensuring that the meter body is in contact with the process fluid at each of its ends. Bonding provides a short circuit by which stray currents can be routed around the magmeter instead of through it. Ensure that the signal wires to the controls are connected properly.

The magnetic flowmeter has been around for many years and has recently been available in either DC or AC/DC versions that require little power and work well on slurries or noisy flows. Special signal conditioning give the DC meter a smooth flow signal and provides excellent zero stability. The choice of meter depends on the product being measured. Use vendors' sales brochures and specification sheets to select which version of the magnetic flowmeter will best suit individual process needs.

GROUNDING

The pulsed DC flowtube is not as sensitive to some stray voltages as other designs. However, significant stray levels could be detected and interfere with the actual flow signal, causing a measurement error. Properly grounding extraneous signals ensures that only the voltage induced in the flowtube's magnetic field is measured.

For conductive, unlined pipe adjacent to the flowtube, grounding the flowtube to the mating pipe flanges is sufficient.

Flowtube-to-liquid grounding with grounding rings or lining protectors is required with lined or nonconductive pipe. Grounding rings are installed between gaskets at each end of the flowtube and contact the process flow. Lining protectors are attached directly to the flowtube and contact the process flow. These grounding devices ensure that the flowtube case is at the same electrical potential as the liquid, establishing a good measurement reference.

To simplify installation, the inside of both flowtube flanges have a ground stud with a ground strap already attached. These straps must be connected to a suitable earth ground having a resistance measurement to earth of less than 1 ohm.

Elaborate or dedicated grounding systems to reach an earth ground are not normally required. It is possible that the process piping itself is adequately grounded, thus eliminating the need for additional connections to an earth ground. To determine whether the process piping is adequately grounded, check the resistance between the piping and good earth ground. If grounding is required, cold water pipes are often available and may be a satisfactory substitute for a dedicated earth ground. Contact the vendor for special grounding applications.

Non-integral grounding rings are available in a variety of sizes and material compositions. They are generally considered suitable for all pipe diameters. Grounding rings require a gasket on each side.

Flowtube lining protectors are optional devices that fit the contour of the flowtube liner and extend a short distance into the tube. Lining protectors prevent several types of liner failure. This option ensures against mechanical damage caused by prying or cutting, flange bolt over-torquing damage to the liner ends, and leading edge wear, which is common with abrasive process fluids. Lining protectors also function as grounding rings by providing a conductive path for the process fluid.

Although this kind of device is needed mainly at the inlet end, it is commonly used at both ends. Lining protectors are assembled to both ends of the flowtube at the plant, and the tube is pressure-tested to ensure seal integrity. Adding lining protectors to flowtubes already in the field is not recommended.

Torquing

If the flange bolts are tightened well beyond what is required to provide a good seal between magmeter flanges and mating pipe flanges, the result could damage the flowmeter. Teflon-lined magmeters are the most susceptible to installation damage.

Maintenance

Very little maintenance is required in magnetic flowmeters. Cleaning and calibration is generally all that is required.

The self-test circuit and modular construction of the magnetic flowmeter system are designed to minimize downtime. The self-test circuit checks a significant portion of the system, allowing troubleshooting to proceed quickly.

Individual calibration allows independent flowtube and transmitter replacement. Independence ensures compatible interface and factory calibration accuracy without additional calibration procedures or equipment.

Usually a unique calibration number, determined at the factory, is loaded into the transmitter. Loading the calibration number into the transmitter enables interface of any given vendor's flowtube with their transmitter. This is done without the need for further calibration.

The smart magnetic flowmeter transmitter performs self-diagnostics on the entire magnetic flowmeter system: the transmitter, the flowtube, and interconnecting wiring. While most of the diagnostics are transmitter- and microprocessor-oriented, some tests diagnose specific flowtube problems.

Troubleshooting

The troubleshooting guide (Table 6-2) and troubleshooting chart (Figure 6-15) are designed to isolate the problem as quickly as possible. Figure 6-16 shows the flowtube circuit diagram referenced throughout the troubleshooting tables.

Checking the coil for continuity and shorts with an ohmmeter will ensure the coil is ok. Also check to see that the electrodes are not shorted to the meter body. Electrical coating in an AC-type magmeter could be manifested by difficulty in adjusting the zero or by sluggish response to changes in flow.

Some transmitters can be field repaired by component or circuit board replacement. Calibration is required after circuit board replacement.

Mass Flowmeters

Mass flowmeters measure mass directly, using the properties of mass, as opposed to those who measure volume or velocity. This type of meter is found in applications with changing fluid parameters. Meters that measure mass directly, in effect, weigh the fluid as it passes through the meter, yielding a highly accurate measurement that is virtually independent of varying process conditions. (See Figure 6-17.) The Coriolis principle flowmeter is a true mass flow flowmeter because it uses the properties of mass to measure mass.

Although true mass flowmeters are available, they are rarely applied due to their high costs. Therefore, two variables will be measured—fluid velocity and fluid density. These two variables are to be multiplied and their product will be the mass flow.

$$M = APV$$

M – mass flow rate
A – cross-sectional area of pipe interior density of fluid
P – density of fluid
V – velocity of fluid flow

Theory of Operation

Coriolis meters operate on the basic principle of motion mechanics. The fluid in motion through a vibrating flow tube is forced to take an acceleration as it

Test A—Electrode Shield Voltage

Connect power to the magmeter and connect all wires. Take all readings at the transmitter first. If you find a faulty reading, repeat the test at the flowtube to determine if the flowtube or cabling is at fault.

→ Measure DC volts from terminals 17 to 18 and from 17 to 19. Each reading should be between 0.1 and 4 V DC, and the difference of the two readings should be less than 0.5 V DC.

→ A reading of less than 0.1 V DC may indicate a shorted electrode.

A reading above 4 V DC may indicate a faulty process ground, an electrode not in contact with the process, a coated electrode, or a nonconductive process.

Test B—Flowtube Coil

Disconnect power from the transmitter by removing the fuse. Disconnect wires 1 and 2 from the transmitter. Take all readings across the wires, not across the transmitter terminals.

→ Measure the resistance across wires 1 and 2 going to the flowtube, using the lowest ohms scale. The reading should be between 2 and 18 Ω.

→ A reading outside this range indicates that the coils or cables are open or shorted.

Test C—Coil Shield to Coil

Disconnect power from the transmitter by removing the fuse. Disconnect wires 1, 2, and ground from the transmitter. Take all readings across the wires, not across the transmitter terminals.

→ Measure the resistance from the coil shield (ground) to wires 1 and 2 using the highest scale. Both readings should be overrange.

→ Any reading on the scale indicates that the coils are shorted to the housing.

Test D—Electrode Shield Resistance

Note: Certain models have a 50-kΩ resistor in series with the electrodes. Any reading near 50 kΩ on these units indicates a shorted electrode.

Disconnect power from the transmitter by removing the fuse. Disconnect wires 17, 18, and 19 from the transmitter. Take all readings across the wires, not across the transmitter terminals. Test with process in the flowtube (either flow or no flow).

→ Measure the resistance from wire 17 to 18 and 17 to 19. This reading will change as you hold the leads on the wires, so use the initial reading. These readings should both be between 100 kΩ and the low MΩ and close to each other.

→ A low reading indicates a possible shorted electrode. A high reading indicates a coated electrode, bad process ground, nonconductive process, or electrode not in contact with process.

Test E—Positive to Negative Electrode

Note: Certain models have a 50-kΩ resistor in series with each electrode, which adds 100 kΩ to this measurement.

Disconnect power from the transmitter by removing the fuse. Disconnect wires 18 and 19 from the transmitter. Take all readings across the wires, not across the transmitter terminals. Test with process in the flowtube (either flow or no flow).

→ Measure the resistance between wires 18 and 19. This reading should be in the low to mid kΩ range.

→ An overrange reading indicates a coated electrode, nonconductive process, or electrode not in contact with the process.

Figure 6-15. Troubleshooting Chart.

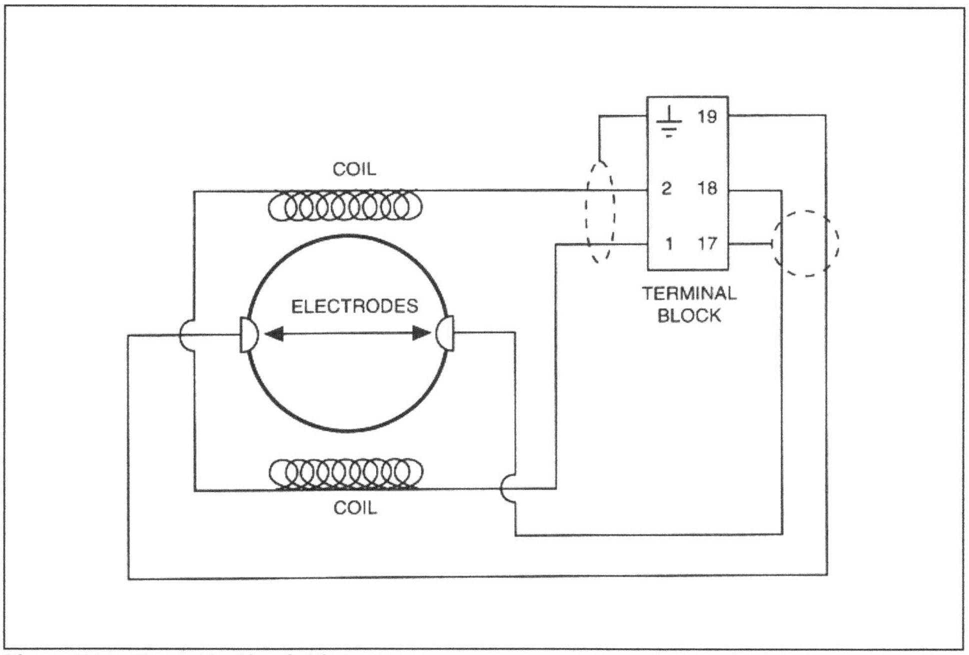

Figure 6-16. Flowtable Circuit Diagram.

Table 6-2. Troubleshooting Table.

Symptom	Potential Cause	Corrective Action
Coil drive open circuit displayed.	Faulty terminal connection.	Isolate the break (faulty connection). Perform: Test B—flowtube coil.
Indicated flow equals half of expected flow.	One signal is being drawn to ground, or is open.	Perform: Test D—electrode shield resistance. Consult your vendor's service center for further instructions.
Indicated flow is erratic.	A less than full flowtube or a non-homogeneous process fluid.	You may need special transmitter features to process the signal correctly.
	Improper grounding.	Make sure the electrode and coil drive shields connect to both the flowtube and the transmitter. Perform: Test D—electrode shield resistance. Perform: Test E—positive-to-negative electrode.
	An inherently noisy process fluid.	Contact your vendor for information regarding the high-signal magnetic flowmeter system.
Reverse flow detected.	Inverted connections at one of the four terminal sites.	Reconnect terminal sets correctly.
	Flow direction is opposite of flowtube arrow.	Reverse the wiring at flowtube terminals 18 and 19; there is no need to invert flowtube.
No flow indicated.	The valves, positioners, or actuators of the physical piping are not properly set.	Perform: Test A—electrode shield voltage. Perform: Test D—electrode shield resistance. Perform: Test E—positive-to-negative electrode
Insufficient process fluid conductivity.	Process is a hydrocarbon.	Perform: Test E—positive-to-negative electrode.

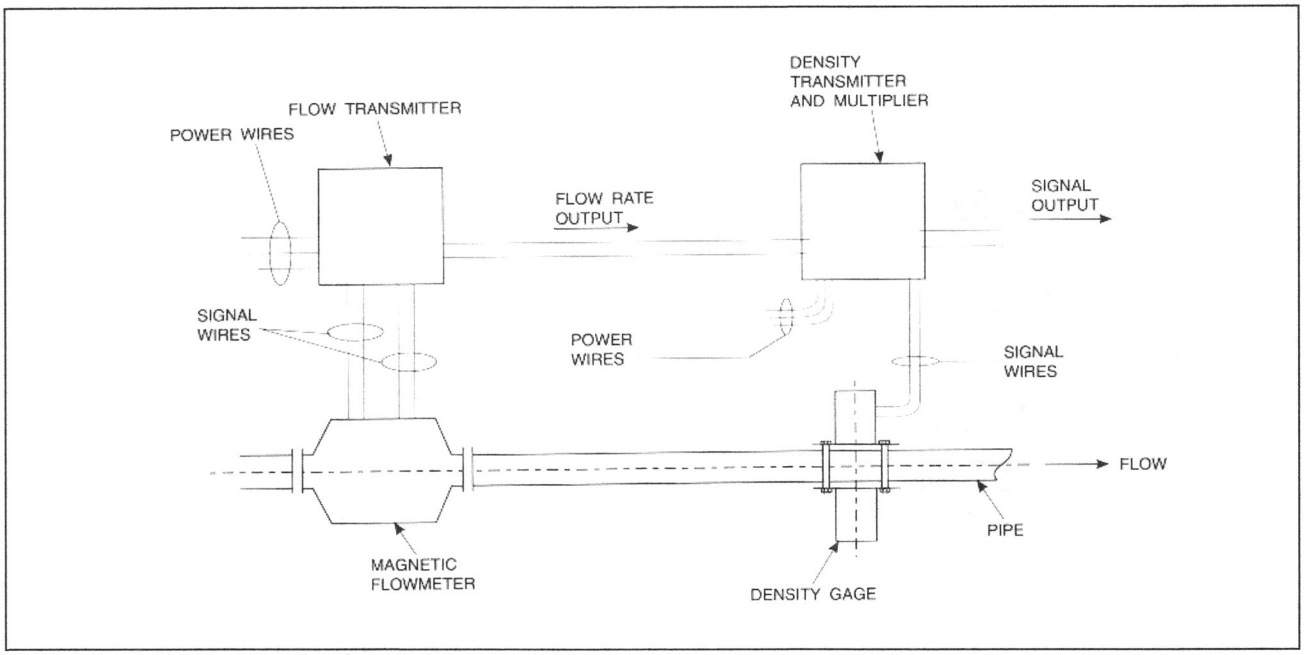

Figure 6-17. Mass Flow Metering System Schematic.

moves toward the point of peak amplitude of vibration. Conversely, the fluid decelerates as it moves away from the point of peak amplitude as it exits the tube. The moving fluid exerts a force on the inlet side of the tube in resistance to this acceleration, causing this side of the tube to lag behind its non-flow position.

On the outlet side, the force-exerted by the flowing fluid is in the opposite direction as the fluid resists the deceleration. This force causes the outlet side to lead ahead of its no-flow position. The result of these forces is a twisting action of the flow tube during flow conditions as it traverses each vibrational cycle. This is demonstrated with a U-tube in Figure 6-18.

Installation

For liquid applications, it is best to install the meter such that it remains completely full of liquid and not allow the trapping of gas bubbles at all times during the measurement. It is preferred that the flow element be mounted in a vertical pipe with flow upward. The flow element should not be the highest point in the system, in a location where the liquid can be siphoned out, or in an orientation that allows gas to become trapped in the flow element.

In gas or vapor applications, the orientation of the flow element should be such that condensate or other liquids are not trapped in the flow element. This may require installing a trap upstream and/or downstream of the flow element.

A downstream shutoff valve is recommended to ensure the ability to obtain zero flow conditions at pressure when making the initial zero adjustment. Temperature extremes should be addressed using specific recommendations by the manufacturer.

Maintenance

One of the major benefits of Coriolis mass flowmeters is the lower maintenance requirement. The benefits of no moving parts and low maintenance have, in many cases, been the primary justification for its higher capital expenditure.

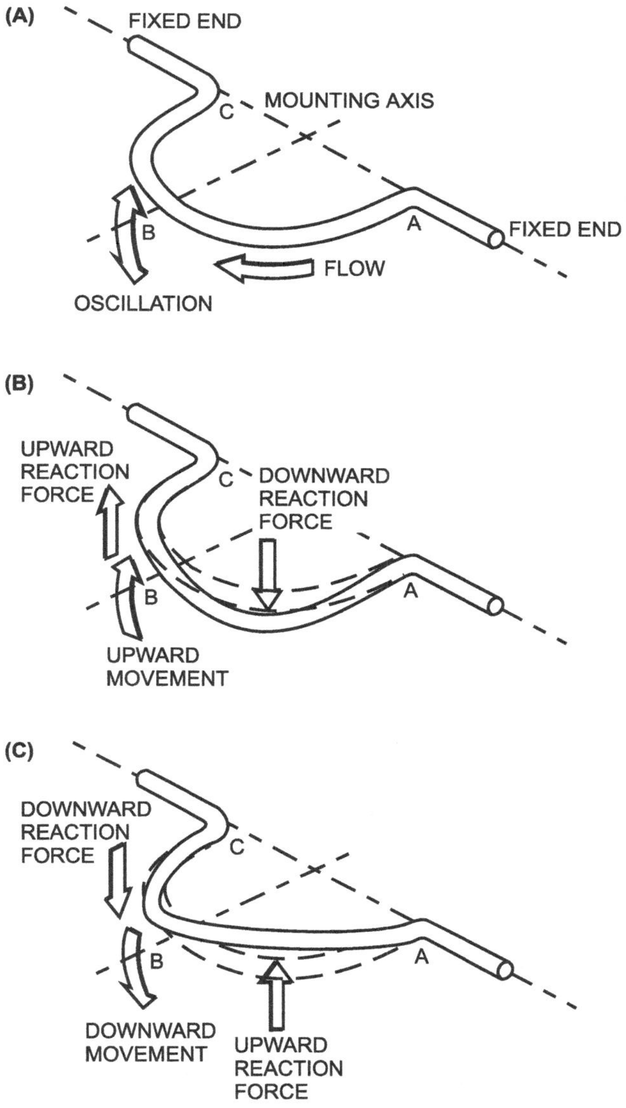

Figure 6-18. Coriolis Principle.
(Courtesy of K-Flow)

The complete functional check of the transmitter portion of the system should be performed on a semi-annual basis. Flow element repair cannot usually be performed in the field and requires replacing the flow element should some type of failure occur.

Coriolis Meters

Coriolis meters are relatively new, and mounting methods have changed since the very early models. Originally, many units had to be mounted on very sturdy platforms in an effort to reduce external vibration effects, temperature expansion effects, and other so-called external effects. The newer devices, in most cases, can be suspended from or mounted in the flow piping. The preferred mounting is in a vertical line, as the meter is always full when a fluid is flowing through the meter, and is self-draining. On gas flows, mounting the case vertically above a horizontal line is the preferred method for the U-shaped meters. Since there are many different commercial shapes of the Coriolis meter, it is best to thoroughly read the installation instructions before selecting a location for the meter.

Some advantages over the magnetic flowmeter are that the fluid does not have to be conductive, is not affected by chemical reactions, and the device measures mass flow as well as density. The measurement is linear. Disadvantages include a maximum size of 4 to 6 inches, higher pressure drop, more critical installation than a magnetic flowmeter, and a higher price in the larger sizes. The choice of meter depends on exactly what is being measured and whether flow or mass flow is required.

As a rule of thumb, in most cases, the Coriolis meter can be mounted wherever a magnetic flowmeter can be mounted. As with any instrument, always check the instruction book for details on mounting. When troubleshooting a Coriolis meter, look at the following areas:

(1) Proper piping to and from the meter.

(2) Proper mounting position for gas or liquid.

(3) Can the tube vibrate freely?

(4) Is the product temperature reasonably constant? Rapid temperature extremes can cause temporary internal binding or rubbing, in some cases.

(5) Is product stickiness causing a buildup on the interior walls of the meter, reducing the effective cross-sectional area of the meter tubing or changing the mass of the tubing?

(6) Is the flow too near the very low end of the range, where the accuracy is less than at a normal flow?

(7) Is the tube material compatible with the product?

(8) Is air entrained in the fluid? Is air slugging occurring?

(9) Is the correct transmission cable being used?

(10) Read the manufacturer's manual and instruction book for more details on a specific brand.

More detailed information on mass flow may be acquired in *Flow Measurement, 2nd Edition*, Chapter 10, published by ISA.

Turbine Flowmeters

Turbine flowmeters are designed to accurately measure the flow of liquids and gases in pipes. They are a volumetric flow-measuring device. The range of flow is from 0.001 gpm to over 25,000 gpm for liquid service and 0.001 acfm to over 25,000 acfm for gas service. The flowmeters are typically manufactured from stainless steel, but also available in a variety of materials, including plastic. Turbine meters are applicable to all clean fluids over a pressure range from sub-atmospheric to over 60,000 psi and temperatures from cryogenic to about 800 degrees C (1500 degrees F).

The turbine flowmeter is perhaps the most accurate type of meter available. A turbine meter has only one moving part — a rotor. The output signal from the flowmeter and sensor is an electronic pulse, but other output signals such as analog, visual, or digital are available.

Turbine Meters

Turbine meters are used on clean fluids or gases when a very accurate measurement is required. The device has a threshold velocity but becomes a very accurate flow measurement once the threshold velocity is exceeded. Since blade configuration is important, any form of trash must be kept from entering the bladed area. For this reason, filters are sometimes placed upstream of the meter to prevent damage to the blading. Bearing and blade material must be compatible with the process material, and there must be sufficient upstream and downstream straight pipe for accurate measurement. Flow straightening tubes are sometimes used when there is not sufficient straight upstream piping. The meter can be damaged by surging or sudden starting or stopping of the product. Design circuitry to allow for gradual starting and stopping of the process fluid. A ramping algorithm is usually used.

Theory of Operation

A turbine flowmeter consists of a rotor mounded on a bearing and shaft in a housing. The fluid to be measured is passed through the housing, causing the rotor to spin with a rotational speed proportional to the velocity of the flowing fluid within the meter. A device to measure the speed of the rotor is employed to make the actual flow measurement. The rotational speed of the sensor shaft and the frequency of the pulses are proportional to the volumetric flow rate through the meter (See Figure 6-19).

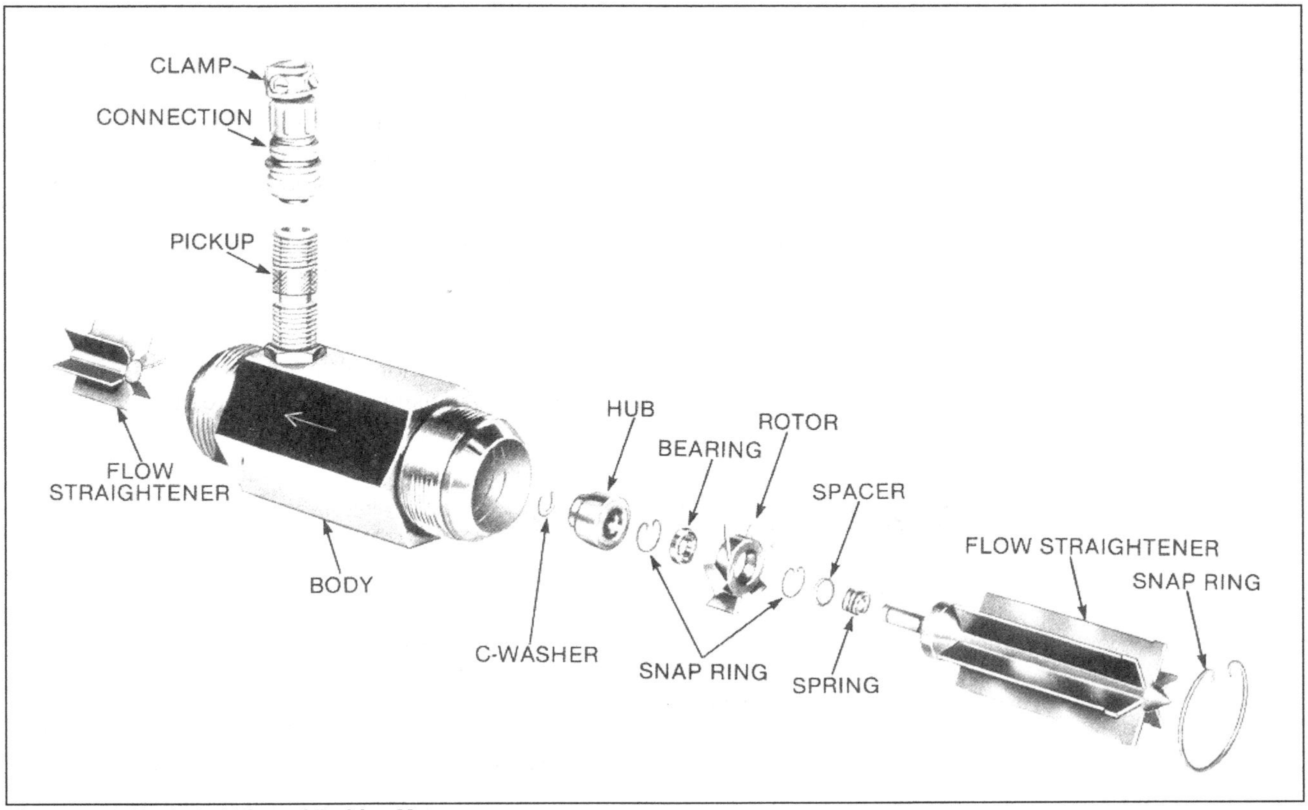

Figure 6-19. Exploded View of Turbine Meter.
(Courtesy of Shute & Koerting's Cox Instrument Division)

Installation

Check the rotation of the rotor by lightly blowing through the meter. It is not necessary to blow hard as it is relatively easy to over-speed.

The most sensitive external influence on a turbine flowmeter is swirl in the pipeline or other plumbing. Common practice is to specify 10 pipe diameters upstream of a meter. It is recommended that flow straighteners are placed upstream of the meter. Flow should be installed in accordance with the directional arrow on the meter. Pulsating flow is difficult to measure accurately. The only electrical connection to the meter itself is to the pickoff.

Maintenance of a turbine flowmeter consists of periodic inspections to ensure the internal parts are not fouled with debris or have suffered any corrosion by the fluid in the system. The bearing should be inspected to ensure it rotates freely.

The internal assembly of the meter may be removed for inspection and cleaning. Most typical meters use a snap ring arrangement to hold the internal assembly into the body. Once removed, the internal assembly may be cleaned with solvent or alcohol.

Contamination in the bearings of a turbine flowmeter is the largest single source of poor meter performance. Flushing the internal assembly of the meter with an appropriate solvent will remove buildup. Blow lightly through the meter to see how freely the rotor spins. And how abruptly it stops spinning. It should coast slowly to a stop.

Calibration

The correct correlation parameter for the turbine flowmeter is Strouhal number vs. Reynolds number. Turbine flowmeter calibration data are typically displayed or presented as flow rate as a function of frequency. This is satisfactory as long as the meter is calibrated in exactly the same fluid and at the same temperature (and pressure for gas applications) as the application conditions.

A better and more commonly used method to present the data is K factor as a function of frequency. K factor is defined as frequency/flow rate (pulses/gallon). This dimensionalized presentation increases the resolution of the data, but it is still limited to the kinematics viscosity (and temperature and pressure) as in the calibration. However, for applications in which the meter is calibrated at the same conditions as the application, it is a good presentation of the data. Find a fluid that can be used in the calibration process that has the same kinematics viscosity at the operating temperature, and perform the calibration at the operating temperature. No corrections to the data are required when using this procedure, and the calibration will be correct.

Open Channel Flowmeters

Theory

High volume flow of liquids in industrial waste systems, waterworks, and irrigations systems are commonly carried in open channels, which are characterized by low system heads and high volume flow rates. Open channel flow is flow in any channel in which the liquid flows with a free surface. Examples of this are rivers, irrigation ditches, canals, flumes, and other uncovered conduits. Certain closed channels, such as sewers and tunnels when flowing partially full and not under pressure, are also classified as open channels.

Methods of Open Channel Flow Measurement

In the timed gravimetric method, the entire contents of the flow stream are collected in some type of container for a fixed length of time. The weight of the fluid is then determined and the flow rate is calculated. The timed gravimetric and dilution methods are generally not used for routine industrial flow measurement because they are not suited to provide a continuous record of flow rate. They are more often used for occasional flow rate measurement at a particular time and place, for calibrating some other type of device, or for developing a liquid level-flow rate curve for a particular location.

In the dilution method, the flow rate is measured by determining how much the flowing water dilutes an added tracer solution. Although brine tracers have been used, radioactive and fluorescent dye tracers are more commonly used today.

In the velocity-area method, the flow rate is calculated by determining the mean flow velocity across a cross section and multiplying this by the flow area at that point. In the open channels, this will generally require two separate measurements, one to determine the mean velocity and the other to determine the flow depth. The velocity-area method is often used like the timed gravimetric and dilution techniques to determine the flow rate in a stream at a particular time of recalibration purposes.

In the hydraulic structures method, some type of hydraulic structure is introduced into the flow stream. The function of the hydraulic structure is to produce a flow that is characterized by a known relationship between a liquid level measurement at some location and the flow rate of the stream. Thus, by knowing the liquid level at the hydraulic structure, the flow rate in the open channel may be determined. The hydraulic structures method differs from the first three in that, provided a standard type of structure is used and certain installations and application rules are followed, no field calibrations or measurements other than a continuous measurement of liquid level are required to obtain a continuous record of flow rate. Because of this, the hydraulic structures method is widely used for industrial open channel flow measurement.

The hydraulic structures method is the most common type of open channel flowmeter used today.

In the slope-hydraulic radius-area method, measurements of water surface slope, cross-sectional area, and wetted perimeter over a length of uniform section channel are used to determine the flow rate, using a resistance equation such as the Manning formula. The Manning formula requires knowledge of the channel cross-section, liquid depth, slope of the water surface, and a roughness factor dependent on the character of the channel. Because of uncertainties associated with the Manning formula, the accuracies obtainable are not as good as those achieved with hydraulic structures and are not commonly used in industrial open channel flow measurement.

Primary Measuring Devices

The most commonly used method of measuring the rate of flows in industrial open channels is that of hydraulic structures. In this method, flow in an open channel is measured by inserting a hydraulic structure into the channel, which changes the level of liquid in or near the structure. By selecting the shape and dimensions of the hydraulic structure, the rate of flow through or over the restriction will be related to the liquid level in a known manner. Thus, the flow rate through the open channel can be derived from a single measurement of the liquid level in or near the structure. The hydraulic structures used in measuring flow in open channels are known as primary measuring device and may be divided into two broad categories — weirs and flumes, shown in Figure 6-20.

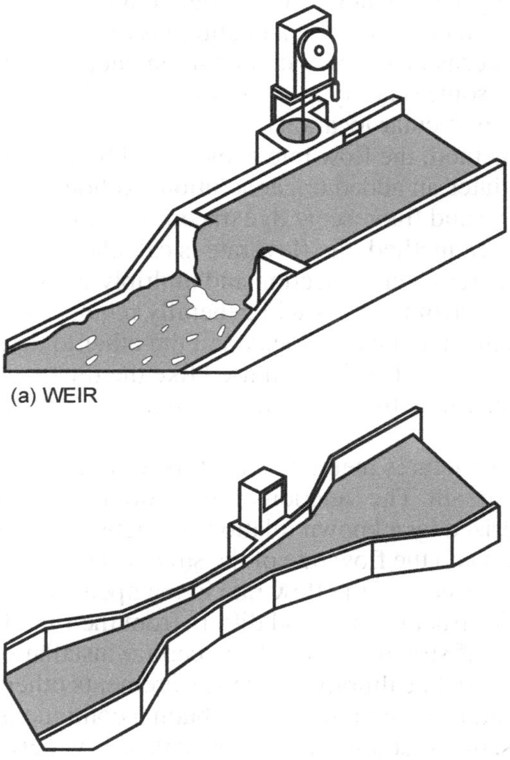

Figure 6-20. Primary Measuring Devices: Weirs and Flumes.

WEIR

A weir is essentially a dam built across an open channel over which the liquid flows, usually through some type of an opening or notch. Weirs are normally classified according to the shape of the notch, the most common types being the triangular (or V-notch) weir, the rectangular weir, and the trapezoidal (or Cipolletti) weir. Each type of weir has an associate equation for determining the flow rate through the weir.

FLUME

A flume is a specially shaped open channel flow section with an area or slope (or both) that is different from that of the channel. These result in an increased velocity and change in the level of the liquid flowing through the flume. A flume normally consists of three sections — a converging section, a throat section, and a diverging section. The flow rate through the flume is a function of the liquid level at some point or points in the flume.

Secondary Measuring Devices: Open Channel Flowmeters

The flow rate or discharge through a weir or flume is usually a function of the liquid level in or near the primary measuring device. A secondary measuring device (or open channel flowmeter) is used in conjunction with a primary measuring device to measure the rate of liquid flow in an open channel. The secondary measuring device has two purposes: (1) to measure the liquid level in the primary measuring device, and (2) to convert this liquid level into an appropriate flow rate according to the known level-flow relationship of the primary measuring device.

Thus, a combination of a weir or flume (primary measuring device) and an open channel flowmeter (secondary measuring device) is necessary to measure flow in an industrial open channel. The flow measurement requires both primary and secondary devices to be complete.

INSTALLATION

The approach section should be straight upstream from the weir for a distance of at least 20 times the maximum expected head of liquid and have little or no slope. The device for measuring the head (level transmitter or flowmeter) should be placed upstream at a distance of at least 3 times the maximum expected head on the weir and should be located in a quiet section of the channel away from all disturbances. Also the zero point of the head-measuring device must be set exactly level with the weir crest.

CALIBRATION

The only calibration required for a weir is to adjust the zero of the level transmitter or flowmeter being used with the weir. The transmitter or flowmeter must be adjusted such that the indicated liquid level matches the actual level of the flow stream, relative to the weir's zero point.

Adjustment of zero and the span of the level transmitter or flowmeter should be performed per the manufacturer's instructions, which vary significantly with the manufacturer, as well as with the technology that is used to measure level.

Level transmitters and flowmeters must also be calibrated to compensate for any hydrostatic heads that result from elevating the transmitter or flowmeter relative to the crest or the bottom of the weir notch. Non-contact measurement must also be made to account for elevating the transducer above the crest level.

MAINTENANCE

Open channel flow measurement devices require periodic maintenance. Make sure all foreign material is cleaned out of the weir or flume. Also the calibration of the transmitter or flowmeter should be checked.

Vortex Shedding Flowmeter

Theory

When a flowing medium strikes a non-streamlined object or obstruction, it separates and moves around the object and passes on downstream. At the point of contact with the object, vortex swirls or eddy currents separate from the object on alternating sides. When this occurs, the separating or shedding causes a local increase in pressure and a decrease in velocity on one side of the object, and a local decrease in pressure with corresponding increase in velocity on the other side of the object. Vortex swirls are shed continuously – 180 degrees out of phase with each other. The frequency of the shedding process is proportional to the velocity of the material flowing past the object.

Vortex Shedding Meters

Vortex shedding meters are excellent for linear flow measurement of liquids and steam. The device has a tolerance for higher temperatures and some solids. The device is simple and does not require a separate transmitter, as the transmitter is an integral part of the sensing device. Mount as you would an orifice plate and troubleshoot the electronics before moving the sensor for inspection.

A damaged densing device or a worn or eroded vortex shedding bar would cause errors in the reading. If the measured flow is hot, check the meter specifications for the maximum operating temperature allowed. Check that the vortex bar is free of trash or buildup.

Installation

The vortex shedding flowmeter is not attitude sensitive; flow can be upward or horizontal without affecting performance. When mounting this meter in the pipeline, the mating pipe should ideally be the same diameter as the meter bore. The pipe must be kept full of liquid at all times when measuring. Straight run requirements vary (see Figure 6-21).

Calibration

If the user provides the correct operating information about how the meter will be used, the manufacturer will calibrate the meter to those conditions. In most cases, the factory calibration includes wet flow calibration whether the meter is installed in a calibration standard where water or air can be used to calibrate the flowmeter element. The electronics are then adjusted so the output signal is correct for the application.

Maintenance

Since the vortex shedding flowmeter has no moving parts, little regular maintenance is required. Many designs permit the sensor to be changed without removing the meter from the pipeline. If it is suspected that the meter is not reading correctly, the installation should be checked first. Experience has shown an incorrectly installed meter is more likely to be the culprit than an actual problem with the meter itself.

Positive Displacement Flowmeters

Positive displacement liquid flowmeters have long been used to measure liquid products. Over the years, numerous design improvements have resulted in an expanded product line that now serves industrial as well as petroleum and domestic applications.

Theory

A liquid flowmeter is, in essence, a hydraulic motor with high volumetric efficiency that absorbs a small amount of energy from the flowing stream. This energy is used to overcome internal friction in driving the flowmeter and its accessories. It is reflected as a pressure drop across the internals of a positive displacement flowmeter that actually creates a hydraulically unbalanced rotor, which caused rotation.

A positive displacement flowmeter is one that continuously divides the flowing stream into known volumetric segments, isolates the segments momentarily, and returns them to the flowing stream while counting the number of displacements.

Flowmeter Size		Liquid Flow			Gas and Steam Flow		
mm	in.	Sched 10	Sched 40*	Sched 80	Sched 10	Sched 40	Sched 80
50	2	+0.4%	ref	-0.8%	+0.4%	ref	-0.8%
80	3	+1%	ref	-0.8%	—	ref	-0.8%
100	4	+1%	ref	-0.2%	—	ref	-0.2%

* K Factor stamped on data plate is based on calibrating with schedule 40 pipe. The flowmeter bore diameter is equivalent to the inside diameter of a schedule 80 pipe.

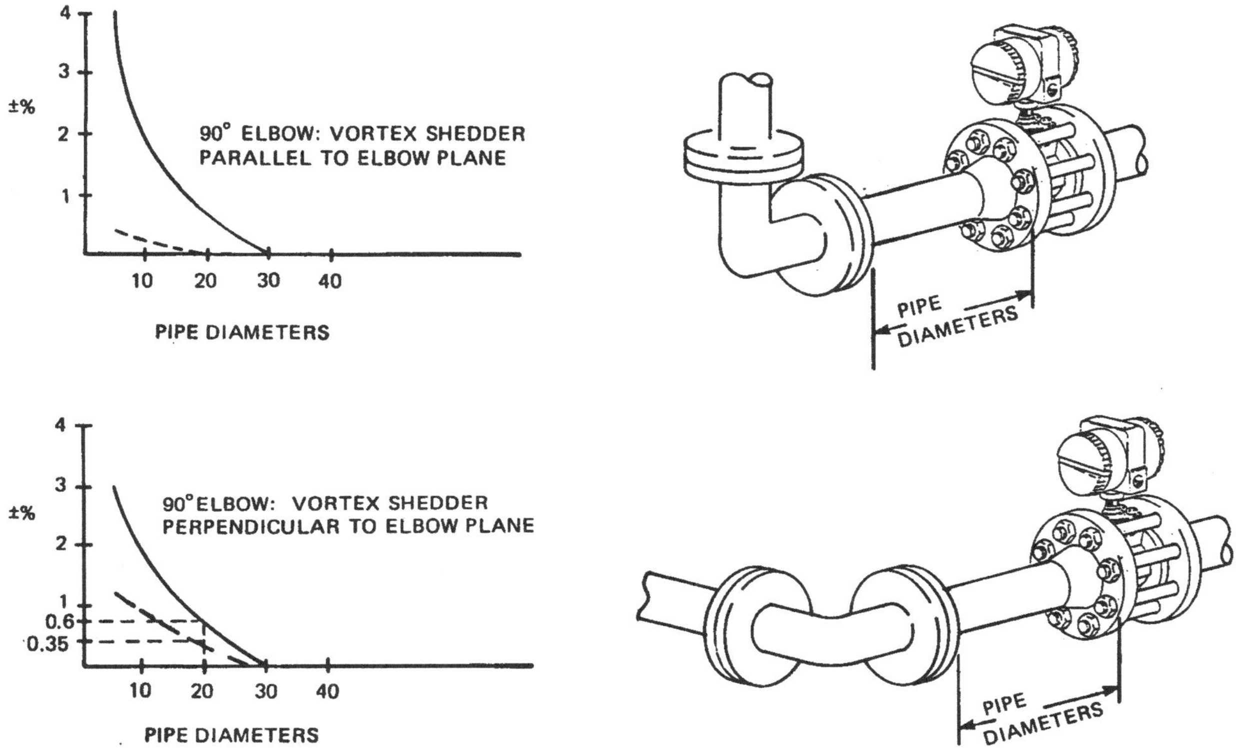

Figure 6-21. K-Factor Offset.
(Courtesy of Inveneys Process Systems/Foxboro)

A positive displacement flowmeter can be broken down into three basic components: the external housing, the measuring unit, and the counter drive train. The positive displacement flowmeter can offer excellent accuracy, repeatability, and reliability in many applications. Positive displacement flowmeters have limitations on applications with extremely dry (non-lubricating) liquids and liquids that contain solid particles.

Positive displacement flowmeters can offer excellent accuracy, repeatability, and reliability in many applications. They can measure varying and high viscosity products over a broad range of flow rates.

The output signal is obtained through the counter drive train and it available in either mechanical or electrical form with a digital or analog signal. The rotary motion of the flowmeter internals can be converted into an output signal capable of driving a broad lien of accessories.

Positive Displacement Meters

Positive displacement meters measure flow volumetrically. The liquid fills a fixed-volume chamber, and each filling is counted. The number of oscillations or revolutions of the device is an indication of the total volumetric flow over time.

These meters are used mostly on batch operations. The normal sizes range from ½ inch or smaller to 2 inches, but some meters can be 6 inches or larger. Body and internal material, including bearings, must be compatible with the fluid, and the fluid should be free of solids.

Follow vendor installation and operating instructions, since there are many versions of these types of meters.

The most frequent problems occur from solids jamming the moving parts of the meter or a mechanical failure of some internal part. Check for air entrainment or slugging if the calibration is suspect.

Target Flowmeters

Versatility and low installation cost make the target flowmeter a viable candidate for many difficult flow measurement applications. Full bore, in-line meters are available in sizes ½ inch to 6 inches. Target flowmeters are used on liquids, gases, vapors, dirty fluids, light slurries, and high viscosity fluids, particularly when fluid characteristics preclude the use of flowmeters with pressure taps or moving parts.

Theory of Operation

Whenever there is flow past an obstacle in a pipe, a force, commonly referred to as drag, is generated to push or drag the obstacle in the direction of flow. Such an obstacle left unsupported would be carried away with the fluid. If, on the other hand, the obstacle was constrained by a force equal and opposite to the drag, the magnitude of that force could be used to determine the rate of flow. This is the underlying principle behind the target flowmeter.

A strain gage (Wheatstone bridge) or mechanical force gage measures the force of the flow (see Figure 6-22).

Target Meters

A target meter is essentially a circular target centered in the meter body or in the pipe line that senses the pressure differences between the front and the rear of the target plate. It can be though of as an "inside-out" orifice plate. The signal produced is the square root of the sensed pressure difference. A feedback mechanism or strain gage is used as the force detector in the transmitter. The device can be used on viscous fluids and also high temperature fluids as well as gases and, in some cases, steam. When the device looses accuracy, drifts, or does not produce a reasonable signal, the problem is usually a damaged target or sensor.

Installation

To obtain the best performance from a target flowmeter, straight runs of upstream and downstream pipe having the same diameter as the meter bore are required. The length of straight run will depend on the type of flow disturbance that precedes the flowmeter. The length can be reduced by the use of a flow conditioner. Typically the straight run requirements are the same as for an orifice plate flowmeter having the same blockage.

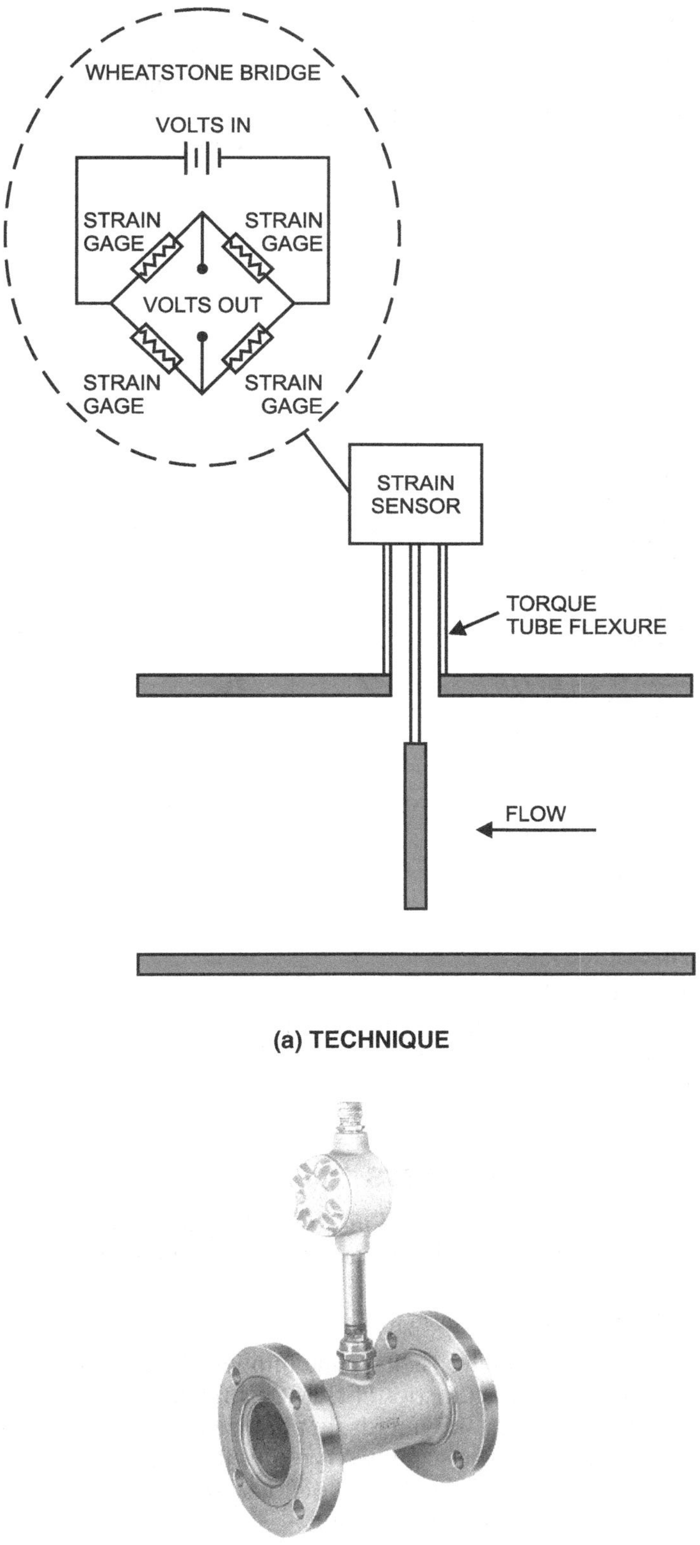

(a) TECHNIQUE

(b) Courtesy of Hersey

Figure 6-22. Strain Gage.

Calibration

High accuracies can be achieved by performing a wet flow calibration of the meter installed in the meter run. Since target flowmeters are generally used where accuracy is not the primary concern, a lower accuracy bench calibration is more often used. After zeroing the transmitter, apply a weight equal to that calculated when full-scale force is suspected form the target support (see Figure 6-23).

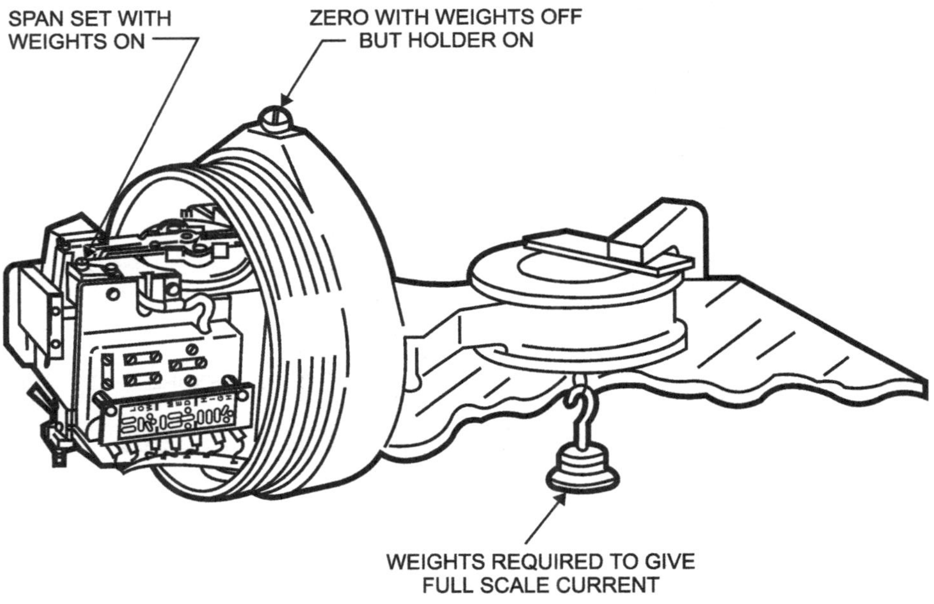

Figure 6-23. Bench Calibration.

Maintenance

Since meter factor depends on target geometry, the target and pipe should be periodically checked for coating or erosion. Any buildup should be cleaned. Do not scour the surface or edges. Replacement of the target is recommended if the edges become rounded.

Thermal Mass Flowmeters

Thermal mass flowmeters came of age when they accurately measured low gas flows such as leak testing and medical analytical equipment.

Theory of Operation

The thermal mass flowmeter, as its name implies, depends on the variation of one or more of the heat characteristics of fluid as a function of flow. While in theory such flowmeters will measure liquids, in practice the commercial versions are limited to measuring gases. There are two types of thermal mass flowmeters: those that measure the "rate of heat loss to the flow stream" and those that measure the "temperature rise of the flow stream."

Installation

Most thermal flowmeters can be mounted in any position, but there is usually a preferred orientation, because drift and pressure effect are somewhat dependent upon position. Unless the gas and piping are very clean, a filter upstream of the flowmeter is recommended.

Calibration

Most thermal flowmeters can trace their calibration back to NIST through some type of volumetric calibrator. Most manufacturers offer calibrations as standard.

Many users who have a significant installed base of flow devices have constructed their own flow calibration capabilities within their facilities.

Maintenance

Thermal mass flowmeters have no moving parts, and when used on clean, non-corrosive gases, they can give years of dependable service. If a sensor plugs, it may be possible to clean it by running a fine wire through it. Always run the wire from the outlet to the inlet, if possible, in case the blockage is confined to the sensor's entrance. Many flowmeters have replacement sensors. Similarly, the electronic circuit card is replaceable.

Ultrasonic Flowmeters

Theory of Operation

Two types of ultrasonic flowmeters are in general use for closed pipe flow measurement: Transit time, which uses pulse transmission and makes use of difference in time for a sonic pulse to travel a fixed distance, and Doppler, which uses continuous wave transmission. Doppler uses a frequency shift by sound scattered or reflected from moving particles in the flow path.

Ultrasonic flowmeters are usually noninvasive and, therefore, can be used on many flows that would damage other meters. Materials of the sensor do not have to be compatible with the fluid measured, and the device can be installed without shutting down the process or cutting into the process piping.

There are many versions of ultrasonic flowmeters. Some use one sensor and others use from two to four and even more sensors. Some sensors must be mounted through the wall of the pipe and contact the process fluid, while others can simply be strapped around the pipe. Essentially, the more complex the installation, the more accurate the measurement. Accuracies are not as absolute as some other measurements; however, the device has excellent repeatability and also has many specialized applications that make it a viable flow measurement device.

Some readouts have built-in signal strength indicators that are used to determine if the measurement is possible or are used to troubleshooting the application if a problem exists.

A strong sensing signal depends on a tight fit of the sensor to the tube. Usually a petroleum jelly product is used to obtain the tight fit when testing for usability. A permanent installation requires gluing or strapping the sensor to the process piping. A lack of sufficient sealing medium can be the cause of a poor signal.

Installation

CLAMP-ON TRANSDUCERS

Many configurations of clamp-on transducers are available, but all share some common requirements:

(1) The pipe wall must be capable of passing sound.

(2) The mounting surface must be clean and smooth.

(3) An acoustic couplet such as oil, grease, or epoxy must be used.

DOPPLER FLOWMETERS

Doppler flowmeters have other requirements:

(1) The inside of the pipe must be free of sound-absorbing material such as scale or dirty grease.

(2) Provide adequate straight section for radial meters.

(3) Do not try to measure the flow of highly aerated materials.

(4) Do not mount the meter at the discharge of the pump.

(5) Do not meter a fluid immediately after a free fall.

(6) Be in a full pipe/conduit.

(7) Be continuous, not pulsating flow.

(8) Contain no material to deposit on the pipe wall.

(9) Require something to reflect or scatter sound.

Calibration

Transit time meters are generally calibrated in a flow lab. Transit time meters are "percent of full scale" devices. Improper installation will cause errors, so try to duplicate the conditions of the flow label when installing. Doppler meters are lists as "percent of rate" devices.

Maintenance

When a sonic meter fails because of low signal strength, the transducers are often blamed. There are several simple tests to check the conditions of the transducers before new ones are installed. If they are clamp-on Doppler transducers, remove them from the pipe. Hold them several inches apart with the radiating surfaces facing each other. Rapidly move them toward and away form each other over a distance of several inches. If a Doppler signal results, the transducers are functioning and the problem is probably in the transmitter or application. Transit time transducers can be tested by connecting them to an oscilloscope and tapping the transducer face with a hard substance such as the edge of a coin. A good transducer should produce a pulse of several hundred millivolts when tapped.

Variable Area Flowmeters

The variable area flowmeter is a special type of differential pressure flowmeter. On the differential pressure flowmeter, the area of the opening is fixed and flow rate is measured as a function of differential pressure across the opening. In the variable area flowmeter, the differential pressure across the opening is constant, and the flow rate is measured as a function of the area of the opening. This areas is generally displayed as the position of a "float" or obstruction that is free to move to produce the varying area (see Figure 6-24).

Variable Area Meters

Variable area meters have been available longer than most flow measuring devices. They are almost universally used for purges and where visibility of

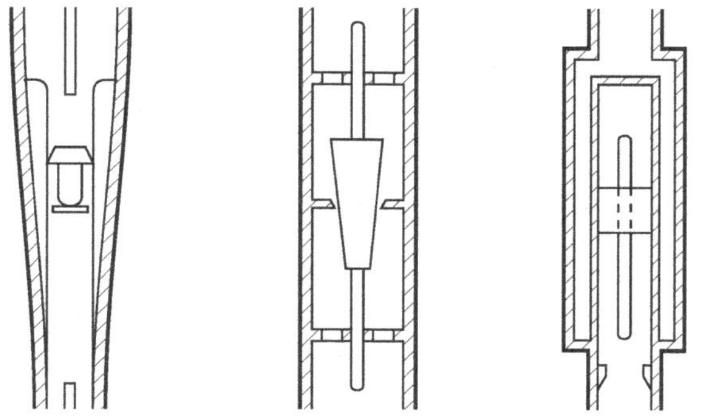

Figure 6-24. Types of Variable Area Flowmeters.

flowing medium is essential. They are restricted to the smaller sizes but serve their flow niche well. Transmitting variable area meters must be mounted so piping does not create stresses on the body of the meter. The float and plunger must be free of any stress or rubbing in order to provide continuous, accurate measurement.

Rotameters – A float made of a material of a density greater than that of the fluid is contained in an upright conical tube whose smaller end is at the bottom. The float, free to move vertically in the tube, is lifted to the position of equilibrium between the upward force of the fluid flowing past the float and the downward force of gravity on the float. In its simplest form the tapered tube is made of glass that is graduated, allowing the flow rate to be read directly by observing the position of the float.

Orifice and Tapered Plug Meters – Equipped with a fixed orifice inside an upright cylindrical chamber. The float has a tapered body with the small end at the bottom and is free to move vertically through the orifice. The flow rate is indicated by the positions of the tapered float.

Piston Type Meters – Piston is closely fitted to a cylindrical sleeve with vertical slots or a series of ports that are progressively uncovered as pressure differential across the piston raises. Flow through the ports exits the meter through an outer chamber. The flow rate is indicated by the position of the piston.

Installation

Variable area flowmeter can be used to measure the flow rate of nearly any liquid or gas. The greatest application of variable area flowmeter is flow rate indication in relatively small pipe sizes. They represent one of the least expensive methods of measuring rate of flow in pipes 2 inches or smaller. Variable area flowmeters generally have fewer installation restrictions than do other types of flowmeters. With the exception of the spring-loaded type, they must be mounted in a vertical position.

Calibration

Most rotameters are originally supplied with a scale for a specific application. Because of the variable area, flowmeter readings are affected by fluid density, it is necessary that the correct scale is used for the application.

Insertion (Sampling) Flowmeters

An insertion flowmeter consists of three basic components: a sensor, a probe assembly, and electronic circuitry. The sensor provides a signal proportional to the velocity or flow rate as a point in the flow stream from which total flow rate is inferred. An application is measuring flow in large line sizes.

Installation

First select the appropriate location, considering flow profile. Install the meter in the orientation recommended by the manufacture. In liquid applications, the pipe must be full.

7

Maintenance Engineering

Overview

Maintenance engineers and technicians now live in a more complicated world than that of their predecessors. When control systems first began with simple field controls, both operators and maintenance personnel were close to the process. They could smell it, see it, touch it, and sometimes taste it. They could feel rumblings from the process as pressures or temperatures were changed. A oneness with the system was felt by all those who were related to its operation and control.

Control began with manual field devices such as hand valves and levers. Gages were used for readout of the process changes. This evolved into field local automatic controls, followed by remote automatic controls in central control rooms. Figure 7-1 provides a historical overview of process control as applied to the pulp and paper industry. Other industries such as chemicals and petrochemicals were more advanced in the application of control systems; the result would be a shift of this curve to the left in some cases.

Each technology advancement has moved operators and maintenance people farther away from the process pulse. Direct use of four of the five human senses has become almost obsolete. It became necessary instead to become accustomed to the pulse of panel instruments that represented the process health and performance.

Each process variable (flow, pressure, temperature, density, pH, conductivity, consistency, etc.) had a discrete place on the panel somewhere. Whether a gage, a recorder pen, or a digital readout was involved, the operator could always find it after learning the panel layout relative to the process layout. The only thing missing was the process touch, feel, smell, and taste, and some method of alarm to replace "being there."

Remote alarms, therefore, became necessary when people were removed from the side of the process into remote rooms. Distributed control systems have made these alarms very easy to implement. It is so easy, in fact, that most systems have too many. The result is that critical alarms are often ignored because some alarm is always present. As with most things, familiarity tends to breed apathy — or even contempt.

In today's distributed control systems (DCS) environment, there is no longer a single monitoring indicator for each process variable. One must have the additional ability to page through cathode ray tube (CRT) displays to find desired information. Frequently, this means overcoming the intimidation factor that a video screen presents to some people.

Maintenance personnel of today must know not only the process and control of its variables, they must also know how to help operators interpret what they can't physically see. With the loss of line-of-sight benefit, remote sensors and transmitters provide readouts in an air-conditioned environment.

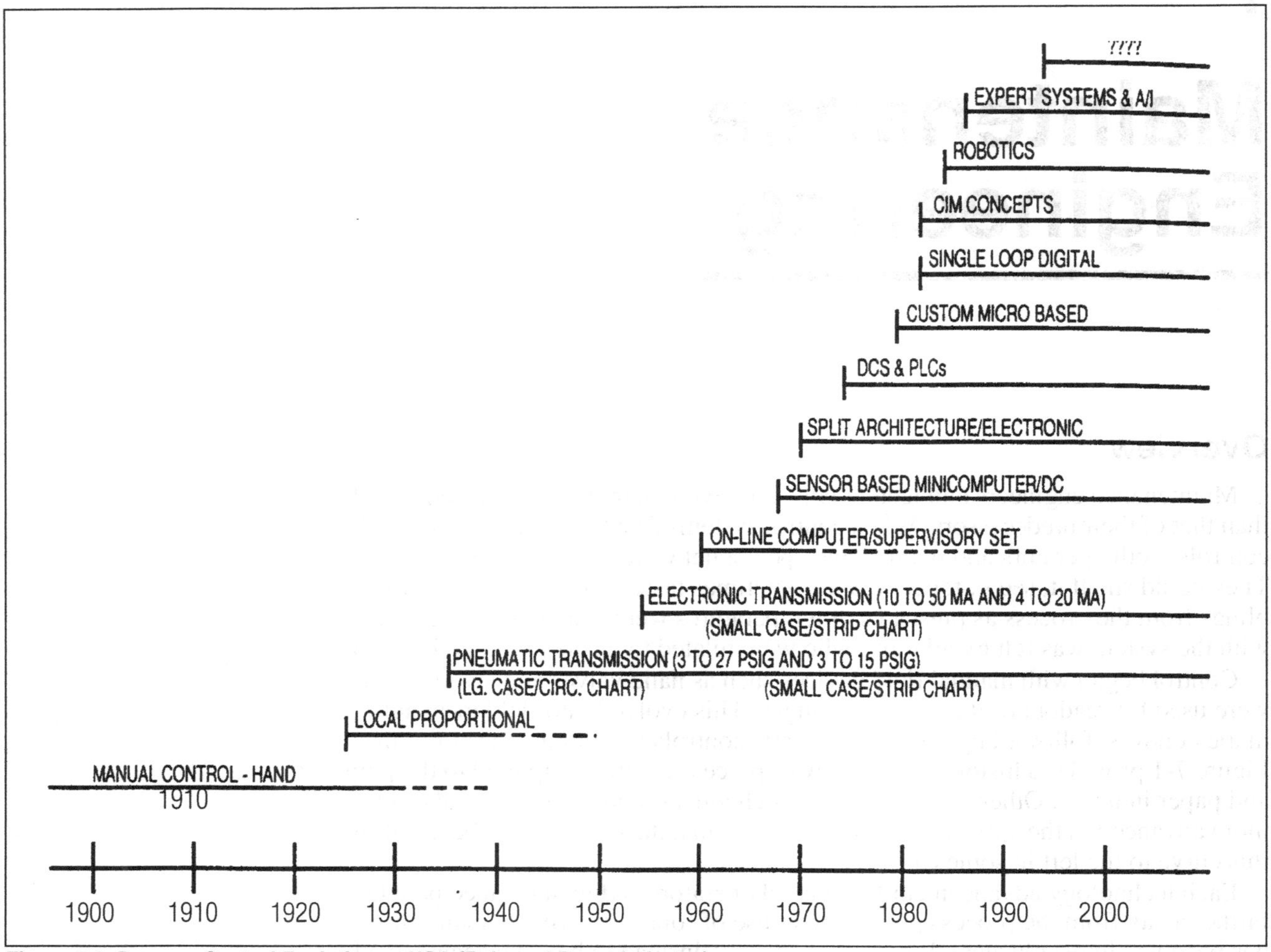

Figure 7-1. Process Control Evolution — USA (Pulp and Paper Industry).

The environment has been produced for the comfort of the computer and controls, but it is also a nice environment for human life. The "up" side to this environment is that physical discomfort brought on by heat, humidity, noise, and dirt has virtually disappeared. The "down" side is that with operator seniority goes control room residence. Fewer senior personnel make the rounds of the field process; the less experienced who are "in the field" may not recognize problems while they are still in a state of infancy.

It is necessary for the maintenance staff to be trained and to train others. This training is not an event, it is a journey — a journey that is more expensive than real travel and even more difficult because of continued demand on discretionary time. This demand is precipitated by the need for more maintenance help, much of which would disappear if more and better training could be implemented. The dilemma is similar to that of the person who would like to take a speed reading course but can't because he or she has too much to read.

Where do skilled maintenance people come from? Control system engineers used to come without formal training in process control. They came from the ranks of mechanical, electrical, and chemical engineering graduates. Many came from physics backgrounds. The technicians came from the ranks of pipe fitters, electricians, and millwrights. With the intensive technical training for military maintenance people beginning with World War II, another resource became available.

The control systems engineer (CSE) should come to the job with a solid background of engineering, physics, and mathematics. As important as the technical skills are the interpersonal skills of written and oral communications. Just as receiving training is important to improve those skills, serving as a trainer is a duty that goes with the territory.

Engineering Assistance

As discussed in the previous section, minimum technical skills are required to become a proficient control systems engineer (CSE). However, most project and operational failures are due to nontechnical rather than technical reasons. Critical factors in the nontechnical category are communications, scheduling, and accounting functions such as budgeting and cost tracking.

In a simplified form, system integration is knitting together the many unit operations and subcontrol systems present at a plant site. It is also necessary to knit together the many human factions within a site and within a company to make advanced control systems workable. The scope goes beyond the plant and company battery limits as just-in-time (JIT) relationships develop with suppliers and electronic data interchange (EDI) develops with customers and suppliers. There is no place for combat in today's marketplace. All parties must work together.

Some of the internal entities the CSE will see are subcultures in operations, maintenance, and other engineering disciplines. They can become friends or foes, depending on attitudes, "turf" issues, and personalities. The CSE can serve as a catalyst to bring the total group together, because controls have an impact on all units of a plant.

A de facto leadership responsibility falls on the CSE's shoulders because others in the chain may not understand the systems or their intent. The CSE must use a value-added approach, taking great care not to appear as a self-appointed guru. It is necessary to reduce terms to their simplest form and not fall into the trap of speaking in acronyms. Writing should be clear and easily understood by readers not well versed in control terms. The task is a difficult one, and not every CSE will be able to pull it off. The successful ones will.

Having established the level of communication language, what is there to communicate about? Plenty! The CSE must know what the process is intended to do before control systems can be modified or selected. This means that close and early relationships must be established with process engineers, project engineers, piping and mechanical engineers, operations management, and operators.

Once the desired future state (scope) has been established for an existing process or for a new project, details of budget and schedule must be worked out. A review of the impact to the site and the business should be frequently made to ensure the exercise doesn't result in corporate hernias! If all is still go, the task of performing detailed engineering is ready to be done. The maintenance side of engineering has profound knowledge that should be integrated into all operations.

It is easy for maintenance needs to be overlooked during project development. All field devices must be in the proper functional location to properly measure and control the process. If the sensor, the controlled variable, the final control element, or any of a myriad of other details is in the wrong location, the success of the system is in jeopardy. If control valves are in the air without platform access, a safety situation has arisen that will preclude proper maintenance.

The maintenance engineer (ME) must have early input and must have understandable reasons for that input. Respect will be evident if he or she is already performing a good job on existing processes. If in a new role or a new plant and yet unproven, the maintenance engineer must work hard at communicating practical needs. This is not to say that others on the team would be hard to deal with, but they may not know any better. This is especially true in a control system upgrade or a new process. If the result becomes too elusive, perhaps a break for some team building would be in order.

Another cause for difficulty can be lack of critical mass. Are there enough resources to do all that has to be done in the time available to get it done? This is where the team players have to come forth with that news, and where management must make some decisions — decisions that relate to priority setting and assignments in keeping with what must be done.

Control systems engineering success probability can be shown best on a matrix (see Figure 7-2). There are no sure ways to guarantee the success of any process improvement or new project. However, success probability can be improved by having a high level of development in the items listed in the left column. These factors relate to strategy, scope, relationships, competency, critical mass, and CSE entry on the project.

The matrix can be used to avoid trouble. If there is doubt, a postmortem should be performed on projects or relationships that didn't turn out too well. The gap will probably be on the list and will be identifiable.

Maintenance Involvement in New Projects

Now that the CSE has been assigned a responsible role on the project, how can he or she make a difference? The most important ingredient for any project is a well defined scope. To accomplish that end, clear understanding must exist regarding what is to be done and what is required from a control systems standpoint.

The CSE should begin with the process engineers and operations people (including key operators) to be sure they are on converging paths. The project should be reduced to a functional specification that is readily understood by all parties. This plateau can be difficult to reach unless everyone works from a common document — the process block diagram. The functional specification and the block diagram are reduced to a process flow sheet. Later conversion of the process flow diagram to a process and instrumentation diagram (P&ID) brings and cements the CSE needs into the picture.

If the company is small, the load of adding controls to the process falls on the ME/CSE. If the company is a large organization that has a separation between control systems engineering and control systems maintenance, this may already have been done before the ME/CSE's involvement. If so, the ME/CSE should thoroughly review and buy into it, or immediately speak up with any concerns and get them resolved. The time is now.

The P&IDs must be shown in accordance with the latest version of ISA-5.1-1984-(R1992)-Instrumentation Symbols and Identification. It is a universally understood method that will make communications with suppliers, consultants, contractors, and the maintenance staff as meaningful as possible.

Terminology should be in accordance with the latest version of ISA-5.1 again for universal communication. The main goal of the P&ID is to develop a complete shorthand depiction of the process that will prevail as further details are fleshed out.

Once the process has been established, installed cost estimates should be made and compared with available funding. It is necessary to develop good estimating methods so the ME/CSE can be comfortable and can defend the resulting numbers. If funding is inadequate, the scope should be reviewed to see what can be removed or deferred to allow the project to fall within the money available. It is easy to fall into the trap of saying, "We can do it for less." This can be a mistake.

	LOW ◄— LEVEL OF DEVELOPMENT —► HIGH				
	1	2	3	4	5
BUSINESS STRATEGY					
MILL STRATEGY					
PROCESS DESIGN CRITERIA					
CONTROL STRATEGY					
SCOPE DEFINITION					
STANDARDIZATION					
HARDWARE					
SOFTWARE					
DESIGN METHODS					
CSE RELATIONSHIPS WITH:					
ELECTRICAL					
MIS					
OTHER DISCIPLINES					
PROCUREMENT					
SYSTEM USERS					
UNDERSTANDING OF CONTROLS					
CSE GROUP					
OTHER DISCIPLINES					
PRODUCTION LEADERS					
OPERATORS					
MAINTENANCE					
CRITICAL MASS					
CSE ENTRY INTO PROJECT	LATE				EARLY

LOW ◄— SUCCESS PROBABILITY —► HIGH

Figure 7-2. Control System Engineering (CSE) Success Probability Matrix.

The project will be one of misery and insufficiency. The scope has to be the driver for approved funding, and funding must be adequate or the scope must be changed.

Concurrent with costs estimates are (1) schedule needs and (2) human resource requirement definitions so staffing can be determined. This returns to the critical mass topic. Four thousand human effort hours mathematically equates to 4000 humans for one hour, but that is not logistically possible. Logistics are important as well. For instance, the process control effort hours tend to cram into the later end of the project, even when early involvement is accomplished.

Equipment layout, piping layout, and HVAC duct layout are all space-consuming entities. It is ideal to establish major raceway corridors for all disciplines so each can proceed, but controls tend to have to follow other actions. This means that when the schedule has been established, staffing will have to be adequate to do the work in the time allotted.

If the ME/CSE's responsibility is also to keep existing processes running during the project, he or she should not forget to account for those hours in developing the resource requirement list. It is also pertinent not only to have a body count, but to recognize skill levels for each function. If it now takes the best available person to keep critical processes maintained, that won't change to accommodate the project. It's training time again.

It is now time for technical maintenance skills to be brought into play. Look at the existing facilities to see what it would look like if a chance arose to do it all over. One shot exists to make the new project optimally maintainable. The maintenance crew should be involved in a brainstorming session to pick locations that are easily accessible, with enough room to do the job without danger to humans or the process.

The operators should be involved in the location and layout of the control room. Is it acoustically acceptable and with safe egress in case of an emergency? Are there dual exits with access to the process area, the equipment, and stairways? Have provisions been made for both high-level lighting for use during shutdowns and variable low-level lighting for use during normal operations? Is there room to pass the largest component in the room through the doors after room closure or must a removable wall panel be provided?

Is there room to spread drawings and instruction manuals for troubleshooting when operations is making demands during an emergency shutdown? Can the system be designed to have on-line, on-screen documentation — on a maintenance screen away from operation screens? Is a communications link possible between the control room and all levels of the operation?

Are signal levels properly isolated in raceway runs? Has a design criterion been developed that determines whether closed wire ducts, open tray, or conduit will be used for raceways? Is there a plant electrical power and grounding plan, or must one be developed? Is there a selected supplier of hardware from a previous partnering arrangement, or must vendors be selected? Is a specification already in place for each type of hardware? Are hardware and software bids ready to be solicited from well written specifications, or must they be generated?

Has it been established who will be responsible for design, installation, checkout, start-up, and final acceptance? The maintenance CSE should be a key player in all phases of the project at some responsible level of authority, or it may be very difficult to keep the resulting process operable.

All these questions — and more — should be asked as the project develops. The best way is to keep a running list of maintenance factors to be implemented on the "next" project, long before the project has even been conceived. With a running plan in place, maintainability can become much easier to implement for all new work. It also helps in the checkout and start-up of a new or modified process.

Installation details regarding location and mounting of field devices should be clear and specific. If left to the installer, one may find surprises to be lived with forever.

There should be a planned sequence of checkout. The installation contractor should have a list of things to be completed before announcing readiness for checkout. When ready, a checkout team made up of engineering, maintenance, and contractor personnel should work from the checkout plan, carefully noting acceptance and rejection details. This begins with dry checks and, depending on the process, ends with water for the process to make sure all loops are capable of throughput control. Control tuning is given initial settings and recorded.

A frequent installation error is the mounting of field devices on brackets that allow vibration from the process. Another is lack of maintenance access without use of ladders or installation of special platforms.

When the real process is run, refinements should be made and recorded. This will provide a starting point in case readjustment during start-up puts the process "in the ditch." It is easy for overzealous tuning by the uninitiated during the throes of start-up frustrations to destroy all early settings.

If the new process is controlled by a DCS, it is possible to have all control algorithms in place and simulation of the process completed far ahead of start-up. Many installations still use analog instruments and hard-tubed or hard-wired signal transmission. Caution should be taken not to lose the techniques (which have now been built into computer control) for logical order in process control.

Documentation, whether manual or electronic, is the key to providing logical and rapid troubleshooting. It is also the vehicle for effective preventive and predictive maintenance. A plan should be in place to keep getting the most out of an installed control base.

Successful Maintenance

It is a rare plant that provides adequate attention to keeping installed instrumentation at optimum performance levels. The key to success of any instrumentation control loop is the basic measurement of the process variable. The sensor must reside in horrible in-process environments and requires considerable attention. This effort is rarely recognized as critical unless the measurement is critical to safety, such as recovery furnaces or chemical towers.

Sensor Verification

Sensors are inherently subject to calibration shift. This has precipitated the drive for "smart transmitters." They go all the way from being self-checking to being capable of remote on-line recalibration by a hand-held calibrator to being adjusted from a remote central control room. Smart transmitters are higher in cost than standard transmitters; therefore, they are normally associated with an extensive rebuild or a new installation. This trend will change as prices drop and retrofits or failure replacements allow for relatively painless upgrades.

In the average existing process, routine calibration checks against original specifications are often of low priority because of the firefighting life style of maintenance. Preventive instrumentation maintenance is usually considered a luxury at budget time and one of the first items to be reduced. This is further augmented by unit operations' need for repeatable measurements rather than accurate measurements. The result is that error shifts are compensated for by control adjustment; and less than optimum performance is accepted because, "That thing has never worked right so just ignore it. Make it read at this value here."

While a plant can operate under such a relaxed attitude toward process control, it rarely can perform with predictability and excellence. In general, U.S. companies tend to obtain less than optimum performance from their control systems. The better performing plants devote proper resources to maintenance and the application of control loops. This requires knowledgeable and skilled control systems engineers, maintenance technicians, and process engineers, all with the motivation to optimize process performance. It requires an operations management team that wants to improve performance.

Process Databases

A process database provides a tool with which managers and operators of unit processes can better run their segments of the plant. They can window into their supply side and output side processes to work toward goals of plant optimization. The key to proper process database tool performance is accurate and repeatable input data on a common, relational database. The application of relational databases to sensor data can provide that capability, but the key is still the need for true process variable input.

It is necessary for existing measurements and control loops to be properly calibrated and tuned to realize the most from a process database. A process database can help diagnose improperly calibrated and tuned control loops, but that should be considered a side benefit and not a prime purpose.

PREREQUISITES

To obtain the benefits of application of a process database, the following systems should be in place:

(1) An active sensor verification program, with systematic record keeping and schedules.

(2) An active process control loop verification and tuning program, with systematic record keeping and schedules.

(3) A proper work system team to leverage the best of the system and the tools, with unit and plant optimization goals clearly defined and operative.

It sounds too simple, but success really boils down to implementating what is in place to make the most of control system benefits. It is the means to obtain plant optimization through use of existing instrumentation. A serious maintenance plan also provides identification of the additional instrumentation needed to improve the system by demonstrating where more is needed to provide the necessary process interfaces. It will also identify obsolete devices.

The High Maintenance System

Implementation of the above prerequisite plan will soon identify which sensors and loops have stability. Frequency of calibration checks can be extended for the stable loops. It will also identify which have instability. The unstable list probably will confirm only known maintenance history that was developed from emergency shutdowns and repair parts records. So a high maintenance control system exists. How can it be rectified?

Some measurements are inherently high maintenance items. A few are pH, oxidation reduction potential (ORP), stack gases such as sulfur compounds and nitrous oxides, and consistency of paper stock. Peer discussions will sort out possible solutions or verify that the problem is not unique.

For the real problem system, the maintenance log should help tie the failures to differing plant conditions. Possible causes might be raw material change, throughput rate changes, ambient temperature changes, operating pressure changes, or any recurring change in concert with the maintenance problems. Is vibration a factor?

The true answer might be that the control system is misapplied and being asked to do something beyond its capability. Are valves properly sized? Is the process variable value compared to the full-scale range allowing enough control resolution? Is the loop stable on manual?

If best detective efforts don't uncover the problem, any simple modifications that can be made to sample points, control points, or environmental protection should be examined. If all fails, it is time to take a fresh look at the process and the control system. The process engineers and the operators should be consulted to see if the process can or should be modified. With the process locked in, with no control system in place, how can it be controlled?

The simple things should be looked for first.

The best solution must then be determined, and the life cycle maintenance costs projected from the log must be compared to life cycle cost of the modifications. Process and product change costs and benefits must be determined. Buy-in to the numbers and agreement with the need for change must be obtained from process engineers and operations.

A request for funding to make the change must be initiated. If approved, a profit improvement project can be begun. If disapproved, the problem must be lived with for a while. Other ways to reduce the maintenance effort must be found, but management should be made aware that the price of labor and materials must be paid when budget time rolls around.

Documentation Control

The backbone of the maintenance effort is documentation. The modern method of documentation is on screen, with ready access to any document. It will be some time before all new installations make use of that method, and much longer before all plants have it for all processes.

However, all plants are probably equipped with personal computers and capable spreadsheet and database software. Even if existing documentation is strictly manual, time would be well spent to enter all pertinent maintenance documents on the computer. The ability to do a location search by drawing number or title or instrument tag number by computer is savings enough. It will also result in less wear and tear on actual drawing files as the frantic three-in-the-morning panic to find a detail takes place.

Upgrade drawings and other change data should be maintained for all instrument indexes, loop drawings, one-line diagrams, and process changes.

The method for making these document changes should be well known by all in the department. Responsibility should lie with one individual to see that the changes actually are made. The responsibility for getting the change in the hands of that person has to lie with each person who makes a change. It is not unusual for a sketch of a change to lie dormant in someone's tool box. It might be forgotten until a crisis strikes again and someone else is troubleshooting with obsolete drawings.

The change procedure should include reasons for and benefits from the change: who made it and when was it made? Sometimes changes are made to satisfy a process condition that is not long term.

The same criteria should apply to hardware and software changes. If a software change is valid, backup disks should be generated with the change incorporated, and copies of the original state should be archived for future reference. It is very easy for a re-boot to be necessary. It does not help to find that the re-boot disk still contains an earlier software version, or that the earlier version is needed but has been overwritten.

In a like manner, changing the hardware might solve a present problem, but repair or replacement action should be started immediately. If this is not done, the next time that component is needed elsewhere it might still be in its defective state. It is important to properly track spare components manually *and* by computer.

Again, it is important to keep careful logs about what has been done in the maintenance cycle. These logs should be repeatedly scanned for recurring malfunctions. How fast a problem has been solved is not as important as how infrequently failure occurs. Predictive and preventive maintenance have a much higher long-term priority than a fast fix, but short-term priorities remain to get systems up rapidly after a failure.

Alternative Methods of Maintenance

The balance between how much maintenance is done internally and how much is done externally can be delicate. Much of the decision relates to the operating philosophy of the site and whether available outside resources are in close proximity. If the site is remote, travel time by an outside maintenance team might be prohibitive. It might also be difficult to secure competent internal staff, in which case it becomes necessary to train those already in place.

If a plant wants to be autonomous, the decision is easy. Staffing with competent people who have the same or higher capabilities as found outside becomes necessary. Some benefits are that all learning remains in house, and turnovers in outside resources do not have an effect. The down side is that it may be necessary to staff for the peaks, which would result in an overstaff condition for normal periods. However, this discretionary time can be put to good use by improving existing systems and working toward unit process and plantwide optimization.

A decision to have a skeleton staff for emergency work, outside help for shutdown work, and specialized technology when necessary might provide a lower overhead burden. There could be a high turnover of outside help. The decision will depend on competing needs and the frequency of needing help. This may not be a problem, but it is always disconcerting to have the feeling that the training of outside people is being done at one's expense. Agreement with the provider of outside service might eliminate this perception, but retention of good people requires billable time. The outside service won't be able to carry a staff any more easily than the companies it serves if the workload is not there to justify it.

Specialized outside help is always a potential need. Complex systems usually come with an offering for resident technical staff. Rates for these people are usually high, but if a plant doesn't have the capability, the outside help is necessary. Careful analysis should be made before making a decision to keep the specialized help after a system is up and running well. The more complex control systems at a given site become, the more viable is retaining in-house talent to maintain them.

Areas of savings can be found in contract personnel costs, spare parts costs, and on-site repair costs. The system vendor is going to provide parts from a third party with a markup to cover overhead costs. Many of the parts needed are standard components that are available from subvendors directly — at a much reduced cost. Another area of savings can be found in repair of boards on-site versus returning them to the vendor on a repair/exchange basis.

Service/Contract Maintenance

Service or contract maintenance, these days more commonly referred to as "outsourcing", is an alternative for maintenance staffing requirements for manufacturing facilities.

Outsourcing can take the form of service or repairs performed by the original equipment manufacturer (OEM) or the OEM's local vendor, or service or repair performed by a third party not associated with either the equipment supplier(s) or the end user. The service or repair function can be performed on a full-time, call-in, or regularly scheduled service basis.

Historically, outsourced maintenance has been in use for decades in various industrial and nonmanufacturing sites, such as airport terminals, office buildings, distribution centers, laboratories, and light manufacturing plants. The skills level of the maintenance people tends to be geared towards the general building maintenance tasks with some HVAC, minor electrical and electrical distribution, and general mechanical skills, as required.

This section will focus upon the outsourcing concept as applied to process industries, such as pulp and paper, where the manufacturing and the maintenance skills levels are much more demanding and require extensive knowledge of the process as well as instrumentation and control logic.

The use of outsourced maintenance services plays a role in the capacity of these industries to provide effective maintenance in their manufacturing locations as an alternative to dedicated in-house maintenance.

In-House Maintenance versus Contract Maintenance

Economics

Economic considerations are reason to consider the use of outside services and contractors. Some of these costs are apparent:

(1) *Training.* In many cases, even those in which the use of outside services may be abhorrent not only to the workforce but to the management of the maintenance function as well, the use of outsourced resources will benefit the operations of the complex. Immediate relief for a non- or undertrained staff can be readily obtained from a contracting service while long-term provisions for improvements to the internal maintenance function via retraining and/or restaffing are being made. In fact, one of the dangers for "outsourcing" suppliers/contractors is loss of skilled technicians in a mill that is restructuring its internal maintenance function.

(2) *Personnel.* The cost of an organization to maintain the mill's equipment can be very high as a portion of operating costs. In the paper industry this expense runs as high as 25% to 30% of the total personnel costs. Outside sources usually claim to be able to provide the personnel necessary to maintain a facility for far less than in-house maintenance organizations.

(3) *Material.* Depending upon the maintenance strategy followed, the material costs can become exorbitant in relation to the total operating costs. For example, if caretaker or "hey boy" strategies of urgent, corrective-only maintenance are the norm, the material costs will be high and will remain high; however, if professional preventive maintenance strategies are followed, then the material costs can be reduced over time.

Other economic issues, not so apparent, include the productivity of the workforce (in-house versus outside) and the overall operating efficiencies of the production equipment.

One economic consideration very often overlooked is the preservation level of the facilities, i.e., floors, roofs, walls—the infrastructure of the plant. These "assets" play an important part in providing equity when the company is borrowing money for operating, yet these same assets are forgotten when their maintenance costs are allocated and budgeted. Unless specifically addressed, the issue of facilities preservation is neglected. The cost to renovate neglected facilities far exceeds the cost of normal preservation.

COST COMPARISONS

This is a difficult issue to analyze since the variables are not consistent from area to area, craft to craft, process to process. When the analysis includes other similar manufacturing facilities but at different locations within the same parent organization, the analysis breaks down because of the accounting practices, procurement practices, employee training, company politics, and so on.

Suffice it to say that any analysis of cost comparisons between outsourcing and use of local in-house employees must be evaluated with both scenarios on a level playing field.

A plant in the Houston, Texas, area reversed its use of outside contracting personnel (in place since start-up) simply because the outside contractor had failed to undergo the process of "continuous improvement" of personnel, tools, and capabilities and, therefore, was seen as a major contributor of the plant's problems rather than as a solution. The decision to go with a traditional in-house maintenance effort was costly, but the expected benefits of improved runnability of the equipment far outweighed the staffing costs.

Never forget the maintenance facilities in cost comparisons and justification.

In most manufacturing plants the maintenance effort is a cost or a profit center and must include the cost of maintaining its own facilities, including tools, equipment, and personnel, as part of its annual budgeting projections. Therefore, unless the annual costs of the building, equipment, and tools for maintenance (including depreciation) are included in the cost evaluation, the outsourcing option will almost always show positive. However, when those costs are included, outsourcing becomes less attractive.

Organizational Options

Several companies are in business today to offer complete facilities and equipment maintenance for a mill or plant. This option not only is a managed process that is available immediately following an assessment of the needs (i.e., preventive, predictive, or reactive), but one of the staffing options includes assimilation of the existing workforce into the employ of the contracting company, much like the recent innovation of contracting security for the plant site. Several organizational options exist, from shared management to complete turnkey-style use of outsourced maintenance.

Option two is a sharing of responsibility between the contracting company and the owner; for example, the company provides management and the contractor provides skilled labor and first-line supervision.

A third option is the supervision and direction of the owner's personnel by contracting planners and supervision, using the contracting company's management plan.

These options are extremely attractive for the facilities of a plant because the equipment for facilities is generic, i.e., air compressors, water systems, effluent systems, HVAC. These applications do not require the specialized expertise deemed necessary for the maintenance of process equipment, i.e., papermaking, packaging, steel converting, assembly, communications, and chemical equipment.

Of course, the decision to undertake an outsourcing strategy requires a knowledge of the "soul" of the plant as well as the body. Does the plant's union contract allow or is management willing to follow through on the conflict and stress that such a change would create? The strain created may very well be artificial, such as that from a strike, but to the principals at the time, it is a real problem, one that requires sensitivity and communications with understanding the underlying problems and concerns.

Resource Availability

Don't forget the "soul" of the plant when evaluating the options.

GEOGRAPHY

Geographic location has a significant impact on the availability of outside skilled personnel. As an example, urban plants (i.e., recycled paper mills and recycled fiber mills) will probably suffer less than others within the pulp and paper industry because of their location near metropolitan areas and their employee lifestyles, where having outsiders "taking away their overtime" is not as much a factor in the decision to use outsourcing. If, on the other hand, the plant is located in a remote rural area, the outsource supplier must make adjustments for employees' living accommodations. This can be an expensive (prohibitive) proposition.

FULL TIME

Decisions must include consideration for the extent of the coverage for the plant, i.e., will the outsource include full time coverage three shifts per day? Shall it include all the craft skills normally associated with the process, and shall that skills level be provided full time? Can the outsource provide full-time, all-skills coverage?

Perhaps only a few or a single skill is required by outsourcing on a full-time basis. Some vendor companies that supply process instrumentation offer this service as a part of their overall marketing program. Unless responsibilities and accountabilities are clearly defined, the cost for this service can be prohibitive compared to on-site, in-house personnel.

ON CALL

An option to full-time coverage by an outsource is on-call coverage. The supplier/vendor is called whenever a problem arises. One problem with this scenario is response time by the supplier/vendor service personnel. Unless the service capability is located within a relatively short distance from the plant site, response time could be lengthy and costly. On the positive side, in-house personnel do not have to be trained for specific equipment that may not require regular, routine maintenance. Without regular, routine maintenance on some specific process equipment, the maintenance personnel lose their technical knowledge and expertise.

SCHEDULED VISITS

Another option, perhaps in combination with the on-call option, is to provide a scheduled visit on a periodic basis, at which time the service person performs periodic diagnostic and preventive maintenance. Again, many vendors of hardware will offer this option as an alternative to training in-house personnel who may have infrequent use of the knowledge they receive in a training session. "Use it or lose it" is a real-world concern, especially with the more technical requirements of electronics, instrumentation, and process control.

Resource Expertise

VENDOR MAINTENANCE

Vendors of equipment will have the best knowledge of their own equipment and, therefore, can provide the best quality in terms of equipment maintenance. The vendor service/maintenance personnel will have immediate access to equipment specifications and the manufacturing documentation, engineering, and design. They will have an immediate supply of spare parts; they will know the necessary troubleshooting techniques; and they will be able to perform the most timely repairs.

The downside is the vendor's availability and cost. Usually the vendor will have a response (to trouble) time exceeding a day (for travel to the site), and the cost will be a factor of the travel to and from and the time spent at the site.

THIRD-PARTY MAINTENANCE

This may be the most controversial outsourcing resource. It cannot be denied that wherever a third-party maintenance organization is effective, a similarly managed and equipped in-house dedicated staff would work as well, perhaps better, and probably for less cost (assuming profit is the only difference in costs).

In terms of expertise, the third-party maintenance organization has to acquire the specific knowledge and skills required for the equipment as does an in-house function. The difference lies then with the learning ability of the individual workers and the time permitted for learning.

PROCESS KNOWLEDGE

Depending upon the resource pool from which the workers are drawn, the edge in this concern has to go to the in-house staff, which, in general terms, will draw their maintenance personnel from within the organization, where process knowledge is likely present in the individual's past experience and training. However, a third-party or vendor-supplied service/repair person can also apply knowledge from similar industrial problems, not familiar to the in-house staff, in solving problems.

SYSTEMS KNOWLEDGE

Systems knowledge is entirely dependent upon the level and degree of training received by the personnel assigned the task of service/repair.

Systems supplier personnel will, of course, have the most knowledge and experience. End users, with a history of the use of specific systems, may very well have the necessary knowledge (with some updating) and experience to service or repair newer technologies from the same system supplier (they will certainly know the supplier corporate standards and some of the personnel).

Obviously, to be effective in the maintenance, service, and repair of systems, the personnel assigned must have specific knowledge and experience with the system hardware.

If the industrial user of a system is inclined to continuously upgrade or update the systems technology (to remain competitive, etc.) and the user is not particularly attached to a particular system vendor, in-house maintenance capability is severely hampered due to knowledge and experience required because of the variety of systems that may be present. The service or repair might better be handled by using the vendors' personnel on an outsourcing basis.

New Systems Installations and Checkout

The installation of new systems in a manufacturing facility is another highly controversial and, sometimes, emotional issue.

In-House

In most locations an existing maintenance function will resist the use of any outside personnel for installation, start-up, and poststart-up service. It is their genuine belief that they, and only they, can provide the best possible installation and follow-up service because of their vested, long-term concern. Supporting that belief is the concept that installing equipment will provide invaluable hands-on knowledge of the equipment for poststart-up servicing and will assure that the equipment has been installed with the utmost care available. The fundamental belief is that since they will, in the future, be responsible for the performance of the equipment and its impact on their own personal well-being, they will be the most responsible for its quality installation, which is a solid argument.

Commonly, the counter argument by mill management and engineering is the time factor. Supposedly, the argument goes, an existing maintenance department is only large enough by economic necessity to maintain the existing equipment and, therefore, will not have sufficient personnel nor time to devote to the installation of new equipment. The new equipment (system) was purchased, after all, based upon an economic return, justified, obviously, upon a timely installation and start-up.

Third Party

Third-party installation is the mostly commonly used approach to new systems installation and checkout in American industry. The use of a third party for these projects is based upon a simple concept: focus and concentration.

This resource provides adequate manpower with specific knowledge and skills, the right tools, and installation know-how on a clearly defined schedule (usually with delay clauses and penalties), and a complete supervisory staff. The supervisory staff includes planning and engineering functions exclusively assigned to the project. The material requirements are known; the time elements are known; and the manpower skills requirements and utilization schedules are known. The project team does not have to be concerned with other events in the plant that might normally influence any other resource, such as in-house crews. The project team, supplied by the third party, is focused.

The quality of workmanship of the installation crews is purely a matter of the competency of the supervisory staff assigned to the project and the company liaison personnel.

Although this option may be more costly in terms of "up-front" costs, the timeliness of the installation will soon pay back any additional costs.

Vendor

Vendor installation, so-called "turnkey" installation, is gaining in popularity with many industries because it combines the knowledge and experience factor of the vendor with a vested-interest commitment by the vendor. Because the vendor is interested in nurturing a lasting relationship with the customer, the vendor will tend to be more concerned with the customer's needs and desires during the project cycle than will a third party, yet will be able to complete the timely installation not provided by in-house staff.

Preventive Maintenance

The need for an organized, plant-wide program of preventive maintenance (PM) is not universally recognized within the manufacturing segment of American business. Although the popularity of total quality production, quality circles, work teams, and so on has been increasing, few seem to grasp the concept that maintenance, as it is practiced today, can be largely prevented. This will reduce the overall need for the highly specialized and skilled personnel that exists in many operations today and will ultimately reduce the cost of maintenance.

The task of preventive maintenance, when analyzed, consists of three basic efforts. Maintenance work is fundamentally *anticipatory* (predictive), *reactive* (corrective), or a combination of both, which could be termed *preemptive* (predic-

It is common practice to hire outside firms during planned shutdowns to supply personnel and tools to replace faulty equipment, repack valves, open equipment for inspection by plant personnel, run performance evaluations, or for equipment suppliers to provide expertise to supervise local maintenance personnel during planned outages of the OEM equipment.

tive followed by corrective). Another group of tasks has little relationship to pure maintenance yet requires the resources, by tradition, of the maintenance effort. These can be termed *obligatory*.

The combination of reducing unplanned maintenance and fully utilizing equipment down time to perform essential preventive maintenance routines promotes increased productivity. Outsourcing can play an important role in the success or failure of preventive maintenance within the manufacturing site.

Many outside firms specialize in predictive (anticipatory) maintenance routines. It is not uncommon for specialized outside crews to be hired just prior to planned shutdowns to perform such tasks as vibration monitoring of rotating equipment, infrared evaluation of electrical distribution equipment, pinpointing location and identification of process leaks, determination of faulty steam traps, and so on.

Limited PM routines can be performed while the equipment is running, requiring less manpower, while the bulk of the PM routines fall during equipment outages and necessitate the use of significantly more manpower. Identification of these routines and when they can be performed is essential for determining the best possible method for accomplishing these tasks.

Many combinations of crews, either in-house, outsourced, or a combination of the two, are feasible. For example, it may be determined that the operations seeks only in-house emergency reactive crews to fight fires, while supplementing with outsourced personnel for PM routines or on major turnarounds or overhauls. It may be decided, on the other hand, that only facilities maintenance will be performed in-house, while all other specialized maintenance will be sought outside.

Power, Grounding, and Isolation Requirements

Power Quality

The importance of a good power system cannot be overemphasized. Poor design of the power system and/or poor maintenance of the power system can substantially add to the cost of maintenance and the down time of a facility. Therefore, power and grounding is covered in detail, to emphasize its importance.

Instrument power system general requirements are as follows:

(1) Reliable power for safe and reliable operations.

(2) Voltage level at all normal operating conditions within the manufacturer's requirements, typically ± 10%.

(3) Protection against transients both externally (lightning, utility system transients, etc.) and internally (motor starting, SCR devices, switching transients, etc.) generated.

(4) Designed to meet the National Electrical Code as a minimum.

(5) Well documented for installation, design, and maintenance purposes.

The majority of electrical transients are internally generated. They come from things such as switching transients, arcs, poor grounding, welding, power faults, and nonlinear loads such as switching power supplies, variable speed drives, SCR drives, etc.

Reliable Power

Reliable instrument power is essential to a good instrument and process control computer system. Reliable power is a combination of good power system design and installation, good operating practices, good maintenance, and a reliable power source(s).

Power System Design and Installation

Good power system design and installation is a team effort between engineering, maintenance, and operations. The system design requirements must be jointly determined and the installation must reflect the design requirements. Obviously these design requirements are very important to a successful installation. Experience indicates that failure to get the design specifications and requirements right represents a major factor in poor or failed designs. One of the problems causing this is the failure of the groups involved to preplan and obtain all the necessary data in order to set the design requirements. This is probably due in part to other priorities such as keeping the place running and "fighting fires." This is really part of a vicious circle: the group doesn't have time to get the design requirements right because they are fighting fires that result from a poor design, which resulted from not getting the design requirements right. The moral of the story is to take the time and do it right the first time, every time, when determining design requirements in order to avoid fighting future fires.

A good power system design provides power overcurrent protection that isolates any fault at a point closest to the fault to avoid tripping the whole power system or a major portion of it. Coordinated fusing should be used to ensure this.

Blown fuse indicators can reduce the repair time substantially. Fuses should be located where they are easily serviced with the size and type of the replacement fuse indicated at the fuse site and spare fuses located conveniently near where they will be used.

An instrument tech was working on a flow switch when he inadvertently shorted to ground the 120 V AC circuit feeding the instrument. This shut down a whole section of the plant. It was determined that after a recent modification, the in-house contractor installed the right size of fuse but the wrong type. These fuses failed to coordinate with the rest of the fuses, and, as a result, an upper level fuse blew and shut down a section of the plant. Proper inspection of the contractor's work would have caught this.

Maintenance Practices

Good maintenance is required to ensure reliable power. Preventive and predictive maintenance programs must be used to minimize major maintenance activities. Good maintenance records to detect high maintenance equipment are required to target the equipment for modification or replacement to improve the up time of the equipment. Good maintenance records also help in troubleshooting and the determination of spare parts stocking. Quality and timely repairs require qualified technicians, properly trained with proper and available documentation, and appropriate spare parts. Staffing must be sufficient that the maintenance department is in the repair business and not the fire fighting business.

Operation Practice

Operation practices must be in accord with the design of the power system, or the power system should be modified to reflect the actual operating requirements. Abuse of the power system by operation practices leads to an unreliable power system. Operating equipment beyond its operation specifications, failure to maintain equipment due to operation requirements, patch work maintenance to meet operation requirements, and use of improper equipment for the application are some of the operation abuses that occur.

Reliable Power Sources

A reliable power source is also required for a reliable power system. Utilities generally provide reliable power for many operations. Instrumentation, however, has additional requirements past normal operations to ensure continued and safe operation of the facility.

Instrument power systems must be fault-tolerant and generally must have the ability to ride through power disturbances and to continue to operate for a period of time during a power outage. Uninterruptible power supplies, power conditioners, and standby generators are some of the methods used to provide reliable sources of power.

Fault tolerance in a power system means that the system is tolerant of failure. For DC systems, high quality power supplies in redundant configurations are used to provide fault tolerance. In AC systems, alternate power sources or feeds such as uninterruptible power supplies (UPSs), standby generators, and alternate utility or power feeders are some of the means used to provide fault tolerance.

Uninterruptible Power Supplies

Uninterruptible power supplies (UPSs) are a common method of providing a reliable source of power. UPSs generally come in two types: on-line and off-line. In on-line UPSs, the batteries are on-line at all times; in off-line UPSs, the batteries are off-line under normal conditions and switched on when the input power is lost. While more expensive, the on-line UPSs are generally considered to provide better and more reliable power. The modern loads from computer-type loads have shifted the electrical load profile for instrumentation systems from linear toward nonlinear. These types of loads have high crest factors (ratio of peak voltage to voltage) as high as 3. In addition, odd harmonics are generated that can cause high neutral currents in 3-phase circuits and other nonlinear effects (excessive heating, for example). These considerations make the selection and sizing of a UPS much more complex and important to ensure reliable, high quality power. UPSs that serve nonlinear loads should have a high crest factor rating and should handle 100% nonlinear load and at least a 50% phase imbalance without providing voltage out of its specified range with no more than 5% total harmonic distortion of the output. Noise-generating loads such as motors, SCR drives, etc., should not be on instrument UPSs. In addition, a UPS should have a maintenance bypass, and a complete bypass around the UPS should be considered. If bypasses are provided, unless properly designed, the UPS may not be providing "isolated" power under all conditions. If "isolated" power is required, the UPS output must provide a separately derived power source (see NEC 2002 Handbook, Article 250), and any reserve and/or bypass power sources must be isolated using transformers. Proper grounding is required to provide a separately derived "isolated" system. Bypasses can, however, provide noise paths that bypass the UPS. Even when the UPS is not on bypass, if a solid-state static switch is used, it may have snubber circuits, which may allow noise into the system. Power conditioning on the reserve and bypass power feeds should be provided to ensure quality power when on reserve or bypass. If the UPS feeds circuits outside the building it is installed in, additional transient protection should be provided for the circuits leaving the building.

Off-line UPSs must switch to battery power when the main power is lost. This switching causes a switching transient of varying time duration, depending on the UPS. Most modern equipment has enough hold-up time to ride through this switching. For some equipment, however, this can be a problem. Before purchase of this type of UPS, a test may be in order.

Some manufacturers recommend for 3-phase computer loads that the neutral be sized to twice the normal size and the supplying transformer be de-rated. This is due to possible additive harmonics in the neutral and the transformer. Neutral current can be as high as 1.73 times the phase current and, in some unusual conditions, as much as 2.73 times the phase current. Manufacturers have reported fires caused by overheated neutrals. While the neutral and transformers are receiving considerable attention, all the electrical equipment is affected by nonlinear loads. The distortion caused by nonlinear loading can affect the operation of other instruments. Transformers are available to handle nonlinear loads.

UPSs are soft or high impedance power sources. This means that they are not able to supply power as freely as a low impedance source such as a transformer. A transformer with 5% impedance can supply up to 20 times its full load current on a momentary surge; a UPS is typically limited to 1.25-1.5 times its full load current. This requires careful consideration when evaluating inrush currents from starting loads.

Maintenance of UPSs is important in ensuring reliable power. UPSs should be regularly exercised and proper battery operation and charging verified. Regular maintenance for batteries is a requirement to ensure proper operation. Upon installation, the UPS voltages, currents, and waveforms should be recorded for all phases. Currents should be measured with a true amp meter. Voltages, currents, and waveforms should be checked regularly (quarterly is a good schedule).

The new UPSs are microprocessor-based and provide considerable ability for diagnostics. UPSs with data monitoring and collection can offer many advantages for maintenance and operations. Also, some come with a modem connection so the manufacturer can troubleshoot problems. Others have the ability to network with computers to provide enhanced monitoring and data acquisition. Careful selection of these features can enhance the maintenance of the UPS for cost effectiveness.

A power line monitor is a good maintenance investment.

The failure of several UPS systems to provide power during an emergency at a nuclear power plant was attributed in part to improper UPS battery maintenance. In another incidence, a UPS had a design flaw in which the electronics operated off the reserve power; when the primary power came back up (but not the reserve after a power outage), the UPS would not come back on line. Proper testing of the UPS would have caught this flaw.

A true RMS amp meter should be used to measure any current where harmonics or nonlinear loading is suspected. A measure of the problem comes from a comparison of a regular clamp-on amp meter (not true) to the reading of a true meter. Hot temperatures are also a symptom of harmonic problems. Generally, the external surface of a piece of equipment is too hot if it cannot be held in the hand for a count of ten (approximately 50°C or 122°F). For terminals under 100 amps, the temperature should normally be no greater than 60°C. Temperatures are easily monitored by an inexpensive infrared temperature detector.

Power Conditioning

Power conditioning provides protection against electrical transients and noise entering the power system. In addition, some types of power conditioners can provide limited voltage regulation. Power conditioners range from UPSs to transformer types to simple types that contain metal oxide varistors (MOVs). Dozens of manufacturers sell these devices. Unfortunately, while there are standards for these devices to be tested to (such as UL 1449, UL 1283, and IEEE C62.41.1 – 2002 and C62.41.2 – 2002), the manufacturers' literature and sale pitches provide a murky picture of these devices and their abilities. The following may provide some guidance in selecting these devices:

(1) Does the application require any ride-through capability?

How long? Milliseconds, seconds, minutes?

(2) Can the application stand a switching transient when the alternate supply comes on line?

(3) Does the application require voltage regulation? Many modern loads that use switching power supplies can operate on a wide range of voltages. Linear power supplies are much more sensitive to voltage variation. Many salesmen will tell you that you don't need voltage regulation any more (the ones selling nonvoltage-regulating equipment) and others will tell you that you do (the ones selling voltage-regulating equipment). The facts are somewhere in the middle, and an analysis of your installation is required to properly apply this equipment. When in doubt, regulate!

(4) Where are the transients/noise coming from? Are they external or internal? Generally, the majority of the transients/noise in a facility are internally generated. These come from motor starts, switching transients, relays, arcs, SCR drives, variable speed drives, faults, welding, etc. External transients come from lightning and utility transients. Generally, internal transients are smaller, but the source is closer than external transients, which are typically larger but the source is farther away. Power conditioning location is dependent on the points of entrance of the transients. The type of conditioning depends on the size of the transients, the frequency, the equipment that is being protected, and the physical arrangements. Generally, power conditioning consists of staged protection with some transient source suppression. For example, a facility might be protected at the service entrance and at the distribution panels with parallel (shunt) transient suppressors. The service entrance suppressor reduces external transients entering the system from the service feed. The suppressor at the distribution panel further reduces external transients and also reduces any internal transients entering the panel. Past the distribution panel, an analysis of the loads must be done. Loads are divided in four general types: those that are generally insensitive to transients, those that are sensitive to transients, those that generate transients, and those that are combinations of the other three. The insensitive loads get no further protection. The loads that are sensitive to transients must be evaluated for cost versus criticality. No further protection or additional low cost protection may be added for noncritical loads. More critical loads may have series protectors, transformer-based conditioners, or even a UPS. Loads that generate transients should have transient suppressors, such as RC networks, diodes, etc., installed. Many surge suppressors work both ways and can be used on transient-generating loads to protect the power system. For loads that do both, bidirectional suppressors should be used to protect the power system. In some cases, isolating devices (such as isolation-type transformers) can also be used.

(5) What kind of transients are appearing? Size, shape, energy, frequency?

(6) What types of noises are occurring, and where are they coming from? Amplitude, frequency, common mode, normal mode, coupling mechanism?

(7) Are the wiring and grounding of the facility contributing to the problem? Creating the problem? It should be remembered that most manufacturer's power conditioning equipment specifications are based on tests under laboratory-type conditions and are presented in the most favorable light. In

practice, the equipment is seldom installed in a manner similar to the test installation and seldom performs as well under real-world conditions. In addition, the vendors, in an attempt to differentiate their products, provide a myriad of specifications that make comparing products a difficult task. Vendors should be held to task to prove they have met their specifications and the power conditioning requirements. Proper selection, location, and physical installation are critical to providing high quality power conditioning.

One vendor makes claim to having the best attenuation of a particular type of transient wave. When requested to provide evidence of what this claim was based on, the vendor said it was based on comparison of their specifications with other vendor specifications. Some proof! Let the buyer beware!

Surge protection devices such as metal oxide varistors (MOV), gas discharge tubes, and silicon avalanche diodes that shunt transients must be installed with leads as short as possible. As the frequency goes up, the inductance of the lead length can become the dominant impedance and a shunt becomes a high impedance, which defeats the purpose of the surge suppressor.

Grounding

Proper grounding is one of the most important factors in ensuring a good power system. Grounding is necessary for several reasons: safety, overcurrent protection as a system reference, lightning protection, static electricity reduction, and noise reduction. The National Electrical Code (NEC), specifically NEC Article 250, sets the minimum code requirements for grounding.

A ground is defined by NEC Article 100 as a conducting connection, whether intentional or accidental, between an electrical circuit or equipment and the earth or to some body that serves in place of the earth. NEC Article 250 provides the code basic requirements for grounding electrical circuits and equipment. Article 250 also defines what electrical systems are required to be grounded. For instrument systems Article 250 states that two-wire DC systems operating at 50 volts or less are not required to be grounded. Most instrument power sources are 120 or 120/208-V AC systems. 480-V AC systems are sometimes used to provide source power for 120/208-V AC instrument systems via step-down transformers. Article 250 requires that any system that can be grounded such that the maximum voltage from ground to the ungrounded conductors does not exceed 150 V AC are required to be grounded (for example, a 240-V AC circuit fed from a center tapped transformer). Three-phase, four-wire wye connected circuits where the neutral is used as a circuit conductor are required to be grounded. Article 250 also applies to instrument circuits by allowing separately derived systems used exclusively for rectifiers that supply only adjustable speed drives not to be grounded.

If you are having strange problems in your instrument power systems, it is probably due to improper grounding.

The common view that the NEC requirements for electrical systems are a "maximum" and sometimes unnecessary is incorrect. The NEC requirements are primarily experience-based practices with some engineering basis and are considered to provide a safe electrical installation. The NEC is primarily concerned with personnel and electrical equipment and device protection. As such they do not always cover all installations fully and do not always meet all the electrical requirements for an installation. The NEC must, therefore, be viewed as the electrical minimum requirements for an installation.

While the NEC allows some instrument circuits to be ungrounded, it is good practice to use grounded circuits. Properly grounded circuits provide a solid reference for electronic equipment to operate from and helps minimize electrical noise problems.

One of the most controversial subjects is the concept of a "quiet" or computer-grade ground. Some instrument, equipment, and computer manufacturers state that they require a "quiet" ground for proper operation of their equipment. This "quiet" ground is usually defined as a separate, independent ground that is dedicated to the instrument or computer system and is separate from the "dirty" power system ground. The problem is that this is a violation of NEC Article 250-81, which requires all grounding electrodes to be connected together to form the grounding electrode system. This requirement is a safety issue, and failure to follow the NEC can result in an unsafe system. Actually, the "quiet" ground is not all that "quiet," and the use of the isolated ground can sometimes cause more problems than it cures. A properly designed ground system can meet the NEC and provide reliable operation. A government standard on this is FIPS 94.

The "quiet" ground requirement probably arises because these systems are designed and many times tested in a laboratory type of environment far removed from a local refinery or petrochemical plant. These systems work well in the laboratory but they don't always work as well in the real world. One of the reasons is there is a "quiet" ground at the test site and a "dirty" ground at plant. The problem really isn't that the ground at the plant is "dirty," but rather the electrical environment (including the grounding system in the plant) is far more complex and subject to many more variabilities than the equipment test environment.

Vendors must design their equipment to deal with the real world, and the users must provide high quality power systems with good grounding systems.

One of the most common problems with grounding in a plant is the double grounding of the neutral. This generally occurs when a bonding jumper is installed in a distribution panel past the first panel or transformer where the circuit has been properly grounded. The problem is that all grounds are not created equal. The earth can be viewed using an analogy to choppy seas. This choppy, ever-changing sea of electrons has different amounts of electrons available at each point of the earth. If one connects two points of the earth (two separate grounds), the two points will be at different potentials to each other, and current will flow between the two points to equalize the charge at both points. Grounding the neutral in two different points will give the same effect, and ground currents will be circulating between the points even if they are relatively close together. These circulating currents can cause operational problems.

A plant had a 240-V AC center-tapped grounded circuit with no neutral to a machine. An electrician was asked to provide a 120-V AC circuit for some instrumentation. There was no 120-V AC distribution panel near by. The electrician had a bright idea and installed a receptacle between one of the phase wires and the ground wire, providing a 120-V AC circuit. This is obviously a violation of the NEC. It is also potentially unsafe since an interruption in the ground wire could energize the machine frame. This problem was detected when unreliable operation of the instrumentation occurred.

In DC power systems, the most common problem is the so-called "ground loop." This term can be misleading as sometimes earth ground is not involved. A "ground loop" is an undesirable current return path in an electrical circuit. Some common causes of "ground loops" are the use of internally grounded instruments inside a loop, shields grounded at both ends, grounded thermocouples with non-isolated transducers, and four-wire grounded transmitters as input to a receiver instrument grounded to a different ground connection. The symptoms of ground loops vary, with many of them being strange. One common one is to have strange voltage or current readings but when the loop is opened, the voltages are OK and the instruments check OK. Another is that some of the loops instruments read correctly and others do not but the individual instruments check OK.

Rack-mounted instruments that share a common power source are a good place for a ground loop to occur.

An instrument tech was called out because several temperature instruments were reading incorrectly. The instrument tech made some measurements and got some very strange readings, but when the individual instruments were checked out, they were OK. An instrument engineer was called in and further tests were made. It was discovered that the thermocouple transducers were non-isolated and the thermocouples were to be ungrounded. It was also discovered that one of the thermocouples was grounded. At first it was thought that the thermocouple had malfunctioned, but it was later determined that the thermocouple had just been recently replaced. The tech who had replaced the thermocouple had not been able to find an ungrounded one, so he had used a grounded one. When the thermocouple was installed in the reactor, it was cold, and the tech verified only that the thermocouple was reading ambient OK. When the thermocouple was heated to operating temperatures, the "ground loop" effects became noticeable. The moral of this story is that isolated transducers are good practice and, many times, are worth their weight in gold.

AC circuits can also have ground loops when there is ground leakage in equipment. Leakages can occur due to leakage through transient protection devices or through insulation such as with nichrome heaters.

By placing the source and return wires or the input and output wires in a clamp-on amp meter in such a manner that the input and output wires have the current flowing in the opposite directions, the ground loop current can be detected. If there is no leakage, the meter should read zero. If current is read, it is equal to the leakage current. This technique is used in many ground fault detectors and is similar to the one that is used for ground fault interrupters.

Ground leakage or loops can detected many times using a clamp-on amp meter.

Noise

Noise is defined as unwanted or undesirable electrical signals in an electrical circuit. Every electrical circuit has noise in it. Noise in a circuit becomes noticeable when the noise level approaches the wanted or desired signal level. Noise in a circuit may come from the outside via conduction or field radiation or from the circuit itself, which also generates noise. The key is to reduce the noise to a level that does not adversely affect the desired signal. The reduction of noise uses two general methods: reduction of noise from the source and reduction of intrusive noise into circuits. Fortunately, many of the techniques available for the reduction of noise work for both methods.

Noise generation can be reduced by preventing the noise from leaving the equipment that is generating the noise or the actual suppression of the noise. The prevention of noise from leaving a noise generator uses the same shielding techniques used for prevention of noise intrusion. These methods put some form of grounded metal shield between the noise generator and the outside world or some device that prevents noise from being conducted via wires to the outside world. Suppression of noise methods include arc suppression, arc removal (replacing electromechanical relays with solid-state relays, for example), separation, circuit redesign, equipment selection (use of zero-crossing firing SCRs instead of phase angle firing, for example), filtering, and isolation. Suppression of the source of noise should be looked at first because it reduces the total amount of noise propagating in the system and will generally be the most cost-effective method if the noise is affecting a number of instrument systems. Suppression of the noise may also reduce the mysterious or "things that go bump in of the night" types of problems.

Noise can enter a system by being conducted through wires or metal or coupled through electromagnetic fields (capacitively, inductively, or plane wave (radiated) coupled).

Interrupting the noise path is a common troubleshooting method for determining whether the problem is conducted noise. Changing the power source for a device, disconnecting a signal or communications path, or disconnecting a ground (note that a safety ground should not be disconnected without providing equivalent protection) are some of the ways to break the conducting path.

CONDUCTED NOISE

Conducted noise is noise that is conducted into the system via wires or metal surfaces and comes from sources such as switching transients, nonlinear devices such as switching power supplies and SCR drives, and noise coupled into the external system that enters the system through the system's connections with the external world (power, signal, communication circuits, etc.). Conducted noise is reduced by filtering or suppressing the noise that is entering on the incoming or outgoing wires. The filter may be as simple as a shunt capacitor or as complex as an active filter. Ferrite beads, common mode chokes, isolating transformers, power conditioners, optical isolators, and fiber optics are just a few of the methods that are used. Distance from the noise source also affects the strength of the conducted noise; i.e., the further from the noise source, the more the noise is reduced by the impedances between the source and the circuit in question. Many times, separate branch circuits are provided for sensitive equipment to take advantage of this effect. The key to successful reduction of conducted noise is to identify the type and strength of the noise that is causing the problem. If the noise generator(s) cannot be suppressed, the most cost-effective method of prevention of the noise entering the sensitive system should be supplied.

One of the computers in a system was experiencing mysterious resets and lockups. Investigation showed that one of the printers was connected to the same UPS that the computer used. Removing the printer from the UPS solved the problem, which was conducted noise from the printer into the UPS system and on to the computer system.

COUPLED NOISE

Coupled noise enters the system via the electromagnetic fields by capacitive, inductive, or plane wave coupling. Capacitive coupling can be viewed as being similar to a capacitor and is electric field-based. Inductive coupling can be viewed as being similar to a transformer and is magnetic field-based. Plane waves are both electric and magnetic field-based and are waves (for example, radio waves).

Capacitive coupled noise can be reduced by placing a grounded metal shield between the noise source and the protected circuit. Aluminum and copper are commonly used materials.

Inductive coupled noise can be reduced by placing a magnetic material shield between the noise source and the protected circuit. Carbon steel is a commonly used material, and a number of special magnetic materials are available for this purpose.

PLANE WAVE NOISE

Plane wave noise or radiated noise occurs when the coupled noise source is more than 1/6 of a wavelength away from the circuit in question. Common sources of this are walkie talkies, radio base stations, and high speed and high frequency circuits.

Plane wave noise can be reduced by placing a metal shield with no hole larger than 1/20 of a wavelength between the noise source and the protected circuit. It should be noted that antennas that are 1/4 wavelength long serve as effective antennas for that frequency, and as frequency goes up, wavelength and, hence, the antenna length goes down. Wires that are just wires at low frequencies become more effective antennas at high frequencies. Generally, the effectiveness of shielding is related to how good a conductor the shield is and the thickness of the shield. Generally, any metal instrument enclosure provides sufficient thickness for protection against plane waves.

Note that aluminum foil or copper braid shields used for cable shields do not provide any protection from magnetic-based noise. Aluminum conduit, aluminum conduit fittings, aluminum or 304 or 316 stainless steel enclosures, and copper MI cable also do not provide any protection against magnetic-based noise.

A cable shield for high frequencies was run into an enclosure, and the shield was opened and the drain wire run down two feet to a ground bus. The shield was ineffective because the drain wire served as an high impedance inductor at the frequencies involved. The drain wire should have been connected, with a path as short as possible, to a low inductive ground return path.

SHIELD GROUNDING

Shield grounding, many times presented as a complex subject, can actually be simple in practice. By following these rules, most loops can be protected by shielding:

Rule #1 — Shields at frequencies below 1 MHz should be grounded only at one point: the zero potential point of the circuit.

Rule #2 — Shields at frequencies above 10 MHz should be grounded in multiple points.

Rule #3 — Nonmagnetic materials such as aluminum foil are effective only against electrostatic (capacitive coupled) noise and provide limited protection against plane wave noise (a function of thickness of the shield).

Rule #4 — Magnetic material is effective against both electrostatic and magnetic (inductive coupled) noise.

Rule #5 — Twisted pairs are effective against magnetic noise and are a general good practice. The tighter the lay (more twists per inch), the more effective the twist.

Rule #6 — Standard grounded metallic instrument enclosures with no opening larger than 1/20 of the wavelength for the highest frequency involved will provide adequate protection against plane wave noise.

Rule #7 — Inductive effects become significant when the length of the wire is longer than 1/20 of the wavelength of the frequency in question.

Rule #8 — In an ungrounded instrument loop, Rule #1 should be followed. Earth grounding the shield does not improve the effectiveness of the shield; if anything, it decreases the effectiveness of the shield.

Rule #9 — The vulnerability of a circuit to coupled noise is related to the circuit loop exposed to the noise. Reduce the loop, reduce the noise.

For multiconductor shielded cables, the shield drain wire length should be kept to a minimum. This requirement, however, many times causes logistics problems for cables that are connected to a single ground or common bus in a cabinet. This may be overcome by connecting the shield drain wires together at the cable (using a compression terminal block) and then running the drain wires to the ground or common connection located as close as possible. This method creates a lower inductive path; i.e., multiple wire returns instead of single wire returns.

Twisted pair cable with individual foil shields and an overall shield run in steel conduit provides excellent noise protection.

Isolation

Isolation is a technique that is used as a means to segregate and protect sensitive electronic equipment. Transformers are used to provide power isolation. These can be standard transformers or can be specially designed isolation transformers that have additional shielding built in. These isolation transformers are used when the isolation is also used to minimize noise and transient intrusion. Isolation of ground return paths via a separately derived system can be accomplished with a standard transformer.

Signal isolation can be very important. Signal isolation is used to eliminate ground loops or to isolate one system from another. Common uses are to isolate grounded thermocouples, to isolate computer systems from the instrument system, to isolate two different systems that are communicating with each other, and to isolate two separately grounded systems. Experience indicates that using isolated transducers is a good practice and is seldom regretted.

A computer was talking via an RS-232 serial port to an RS-232/RS 485 converter and then to a group of controllers. The system exhibited intermittent bizarre behavior. A check with a line monitor indicated that the computer was sending and receiving correctly. A grounding problem was suspected, but the system did not appear to be incorrectly grounded. A check of the RS-232/RS-485 converter showed, however, that it was a non-isolated type. The converter was replaced by an isolated type and the problem cleared up. The computer was providing a ground, and the controllers provided a different ground, creating a ground loop that caused the problem. Moral of the story: since you have little control over how computers are built, it is good practice to provide isolation for your computer I/O from the real world.

Instrument Air Requirements

With all the electronic and electrical controls in place or on their way, one can get rid of the pesky high maintenance air compressors, right? Wrong! The prime movers for most control valves, damper drives, and fan louvers are still pneumatically operated. They are likely to be so for some time. Many current-to-pneumatic (I/P) and pneumatic-to-current (P/I) transducers are still in place, and many pneumatic control loops are still in service. Compressed air as a control necessity is going to be around for a while.

The two worst enemies of compressed air for control service are moisture and oil. Moisture is always a potential problem because the cooling of compressed air below its dew point causes moisture to condense. Oil can be dealt with more easily because of nonlubricated compressors. Precautions must still be taken to condition compressed air to be of instrument quality.

Types of compressors found in service for instrument air are rotary vane, rotary liquid ring, diaphragm, rotary screw, and centrifugal. Some reciprocating oiled piston compressors are still in service, but it would not be advisable to start a new installation with one because of the added concern for oil.

Compressor intakes should be well filtered and located away from contaminants. Each installation should be addressed for possible contaminants and proper scrubbers or absorbers installed to reduce ambient-imposed problems. Another source of internal corrosion is synthetic lubricants that attack seals and diaphragms.

The compression process causes heat. Inner coolers and after coolers are used to bring the temperature down, causing moisture to drop out as the temperature drops below the dew point. It is, therefore, necessary to provide air drying downstream of the compressor system. Types of air dryers are deliquescent, mechanical refrigeration, and regenerative desiccant.

The deliquescent type requires no utility but is limited in capacity (1 to 5000 at 100°F and 100 psig), and outlet dew point is 30°F or higher. Refrigeration types require electric power. They have a higher flow range of 0 to 10,000 at 100°F and 100 psig but are good only for dew points of only 35°F and higher.

Regenerative desiccant types come with heaters and without, and the motive utility is air; they can reach dew points of -40°F but can go lower if properly designed. Capacity is 0 to 10,000 at 100°F and 100 psig.

The dew point at line pressure should be at least 18°F (10°C) below the minimum temperature to which any part of the compressed air system will be exposed during any season of the year. In no case should the dew point at line pressure exceed approximately 35°F (2°C).

The selection of the proper type and size of air dryer must be based on the actual inlet and flow conditions under which the dryer is expected to operate. To assure continuous delivery of instrument air in compliance with the referenced standards, it is necessary to verify the quality of the system air. Tests would include oil content, dew point, particulates, and other contaminants. Tests should be conducted on initial start-up and retests routinely made at intervals depending upon local conditions.

Maximum total oil content, exclusive of noncondensibles, should be as close to zero as possible. Under no circumstances should it exceed one part per million on either a weight or volume basis.

Once a quality instrument air system has been established, it becomes the most expensive utility in the plant. It is, therefore, necessary to make sure the distribution system is leak-free. One report regarding leakage in industrial plants states that the average industrial plant has a 17% leak rate on compressed air fittings. There are two sides to the leakage problem: (1) piping supply headers and (2) tubing that is used to feed instruments off the header and signal tubing.

Care should be taken in the plumbed makeup of the air system. Screwed fitting technology has been around for over 150 years. It relies on the taper between the male and female threads, with a lubricant between them. Proper lubricant type and application determine success or failure of the joint.

Older tube fittings were of the flared type. Improvement of tube connections came with the development of compression tube fittings. They require proper tube dimensions and close tolerance on the fitting machining. The tube is inserted into the fitting and the compression nut is tightened, causing a ferrule to swage into the tube. Care must be taken to cut the tubing squarely and to remove any burrs caused by the cutting. It is also critical not to excessively tighten the compression nut. Compression fittings are available for stainless steel, copper, and plastic tubing. They are not interchangeable.

Communication Requirements

Communication requirements come in several forms. Two of the best understood but most troublesome are verbal and written communications between people in key roles. They were touched on earlier in the chapter, with emphasis on the importance of communicating in non-intimidating terms. The important mission here is to reduce technology to its simplest forms so true communication can take place. A third requirement relates to being able to communicate throughout the plant as maintenance functions are performed.

The fourth and newest communication requirement relates to the connectivity between hardware and software components. When life was simpler, this meant standardizing on a 3-15 psig pneumatic signal level, a 4-20 or 10-50 milliamp electric signal level, or a 1-5 volt signal. Those were the "good old days" in terms of simplicity, but these were not nearly as powerful as communication linkages today.

The communication methods of today are a multiplicity of network protocols, many of which are similar—but not exact. There are terms like Ethernet™, ControlNet, token ring, star, TCP/IP, 10 BaseT, local area networks, wide area networks, thicknet, thinnet, bridges, gateways, fieldbus protocol, open system architecture, and more—all of which have an application and a meaning. Volumes are written about them and some are still being written.

Heating, Ventilating, Cooling, and Air Conditioning Systems

Maintenance for Equipment Cooling Systems

Proper maintenance techniques can best be classified as the art of maintenance. As in all forms of art, some fundamental requirements are necessary to understand your subject and the tools of your trade. The maintenance mechanic must have a conceptual understanding of the proper operation of the maintained systems. In addition, he must have proper diagnostic equipment and tools at his disposal. Regular training must also be part of a maintenance program.

The goal here is to provide an overview of the maintenance knowledge required for air conditioning used for equipment cooling. Additionally, some typical maintenance checklist items for various types of equipment found in the industrial setting are presented.

To begin, let us review the vapor compression refrigeration cycle.

The evaporator absorbs heat and the condenser rejects heat. The refrigerant is a gas on the left side and at low pressure. The refrigerant of the right side is at a much higher pressure and is a liquid phase.

A metering device (orifice) allows the refrigerant to go from high pressure liquid to low pressure gas. According to the rules of physics, when a liquid changes state to a gas, heat is added to the process. All of us have experienced this vaporization process, when an alcohol swab is used on the arm before an injection or blood sample. That cool feeling where your arm is wiped is not because the alcohol is cold, as the alcohol is at room temperature in the closed container, but rather, the process of your arm giving heat (reducing your arm temperature) to the alcohol vaporization process.

We sustain the vaporization process by compressing the vapor refrigerant into liquid with the compressor. (Heat quantitatively is referred to as Q in the following equations.) The evaporator and condenser are similar in construction having tubes with larger fins attached to increase the surface area of the tubes. The ability of the evaporator to gain heat is a function of the surface area of the evaporator, the temperature of the air blowing across the evaporator and the volume of air passing over the evaporator (mass flow rate). Temperature differential is the temperature difference between the air passing over a cooling coil and the temperature of the cooling coil.

$$Q \text{ heat gain} = \text{Surface area} \times \text{Temperature differential} \times \text{Mass flow}$$

The condenser operates at higher temperature than the ambient air temperature and rejects heat to the air. The ability to reject heat relies on the same criteria as the evaporator, surface area, temperature difference of entering and leaving air, and the mass flow of air.

Schematically now we have three heat flow systems in series from the evaporator to the condenser.

When designing cooling systems, the designer has to know the cooling load Q input and the air temperature. (This information is a part of the design data.) Small air conditioning systems are rated in Btu's (British thermal units). (One Btu is the heat energy required to raise one pound of water 1 degree Fahrenheit at room temperature.) Once the design heat flow, Q, is known, the designer must know the design condition for the air receiving the rejected heat. From the difference between the entering evaporator air temperature and the entering condenser air temperature the designer determines temperature differential. This differential is how much temperature increase the refrigerant system must impart to the condenser to cause heat flow away from the condenser, the rise in energy level is "thermodynamic lift."

The compressor is sized to impart sufficient energy to the gas to change the phase to liquid at high temperature and pressure. The compressor is selected to meet a specific work load of refrigerant flow and thermodynamic lift. Why is it important to know about compressor selection? If we ask the compressor to work harder than the design ability the compressor will fail. How do we ask the compressor to work above design? When the condenser fins become loaded with debris, air flow through the condenser is reduced. For proper heat flow to continue the compressor must provide higher temperatures requiring higher pressure and more compressor work.

Other typical problems are a dirty evaporator that reduces cooling available. In the case of an equipment cooling system, this reduced air flow means higher operating temperatures for your equipment and more heat related stress and failure.

Loss of refrigerant will reduce the cooling capacity. Typical results include freezing of the evaporator and if the refrigerant charge is sufficiently low, the loss of lubrication will cause compressor failure (the refrigerant has small amounts of lubricant suspended in the refrigerant). Lack of design refrigerant flow will cause the oil to settle out of the gas and remain in the evaporator and condenser.

If your cooling equipment operates when outside air temperature conditions are below 40 degrees Fahrenheit, the air conditioning should be fitted with a system of "low ambient control." This system may turn off the condenser fan or operate dampers to restrict air flow through the condenser. Make a thorough checkout of the low ambient control early in the cooler seasons.

Hot gas bypass is a technique used to keep the compressor running when only small cooling loads are present. The refrigerant bypasses the evaporator and returns to the inlet of the compressor. This technique allows the air conditioning to operate without damage to the compressor. However, energy costs will be higher as the compressor runs under light load.

There are numerous other technical innovations used in the cooling systems for equipment cooling. The purpose here is to provide an overview of the concepts for cooling system operation.

You should now understand the critical nature of the design for heat flows through the cooling system. Anything that impairs the ability of the airflows to the cooling units will degrade the performance of our systems. Likewise, anything that impairs the ability of the finned coils to pass the required heat flows will also degrade the system operation. The compressor motor is sized to produce a required level of work. If we require the compressor to deliver work in excess of design we begin to experience random compressor failures.

The biggest source of cooling equipment failure is the failure to keep the cooling system within the parameters of the designer. The following are some recommended procedures for proper maintenance of typical cooling equipment. Specific guidance for maintenance can be found in the installation documents for your cooling equipment. Should the installation documents not exist, call the vendor and ask for another copy.

The discussion here will not enable you to perform major maintenance on cooling systems. The purpose is to expose you to the concepts of maintenance. Proper maintenance begins through system knowledge.

TYPICAL MAINTENANCE TASKS

COMPUTER ROOM UNIT

SAFETY NOTE: BE SURE TO WEAR ADEQUATE EYE PROTECTION ANY TIME YOU ARE WORKING AROUND COOLING SYSTEMS.

CHECK AND RECORD COMPRESSOR AMPERAGE AND VOLTAGE

COMPRESSOR #1 Ll________ L2________ L3________
L1-L2________ L2-L3________ L3-L1________

CHECK SIGHT GLASS, IF APPLICABLE, FOR PRESENCE OF MOISTURE AND PROPER GAS CHARGE (NO GAS BUBBLES)
CHECK COMPRESSOR OIL LEVEL SIGHT GLASS. IF INSTALLED.
CHECK OUT SAFETY SHUT DOWN SYSTEMS, SUCH AS SMOKE DETECTION.

SHUT DOWN UNIT AND LOCKOUT ELECTRICAL CIRCUIT. THEN MAKE THE FOLLOWING TESTS:
CHECK AIR FILTER, REPLACE AS NECESSARY
CHECK REFRIGERANT LINES AND CAPILLARY TUBES FOR SIGNS OF WEAKNESS CAUSED BY VIBRATION OR RUBBING.
CHECK DISCONNECTS AND STARTERS, TIGHTEN ELECTRICAL CONNECTIONS.
INSPECT MOTOR STARTER CONTACTS FOR EXCESSIVE WEAR, PITTING, DISCOLORATION AND FOR PROPER ALIGNMENT, REPLACE AS REQUIRED. CHECK COMPRESSOR MOUNTS OR ISOLATORS FOR TIGHTNESS OR DEFECTS.
INSPECT REFRIGERANT LINE INSULATION AND REPAIR WHERE POSSIBLE. BE CERTAIN TO DISCONNECT HIGH VOLTAGE BEFORE SPRAYING EQUIPMENT WITH WATER.
INSPECT COILS FOR DAMAGE, STRAIGHTEN ANY DAMAGED FINS WITH A FIN COMB. CLEAN COIL WITH A POWER CLEANER AND CLEAN OTHER DEBRIS OR GREASE FROM THE UNIT. TREAT OR REPAIR ANY AREAS OF CORROSION.
INSPECT COMPRESSOR AND TREAT FOR CORROSION AS REQUIRED. INSPECT EXTERIOR AND INTERIOR OF FAN SYSTEM.
INSPECT INSULATION AND REPORT CONDITION.
CLEAN FAN BLADES, CHECK TIGHTNESS OF SET SCREWS ON FAN HUB. CLEAN MOTOR VENTILATION PASSAGES.
LUBRICATE FAN AND MOTOR BEARINGS PER MANUFACTURER'S INSTRUCTIONS. INSPECT SEALS, RETAINERS AND SLINGS.
INSPECT MOTOR MOUNTS, SUPPORTS, ISOLATORS AND COUPLINGS FOR TIGHTNESS OR DEFECTS.
INSPECT BELTS FOR INDICATION OF OVERLOADING, MISALIGNMENT OR IMPROPER BELT TENSION. ADJUST OR REPLACE AS REQUIRED.
CHECK PULLEYS FOR WEAR AND TIGHTNESS. NOTE ABNORMAL CONDITIONS.
RESTORE POWER AND ALLOW THE UNIT TO STABILIZE RUNNING CONDITIONS.
MEASURE MOTOR AMPERAGE AND RECORD:

L1 ________ L2 ________ L3 ________

CHECK VOLTAGE ACROSS PHASES

L1-L2 ________ L2-L3 ________ L3-L1 ________

COMPARE WITH NAMEPLATE RATING.
INSURE ALL PANELS ARE CLOSED AND VERIFY THE UNIT IS OPERATIONAL.

8 Temperature Devices

Overview

Temperature may be measured with the basic thermometer to the more advanced nuclear temperature measurement. In most cases, the temperature element voltage or resistance has to be amplified and converted to a current for process measurement. The actual temperature, the accuracy, the application, and the speed of response will be the determining reason on what type of temperature measurement is used.

Temperature Sensors

Temperature measurement is not always what is expected. Although the sensing element can be accurate to within a tenth of a degree, the measurement of a gas or liquid temperature may be off by many degrees. The mounting and location of a temperature sensor is extremely important. If a temperature reading does not agree with a calculated temperature or simply does not agree with other temperature sensors in a system, the following should be considered: check the calibration of the transmitter. If the transmitter uses a filled temperature bulb, check actual sensor calibration with a field "dry block" calibration device or a liquid bath. If the sensor is an RTD or thermocouple, first check the output of the transmitter by using a temperature simulator/calibrator, and remember to correct for the reference junction compensation if the device is not automatically compensated. If the loop calibration is still incorrect, check the actual sensor with a field "dry block" calibration device.

If the results of this calibration do not provide a correct reading, consider the following suggestions:

(1) Is the sensor compatible with the process? Some thermocouples deteriorate in oxidizing atmospheres but not in reducing atmospheres and vice versa.

(2) Is the sensor installed in a location that is conductive to temperature pickup? Any wall effects when used on heated or cooled tanks?

(3) Is there a thermowell effect? Lag due to a thick thermowell?

(4) Is the temperature sensor bottomed out properly against the end of the thermowell? Does the sensor fit too loosely in the well?

(5) Are the well and sensor inserted far enough into the line or tank for proper temperature pickup? Insertion length is very important. These dimensions are listed in most temperature books or in vendor's installation pamphlets.

(6) If the sensor is a filled-tube system, is there a crimp or leak in the capillary, causing errors or loss of signal? Check spring linkages in the transmitter or receiver.

(7) If the measurement is electrical, are all the connections tight? Is the correct thermocouple type being used and does the transmission wire match the thermocouple type? Check the color code of transmission wire and sensor wires. The red wire is always negative. Is 3-lead extension wire being used when long runs are used? Is the element grounded or ungrounded? Is the correct calibration curve being used for the type of T/C or RTD?

(8) Check for loose tubing fittings or electrical wiring problems. Follow normal transmission line troubleshooting routines.

(9) If a wire-wound ceramic tip type of RTD is used in an agitated vessel or where vibration is a problem, the wire-wound sensor tip can easily be broken. Try etched or ribbon-type sensor elements that can withstand severe vibration.

Special thermocouple calibrators have built-in reference junction compensation, but using a millivolt meter for calibrating thermocouples connected to ordinary receiving instruments having copper input terminals requires reference junction compensation. Add the equivalent millivolt value of the terminal temperature to the millivolt reading of the thermocouple at this connection. This value for the terminal temperature may be taken from the mV/temperature tables for the type of thermocouple being used. Conversely, when a nonthermocouple instrument is being used as a millivolt source, a value corresponding to the temperature of the instrument terminals must be subtracted from the setting of the millivolt source in order to arrive at the actual input to the thermocouple receiver. (A simple way to remember to add or subtract is to remember "subtract when sending" or "S = S.")

There are several types of noncontacting temperature measurement devices; infrared and pyrometry are the most common. The most common problems with these devices is with selection of the proper unit for the measurement being made. Emissivity, lens focus, and distance between the source and the measurement must be considered. The noncontacting temperature-sensing devices can be very handy for applications in which a probe is not practical. Temperature ranges can be quite high and accuracies are excellent when properly used.

Types of Temperature Sensors

Noble Metal Thermocouple
Base Metal Thermocouple
Resistance Temperature Detectors
Thermistors
Integrated Circuit Sensors
Infrared Temperature Transducers
Optical Fiber Thermometry
Thermometer
Radiation Pyrometer
Filled Thermometers
Bi-metallic
Temp-Tips

Selection of a Type

The selection of a type of transducer is determined by the application. The following questions should be considered.

- Upper temperature limit
- Temperature range
- Atmosphere
- Tolerances/accuracy
- Speed of response
- Resolution (change per degree)
- Space availability
- Heat transfer area
- Grounding considerations
- Velocity and properties of the fluid or atmosphere
- Fluid or gas pressure
- Lead length (primary element to indicator or transmitter)
- Number of connections
- Type of output (resistance, current, voltage) into what impedance
- Averaging or difference applications
- Linearity required
- Costs
- Simplicity acceptable

Placement of Sensor (Primary Element)

A desirable measuring location must be selected. It should see exactly the same heat as does the process, follow all temperature fluctuations at exactly the same rate as the process, and should in itself cause no disturbance in the process temperature.

- Tip must be in the hot zone (atmosphere, liquid or contract to be measured).
- Primary element in a well must have contact with the end of the well. A spring loaded head should be used. Response will be slower because of the mass of the well.
- The connection between the sensor and the extension wire or transmitter is done in the gradient area. If the gradient area is quite warm, cast iron heads should be used and transmitters should not be placed in the head.

Speed of Response

Speed of response must be considered in processes where immediate temperature changes can affect the product or safety. The following affect the speed of response:

- Size of the wire at the TC junction. The type of junction (exposed, grounded or ungrounded) and the size of the sheath determine the speed of response of thermocouples.
- Size of the resistor, type of tip and size of the sheath determine the speed of response of RTDs. Generally RTDs are 1/3 slower responding than thermocouples or thermistors. Thin film RTDs are generally faster responding than wire wound RTDs. If the sensor is placed in a well, response time will be slowed. The junction must press firmly on the base of the thermowell since RTDs average temperature over the bottom two or more inches of the sheath, response will be slower than thermocouples.

- A thermistor is very sensitive and will measure small changes with very fast response, but have a very narrow temperature range.
- Integrated Circuit Temperature Transducers are slow to respond.
- Infrared Temperature Transducer is a non-contact measurement; output response time is 10 ms to 10 sec.
- Radiation Pyrometers are non-contact measuring devices and therefore are slow in responding.
- Fiber Optic Infrared Sensors are slow in responding.
- Fiber Optic Thermometry is capable of responding to temperature changes up to 50,000 times per second.
- Thermometers are generally slow in responding because they are put in large areas where the ambient temperature is slow changing or in large volumes of liquid.
- Bi-metallic temperature devices are very slow in responding.
- Tem-tips are a temperature device, which is inserted in hot molten metal and sent to a recording device. The speed of response is very fast.

Temperature Range

The temperature of the process determines the type of sensor that should be used. We must measure sub-zero temperatures to temperatures above 2000°C.

- Very high temperatures (above 1000°C.)
  Infrared Temperature Transducers (Pyrometers)
  Optical fiber thermometry
  Noble Metal Thermocouples (Type B, R, S)
  Tungsten Thermocouples (Type W, W3, W5)
- High Temperatures (200°C to 1200°C)
  Infrared Temperature Transducer
  Type R or S Noble Metal Thermocouples
  Type K or N Base Metal Thermocouples
- Medium High Temperature (100°C to 800°C)
  Type E, K or N Base Metal Thermocouple
  Type R or S Noble Metal Thermocouple
- Medium Temperature (0°C to 700°C)
  Type E, J, K, or T Base Metal Thermocouple
  Platinum RTD 100 ohm
- Medium Low Temperature (0°C to 350°C)
  Type E, J, K or T Base Metal Thermocouple
  Platinum RTD 100 ohm
  Platinum RTD 1000 ohm
- Low Temperatures (-50°C to 200°C)
  Type E or T Base Metal Thermocouple
  Platinum RTD 100 or 1000 ohm
  Balco RTD
  Thermistor
  Semiconductor Devices

- Subzero Temperatures (-200°C to 0°C)
  Type E or T Base Metal Thermocouples
  Platinum RTD 100 or 1000 ohms

Accuracy

Accuracy of the primary element (sensor) is generally the limiting factor on the accuracy of the measurement. In order to improve accuracy of the system, the following ideas are presented:

- Purchase the proper type of sensor for the application.
- Use the proper extension wire.
- Run extension wire in conduit and raceways reserved for low level signals.
- Parallel and average multiple selected thermocouples.
- Limit the distance and number of connections of the extension wire.
- Location of the reference junction must be where the thermocouple or thermocouple extension wire meets the copper-copper extension wire. This temperature must be stable and measured.
- Use high impedance measuring devices.
- If RTD's are used, insure use of low transducer excitation and use only 3 or 4 wire circuits designed for lead length compensation.
- Infrared temperature transducers (pyrometers) must be pointed at black body (non-reflective surface).

Connections

Connections induce errors especially when using RTDs. Lead length, small wire size and poor connections add resistance to the circuit. Changes in the resistance of the RTD loop will give improper temperature readings. Connections loosen and corrode after time. The differences in materials (wire and screws) cause corrosion, especially in harsh atmospheres. Soft metals connections loosen, especially gold, aluminum and copper.

Transmitters

Transmitters are used when the measurement is to be transmitted a long ways or when the type of signal has to be changed (millivolt to current, resistance to current). Transmitters should be located near the primary element without being in a hot gradient area. The reference junction compensation is normally done at the transmitter input. The temperature must be stable and measured.

Transmitters have specifications and tolerances and add error to the loop. They could add less error than error caused by noise and resistance. Temperature transmitters should have high input impedance. RTD transmitters should have low transducer excitation voltage and still supply enough power. RTD transmitters should accept three or four lead RTDs and have a three- or four-wire bridge with lead length compensation network or a four wire potentiometric circuit with lead length compensation.

Input Devices

Input devices, controllers, recorders, indicators, and DCS systems have specifications and tolerances, as do primary element sensors, and contribute to the total accuracy of the loop. Input devices must have high input impedance. For RTD devices, they must have lead compensation and accept 3 or 4 wire sensors and low excitation voltage.

Pneumatic devices are generally the least accurate, followed by analog devices, with microprocessor input devices and optical fiber thermometry being the most accurate.

Thermocouples

Thermocouples are simple, inexpensive, durable, and relatively accurate sensors that can be used in a wide variety of applications and environmental conditions. They have a wide measurement range, covering temperatures from below –183°C to about 2500°C with an accuracy of 0.1 to 1% of absolute temperature. Most thermocouples are made up of metal wires assembled in a variety of styles to provide direct or differential temperature measurements.

Thermocouples have a rapid dynamic response to temperature change. However, for protection and ease of calibration and removal, most thermocouples are installed in insulated thermowells, which delays dynamic response considerably. One thing to remember in installing thermocouples is that red is the negative lead.

Types

About a dozen types of thermocouples are commonly used in industrial application. Seven of these have been assigned letter designations by ISA. By convention, a slash mark is used to separate the materials of each thermocouple wire. For example, copper/constantan identifies a thermocouple with one copper wire and one constantan wire. The order in which there wire materials are listed identifies the polarity of the wires. The first wire, on the left of the slash, has a positive polarity when the measuring junction is a higher temperature than the reference junction. To assist in visual identification of wires and to avoid inadvertent cross wiring, many thermocouple wires and extension wires are color-coded. The color red is always used to identify the negative lead. (See Table 8-1).

Table 8-1. Color Code — Duplex-Insulated Thermocouple Wires

Thermocouple			Color of Insulation		
Type	Positive	Negative	Overall*	Positive*	Negative
E	EP	EN	Brown	Purple	Red
J	JP	JN	Brown	White	Red
K	KP	KN	Brown	Yellow	Red
T	TP	TN	Brown	Blue	Red

* A tracer color of the positive wire code color may be used in the overall braid.

Thermocouples can be divided into three functional classes; base metal, noble metal and refractory metal.

Base metal thermocouples are useful for measuring temperatures under 1000°C. This class includes thermocouples made of iron/constantan (Type J), copper/constantan (Type T), Chromel-Alumel™ (Type K), Chromel™/constantan (Type E), and alloys of copper, nickel, iron, chromium, manganese, aluminum, and other elements.

Noble metal thermocouples are useful to about 2000°C. This class includes tungsten/rhenium alloy thermocouples as well as those made of tantalum, molybdenum, and other alloys. Reference temperature tables represent the thermoelectric voltage in millivolts. (See Table 8-2).

Table 8-2. Typical Temperature Table

TEMPERATURES IN DEGREES CELSIUS (IPTS-68) REFERENCE JUNCTIONS AT 0°C

THERMOELECTRIC VOLTAGE IN MILLIVOLTS

DEG C	0	1	2	3	4	5	6	7	8	9	10	DEG C
-210	-8.096											-210
-200	-7.890	-7.912	-7.934	-7.955	-7.976	-7.996	-8.017	-8.037	-8.057	-8.076	-8.096	-200
-190	-7.659	-7.683	-7.707	-7.731	-7.755	-7.778	-7.801	-7.824	-7.846	-7.868	-7.890	-190
-180	-7.402	-7.429	-7.455	-7.482	-7.508	-7.533	-7.559	-7.584	-7.609	-7.634	-7.659	-180
-170	-7.122	-7.151	-7.180	-7.209	-7.237	-7.265	-7.293	-7.321	-7.348	-7.375	-7.402	-170
-160	-6.821	-6.852	-6.883	-6.914	-6.944	-6.974	-7.004	-7.034	-7.064	-7.093	-7.122	-160
-150	-6.499	-6.532	-6.565	-6.598	-6.630	-6.663	-6.695	-6.727	-6.758	-6.790	-6.821	-150
-140	-6.159	-6.194	-6.228	-6.263	-6.297	-6.331	-6.365	-6.399	-6.433	-6.466	-6.499	-140
-130	-5.801	-5.837	-5.874	-5.910	-5.946	-5.982	-6.018	-6.053	-6.089	-6.124	-6.159	-130
-120	-5.426	-5.464	-5.502	-5.540	-5.578	-5.615	-5.653	-5.690	-5.727	-5.764	-5.801	-120
-110	-5.036	-5.076	-5.115	-5.155	-5.194	-5.233	-5.272	-5.311	-5.349	-5.388	-5.426	-110
-100	-4.632	-4.673	-4.714	-4.755	-4.795	-4.836	-4.876	-4.916	-4.956	-4.996	-5.036	-100
-90	-4.215	-4.257	-4.299	-4.341	-4.383	-4.425	-4.467	-4.508	-4.550	-4.591	-4.632	-90
-80	-3.785	-3.829	-3.872	-3.915	-3.958	-4.001	-4.044	-4.087	-4.130	-4.172	-4.215	-80
-70	-3.344	-3.389	-3.433	-3.478	-3.522	-3.566	-3.610	-3.654	-3.698	-3.742	-3.785	-70
-60	-2.892	-2.938	-2.984	-3.029	-3.074	-3.120	-3.165	-3.210	-3.255	-3.299	-3.344	-60
-50	-2.431	-2.478	-2.524	-2.570	-2.617	-2.663	-2.709	-2.755	-2.801	-2.847	-2.892	-50
-40	-1.960	-2.008	-2.055	-2.102	-2.150	-2.197	-2.244	-2.291	-2.338	-2.384	-2.431	-40
-30	-1.481	-1.530	-1.578	-1.626	-1.674	-1.722	-1.770	-1.818	-1.865	-1.913	-1.960	-30
-20	-0.995	-1.044	-1.093	-1.141	-1.190	-1.239	-1.288	-1.336	-1.385	-1.433	-1.481	-20
-10	-0.501	-0.550	-0.600	-0.650	-0.699	-0.748	-0.798	-0.847	-0.896	-0.945	-0.995	-10
-0	0.000	-0.050	-0.101	-0.151	-0.201	-0.251	-0.301	-0.351	-0.401	-0.451	-0.501	-0
0	0.000	0.050	0.101	0.151	0.202	0.253	0.303	0.354	0.405	0.456	0.507	0
10	0.507	0.558	0.609	0.660	0.711	0.762	0.813	0.865	0.916	0.967	1.019	10
20	1.019	1.070	1.122	1.174	1.225	1.277	1.329	1.381	1.432	1.484	1.536	20
30	1.536	1.588	1.640	1.693	1.745	1.797	1.849	1.901	1.954	2.006	2.058	30
40	2.058	2.111	2.163	2.216	2.268	2.321	2.374	2.426	2.479	2.532	2.585	40
50	2.585	2.638	2.691	2.743	2.796	2.849	2.902	2.956	3.009	3.062	3.115	50
60	3.115	3.168	3.221	3.275	3.328	3.381	3.435	3.488	3.542	3.595	3.649	60
70	3.649	3.702	3.756	3.809	3.863	3.917	3.971	4.024	4.078	4.132	4.186	70
80	4.186	4.239	4.293	4.347	4.401	4.455	4.509	4.563	4.617	4.671	4.725	80
90	4.725	4.780	4.834	4.888	4.942	4.996	5.050	5.105	5.159	5.213	5.268	90
100	5.268	5.322	5.376	5.431	5.485	5.540	5.594	5.649	5.703	5.758	5.812	100
110	5.812	5.867	5.921	5.976	6.031	6.085	6.140	6.195	6.249	6.304	6.359	110
120	6.359	6.414	6.468	6.523	6.578	6.633	6.688	6.742	6.797	6.852	6.907	120
130	6.907	6.962	7.017	7.072	7.127	7.182	7.237	7.292	7.347	7.402	7.457	130
140	7.457	7.512	7.567	7.622	7.677	7.732	7.787	7.843	7.898	7.953	8.008	140
150	8.008	8.063	8.118	8.174	8.229	8.284	8.339	8.394	8.450	8.505	8.560	150
160	8.560	8.616	8.671	8.726	8.781	8.837	8.892	8.947	9.003	9.058	9.113	160
170	9.113	9.169	9.224	9.279	9.335	9.390	9.446	9.501	9.556	9.612	9.667	170
180	9.667	9.723	9.778	9.834	9.889	9.944	10.000	10.055	10.111	10.166	10.222	180
190	10.222	10.277	10.333	10.388	10.444	10.499	10.555	10.610	10.666	10.721	10.777	190
200	10.777	10.832	10.888	10.943	10.999	11.054	11.110	11.165	11.221	11.276	11.332	200
210	11.332	11.387	11.443	11.498	11.554	11.609	11.665	11.720	11.776	11.831	11.887	210
220	11.887	11.943	11.998	12.054	12.109	12.165	12.220	12.276	12.331	12.387	12.442	220
230	12.442	12.498	12.553	12.609	12.664	12.720	12.776	12.831	12.887	12.942	12.998	230
240	12.998	13.053	13.109	13.164	13.220	13.275	13.331	13.386	13.442	13.497	13.553	240
250	13.553	13.608	13.664	13.719	13.775	13.830	13.886	13.941	13.997	14.052	14.108	250
260	14.108	14.163	14.219	14.274	14.330	14.385	14.441	14.496	14.552	14.607	14.663	260
270	14.663	14.718	14.774	14.829	14.885	14.940	14.995	15.051	15.106	15.162	15.217	270
280	15.217	15.273	15.328	15.383	15.439	15.494	15.550	15.605	15.661	15.716	15.771	280
290	15.771	15.827	15.882	15.938	15.993	16.048	16.104	16.159	16.214	16.270	16.325	290
300	16.325	16.380	16.436	16.491	16.547	16.602	16.657	16.713	16.768	16.823	16.879	300
310	16.879	16.934	16.989	17.044	17.100	17.155	17.210	17.266	17.321	17.376	17.432	310
320	17.432	17.487	17.542	17.597	17.653	17.708	17.763	17.818	17.874	17.929	17.984	320
330	17.984	18.039	18.095	18.150	18.205	18.260	18.316	18.371	18.426	18.481	18.537	330
340	18.537	18.592	18.647	18.702	18.757	18.813	18.868	18.923	18.978	19.033	19.089	340
DEG C	0	1	2	3	4	5	6	7	8	9	10	DEG C

Type B Noble Metal Thermocouple

Type B thermocouples are Platinum–30% Rhodium (+) versus Platinum-6% Rhodium (-). The temperature range is from 0 to 1820°C (0–3300 degree F). A usable temperature range is 870 to 1700°C (1600 to 3100°F). Output is 0.000 to 13.814 mv. ANSI color code is BP (+) grey and BN (-) Red.

Table 8-2. Typical Temperature Table (cont'd.)

TEMPERATURES IN DEGREES CELSIUS (IPTS-68) REFERENCE JUNCTIONS AT 0°C

THERMOELECTRIC VOLTAGE IN MILLIVOLTS

DEG C	0	1	2	3	4	5	6	7	8	9	10	DEG C
350	19.089	19.144	19.199	19.254	19.309	19.364	19.420	19.475	19.530	19.585	19.640	350
360	19.640	19.695	19.751	19.806	19.861	19.916	19.971	20.026	20.081	20.137	20.192	360
370	20.192	20.247	20.302	20.357	20.412	20.467	20.523	20.578	20.633	20.688	20.743	370
380	20.743	20.798	20.853	20.909	20.964	21.019	21.074	21.129	21.184	21.239	21.295	380
390	21.295	21.350	21.405	21.460	21.515	21.570	21.625	21.680	21.736	21.791	21.846	390
400	21.846	21.901	21.956	22.011	22.066	22.122	22.177	22.232	22.287	22.342	22.397	400
410	22.397	22.453	22.508	22.563	22.618	22.673	22.728	22.784	22.839	22.894	22.949	410
420	22.949	23.004	23.060	23.115	23.170	23.225	23.280	23.336	23.391	23.446	23.501	420
430	23.501	23.556	23.612	23.667	23.722	23.777	23.833	23.888	23.943	23.999	24.054	430
440	24.054	24.109	24.164	24.220	24.275	24.330	24.386	24.441	24.496	24.552	24.607	440
450	24.607	24.662	24.718	24.773	24.829	24.884	24.939	24.995	25.050	25.106	25.161	450
460	25.161	25.217	25.272	25.327	25.383	25.438	25.494	25.549	25.605	25.661	25.716	460
470	25.716	25.772	25.827	25.883	25.938	25.994	26.050	26.105	26.161	26.216	26.272	470
480	26.272	26.328	26.383	26.439	26.495	26.551	26.606	26.662	26.718	26.774	26.829	480
490	26.829	26.885	26.941	26.997	27.053	27.109	27.165	27.220	27.276	27.332	27.388	490
500	27.388	27.444	27.500	27.556	27.612	27.668	27.724	27.780	27.836	27.893	27.949	500
510	27.949	28.005	28.061	28.117	28.173	28.230	28.286	28.342	28.398	28.455	28.511	510
520	28.511	28.567	28.624	28.680	28.736	28.793	28.849	28.906	28.962	29.019	29.075	520
530	29.075	29.132	29.188	29.245	29.301	29.358	29.415	29.471	29.528	29.585	29.642	530
540	29.642	29.698	29.755	29.812	29.869	29.926	29.983	30.039	30.096	30.153	30.210	540
550	30.210	30.267	30.324	30.381	30.439	30.496	30.553	30.610	30.667	30.724	30.782	550
560	30.782	30.839	30.896	30.954	31.011	31.068	31.126	31.183	31.241	31.298	31.356	560
570	31.356	31.413	31.471	31.528	31.586	31.644	31.702	31.759	31.817	31.875	31.933	570
580	31.933	31.991	32.048	32.106	32.164	32.222	32.280	32.338	32.396	32.455	32.513	580
590	32.513	32.571	32.629	32.687	32.746	32.804	32.862	32.921	32.979	33.038	33.096	590
600	33.096	33.155	33.213	33.272	33.330	33.389	33.448	33.506	33.565	33.624	33.683	600
610	33.683	33.742	33.800	33.859	33.918	33.977	34.036	34.095	34.155	34.214	34.273	610
620	34.273	34.332	34.391	34.451	34.510	34.569	34.629	34.688	34.748	34.807	34.867	620
630	34.867	34.926	34.986	35.046	35.105	35.165	35.225	35.285	35.344	35.404	35.464	630
640	35.464	35.524	35.584	35.644	35.704	35.764	35.825	35.885	35.945	36.005	36.066	640
650	36.066	36.126	36.186	36.247	36.307	36.368	36.428	36.489	36.549	36.610	36.671	650
660	36.671	36.732	36.792	36.853	36.914	36.975	37.036	37.097	37.158	37.219	37.280	660
670	37.280	37.341	37.402	37.463	37.525	37.586	37.647	37.709	37.770	37.831	37.893	670
680	37.893	37.954	38.016	38.078	38.139	38.201	38.262	38.324	38.386	38.448	38.510	680
690	38.510	38.572	38.633	38.695	38.757	38.819	38.882	38.944	39.006	39.068	39.130	690
700	39.130	39.192	39.255	39.317	39.379	39.442	39.504	39.567	39.629	39.692	39.754	700
710	39.754	39.817	39.880	39.942	40.005	40.068	40.131	40.193	40.256	40.319	40.382	710
720	40.382	40.445	40.508	40.571	40.634	40.697	40.760	40.823	40.886	40.950	41.013	720
730	41.013	41.076	41.139	41.203	41.266	41.329	41.393	41.456	41.520	41.583	41.647	730
740	41.647	41.710	41.774	41.837	41.901	41.965	42.028	42.092	42.156	42.219	42.283	740
750	42.283	42.347	42.411	42.475	42.538	42.602	42.666	42.730	42.794	42.858	42.922	750
760	42.922											760
DEG C	0	1	2	3	4	5	6	7	8	9	10	DEG C

The advantages of Type B thermocouples are:

- Better stability than type S or R
- Increased mechanical strength
- Usable in higher temperature than other thermocouples
- Reference junction compensation is not required if junction temperature does not exceed 150°F
- Can be used in neutral or oxidizing atmospheres
- Can use copper-copper extension wire

The disadvantages are:

- Available in standard grade only
- High temperature limit requires the use of alumina insulators and protection tubes. Do not insert in metal protection tubes
- Easily contaminated in other than oxidizing atmospheres
- Nonlinear output
- More expensive than Base Metal Thermocouples

Type R Noble Metal Thermocouple

Type R thermocouples are Platinum–13% Rhodium (+) versus Platinum (-). The temperature range is –50 to 1760°C (–60 to 3210°F) with a usable temperature range of 0 to 1480°C (32 to 2700°F). The output is –0.226 to 21.006 millivolts. The ANSI color code is RP (+) Black, RN (-) Red.

The advantages of Type R thermocouples are:

- Usable in oxidizing atmospheres
- Provides a higher usable range than type K
- Has high conformity to published calibration data
- Frequently more practical than non-contact pyrometers

The disadvantages are:

- Easily contaminated in other than oxidizing atmospheres
- Nonlinear output
- More expensive than other thermocouples
- Copper and #11 alloy (SX) thermocouple extension wire required between primary element and input device

Type S Noble Metal Thermocouple

Type S thermocouples are Platinum–10% Rhodium (+) versus Platinum (-). The temperature range is –50 to 1760°C (–60 to 3210°F) and the usable temperature range is 0 to 1480°C (32 to 2700°F). ANSI color code is SP (+) Black and SN (-) Red.

Advantages of Type S thermocouples:

- Usable in oxidizing atmospheres
- Provides a higher usable range than type K
- Frequently more practical than non-contact pyrometers
- Has a high conformity to published data

The disadvantages are:

- Easily contaminated in other than oxidizing atmospheres
- Nonlinear output
- More expensive than base metal thermocouples
- Copper and #11 Alloy SX thermocouple extension wire required between primary element and input device

Type E Base Metal Thermocouple

Type E thermocouples are Nickel–10% Chromium (+) versus copper-nickel (-). The temperature range is –270 to 1000°C (–450 to 1800°F) with a usable range of –190 to 900°C (–300 to 1600°F). ANSI color code is EP (+) Purple and EN (-) Red.

Advantages of Type E thermocouples:

- Has high thermoelectric power (EMF output)
- Both elements are highly corrosion resistant permitting use in oxidizing atmospheres
- Does not corrode at sub–zero temperatures

Disadvantages are:

- Stability is unsatisfactory in reducing atmospheres
- Reference junction compensation required
- Nonlinear output
- Limited use in a vacuum
- Nickel Chromium and Constantan (EX) thermocouple extension wire required between primary element and input devices

Type J Base Metal Thermocouple

Type J thermocouples are Iron (+) versus copper-nickel (-). The iron is magnetic and the constantan non-magnetic. The temperature range is –210 to 870°C (–300 to 1600°F) with a usable range of 0–750°C (32 to 1382°F). ANSI color code is JP (+) White and JN (-) Red.

Advantages of Type J thermocouples are:

- Comparatively inexpensive
- Can be used in neutral or reducing atmospheres

Disadvantages are:

- Maximum upper temperature in oxidizing atmospheres is 1400°F due to oxidation of the iron
- Protection tubes must be used above 900°F
- Protection tubes should always be used in a contaminating medium
- High limit of error. Error increases in time in oxidizing atmospheres
- Reference Junction compensation required
- Non linear output
- Not recommended at low temperatures
- Iron and Constantan (JX) thermocouple extension wire required between primary element and input device

Type K Base Metal Thermocouple

Type K thermocouples are Nickel–10% Chromium (+) versus Nickel–5% aluminum-silicon. Chromel non-magnetic, Alumel magnetic. The temperature range is –270 to 1370°C (–300 to 2300°F) with a usable temperature range of 0–1250°C (32 to 2282°F). ANSI color code is KP (+) Yellow and KN (-) Red.

Advantages of Type K thermocouples are:

- Suitable for oxidizing atmospheres
- In higher temperature ranges, provides a more mechanically and thermally-rugged unit than platinum, rhodium-platinum, and a longer life than iron constantan
- Less expensive than platinum, rhodium-platinum thermocouples

Disadvantages are:

- Especially vulnerable to reducing atmospheres, requiring substantial protection when used
- Vented protection tube suggested in reducing atmospheres

- Non linear output
- Reference junction compensation required
- Limited use in a vacuum
- After thermocouple sees 2100°F, accuracy below 600°F is shifted
- Carbide precipitation (green rot)

Type N Base Metal Thermocouple

The type N thermocouples are Nickel–14% Chromium–1.4% Silicon (+) versus Nickel–4.4% Silicon–0.1% Magnesium (-). The temperature range is from –270 to 1300°C (–450 to 2370°F) with usable temperature range from 0°C to 1250°C (32 to 2282°F). Also called Nicrosil-Nisil.

The advantages of Type N thermocouples are:

- Linearity of noble metal thermocouple with cost advantages of base metal thermocouple
- Linear from 500 to 1000°C
- Better stability than other thermocouples
- Better high temperature oxidation resistance than type K

A disadvantage is:

- Not many instruments support type N

Type T Base Metal Thermocouple

The type T thermocouple are Copper (+) versus Copper–Nickel (-). Copper yellow metal, Constantan silver metal. The temperature range is –270 to 400°C (–450 to 750°F) with a usable temperature range of –190 to 370°C (–300 to 700°F). ANSI color code TP (+) Blue, TN (-) Red

Advantages of type T thermocouple:

- Resists atmosphere corrosion
- Applicable in reducing or oxidizing atmospheres below 600°F
- Its stability makes it useful in sub-zero temperatures
- Has high conformity to published calibration data
- Very accurate
- Good where moisture is present
- May be used in a vacuum

Disadvantages:

- Copper oxidizes above 600°F
- Nonlinear output
- Copper and Constantan (TX) thermocouple extension wire must be used between primary element and input device
- Reference junction compensation required

Type W Base Metal Thermocouples

- Letter code not ANSI standard
- Temperature range 0°C to 2300°C
- Useful in a vacuum
- Type W: Tungsten vs. Tungsten 26% Rhenium
- Type W3: Tungsten 3% Rhenium verses Tungsten 25% Rhenium
- Type W5: Tungsten 5% Rhenium versus Tungsten 26% Thenium

Installation

While some thermocouples may be exposed directly to the process medium, they are typically installed in a protecting tube or a thermowell. Proper installation of a thermocouple will provide protection of the thermocouple and help ensure the accuracy of the measurement. To determine the appropriate type of thermocouple, assembly components, and extension wire for a particular application several other factors must be considered when planning a thermocouple installation.

Determine the best location, the one that provides the most representative measurement. The depth of immersion is also critical to the accuracy of the measurement. The position of the thermocouple is also an important installation consideration. Thermowells are usually installed perpendicular to a pipe or vessel wall, at an angle, or in an elbow. (See Figure 8-1). In gas applications, the thermocouple should be located where the mass velocity is as high as possible. In high temperature applications such as ovens or furnaces, vertical installation of the thermocouple through the top of the vessel will prevent the thermowell from bending or sagging.

Thermocouple extension wire should be installed in such a way that it is protected from excessive heat, moisture, and mechanical damage. Whenever practical, it should be installed in conduit. To minimize errors that can be introduced by junction boxes, the extension wire should be run from the connection head to the measuring instrument in one continuous length. To prevent induced extraneous voltages, electrical wires should never be run in the same conduit with extension wires.

On lines at high-steam pressure pipes or high fluid pressure pipes, it is advisable to weld the thermowell to avoid leakage of steam or fluid. Primary elements in a well must have contact with the end of the thermowell. A spring-loaded head should be used.

Checking the Accuracy

One common method for testing the accuracy of a new thermocouple involves placing the reference junction in an ice bath to establish the reference temperature, and placing the measuring junction in a variable calibrating temperature bath. The voltage output of the thermocouple is read with a calibrated potentiometer, high-impedance voltmeter, or thermocouple indicator. Then compare with the appropriate thermocouple reference table to determine whether or not the thermocouple output agrees with the table within the prescribed limits of error throughout its usable range.

Used or installed thermocouples that are exposed to high temperature in various atmospheres may change characteristics. The purpose of checking an installed thermocouple is to determine the temperature error in actual service, not the temperature-emf characteristics of the thermocouple. Therefore, a used thermocouple should always be checked in its normal, installed location.

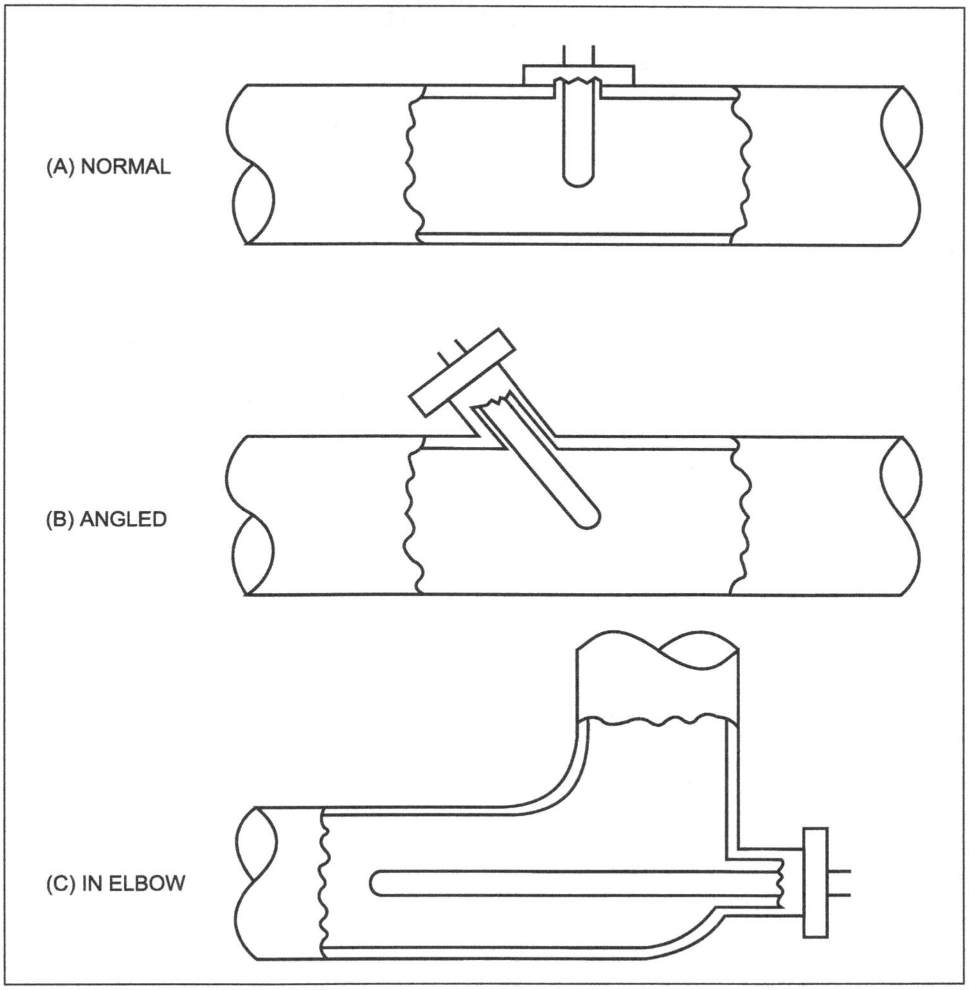

Figure 8-1. Thermowell Installations.

Readout Instrumentation

The low millivolt readings must be changed to a temperature readout. The millivolt signal can either be sent to a temperature gage or a temperature transmitter via thermocouple extension wire. In some cases the temperature transmitter is located on the thermowell, and a 4-20 mA signal is sent to the control system. In other cases the temperature transmitter is located near, but not on the thermowell and is connected to the thermocouple with extension wire, then converted to 4-20 mA and sent to the control system. In some systems, thermocouple extension wire is connected between the thermocouple and a panel mounted temperature gage, or directly into a control systems input block as a millivolt signal.

Resistance Temperature Devices

Resistance thermometers respond to temperature by changing their electrical resistance. Two common types of resistance thermometers are resistance temperature detectors (RTDs), which have metallic sensing elements, and thermistors, which have semiconductor elements.

Resistance Temperature Detector (RTD)

An RTD consists of a sensing element fabricated of metal wire or metal fiber that responds to temperature changes by changing its resistance. The sensor is connected to a readout instrumentation that monitors the resistance to a temperature value. The principle of operation of an RTD is based on the fact that the electrical resistance of some metals varies directly with temperature changes.

RTD Readout

Temperature measurement with an RTD is actually a measurement of the sensor's resistance, using the sensor calibration to convert the measurement into temperature. This is achieved by connecting the sensor to a transducer that has a bridge circuit, typically a Wheatstone bridge or a Mueller bridge. The sensor leads are connected so that the RTD forms one leg of the bridge circuit (See Figure 8-2). Three-wire and four-wire sensors provide compensation for lead resistance (See Figure 8-3).

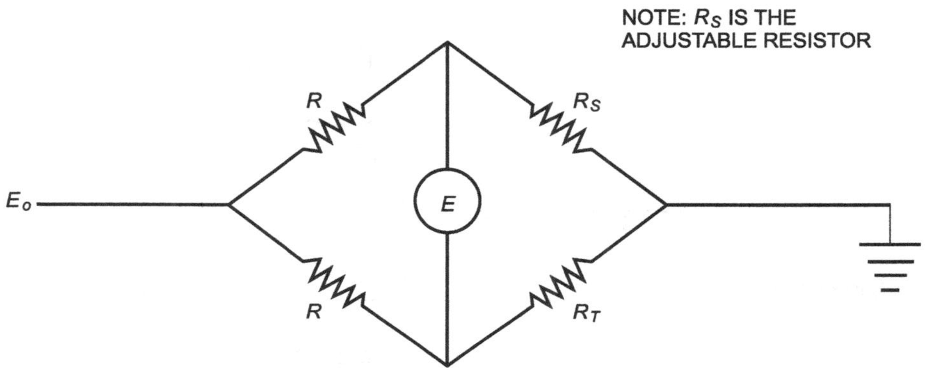

Figure 8-2. Two-Wire RTD and Wheatstone Bridge Circuit.

RTD Accuracy

In addition to lead wire resistance, which can be compensated through the use of a three-wire or four-wire sensor, several other factors influence the accuracy of RTD temperature measurement. One of these is a lack of a single standard for RTDs. Several standards specify different values for the temperature coefficient of resistance for a given metal. In addition, manufacturers observe different tolerances in the manufacture of RTDs. Therefore, accuracy problems can occur when RTDs from different manufacturers are used in the same system, or when an RTD from one manufacturer is replaced with an RTD from another manufacturer. Self-heating caused by current flow can also affect accuracy.

100 Ohm Platinum

Temperature range –200 to 850°C (–325 to 1560°F) with a usable range of –200 to 500°C (–320 to 1000°F); resistance 18.49 to 390.26 ohms. Ice point 0°C (32°F) equals 100 ohms +/–.2 ohms.

Advantages of 100 Ohm Platinum temperature device are:

- More linear than thermocouples
- Long temperature averaging probes available

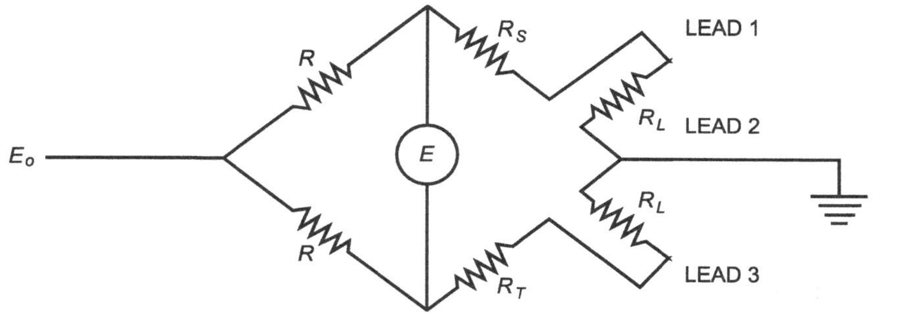

(a) A WHEATSTONE BRIDGE FOR 3-WIRE MEASUREMENTS

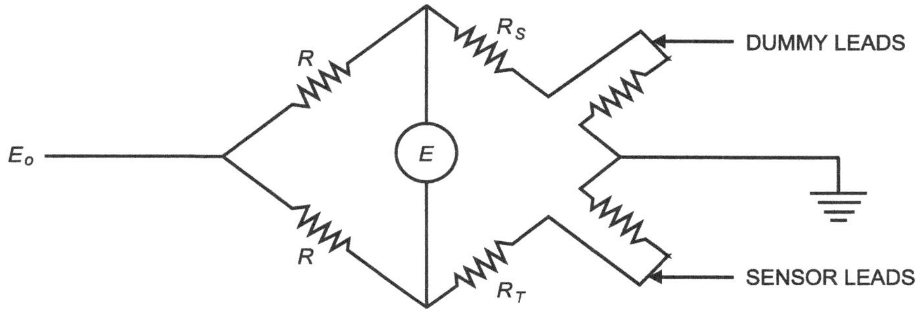

(b) A WHEATSTONE BRIDGE FOR USE WITH A 4-WIRE DUMMY LEAD CONFIGURATION

Figure 8-3. Three-Wire and Four-Wire RTDs in Wheatstone Bridge.

- Not tip sensitive
- High resolution
- Most repeatable
- Stable
- More accuracy over a smaller temperature span at lower temperature than most thermocouples

Disadvantages:

- Slower responding than thermocouples (about 1/3 response time)
- Is not self powered
- Is not tip sensitive
- Must have equal length and same type and gage of lead wires
- A small lead impedance can contribute a significant error in temperature measurement
- More expensive than thermocouples
- Very fragile

- Self-heating. A current must be passed through the device to provide a voltage that can be measured. This self heating appears as a measurement error
- Two-wire RTD very susceptible to errors due to lead impedance
- Three-wire and four-wire RTDs will reduce the affects of lead impedance with proper input circuit
- Not recommended for shallow immersion depth applications

1000 Ohm Platinum

Temperature range –200 to 550°C. Ice point 0°C (32°F) equals 1000 ohms. Resistance 199.4887 to 2913.536

Advantages:

- Lead resistance does not affect reading as much as other RTDs

Disadvantages:

- About the same as 100 Ohm RTDs

25 Ohm Platinum

Temperature range –325 to 850°F. Ice point 0°C (32°F) equals 25 ohms. Not supported by many instruments.

10 Ohm Copper

Temperature range 0 to 120°C. Ice point 25°C (77°F) equals 10 ohms. Very little instrument support.

100 Ohm Copper

Temperature range –300 to 500°F. Ice point 100 ohms equals 0°C. Very little instrumentation support.

120 Ohm Nickel

Temperature range –50 to 500°F. Ice point 0° equals 120 ohms. Very little instrumentation support.

1000 Ohm Nickel at 0°C

Temperature range –50 to 500°F. Ice point 0°C equals 1000 ohms. Not accurate. Little support from manufacturers.

1000 Ohm Nickel at 70°F

Temperature range –50 to 500°F. Lead impedance does not affect reading as much as other RTD's. Little support.

Balco

Temperature range 32 to 400°F. Ice point 2000 ohms at 0°C. Balco is a wire with high resistance, so a relatively small element can be made with a resistance of 2,000 or even 20,000 ohms. Thus, if small temperatures are measured, the high resistance in a proper bridge network will give a potentially higher output. There is no standard curve for this material, and it is not used for precision measurements.

Tungsten

Resistivity 30 ohms per circular mil foot. Reserved for very high temperatures.

Gold

Resistivity 13.0 ohms per circular mil foot. Rarely used because of their low resistivity.

Silver

Resistivity 8.8 ohms per circular mil foot. Rarely used because of their low resistivity.

Thermistors

Like the RTD, the thermistor is also a temperature sensitive resistor. They have a negative temperature coefficient, that is, their resistance decreases with increasing temperatures. The temperature range is –140 to 300°C (–220 to 572°F). The resistance is 2K to 1 Meg Ohm.

Thermistors are made of solid semiconductor materials that have a high coefficient of resistivity. The relationship between resistance and temperature and the linear current-voltage characteristics are the primary importance. Typical thermistors are suitable for temperature as low as –20°C and others as high as 600°C.

Thermistor Accuracy

A thermistor will maintain a stable and repeatable resistance versus temperature relationship when it is used within its specified temperature range. Another characteristic of thermistors is that they are made with two lead wires instead of the typical three-wire and four-wire RTDs. This is because the large resistance values for thermistors virtually eliminates the concern about lead wire resistance.

Thermistor Readout

A variety of circuit designs are used to measure temperature with a thermistor. For example, a thermistor may be connected to a Wheatstone bridge, much like a two-wire RTD circuit, or a series circuit that includes a battery, the sensor, and a micro-ammeter may be used. As long as voltage is constant, the current flow will be determined only by the resistance of the thermistor.

The advantages of thermistors are:

- More sensitive than RTD or thermocouple
- Exhibit by far the largest parameter change with change in temperature
- Thermistor circuits detect minute changes in temperature, which could not be observed with the RTD or TC circuit
- Small and responds quickly

The disadvantages are:

- Very non linear. Linear thermistors are available for a particular range
- No standard thermistor curves
- Because they are semiconductors, thermistors are more susceptible to permanent decalibration at high temperatures
- More fragile than RTDs or TCs and must be carefully mounted

Integrated Circuit Temperature Transducer

Also called monolithic linear temperature sensor. Available in both voltage and current outputs. Temperature range of 0 to 200°C.

Advantages of Integrated Circuit Temperature Transducers:

- Output linearity proportional to absolute temperature
- Useful for thermocouple reference junction compensation circuits
- Inexpensive
- Most linear of all sensors
- Available in both voltage and current outputs

The disadvantages are:

- They are semiconductor devices and thus have a limited temperature range
- Same problems of self heating and fragility as RTDs
- Requires an external power source
- Maximum temperature 200°C
- Slow response
- Limited configurations
- Not supported by many measuring devices

Infrared Temperature Transducers

Temperature measurement through the use of mechanical thermometers, thermocouples, RTD's and thermistors require placing the sensor in physical contact with the medium or object being measured, either directly or within a thermowell. However, in some applications, contact measurement is not suitable. In these situations, non-contacting temperature measurement can be achieved through the use of radiation or optical pyrometers. Radiation pyrometers are capable of measuring temperatures to approximately 4000°C without touching the object being measured.

Principles of Radiation Pyrometers

Temperature measurement with radiation pyrometers is based on the principle that all objects emit radiant energy. Radiant energy is emitted in the form of electromagnetic waves, considered to be a stream of photons that travel at the speed of light. The wavelength of radiant energy emitted by a hot object range from the visible light portion to the infrared portion of the electromagnetic spectrum. Radiation pyrometers measure the temperature of an object by measuring the intensity of the radiation it emits. Since emittance will differ from one object to another, a standard, called a blackbody, is used as a reference for calibrating radiation pyrometers and serves as the basis for the laws that define the relationship of the intensity of radiation and wavelength, having a surface that does not reflect or pass radiation.

Pyrometer Accuracy

In addition to proper calibration, several techniques can help to ensure accurate measurement with radiation pyrometers. It is important that the radiation observed by the pyrometer is actually emitted by the target object, not radiation reflected from its surface. Measurement accuracy can also be affected by the presence of gases or vapors between the target and pyrometer. A film of dirt on the viewing window or lens will also affect measurement accuracy.

Infrared Temperature Transducers are non-contact measurement. You point at a non-reflective process. Temperature ranges of infrared temperature transducers are 0 to 3500°C (0 to 6500°F). accuracy is +/–0.75% or full scale or 3°F with a repeatability of +/–0.25% of full scale. The outputs can be 4-20 mA, simulated TC, or local indication. Cooling can be accomplished with a cooling jacket with air or liquid cooling.

TYPES:

Automatic Optical Pyrometer

Temperature span of 500 to 3000°C (900 to 5000°F). Best choice for most common incandescent applications in such area as iron and steel, foundries, forging, hardening, annealing, all types of semiconductor processes.

Two Color Pyrometer

Temperature span of 700 to 3500°C (1300 to 6500°F). For difficult high temperature applications where the object is partly obliterated, moves in the field of view, or where emittance varies. Molten metal, wire and rod, vacuum furnaces and kilns are typical uses.

General Purpose Infrared Thermometer, Medium Temperature

Temperature span of 250 to 1100°C (500 to 2200°F). Best choice for ferrous and non-ferrous metals.

General Purpose Infrared Thermometer, Low Temperature

Temperature range 80 to 800°C (150 to 1400°F). Very versatile for all types of low temperature uses from thick plastics, rubber and textiles to metals.

General Purpose Infrared Thermometer, Very Low Temperature

Temperature range 0 to 500°C (0 to 1000°F). Ideally suited to all types of very low temperature applications such as print drying, food and textile processing.

Flame Pyrometer

Temperature range 800 to 2400°C (1500 to 4500°F). exclusively for flame temperature, combustion efficiency and pollution measurement and control in boilers, incinerators, chemical reactors, glass melting tanks, kilns, etc.

Glass Infrared Thermometer

Temperature range 50 to 1200°C (100 to 2500°F). Measures glass surface temperature in such operations as forming, bending, tempering, annealing and sealing.

Plastic Infrared Thermometer

Temperature range 0 to 800°C (50 to 1500°F). Ideal for paints and organics such as waxes and oils.

Plastic Infrared Thermometer, Low Temperature

Temperature range 20 to 400°C (50 to 800°F). Best choice for most textiles.

Fiber Optic Infrared Sensors

Applications where direct sighting with conventional infrared instrumentation is difficult or impossible. Used in environments where RF or EMI interference is a problem and high ambient temperature applications where conventional water cooling of the head is impractical or undesirable. Also used in vacuum applications where sighting through the window is difficult or impossible.

Optical Fiber Thermometry

Planck's Theory of Blackbody Radiation

A thin film of metal is sputtered on the end of a single crystal of aluminum oxide (sapphire) to form a blackbody cavity. When the sensor is inserted into a high temperature environment, the metal cavity emits radiant energy. Since the sapphire optical fiber is optically transparent, this radiant output can be transported through it and through a second flexible low temperature optical fiber to a photo-detection device with almost no detectable radiation loss along the optical transport path. Here the radiation passes through a narrow band wavelength filter and strikes a photo-detector where it undergoes a linear conversion into its equivalent electrical energy (one photon input equals one electron output). Since the resulting current has a directly linear relationship to the optical fiber's radiation at the wavelength defined by the filter, the Planck equation can be applied to determine the temperature of the blackbody.

Temperature range 500 to 2000°C (950 to 3600°F).

Advantages:

- Wide optical bandwidth results in less thermal radiation heat transfer
- Heat transfer errors held to absolute minimum over a wide range of gas flow, pressure, and temperature conditions
- Not subject to drift over periods of time
- Blackbody insensitive to contamination, thermal aging, and electro-magnetic interference
- Capable of responding to temperature changes up to 50,000 times per second

Disadvantages:

- Not supported by ANSI as yet
- New to the industries

Thermometers

Filled Thermometer

Filled thermometers contain a gas or a volatile liquid and rely on pressure measurements to provide temperature indications. The components of a typical filled thermometer are a bulb that is exposed to the fluid being measured, a capillary tube, a pressure element such as a Bourdon tube, and a scale. The bulb, capillary tube, and Bourdon tube are filled with a liquid, vapor, or gas. When the temperature changes, the fluid either expands or contracts, which causes the Bourdon tube to move, thereby moving the position of the needle on the scale (See Figure 8-4). Vapor systems have a relatively narrow temperature measurement span. To check the calibration of bimetallic thermometers, place the thermometer in a liquid or gas temperature controlled environment and compare the temperature with a calibrated temperature indicator. Check it in the range that the thermometer is going to be used in. Some have an adjustment screw on the face to adjust the pointer.

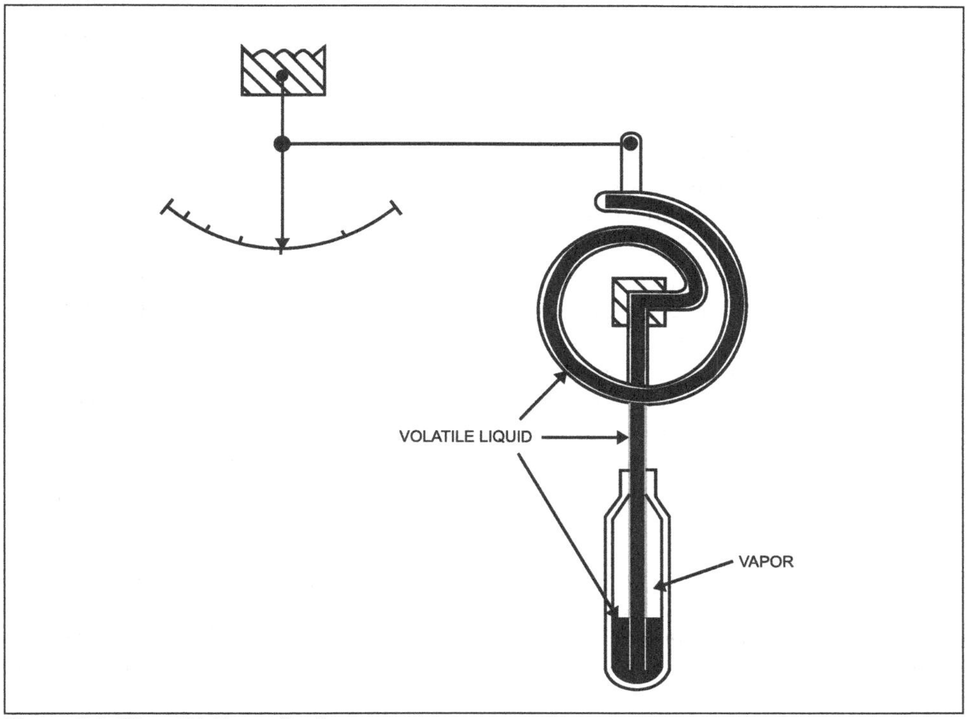

Figure 8-4. Class IIA Vapor System.

Bimetallic Thermometer

Bimetallic thermometers use the differences in the thermal expansion properties of metals to provide temperature measurement capability. Typical bimetallic usually use spiral or helical configuration of bonded metal combination. One end is fixed, and the other is attached to a pointer that indicates temperature on a scale.

To check the calibration of bimetallic thermometers, place the thermometer in a liquid or gas temperature controlled environment and compare the temperature with a calibrated temperature indicator. Check it in the range that the thermometers are going to be used in some have an adjustment screw on the face to adjust the pointer.

OPERATION

The measurement of the temperature of a material requires the insertion of a sensor (usually protected by a thermowell) into a pipe or tank containing the material. Electrical-type temperature sensors such as RTDs or thermocouples produce low-level electrical signals that are proportional to the sensed temperature. Long runs of lead wire between the sensor and signal conditioning device can result in large errors due to noise unless great care is taken with shielding. In addition, extra compensating leads (for RTD inputs) or special extension wire (for thermocouple input) are needed between the sensor and signal conditioning device.

These negative characteristics can be eliminated by mounting a signal-conditioning device close to the point of measurement. Temperature transmitters offer a convenient, reliable, and economical way to do this. The temperature is converted to a standardized high-level 4–20 mA DC signal, which is in turn transmitted to the control room via two copper wires.

INSTALLATION

The transmitter can be attached directly to the sensor or remote mounted. See Figure 8-5.

Mounting stability is an important consideration. The transmitter, though rugged, may have to be supported separately under high-vibration conditions, particularly if extensive thermowell lagging or long extension fittings are required.

The transmitter will typically operate within specifications for ambient temperatures in the range of 25°C to 85°C. It will function, but not necessarily within specifications, over the range of 40°C to 100°C.

Aside from ambient temperature variations, heat from the process is transferred from the thermowell to the transmitter housing. If the process temperature is near or beyond specification limits, thermowell lagging or an extension nipple should be used to protect the transmitter from the overtemperature condition.

The transmitter has been carefully designed to resist attack by moisture or other corrosives. The coated circuit boards are mounted in a compartment that is completely sealed from the conduit entrances. O-ring seals protect the interior when the covers are installed. In humid environments, however, it is still possible for moisture "breathing" to occur in conduit lines. If the transmitter is mounted at a low point in the conduit run, the terminal compartment could fill with water, causing shorting. The transmitter, therefore, should be mounted so that moisture from the conduit will not drain into the housing.

A number of optionally available certifications facilitate use of the temperature transmitters in most hazardous areas. For such installations, care must be taken to assure that the temperature transmitters are installed in accordance with all applicable codes and standards and that the ratings and instructions marked on the transmitter are strictly followed.

The DC power supply need not be well regulated but should provide power with less than 2% ripple.

Various RTD configurations are used in industry, with each configuration offering a specific solution for compensating the effects of leadwire resistance. The typical sensors use 3- and 4-wire designs.

THERMOCOUPLE OR MILLIVOLT INPUTS

In the case of thermocouples, connections between the sensor and the transmitter must be made with thermocouple wire. For process mounting applications, the thermocouple may be connected directly to the transmitter. For installations in which the transmitter is mounted remotely from the sensor, appropriate thermocouple extension wire must be used. As with all low-level signal wiring, shielding is recommended for long runs.

The transmitter will operate with the current signal loop either floating or grounded. However, many types of readout devices are affected by the extra noise in floating systems. If operation appears noisy or erratic, grounding the current signal loop at a single point may solve the problem. The best place to ground the loop is at the negative terminal of the power supply. Alternately, either side of the readout device could be grounded. Do not ground the current signal loop at more than one point.

The transmitter will withstand electrical transients of the energy level usually encountered in static discharges or induced switching transients. However, high-energy transients, such as those induced in wiring from nearby lightning strikes, can damage the transmitter and the sensor. If the transmitter will be installed in an area in which such high transients are probable, a transient protector should be installed at the transmitter location.

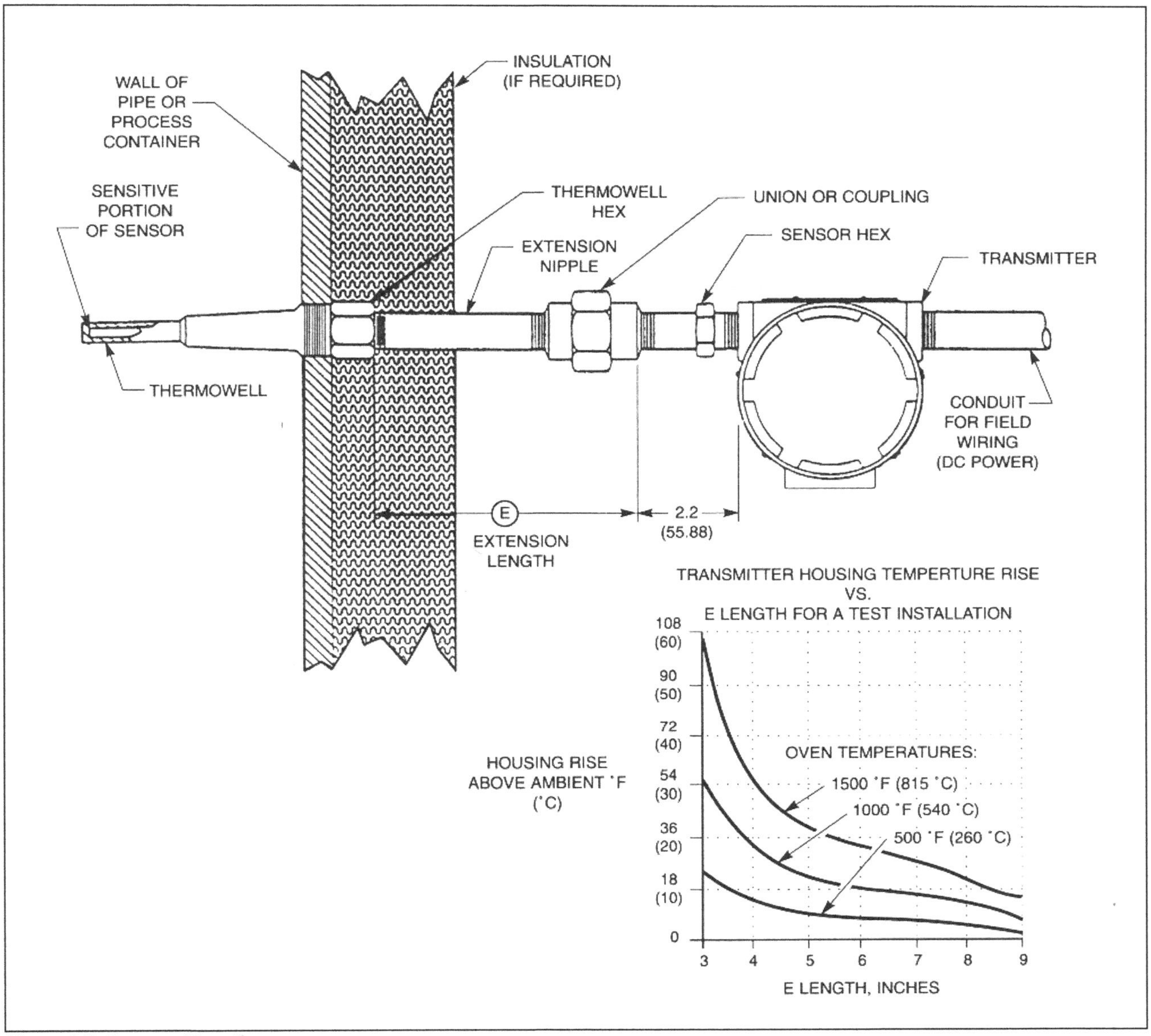

Figure 8-5. Process Mounting.
(Courtesy of Rosemount, Inc.)

Intrinsically safe installations can be used instead of explosionproof installations in hazardous area. The transmitter and sensor are located in a hazardous area, and the current signal leads are connected to equipment in a nonhazardous area through intrinsic safety barriers that limit the voltage and current fed into the hazardous area. Installation must be made in accordance with the barrier manufacturer's instructions for the specific barrier used.

Mechanical installation may be accomplished using the following guidelines:

(1) Install thermowell.

(2) Install any separate extension nipples.

(3) Screw sensor into thermowell.

(4) If the process mounting technique is use, screw the transmitter onto the threaded end of the sensor, carefully pulling the sensor leads into the terminal compartment of the transmitter. For transmitter locations that are remote from the sensor, use conduit between the sensor and transmitter.

(5) Install drain seals if required for severe environments or code requirements (see Figure 8-6).

(6) Install conduit for field wiring to the other hub of the transmitter.

(7) Pull field wiring leads through the conduit into the terminal side of the housing.

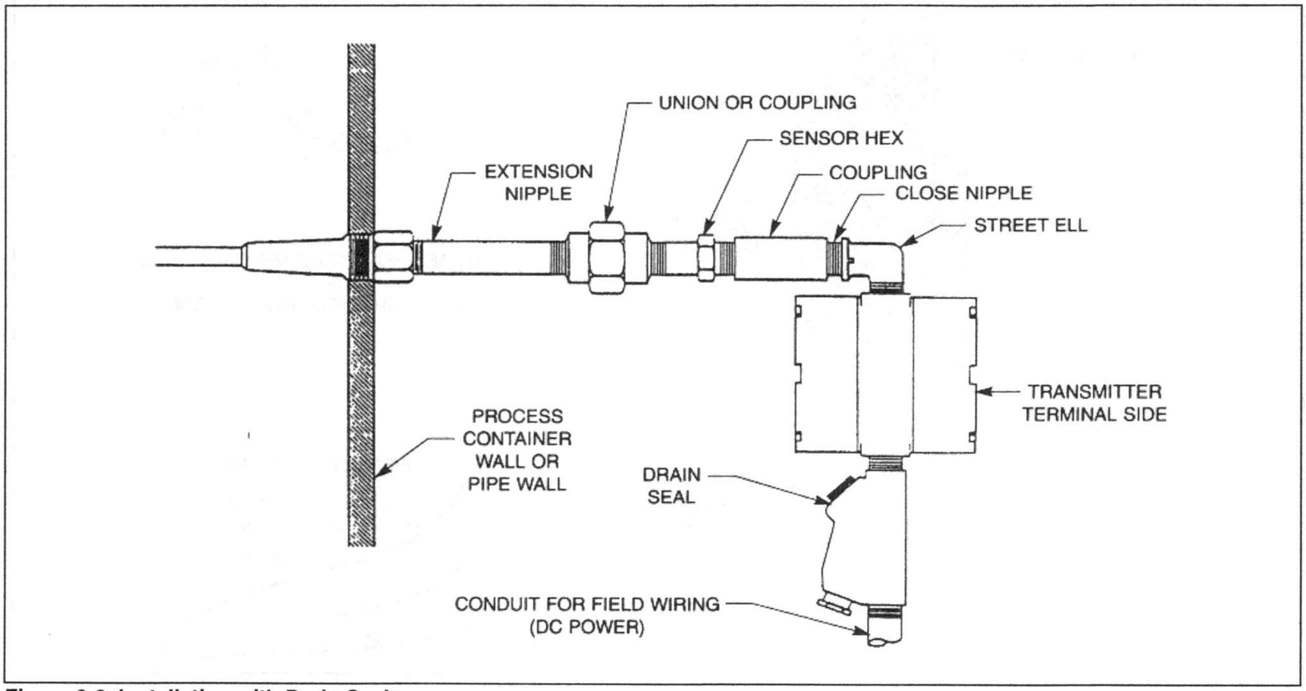

Figure 8-6. Installation with Drain Seal.
(Courtesy of Rosemount, Inc.)

CALIBRATION

The transmitter is factory calibrated to the temperature range shown on the nameplate. Only a few calibration laboratories have the kind of precision temperature baths necessary for accurate direct calibration of a temperature sensor or sensor/transmitter system. As a result, the transmitter is normally calibrated by substituting a resistance decade box for an RTD or a millivolt source for a thermocouple. These techniques are described separately below.

RTD Transmitter, Calibration Equipment Required:

- Voltmeter/Milliampmeter (such as a 5-digit DVM). Voltage rating is dependent upon the test signal (see previous paragraph). Accuracy is ±0.01%; resolution is 1 mV.
- DC Power Supply. Power capability is 24 V DC at 35 mA.
- Resistance Decade Box. Precision type, 5-dial, with large dial providing 100-ohm steps. Accuracy is ±0.02 ohm. The decade box should be periodically calibrated against a 5-dial Wheatstone bridge.

Calibration Procedure:

The steps that follow describe the procedure for calibrating an RTD transmitter (see Figure 8-7):

(1) Remove the transmitter terminal side housing cover.

(2) If an RTD is already connected, remove all RTD lead connections.

(3) Attach the calibration test equipment as shown in Figure 8-9. Terminal connections may be made conveniently using miniature banana plugs.

(4) Determine the RTD resistance at the desired base and full-scale temperatures.

(5) Turn power on.

(6) Set decade box to the resistance that corresponds to the desired base temperature. Adjust zero potentiometer until the output is 4 mA.

(7) Set decade box to the resistance that corresponds to the desired full-scale temperature. Adjust span potentiometer until the output is 20 mA.

(8) Repeat above steps until both 4 and 20 mA readings are obtained without readjusting span and zero potentiometers.

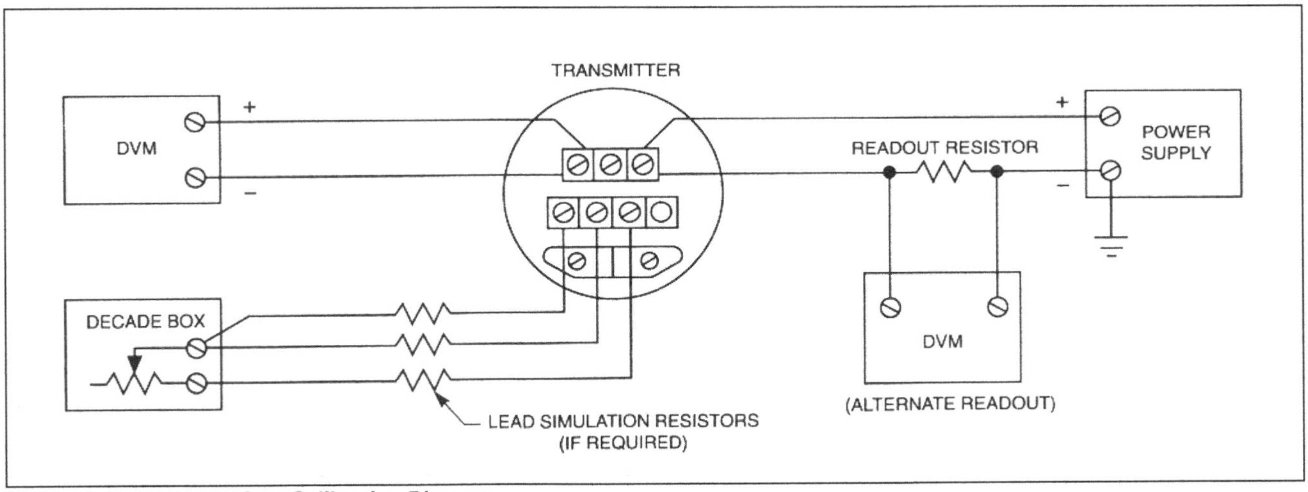

Figure 8-7. RTD Transmitter Calibration Diagram.

Calibration Using a Compensated Thermocouple Simulator, Calibration Equipment Required (see Figure 8-8):

- Compensated Thermocouple Simulator: Precision voltage source providing conformity to NIST Monograph 125 thermocouple curves. Reflect accuracy of simulator to desired calibration span. A simulator accuracy four times better than the transmitter is recommended (0.05% of calibrated span or 0.005 mV, whichever is greater). Simulator inaccuracies greater than this will degrade system accuracy and factory calibration is recommended.

- Voltmeter: Such as a 5-digit DVM. Accuracy is ±0.01%; resolution is 1 mV.

- DC Power Supply: Power capability is 24 V DC at 35 mA.

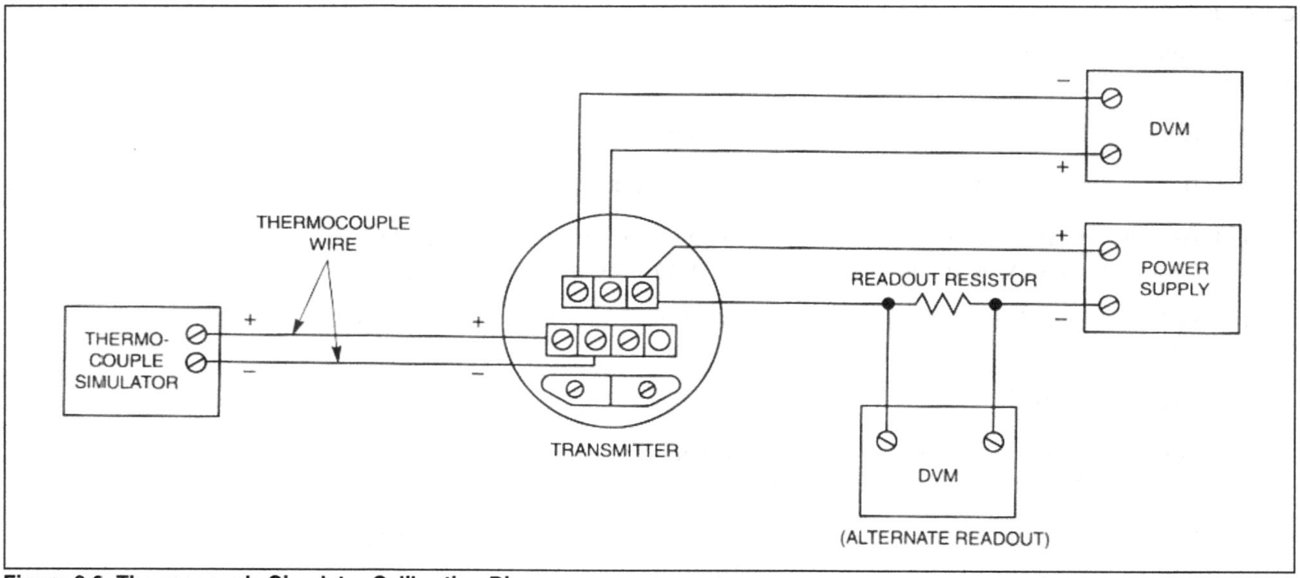

Figure 8-8. Thermocouple Simulator Calibration Diagram.
(Courtesy of Rosemount, Inc.)

- Thermocouple Wire: The thermocouple wire used should be the same type as the wire the thermocouple is constructed of.

Calibration Procedure:

(1) Remove the transmitter terminal side housing cover.

(2) If a thermocouple is already connected, remove all the thermocouple lead connections.

(3) Determine the base and full scale temperatures.

(4) Turn power on.

(5) Refer to the thermocouple simulator instructions for setting the thermocouple type and engineering units. Set the simulator to the base (zero) temperature and adjust the zero pot until the output is 4 mA (or 40 mV at the test terminals).

(6) Set the simulator to the full scale temperature and adjust the span pot until the output is 20 mA.

(7) Repeat steps until both the 4 and 20 mA readings are obtained without readjusting the pots.

MAINTENANCE

In case of a failure, particularly one in which the transmitters output goes to one extreme and stays there, the first step is to determine whether the fault lies with the sensor or the transmitter. Although only a careful calibration can determine sensor shifts, catastrophic failures (such as an open or shorted sensor element) can be checked with an ohmmeter at the time the transmitter is disconnected from the sensor.

TROUBLESHOOTING

This section offers tips for troubleshooting several kinds of potential malfunctions. To determine a malfunction, use pin-like probes to break through the protective coating to make measurements on a circuit board.

Symptom: No output.

System:

- Break in continuity of signal loop; check leads, load power supply.
- Reverse polarity connection.

Circuit:

- Dirty or defective terminals.
- Dirty of defective circuit interconnection pins of receptacles.

Symptom: Output signal too low regardless of input.

RTD Transmitter:

System:

- Supply voltage too low at transmitter terminals.

Sensor:

- One RTD lead shorted to another.
- RTD shorted to ground.
- Incorrect RTD lead connection.
- Open RTD lead on double lead side.

Thermocouple Transmitter:

System:

- Input open (if downscale burnout protection).
- Input shorted.
- Input polarity reversed.

Circuit:

- Zero pot element open.
- Defective transistor.

Symptom: Unit cannot be trimmed to correct base temperature.

System:

- Desired range outside capability of unit.

Circuit:

- Defective zero pot.

Symptom: Unit cannot be trimmed to correct span.

System:

- Desired range outside capability of unit.

Circuit:

- Supply voltage too low at transmitter terminals.
- Defective span pot.
- Defective component.

Symptom: Excessive output shift with ambient temperature. (If suspected, test transmitter in environmental oven at ±50°C from ambient and compare results with specs.)

All models:

- Defective component in amplifier or current control section.

Symptom: Output above scale regardless of input.

Sensor:

Sensing element open

Circuit:

Dirty or defective terminals, interconnecting pins or receptacles.
Component failure.

Symptom: Excessive current (over 30 mA).

System:

- Short between current signal leads.
- Current signal leads connected to sensor terminals (this will damage many components in the unit).
- Sensor or sensor leads shorted to ground when positive side of power supply is grounded.

Symptom: Erratic readings or excessive noise in readout.

System:

- Poor connections in the signal loop at transmitter, power supply, or receiver.
- Signal loop grounded at more than one point.
- Signal loop not grounded at any point.
- Noisy power supply.
- Unusually high noise pick-up in sensor or signal leads. Any run of low-level leads (between sensor and transmitter) should be shielded, with shielding grounded at one point only.

Circuit:

- Intermittently defective component on any board.

Electrical Block Diagrams

RESISTANCE TEMPERATURE TRANSMITTER

The specific operation of the different functional blocks is described below.

RTD

The transmitter is designed for use with a platinum resistance temperature sensor (100 ohms R^o unless otherwise specified). This sensor type has a positive temperature coefficient, with a highly stable resistance versus temperature relationship. (See Figure 8-9).

Resistance Bridge

The bridge converts the resistance versus temperature relationship of the sensor to a millivolt versus resistance signal. A variable resistance network in the leg of the bridge adjacent to the RTD allows zero adjustment. Selectable resistors in the same leg allow a choice of zero regimes for wide zero rangeability while maintaining required resolution.

The two arms of the bridge opposite the RTD consist of a fixed divider network. All of the output signal (except for the sensor current, which is fixed) flows through the feedback resistor, thus rebalancing the bridge output to zero. The bridge is driven by a current source. Span adjustability is achieved by varying the current level through the RTD. The current source also supplies a linearizing signal, so output is linear with temperature.

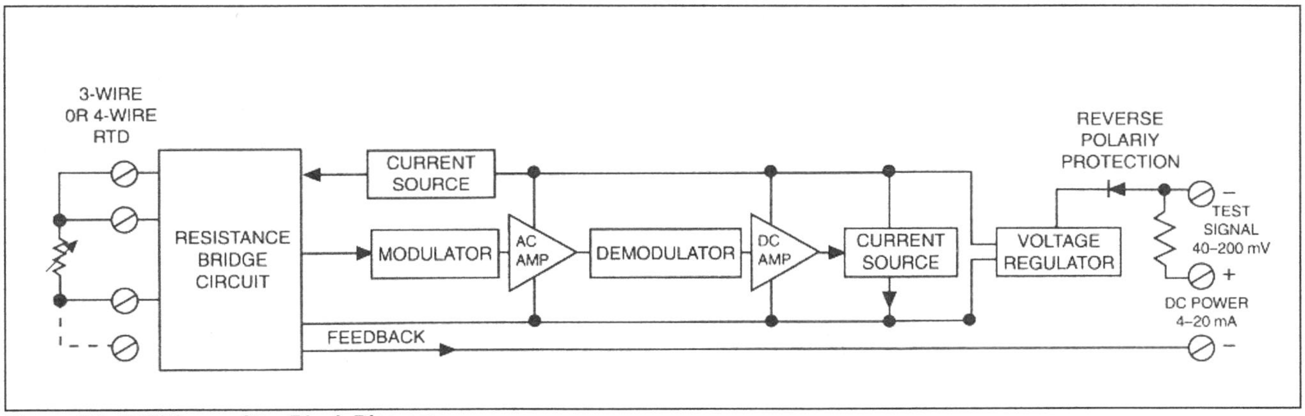

Figure 8-9. RTD Transmitter Block Diagram.
(Courtesy of Rosemount, Inc.)

DC Amplification, Current Control

The differential signal from the bridge is amplified by a precision DC amp to drive a transistor that controls a current signal proportional to sensor temperature. This signal is combined with the current used to operate the circuit, and the result is fed through the bridge feedback resistor. Thus, the transistor passes just enough current so that total 4–20 mA signal is dependent only on sensor temperature. A limiter resistor keeps the circuit from drawing too much current under over-scale conditions.

Voltage Regulation: A voltage regulator circuit provides a stable voltage of 9.1 V DC to ensure the signal is independent of supply voltage and load resistance variations. Protection against damage due to reverse power hookup is provided by a diode.

THERMOCOUPLE TEMPERATURE TRANSMITTER

The specific operation of the different functional blocks shown in Figure 8-10 is described below.

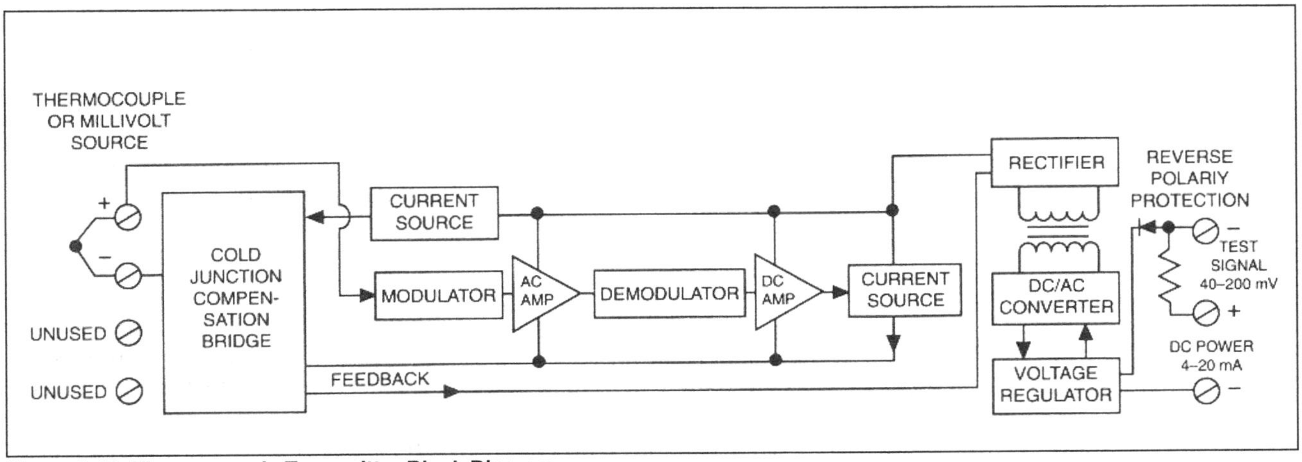

Figure 8-10. Thermocouple Transmitter Block Diagram.
(Courtesy of Rosemount, Inc.)

Thermocouple

The Model 444TE, TJ, TK, TR, TS, or TT transmitter is designed for use with a thermocouple for temperature measurement. A thermocouple consists of two specific dissimilar metals joined at the measurement side. If this "hot" junction is at temperature T_m, and the point at which the thermocouple wires attach to ordinary copper wire (the "cold" junction) is at temperature T_a, the thermocouple produces a millivolt-level signal proportional to T_m minus T_a. Some thermocouple designs use an intentionally grounded hot junction for fast response, while others use an electrically insulated hot junction. Either type may be used with the Model 444.

Bridge

Cold-junction compensation for the thermocouple is provided by two compensation resistors. When the bridge is fed the correct current, the output from the compensation resistors simulates that of a thermocouple at temperature T_a. This voltage is in series with the thermocouple, and cancels the thermocouple's sensitivity to T_a. The resultant output is proportional only to T_m. A variable resistance network in the bridge allows zero adjustment, while a selectable resistor in the same bridge leg allows a choice of zero regimes for wide zero rangeability while maintaining required resolution. Span is adjusted by varying the magnitude of the feedback resistance. Signal current flowing through the resistance rebalances the combined bridge and thermocouple output to zero.

The bridge is driven by a current source that is factory adjustable for correct cold-junction compensation. The voltage divider forces the transmitter output to overscale saturation if the thermocouple opens. This is upscale burnout indication.

Modulation, AC Amplification, Demodulation

The differential signal from the bridge is converted to an AC signal. This signal is then amplified and converted back to a DC signal. This carrier amplifier technique helps ensure the signal will not be affected by ambient temperature changes.

DC Amplification, Current Control

The DC signal is further amplified to drive a voltage-to-current converter that generates a current signal that is proportional to sensor temperature. This current, combined with the current used to operate the circuit, is fed through a feedback resistor rebalancing the bridge circuit. As a result, the 4–20 mA output current is regulated depending only on the sensor temperature. A limiter resistor, in the voltage-to-current converter circuit, limits the circuit from drawing too much current when in overscale conditions.

Isolation

Isolated power to run the unit is provided by a DC-to-AC converter that feeds a transformer. On the secondary side of this transformer, diodes rectify that AC to provide DC power to the circuitry.

Voltage Regulation

A voltage regulator circuit provides a stable voltage of 8.8 volts independent of supply voltage and load resistance variations. This stable 8.8 volts is applied to the DC-to-AC converter, which generates a square wave at the primary of the isolation transformer.

The transformer winding ratio is 1:1, which produces a square wave of equal magnitude at the secondary. This is an important feature because the voltages are equal, and the current drawn by the secondary circuit is accurately reflected in the current drawn by the primary circuit. Then the secondary signal is rectified, thus providing an isolated voltage supply for the 444 circuit. A reverse-polarity protection diode, in series with the DC power, protects the circuit against damage due to reverse power hookup.

Maintaining Temperature Devices

Contamination is the major problem with thermocouples due to oxidation or reducing atmospheres. Replacement of the device is generally required. Another major problem is loose connections, especially in RTDs.

Improper installation, wrong devices used, wrong extension wire used, wire reversals, and wrong placement of the measuring device will give false readings.

Potentiometers, decade boxes and other calibration devices can be used to simulate the input or measure the output.

First check the calibration of each device (primary element, transmitter, and input device) then check the overall loop by simulating the input and checking the final output of the controlling instrument.

Thermocouple calibrating ovens and ice baths are used to check thermocouples and other temperature devices.

Temperature transmitters can be ranged and calibrated in shops and laboratories prior to installation. Field calibration of smart devices can be ranged and calibration checked from the control system. Generally you can use a temperature calibrating instrument of the proper type as a simulated input and check the 4-20 mA output of the transmitter. Utilize at least a three-point calibration, zero, span and midrange.

9

Panel and Transmitting Instruments

Overview

In older plants, control rooms consist of control panels along many sides of the control room. Many instruments and switches are located on these panels. Behind those panels are many wires, terminal blocks, and more instruments and supplies.

Newer plants are designed with consoles, which include CRT screens and keyboards. All control is done by inputting through the keyboard and checking the results on the CRT screens. The computer CPUs and input and output devices are located in another room. In some cases, programmable logic controllers and other remote field controllers are located throughout the plant.

It is better to have the input device close to the measurement point and the output device close to the control point.

By using data highways and networking cables, much information can be sent to the main computer.

Panel and Behind-Panel Instruments

Although the process industries are moving rapidly to distributed control systems (DCS), programmable logic controllers (PLC), and personal computers (PC), conventional control panels still contain a percentage of the existing and operating control systems. Even with the advent of DCS, local control panels and racks used for motor control and other functions still follow the rules used for building and maintaining conventional panels.

When control panels are new, they look clean, neat, and functional. Unfortunately, years later, the face of the panel can look dingy and the rear of the panel can look like a bowl of spaghetti. To keep control panels looking as good as new and functioning properly, every addition or modification should follow the original panel specifications, and all work should be performed in a professional manner. Making repairs or additions on a temporary basis leads to panels that look bad and can be hazardous to maintain. Use ISA-RP60.1 through RP60.9 to review basic panel design and construction requirements.

Panel (Air and Electrical) Distribution and Problems

For good control and maximum uptime, instrument air must be properly filtered and reduced to the required pressure levels. Although this sounds simple, poor air supply systems are the cause of many control and maintenance problems. Dual filters and air sets, with proper valving to allow on-line switching of these devices, is necessary if one does not want to shut down a process simply to change filters or repair a malfunctioning air regulator. A good pressure gage should be mounted on top of the air header, close to the air sets, as an aid in setting the air pressure. The unreduced panel air supply pressure is almost always indicated on the face of the control panel and is alarmed when it goes below a set low pressure. Although the panel supply air pressure is usually 80 to 120 pounds per square inch (psig), it is reduced by the panel air reducing valve to about 20 psig, which is the air supply requirement for most instruments.

The main air header should be sloped towards a drain valve, which should be located at the far end of the air header. Air takeoffs should be located at the top of the air header so that any water that might be in the header is not drawn into the tubing supplying air to the individual instruments. Each air takeoff should include an isolation valve, properly identified and located at the header, to allow closing off the individual air lines for instrument maintenance or removal. This information should be kept in mind when building in-house panels or adding instruments to an existing panel.

Always check the air set and filter when troubleshooting any pneumatic control problem. Open the air header blowdown valve until the air blows free of moisture. If it takes more than five to ten seconds to blow the air header clear, one will not only have panel problems but will probably have an air dryer problem that could be affecting many loops.

Wet air causes many instrument problems. For this reason, instrument air supply tubing should slope away from the user and towards the air header, where any moisture can collect and be periodically drained from the header.

Electrical power to a panel should be properly labeled to identify all wires to and from the main power supply box or subpanel. Similar to the individual air supply shutoff valves used on pneumatic panels, electric or electronic loops also need individual shutoff switches or fusing to allow working on one loop without having to shut off power to the entire panel. Individual fusing also prevent a single short from damaging other power sources or controllers.

When troubleshooting an electronic loop, always identify the loop fuse or switch before attaching test gear or removing any wires. Be sure there are no common power feeds to the loop that, when opened, will inadvertently trip out a dozen other loops and possibly trip the entire plant. When working on critical systems, be sure the loop being worked on does not contain an alarm or trip that will fail to the trip position when power is removed from the loop or the process signal goes to zero.

Most electronic instruments operate internally on DC power but must receive their initial power as 120 volts AC. When battery backup systems are used, the instruments, in some cases, are supplied DC through the battery backup system. Most present day battery backup systems are selected to supply 120 volts AC to the panel instruments, since items that require backup might include devices that require AC power, which is internally converted to DC.

Because there are many ways to supply power to the various instruments in a control panel, it is recommended that the source of power, for the loop being checked, be disconnected before performing maintenance that could involve personal contact with potentially hot circuits. Obviously, power must be left on while checking for voltage, current, or power, but if the maintenance requires removing a wire or the entire instrument, it is best to remove power before performing this type of work.

DC power wiring should be separated from AC power wiring by keeping the two sources separated physically by distance, partitioning, or both. Although standards require this, do not rely on this always being true, especially on old panels. Many technicians have been stung by 120 volts AC or greater when they thought they were working on a DC circuit. Power distribution terminals should be labeled as 120-V AC, 24-V DC etc., to easily identify their source and voltage and should be physically separated by sufficient distance to prevent accidental shorting of the DC power with the AC power source.

Always check that the power is off before disconnecting any wires.

Process and Controller Tubing and Wiring

Panel tubing is usually copper but can be stainless steel, PVC-coated copper, or plastic. Copper is usually used when there are no corrosion problems, since it is rigid but easily bent and presents a neat, orderly installation. Stainless tubing has the same attributes, and it is resistant to corrosion. It is, however, considerably more expensive and is somewhat more difficult to bend. Plastic tubing is used when cost is important or when the panel equipment will be changed often. The use of plastic tubing is tolerated both in new panels and for quick fixes or for corrosion protection; with 90-degree metal corner supports and sufficient tie downs, plastic tubing can produce a credible job.

All metal tubing should be of the seamless type and should be bent to allow the connection to be easily removed from the instrument. Allow room to pull the tubing at least one inch away from the connection; otherwise, the tubing may not be removable after it is initially connected. Fittings should be kept to a minimum to reduce potential air leaks. All bends should be 90 degrees, if possible, and tubes should be parallel and supported sufficiently to prevent sagging. Use a tube bender when bending metal tubes. Bending a tube by hand, without first filling the tube with sand and then slowly bending the tube, will usually cause cracking, pinching, and poor looking bends. If one must use the hand approach, the tube should be blown clean before installing it in the panel.

Any high pressure, hazardous chemical, or direct steam connections to a local panel should have a block and bleed valve in the tubing or piping to allow shutting off the line and bleeding it down before opening any fitting or disconnecting the tubing. For hazardous chemicals, the bleed valve should be piped to the bottom of the panel where it can be collected in the proper container. Any condensate valve should be piped to the bottom of the panel and vented outside the panel to a safe place. Do not vent any hazardous liquid or gas inside the panel. Modern control room panels have only instrument air and electronic or electric signals entering the shell of the panel. Direct process connections are a potential control room operator hazard and, although a possibility in an old plant, should never be allowed in a new installation and should be retrofitted wherever a potential hazard might exist.

Pneumatic or hydraulic tubing fittings should be checked for small leaks. Oil leaks can easily be seen, but air leaks should be checked by using a leak detecting fluid or simply by applying soapy water to the connections or to parts of the tubing that are suspect. Ultrasonic leak detectors are also available that can detect minute leaks without touching the tubing or fittings. Some tube fitting manufacturers offer gap inspection gages that indicate when their fittings are properly tightened. Others mention a certain number of turns from finger tight. In any case, do not overtighten fittings, as this can cause added stress on the tubing or damage the tubing so that rather then preventing one, a leak is created.

If the instrument air is not always dry, be especially careful when installing plastic tubing, making certain that it does not sag or have low points that will collect water and eventually interfere with signal transmission.

Most electronic controllers provide a DC current source for the field wiring to and from the panel. This current 4-20 mA source is converted to a 1- to 5- volt signal, through a dropping resistor at the control panel terminal strip or in back of the controller or panel device. If the loop contains more than one device in the panel, each device is usually wired in parallel across a single calibration or dropping resistor. This allows disconnecting a device from a loop without actually opening the entire loop; i.e., if a recorder is unplugged from a control loop, the control loop is still connected to the process and is active. This wiring method is important for maintenance purposes, as it allows one to perform maintenance on one part of a loop without affecting the rest of the loop. Many modern controllers have built-in

If a pneumatic process signal is unsteady or bouncing and you need some damping, a simple resistive/capacitive network can be built by adding a needle valve and a volume chamber in series somewhere in the input tubing. Some devices have built-in damping adjustments, but if they do not, you can build your own by attaching a ¼-inch needle valve to a short piece of 1-inch pipe, capped at both ends. The needle valve should be a long taper needle valve for good adjustability and should be located on the process side of the tubing. The volume chamber should be between the needle valve and the panel instrument. The size of the volume chamber and the setting of the needle valve will determine the amount of damping performed.

dropping resistors, so it is important to read the installation booklet before working on an unfamiliar instrument.

When planning any critical loop wiring, do not allow more than one wire to be connected to the panel side of the terminal strip. Use jumpers on the field side, as required, to allow removing a wire within the panel without having to loosen a terminal that might contain several wires that, when loosened, might cause a trip.

In order to have only one wire under a terminal, jumpering of the terminal block is required. The use of miniature, modular terminal strips allows space for more terminals and usually provides special terminal arrangements that allow the jumpering without a maze of jumbled wiring. Figure 9-1 shows a conventional terminal strip using a complex wiring scheme to allow the "one wire under a terminal" principle. Figure 9-2 shows a modular terminal system that uses a double row of terminals, in the same space as a single terminal, and allows the jumpering to take place on one row of terminals while the field wires can terminate on the other row. Having two or more wires under one terminal allows the possibility of accidentally opening the circuit while attempting to remove only one wire. Critical loops should use the "one wire under a terminal approach" to avoid accidental trips.

For various reasons, some loops have portions of the loop wire in series, and removing a wire anywhere in that loop will cause the loop to loose power or continuity. Therefore, always refer to a loop drawing and check all of the wiring before doing work on any loop. When adding a new device to an existing loop, the same precautions apply. If adding a device in series with an existing loop, check the resistance being added to determine if the total resistance exceeds the power capability of the loop. Most instruments can handle about 600 to 800 ohms of total resistance before they become loaded down to the point that they do not perform to specifications. Check the vendor's specifications whenever approaching more than 500 or 600 ohms of total resistance in the series portion of the loop.

When a loop exceeds the maximum allowed resistance, the use of one-to-one repeaters or signal amplifiers should be considered. Also, if a ground loop is detected (usually only when interconnected, nonisolated instruments are grounded in two or more places), a signal isolator, one-to-one repeater, or signal conditioner can be used to eliminate the problem. When using loop-powered isolators, check the resistance load they must feed; most loop-powered isolators have lower load resistance capability than separately powered devices.

Loop grounding, in almost all cases, is done at the rack or control panel and not in the field. Since all the wiring terminates in the control or rack room in one location, all the grounding wires can be run conveniently to a single-source earth ground. Since the wiring is inside a building, the connections are protected and grounding will remain solid. Do not ground the loops at both ends, as this will usually cause ground loop problems. Grounding should be connected so as to produce a minimum resistance to ground. When using intrinsically safe circuits, it is required to have a resistance of less than one ohm to ground between the I.S. barrier and earth ground. This requires good grounding techniques and reasonably large grounding wire sizes. Grounding wires should always be green in color.

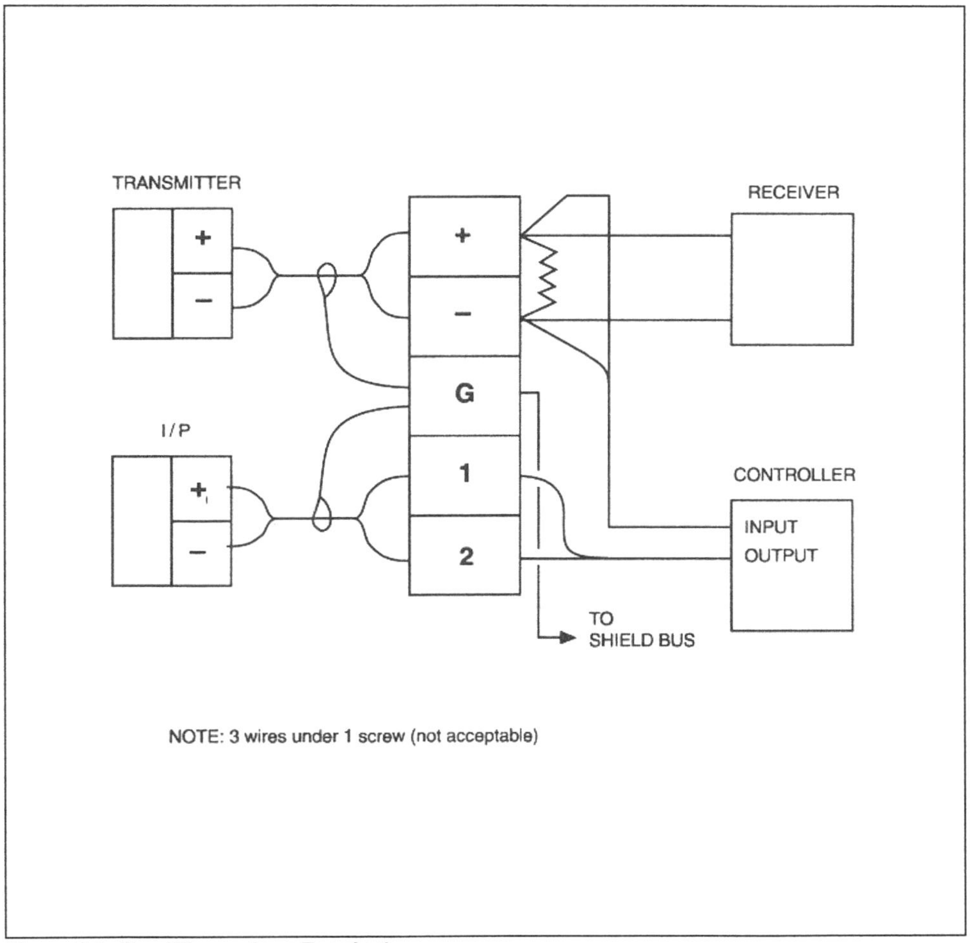

Figure 9-1. One Wire under a Terminal.

When rewiring or adding a loop to a panel, always separate the wiring from any AC source; i.e., do not run DC wiring in an AC cable tray or in an existing bundle of AC wiring; this is unsafe and may cause induced voltages in the signal cables. Intrinsically safe wiring must be run in a separate tray and must be certain designated distances from AC and nonintrinsically safe wiring. Refer to intrinsic safe wiring codes or ANSI/ISA-RP12.06.01-2003, Recommended Practice for Wiring Methods for Hazardous (Classified) Locations Instrumentation Part 1: Intrinsic Safety.

When adding tubing or wiring to an existing panel, try to maintain the same quality of workmanship as existed in the panel when it was new. When changing or removing instrumentation, also remove unused tubing or wiring whenever possible; don't merely disconnect it and let it hang. If it is best to leave disconnected wiring or tubing in a panel, "dress" it up and tag it as "spare." Be certain that unused or disconnected wiring is also disconnected from the power source. Either remove the fuse or disconnect the wire from the power source; don't merely disconnect one end and forget it. Tape it, identify it and neatly secure it to the rest of the wiring. Unused wiring and tubing simply adds confusion and leads to potential future problems. Housekeeping is a continual job. Clean up the work area after each job. Insist on time to do the job properly.

Most panels have special holders for storing drawings. When instrumentation is modified or removed from a panel, changes should be marked on the drawings; they should be dated and signed by the person involved. When this is done, the engineering or maintenance department should update the original and file draw-

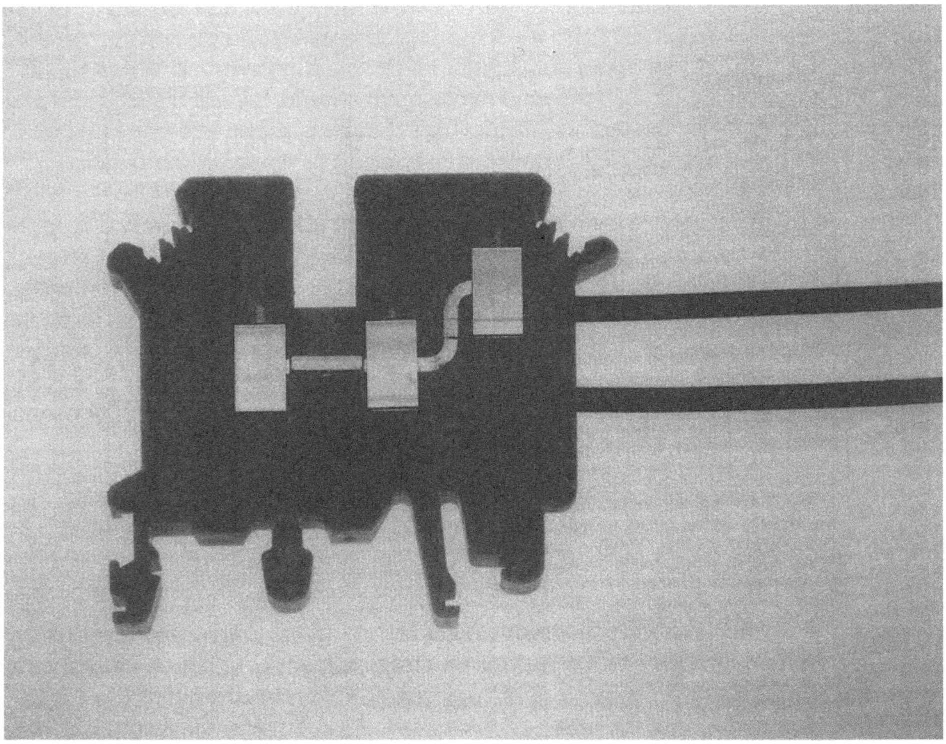

Figure 9-2. Modular Terminal System.
(Courtesy of Phoenix Contact, Inc.)

ings. Only by doing this whenever a change is made can one hope to have correct drawings the next time troubleshooting or work on the panel is necessary.

Modern instruments are mounted in housings that allow their removal from the panel simply by turning a screw, or switch, and pulling the instrument from its housing, automatically disconnecting any flexible tubes or wires that connected the device to the rear of the housing. Not all similar instruments or devices fit into the same "model number" housing. Even instruments of the same manufacturer and type may require that the rear manifold or backplane be altered or replaced with a different "model number" part. Read the installation manual when changing instruments within a housing.

When servicing any controlling instrument, always inform and obtain the operator's permission to place the controller on manual or to remove the instrument from the panel. Do not change tuning or alarm settings without informing the operator. New tuning or alarm settings should be logged into the operator's log book as well as your own log book.

When a new hole must be cut into an existing and working panel, considerable care must be taken to prevent metal filings and dust from falling into the instrumentation located below the area being worked. Also, to prevent damage to the finish of the panel, outline the area to be cut out with masking tape. The masking tape prevents chipping of the paint while making the cut. Place a drop cloth or piece of plastic film below and on both sides of the area to be cut out to catch any metal or plastic shavings. All too often, hours or days after such an operation, problems are caused for instruments below the new addition by some small piece of metal or debris that has fallen into an instrument.

New wiring should follow the same specifications used to construct the original panel. Use existing cable trays wherever possible and bring supply voltage from the main power supply distribution panel. Do not shortcut this procedure and tap into the supply of an instrument close by, as this violates the original panel design and will someday cause unexpected problems when this loop is being worked on and power to both loops might be accidentally killed. Document all changes and additions as soon as the work is completed.

Electric Switches, Alarms, and Gages

Panels contain many auxiliary devices that make up an operating system. Unlike the newer DCS systems, panel alarms, switches, push buttons, totalizers, etc., are individual panel or behind-panel devices that require mounting, wiring or tubing, and commissioning before they can perform their job. Care should be taken not to mount these devices in hard-to-see or hard-to-reach areas of the panel. Adjustment screws and scales should face out so the technician can easily see these when adjustments are required. If adjusting screws must face up, provide room above the instrument to allow for a vertical screwdriver adjustment. A capped pneumatic tee connection or an electrical parallel connection allows the system to be checked without disconnecting tubing or wiring, which simplifies maintenance and reduces set-up time.

When purchasing a pressure switch or relay, specify double throw switches so that they can be wired either normally open or normally closed. An adjustable dead band should be specified when an alarm or interlock is to function without chatter or turning on or off at the same point. Most pressure switches have calibrated adjustment scales, but the switch should be tested with an accurate pressure gage and a meter across the switching contacts. All switches should be mounted so the cover plate is easily accessible and the switch should be oriented per the vendor requirements, since some positions may cause a shift in the setting or produce poor repeatability.

When a large number of I/P transducers or pressure switches are involved, it may be practical to mount permanent test gear in the rear of the panel. Quick disconnects and special terminals with test points encourage more frequent checking and provide for long-term accuracy.

When adding push buttons or selector switches, locate them so that there is little change of accidentally activating them. Provide special covers for trip or shutdown buttons, and code all emergency buttons red or the designated emergency colors for the particular plant or mill. Follow all codes regarding distances between push buttons, terminals, and adjacent devices. Provide insulating barriers where there is a possibility of easily shorting or accidentally touching live, powered contacts. Always follow the National Electrical Code.

Panel alarm annunciators should be checked for burned out bulbs at least once a month. Most annunciators have dual bulbs, but do not wait until both bulbs burn out before replacing them. A burned out bulb is the same as a nonfunctioning alarm. Use the test button on the annunciator system at every opportunity and keep the annunciators in working condition. When alarms are changed, immediately have new windows engraved and/or temporarily mark the old window properly and legibly, so that the operator understands the new alarm.

If the area is noisy and the standard annunciator horn might not be heard, it is better to mount a rotating beacon on top of the panel than to buy a louder horn. Loud horns are eventually stuffed with rags or silenced, but rotating beacons are accepted and work well. Alarms should be heard but should not be so loud as to aggravate the operator when he or she is trying to cope with a problem. Use a different sounding horn for critical alarms.

Direct-connected draft gages should be checked and blown down whenever an operator suggests that the readings are inaccurate or acting strange. Piping should slope away form the gage, and there should be a blow-down valve just after the instrument connection. When troubleshooting draft gages, the first thing to look at is the piping, both in the control panel and also at the boiler. Condensate and dust can plug the sensing line and cause either a loss of signal or an intermittent or fluctuating signal. Sagging sensing or transmission lines can trap condensate and cause the same problems.

When connecting new tubing or cable to an existing panel, openings in the floor of the panel should always be sealed to prevent corrosive or hazardous gases from entering the panel. Even when these gases are not present, if the tubing or cable entry points allow moist or unconditioned air to enter the panel, they should be sealed. This is very evident when a local panel is mounted on the second floor and the connections enter through an opening in the floor. The opening acts as a chimney and draws these contaminants into the panel. Unsealed openings are also a fire hazard.

Special air conditioners, heaters, or filters may be required for panels located in areas where environmental conditions are conducive to instrument deterioration. These can be added, but it is important that they are added before the instruments begin to show signs of deterioration.

Purged panels are required in explosive atmospheres where nonintrinsically safe instrumentation or electrical devices are used in nonexplosion-proof housings. Always check the area with an explosive meter and obtain a safety permit before opening any panel located in a hazardous area. For more information, see ISA RP 12.4-1996, Pressurized Enclosures, and ISA 12.04.01-2003, Electrical Apparatus for Explosive Gas Atmospheres, Part 2, Pressurized Enclosures.

Most panels have an electrical outlet for an extension light or some other electrical use. Unless it is a specially marked outlet for maintenance purposes, do not use this receptacle for your drill, vacuum cleaner, or soldering iron. Many panels have been built with this source tied into the instrument power. When you use your drill on this power source, you may find some weird things happen, and the panel operator may not be too happy.

If your panels are wired this way, it is suggested that you run a separately fused power source to a maintenance convenience plug and label it properly. Then, when you overload this circuit, you will not kill a working panel.

Panels should have good interior lighting. A white panel interior color affords a good background and light reflective surface. Panel manufacturers have various standards. If you have specific needs, always spell them out when going out for bid. Changes to a panel after it is built, or even when only in the design stage, can be expensive; when changes are made after the panel is built, they tend to produce a panel that simply is not as good as a panel built properly the first time.

This section is a general review of simple and well used practices. Each plant or mill has its own standards and specifications that must be upheld if the work is to have quality, be understood by others, and be safe. Good housekeeping and good documentation, with a thought for others who might be the next to work on the panel, will be a help to everyone.

Gages have been used in monitoring processes for many years. Field gages are still seen on pipes and vessels, measuring pressure and temperatures. Some gages are filled so the vibrations do not affect them as much.

Some gages connect directly to the process such as pressure gages on a pipe while other gages are connected by wires to monitor voltages and currents.

Diaphragms, bellows, bi-metallic, and other concepts of measurements are used.

Very little maintenance is required. Most gages have a zero adjustment, which moves the needle slightly. You should set the needle at the verified operational set-point. Gages operate more accurately in the middle 1/3 of the scale. Replacement is generally the most cost effective means of correcting the problem.

Panel Meters

Panel meters can be analog or digital. Panel meters consist of gages which are panel mounted and the information measured is connected by wires or tubing. Panels can be field mounted or in the control room. Many panel meters have alarm points indicated. Newer bar graph panel meters may change color as you get close to an alarm condition or in an alarm condition. Other panel meters have switches that can send a contact closure to another device.

Discrete Switches

Discrete switches are mounted both in the field and also in control panels, racks, and control rooms. Maintenance is highly dependent on the field location and the process being detected. It is preferable to mount any field switch in a protected area or within an enclosed housing of some sort. However, many are simply mounted on process piping or on equipment that is located in a hostile area. When this occurs, and the device becomes a maintenance item due to corrosion, water or liquid damage or vibration, it may be best to simply relocate the switch.

The following is a quick checklist of things to look for and correct when troubleshooting discrete switches.

(1) Is the mounting bracket free of vibration? Vibration can cause premature failure or intermittent closures. Add support or relocate.

(2) Is switch interior wet? Add drip leg to wiring from device to the first junction box. Provide a drainage hole in the low point of the connecting box or mount the switch above the low point so that moisture will collect and drain away from the switch. Add a waterproof connector fitting to the switch housing. Move switch so it is away from any dripping liquid. Inspect housing connection gasket and replace if damaged. Tighten all screws used to attach the coverplate to the switch. If the cover is threaded, check the O-ring seal, apply lubricant if needed, or replace. Check that the cover is tightened sufficiently to seal, but do not over tighten.

(3) Is switch too hot or cold? Move switch or add insulation to protect against the high or low temperature.

(4) Check the locking nuts on zero and span adjustment screws. Tighten and recheck actuating setting.

(5) Check that the process connecting tubing is not plugged or crimped. Also check that the process connecting line is full of fluid, if on liquid service, and free of liquid if on gas service. On liquid service, bleed down or vent the process connection leg if it is thought to contain entrained air.

(6) Inspect and tighten any loose process or air tubing fittings. Inspect for minor leaks.

If the discrete contact is cycling or chattering, look for loose setting screws or nuts, or too small a dead band setting, as well as excessive vibration for the cause of the problem.

If the switch is open-circuited, look for broken wires, loose wiring connections, a bad switch, a wire on the wrong side of the switch, or a blown fuse.

If the switch is closed when it is supposed to be open, look for a fused contact, something across the switch contact or wires, or a wire on the wrong switch terminal.

When specifying a discrete switch, whether simply for maintenance inventory or for a particular project, specify the double-throw type rather than the single-throw. Many times a contact function has to be changed after the switch is installed. The small added cost of a double-throw switch pays off quickly because only one type of switch needs to be stocked and design errors will not require ordering a new switch.

When setting a discrete switch, it is best to make the adjustment in the field so that ambient conditions, location, and the mounting position is taken into account. These variables usually affect the calibration "zero."

Potentiometers

Potentiometers and reostats are used to adjust the output of electrical voltage or current. Potentiometers are also used to calibrate analog devices.

Recorders

Recorders have been used in plants for many years. The first recorders were single point recorders, that is they measured and recorded only one input. These recorders were either round chart or strip chart recorders. As concepts improved, multipoint recorders were introduced which provided recording several inputs at the same time. Switches were added to these recorders for alarm and control outputs.

Different colored ink was used to mark the paper. Some newer recorders utilized thermal paper and thermal print-heads. The very newest recorders utilize strictly digital indication.

Transducers

The term "transducer" applies to a device that is used to convert one type of signal to another type. In most control loops such instruments transduce an electrical signal to a pneumatic signal or a pneumatic signal to an electric signal. Usually the output, whether pneumatic or electric, is a standard 3–15 psi or a 4–20 mA signal and the input is usually a voltage, current, or pneumatic signal. The output of the transducer is used to provide a usable signal to an indicator, controller, valve or valve positioner, recorder, or an input to a DCS, if the initial signal is not compatible with the receiving device.

Current-to-Pneumatic Transducer

A current-to-pneumatic (I/P) transducer (see Figure 9-3) accepts an electrical signal and produces a proportional pneumatic output. Typically, 4–20 mA is converted to 3–15 psi (0.02–1.0 kg/cm^2). Models are usually available in direct and reverse action and are field-selectable with full or split range inputs or outputs.

The most common application if an I/P transducer is to receive an electrical signal from a controller and produce a pneumatic output for operating a control valve actuator or positioner. The device can be mounted on the wall or a pipe stand or directly on a valve actuator. Some industries prefer to mount them directly on the valve and select models that can withstand the normal vibration and accelerations experienced when close-coupled to the piping. Other industries prefer to remote mount the devices on instrument pipe stands to reduce any vibration effects.

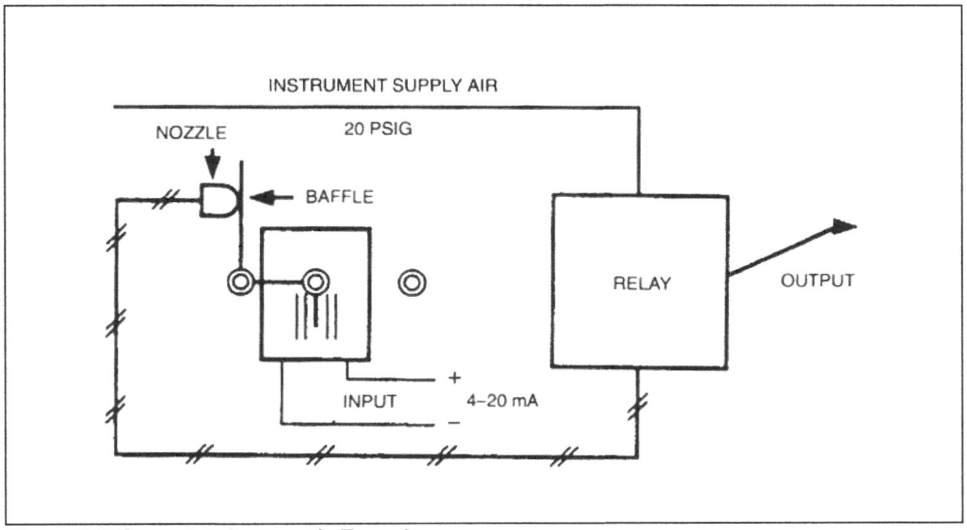

Figure 9-3. Current to Pneumatic Transducer.

When troubleshooting an inoperative or poorly operating I/P transducer, first look at the air supply and check that it is at least 20 psig or 3 pounds above the required output of the device. Slightly open the air set drain and check that the supply air is dry and oil free. If it is not, blow down the air line until the air is dry, open the transducer and clean the pilot orifice and the nozzle/baffle of any dirt or oil buildup. Also, check that any external bleed opening is clear and venting properly. Find the cause of the wet or oily air supply after cleaning the transducer and correct that problem also.

Since excessive vibration can cause problems, always look for loose adjustment nuts and fittings and tighten them when required.

Calibration is much the same for any transmitter. Provide an accurate input and check the output with an accurate test gage or test device. Check zero, span, and linearity; adjust as necessary. Use either an accurate current generator or an accurate voltage generator with a precision 250 ohm, ½ watt resistor. The pneumatic output indicator or test gage should have an output load volume of a minimum of 2 cubic inches. This will provide a stable operation and reading. If the transducer is bench calibrated, always check the zero after the transducer is re-installed in the field, since most devices are somewhat position-sensitive.

When the transducer is used to feed a valve positioner or directly feed a valve, always restroke the valve, through the transducer, to check the system before declaring the unit calibrated.

There are many types and models of transducers. Some have less output volume capacity than others and one needs to know where they are to be used before specifying a new unit or a replacement unit. Do not simply replace one unit with another type or brand unless the specifications are similar.

Pneumatic-to-Current Transducers

A pneumatic-to-current transducer (P/I) (see Figure 9-4) is the opposite or reverse of the type of transducer mentioned above. The P/I transducer converts a pneumatic signal to an electronic signal. The electronic signal can be a 4–20 mA DC, 10–50 mA DC or a few variations of these standard signals. The function is usually to convert an older pneumatic transmitter to a signal that is compatible with newer electronic control systems. Many times a large number of pneumatic signals must be converted to supply information to an electronic DCS system or electronic controllers. In this case, the P/Is are usually of the miniature type and are mounted, closely spaced, on some form of rail with a common air supply header system.

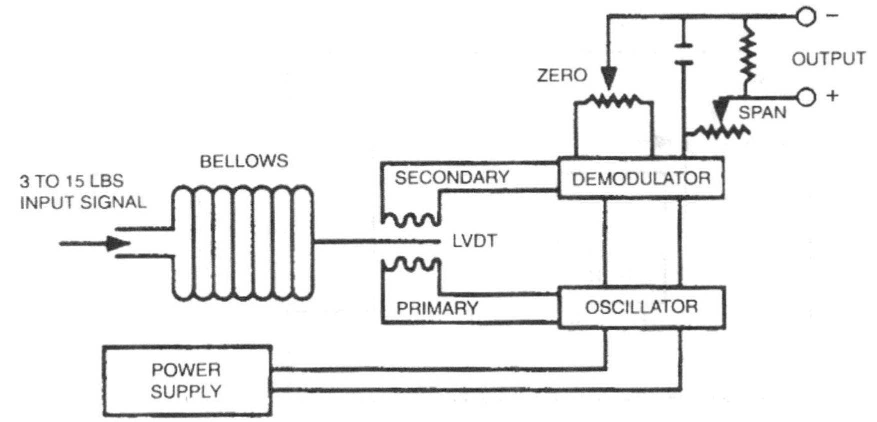

Figure 9-4. Pneumatic to Current Transducer.

Calibration and troubleshooting is very similar to the I/P transducer, except the output, which is electronic, should be checked for proper grounding and total resistance of the output loop. Most P/I transducers can feed up to about 500 to 800 ohms. If the loop resistance approaches or exceeds the manufacturer's published maximum allowable resistance and the output cannot be made to go to full scale, the loop resistance needs to be reduced.

When installing transducers of this type, be careful to allow easy disconnects of the input tubing, since the inputs must be disconnected for calibration or servicing. If metal tubing is used, provide sufficient bends to allow for loosening the tube fitting and pulling the tubing at least several inches away from the connecting point to the transducer.

Before using a P/I transducer to convert an old pneumatic transmitter output to an electronic signal, consider whether it is best to simply replace the pneumatic transmitter with an electronic one or whether it is economical to simply add a P/I transducer. Consider the accuracy of two devices versus one and remember that you will have to maintain and calibrate two devices rather than one.

Smart Transmitters

This section gives the basics of the operation, installation, calibration, and maintenance of smart transmitters. Because of the design and operation differences, smart transmitters require calibration procedures that are different from those of analog devices. The user manual for the particular device should always be consulted in order to achieve the best performance and reliability.

A smart transmitter will be used as an example throughout to maintain a consistent train of thought. Most of the concepts introduced for smart pressure transmitters hold for other types of microprocessor-based field instrumentation as well.

The features of smart transmitters can provide substantial benefits to users in terms of time and labor savings as well as an increase in plant operating safety.

The overwhelming favorite means of communicating a process signal from the field to the control room is via the 4–20 mA analog current loop. While fast, reliable, and an industry standard, it leaves room for improvement in the areas of maintenance capabilities, performance, and diagnostics, to name a few. Analog field instruments can be inaccurate, less reliable, and expensive to install and maintain. Smart field instruments may not only address these concerns but can provide further benefits as well.

Smart transmitters have a microprocessor built into each measuring device, which manages communications and performs diagnostic checks, error correction, and setup functions. Communications is now a two-way street — not only from the field to the control room, but also from the control room to the field. Some smart transmitters provide simultaneous analog and digital communications. Listed below are some features of smart transmitters that can provide advantages to users over analog devices.

(1) Improve accuracy
  (a) Ambient temperature and pressure compensation
  (b) Sensor linearization
  (c) Characterization over entire span of instrument
  (d) No D/A conversion error
  (e) Increased rangeability
  (f) Digitally adjustable dampening
  (g) Noise resistance

(2) Greater measurement stability
  (a) Elimination of loop errors
  (b) Inherent temperature insensitivity of digital electronics
  (c) No potentiometers in electronics that can create drift

(3) Reduced start-up, operating, and maintenance costs
  (a) Self-diagnosis
  (b) Reduced wiring costs through multi-drop installation
  (c) Built-in setup features
  (d) Multiple function devices
  (e) Wider rangeability/reduced inventory
  (f) Reduced record keeping through use of transmitter database

(4) Increased safety
  (a) Fewer trips into hazardous areas
  (b) Built-in security to prevent unauthorized tampering
  (c) Ability to set transmitter gross fail conditions

The functionality of smart field instruments can be thought of as analogous to the computer industry. Not too many years ago the norm for computing in the workplace was having a computer terminal at the desk that was connected to a mainframe. While this setup provided a local interface to computing capability, it was limited in functionality, flexibility, and specialization capability. With the advent of the personal computer and local area network, there is an enormous amount of distributed intelligence in the office area. Individual PCs/workstations can be configured to offer the most in terms of flexibility, functionality, and productivity for the individual user while still being able to share information and capabilities with the rest of the network. Smart transmitters are like these personal computers. They, in effect, distribute the intelligence throughout the process of field measurement to provide additional flexibility and functionality for the user. Just as PCs are slightly more expensive than computer terminals, the additional functionality of smart transmitters sometimes does not come for free. However, many analyses have shown that the additional expense is well justified. In fact, some studies have shown a savings of $2,000 per field device over a five-year time frame.

The rest of this section will focus on the setup, calibration, maintenance, and troubleshooting of smart transmitters. Before discussing these actual procedures, however, it is important to have a rudimentary understanding of the operation of the devices in order to achieve the maximum benefits.

Setup

The theory of operation of a smart transmitter is best explained by comparing it to the operation of an analog transmitter. The block diagrams shown in Figures 9-5 and 9-6 are used to illustrate this discussion.

With an analog pressure transmitter, the signal from the sensor is conditioned by the transmitter electronics, producing an analog 4 to 20 mA output that is proportional to the applied input signal. Zero and span adjustments are made via potentiometers to set the calibrated span of the device.

With a smart pressure transmitter, the input signal from the sensor is converted to a digital format via an analog-to-digital converter for use by the microprocessor (see Figure 9-7). The microprocessor accepts the digital input signal from the sensor module along with temperature and pressure correction coefficients, which are generated during characterization and stored in the nonvolatile memory of the device. Characterization is the process of matching the electronics of the device to a specific sensor. Different devices will have different procedures here, but most involve the application of the measured variable to the device over the entire operating range combined with cycling over the device's rated ambient temperature range. Data generated from these cycles is used to generate correction coefficients that are permanently stored in the devices nonvolatile memory.

Some smart devices are available in retrofit packages that can be used to upgrade existing analog field devices. In this case the characterization is done in the field with known inputs. In either case, this information is stored in the device for use in continuously correcting the output signal of the device based on its ambient conditions. This provides a much more accurate control signal.

After the microprocessor performs the above corrections, a corrected digital signal is generated by the microprocessor. In some devices this signal is available in engineering units, while other devices may present a scaled digital signal that the control system interprets. This digital representation of the process variable is sent to the D/A converter for conversion into a 4 to 20 mA analog signal. The signal is also sent to the modem in the device for conversion into a digital communications signal.

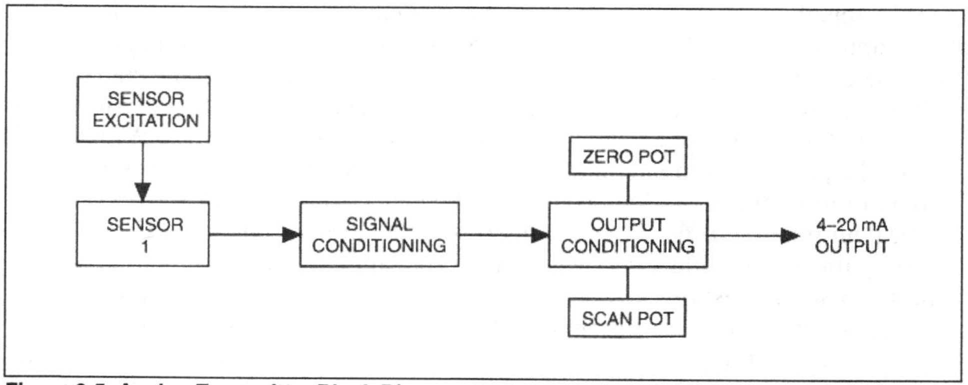

Figure 9-5. Analog Transmitter Block Diagram.

Calibration

Because smart transmitters operate under principles that are different from those for analog devices, the calibration and maintenance procedures required to achieve optimum device performance are different. The calibration of analog

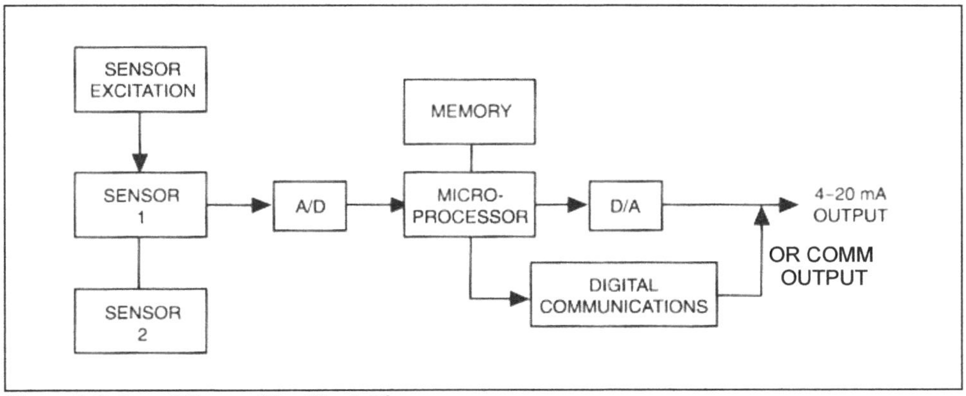

Figure 9-6. Smart Transmitter Block Diagram.

pressure transmitters requires the application of known pressures and the adjusting of potentiometers. Whenever this calibration is changed, or the device is rearranged, another set of known pressures must be applied.

"Calibration" of smart devices is divided into several sections: characterization, digital trim (zero sensor trim, full sensor trim, 4 mA trim, 20 mA trim), and rearranging (4 mA set point, 20 mA set point). Figure 9-7 illustrates the calibration functions that may be performed by the user.

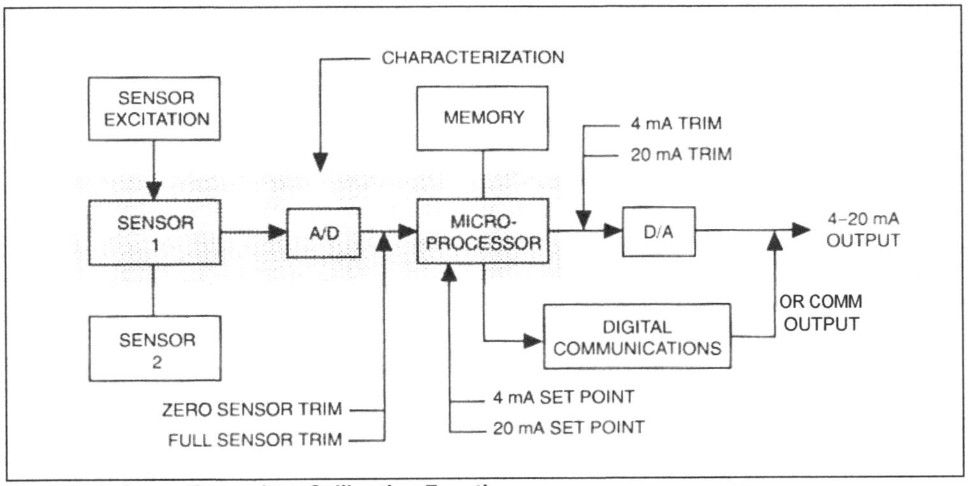

Figure 9-7. Smart Transmitter Calibration Functions.

CHARACTERIZATION

The characterization of a smart transmitter allows for permanent storage of reference information. Known pressures are applied and the device stores information about these pressures and how they make the sensing cell react. In effect, a transfer function of applied pressures versus output is created (see Figure 9-8). It is then possible to range a smart pressure transmitter based on the initial characterization information simply by entering the desired span end points through the hand-held communicator. This moves the span end points along the know characterization curve.

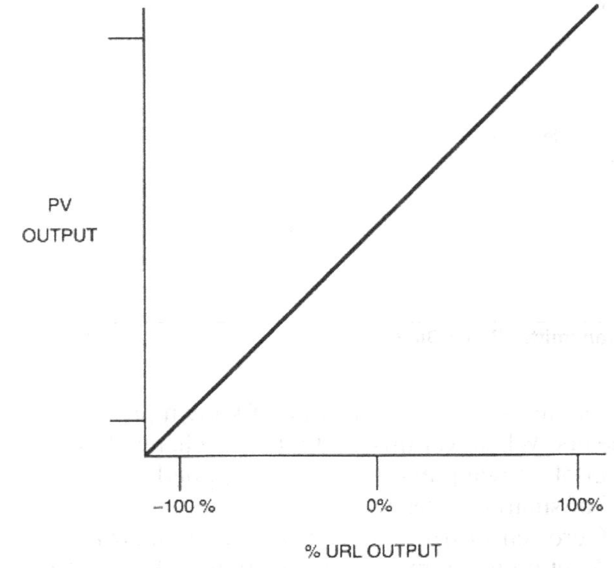

Figure 9-8. Transfer Function.

The method of ranging and calibration chosen for smart transmitters depends on the performance desired. Tradeoffs can be made favoring time over performance.

Adjustments to the initial characterization curve can be made with the digital trim function in order to match plant standards.

The digital trim feature should not be confused with rearranging. Although you can still match a pressure input to a 4 or 20 mA output as in the calibration of conventional analog devices through a rearranging function, you have not affected the transmitter's digital interpretation of that input; you have simply moved the range points within the transmitter's allowed range.

DIGITAL TRIM

In order to understand the digital trim function it is necessary to remember the differences in operation between conventional analog devices and smart devices. The relationship between the input pressure and the transmitter's output is determined by the initial characterization curve.

Digital trim is a calibration function that allows the user to correct the transmitter's digital signal to match plant standards or compensate for installation effects. The term "digital trim" refers to two functions. The first function, called sensor trim, consists of matching the process variable reading of the transmitter to a precision pressure input. The second function, called 4–20 mA trim, consists of adjusting the analog output electronics to match the loop readout meter.

With a smart transmitter, one can alter the transmitter interpretation of the input signal using a digital trim function called sensor trim. The transmitter conversion of that interpretation can also be altered into an analog 4–20 mA output, using a digital trim function called 4–20 mA trim.

A smart transmitter can be rearranged in three ways (the last two are most similar to traditional "calibration" of a conventional analog device):

(1) Using the keyboard of the communicator

(2) Using a sensor input and the communicator

(3) Using a sensor input and the integral span and zero buttons

In each of these methods the transmitter uses its initial characterization curve to interpret the readings given by the sensor at a particular input pressure. It is possible that the stored characterization curve does not match the plant standards for pressure. This discrepancy may be noticed when the plant standard is applied to the transmitter and the value is compared to the digital reading obtained with the communicator. It is most likely that the difference between the readings is caused by the differences in measurement standards and the conditions under which these standards are applied.

Should there be a difference here, one can choose to leave the characterization alone, or alter the characterization in order to match ones plant standard in one of two ways: (1) by recharacterizing the transmitter, or (2) through digital trim.

Once a transmitter has been characterized, the easiest method to change the characterization curve to match a plant standard is through a digital trim. Digital trim is similar to "calibration" of conventional transmitters, but with microprocessor electronics it is a two-step process.

To eliminate the possibility of overtrimming a transmitter, manufacturers may place limits on how much a given device can be trimmed from factory characterization.

There are two ways to perform sensor trim: zero sensor trim and full sensor trim. They vary in complexity, and their use is application-dependent. Zero sensor trim is a simple one-point adjustment, commonly used to correct for mounting position effects and any zero shifts caused by the application of static pressure. It must be zero-based, and it may be performed when an exact pressure source is not available for the second pressure needed in a full trim. The result is very similar to "tweaking" a zero pot on an analog transmitter when the transmitter is in the final mounting position. Since this method maintains the slope of the characterization curve, it should not be used in place of a full trim over the full sensor range. Since it is zero-based, it generally should not be used for absolute pressure transmitters.

A full sensor trim is a two-point process in which two end point pressures are applied and all output is linearized between them. It allows the user to select low trim and high trim end point values to provide a digital calibration over the required measurement range. The low trim value determines the offset or "zero" adjustment to the characterization curve by correcting all points along the curve equally.

The low trim graph shown in Figure 9-9 illustrates an example of low trim correction. The values have, of course, been greatly exaggerated.

Remember that the transmitter can be only as accurate as the equipment used to digitally trim it. For best results, use precise equipment under stable ambient conditions. If such equipment is not available, it may be better to return the transmitter to a manufacturer's service center for verification of the trim values.

Because a low trim does not need to be performed at a true process variable of zero as in a zero sensor trim, it is possible to use low trim to correct for mounting position effects on an absolute pressure transmitter. This allows the "zero" correction to be performed at atmospheric pressure.

For best accuracy, the digital trim points should be set equal to or slightly greater than the maximum 4 and 20 mA set points. Trim points set inside the range points can cause excessive errors outside the trim points.

4–20 mA trim can also be used to make adjustments to allow for peculiarities of a particular readout device in the loop.

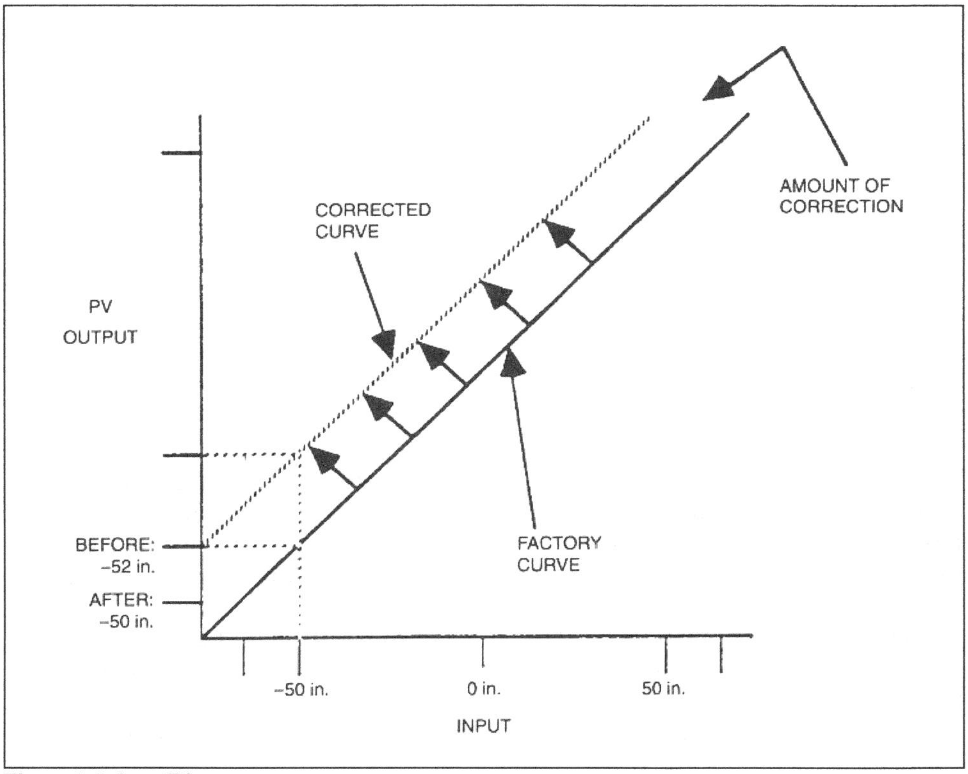

Figure 9-9. Low Trim.

For example, assume that a negative 50 in. H_2O has been applied to the transmitter with a highly accurate pressure source, and the process variable reading shows negative 52 in. H_2O. If corrections are made using the low trim procedure, the characterization curve can be adjusted so that the digital process variable will match the pressure input. The effect of this trim is that the entire characterization curve shifts the same amount at every point.

The high trim value determines the "span" or slope adjustment to the characterization curve based on the low trim value. These values should never be changed without a highly accurate pressure source. The graph shown in Figure 9-10 shows a high value trim correction based on the low trim adjustment just made. Continuing the example, the high trim value has been set at 50 in. H_2O. This has the effect of altering the slope of the characterization curve from the low point to the high point.

The other digital trim function available is 4–20 mA trim. This procedure requires an accurate current meter and is used to match the transmitter's analog output to the loop readout meter. It is, in essence, a "calibration" of the output electronics.

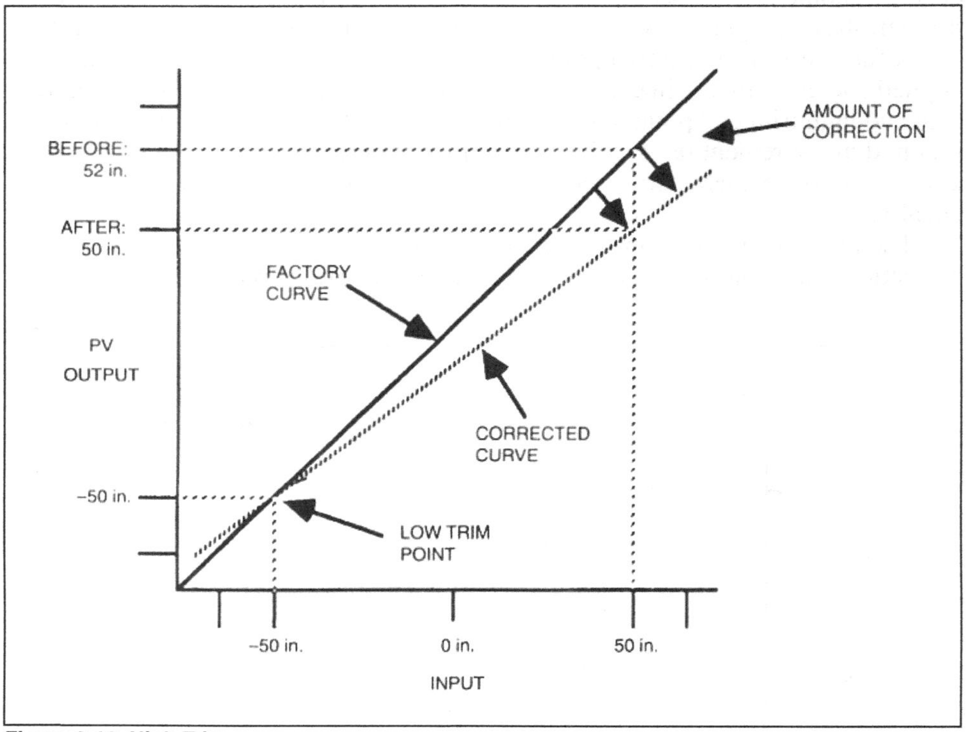

Figure 9-10. High Trim.

To determine whether the output circuitry should be trimmed, connect the transmitter as shown in Figure 9-11. Enter the loop test mode of the transmitter. Drive the analog output to 4 and 20 mA, checking the output against the precision reference meter. Any necessary corrections to the analog output are made during this procedure.

The digital trim functions are available to the user as part of the overall calibration procedure. Performing these digital trim functions allows the transmitter to perform with the highest possible accuracy. It also allows the user to match the device to accepted plant standards. Since such highly accurate calibration equip-

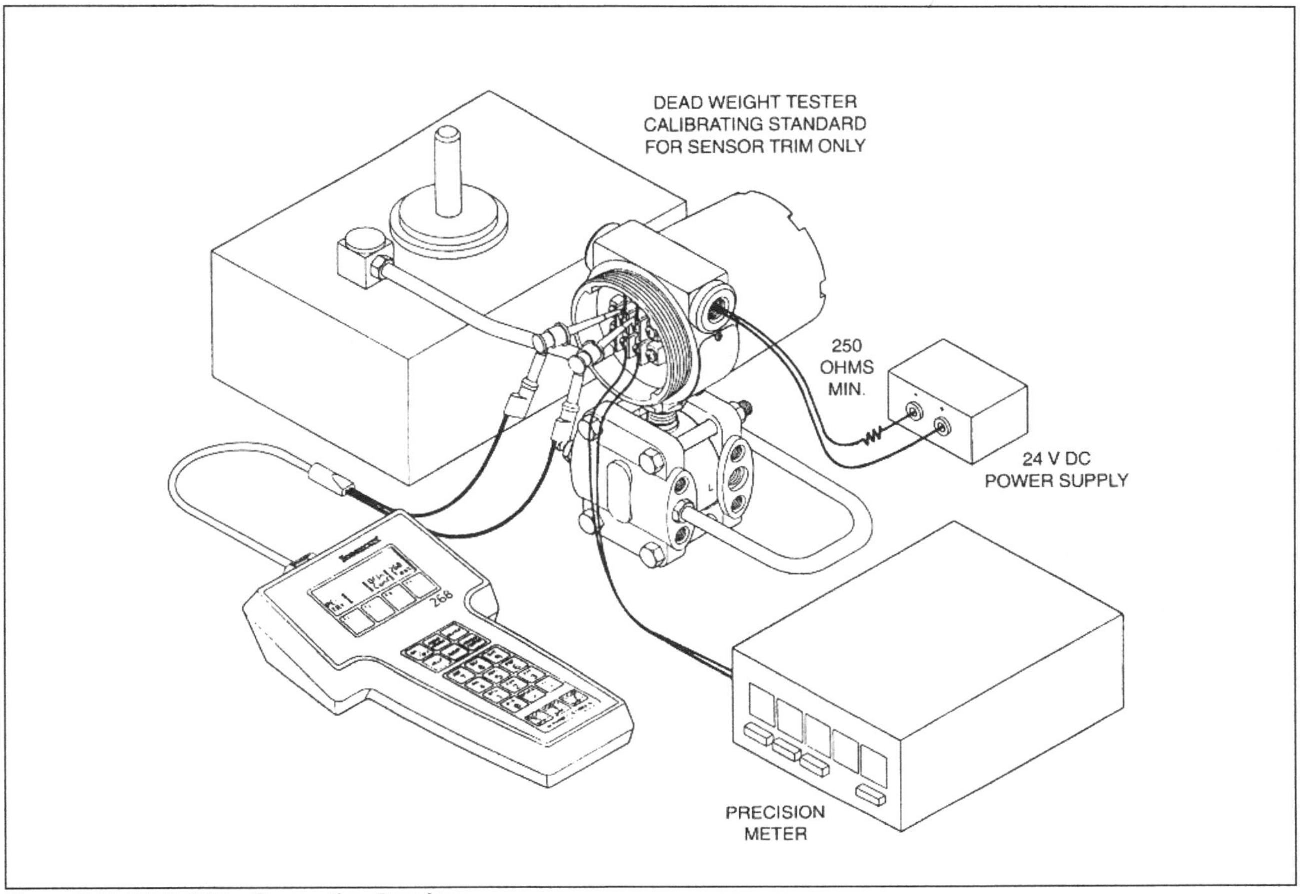

Figure 9-11. Digital Trim Connection Drawing.
(Courtesy of Rosemount, Inc.)

ment is necessary, the digital trim functions may not be routinely performed by users.

REARRANGING

In addition to the digital trim capabilities, smart transmitters also have the ability to be rearranged anywhere along their characterization curve. In essence, range values are simply picked along a predetermined curve. As previously mentioned, there are typically several methods by which a smart transmitter can be rearranged:

(1) With reference pressure input and local zero and span adjustments

(2) With the hand-held communicator via keyboard entry

(3) With a reference pressure and the hand-held communicator

The important differences between these methods may help in choosing which method is best for a particular application.

In rearranging the 4 mA point with a pressure input source (via either the handheld communicator or the local zero button, for example:

If the transmitter is ranged so that

$$4 \text{ mA} = 0 \text{ in. } H_2O$$
$$20 \text{ mA} = 100 \text{ in. } H_2O$$

and the 4 mA setting is changed to 50 in. H_2O by applying 50 in. H_2O, the new settings are:

4 mA = 50 in. H_2O
20 mA = 150 H_2O

The 100 in. H_2O span is maintained.

It is also important to note that when using a pressure source, the 4 and 20 mA set points are based on the transmitter interpretation of the pressures that have been provided during characterization and digital trim.

Using the zero and span adjustments on smart transmitters is similar to using the potentiometers on conventional analog transmitters. Their operation, however, is discreet rather than continuous like that of potentiometers. It is also totally non-interactive, unlike analog instruments.

Although the 4 and 20 mA set points will operate properly within these applied settings, the transmitter's digital output in engineering units may indicate a slightly different value. Recall that sensor trim is used to match the transmitter reading in engineering units to the plant standard, thereby eliminating any discrepancy.

To rearrange a smart transmitter using the local zero and span adjustments, follow these steps:

(1) Using a pressure source with an accuracy three to ten times greater than the desired calibrated accuracy, apply a pressure equivalent to the lower calibrated value to the high side of the transmitter.

(2) Activate the local zero adjustment. Verify that the output is 4 mA.

(3) Apply a pressure that is equivalent to the higher calibrated value to the high side of the transmitter.

(4) Activate the local span adjustment. Verify that the output is 20 mA.

When the local zero/span adjustment is activated, the transmitter saves the current digital process variable reading as the 4 or 20 mA range point.

Rearranging with the keypad on the hand-held communicator typically changes the 4 and 20 mA points independently. When the 4 or 20 mA setting is changed, via the keypad of the hand-held communicator, the span is also changed. For example:

If the transmitter is ranged so that

4 mA = 0 in. H_2O
20 mA = 100 in. H_2O

and the 4 mA setting is changed to 50 in. H_2O, the new settings are:

4 mA = 50 in. H_2O
20 mA = 100 in. H_2O

Note that the 20 mA set point remained at 100 in. H_2O, which changed the span from 100 in. H_2O to 50 in. H_2O.

When using only the keypad and the hand-held communicator, simply type the 4 and 20 mA set points into the communicator using the numeric keypad and send this information to the transmitter. The transmitter again saves these values as the 4 or 20 mA range points.

Commissioning

A smart transmitter should be commissioned with the hand-held communicator before being put into service. Commissioning may consist of testing the transmitter and the loop and verifying the transmitter configuration data. The transmitter may be commissioned either on the bench or after installation.

To commission on the bench, connect the transmitter as shown in Figure 9-12. The equipment required includes a power supply and a current to the voltmeter. A resistor (typically 250 Ω) may be necessary for communications. After connecting the equipment and venting the transmitter to atmosphere, turn on the hand-held communicator or other communicating host to verify communications. Many devices will display the transmitter manufacturer, model number, and tag. If a screen appears indicating transmitter security is on, the transmitter will not accept changes to its memory. It will be necessary to disengage this function if changes are to be made to the transmitter memory during commissioning.

It may be useful to commission the transmitter before installation to ensure that all components are in good working order and to acquaint oneself with the operation of the device installation.

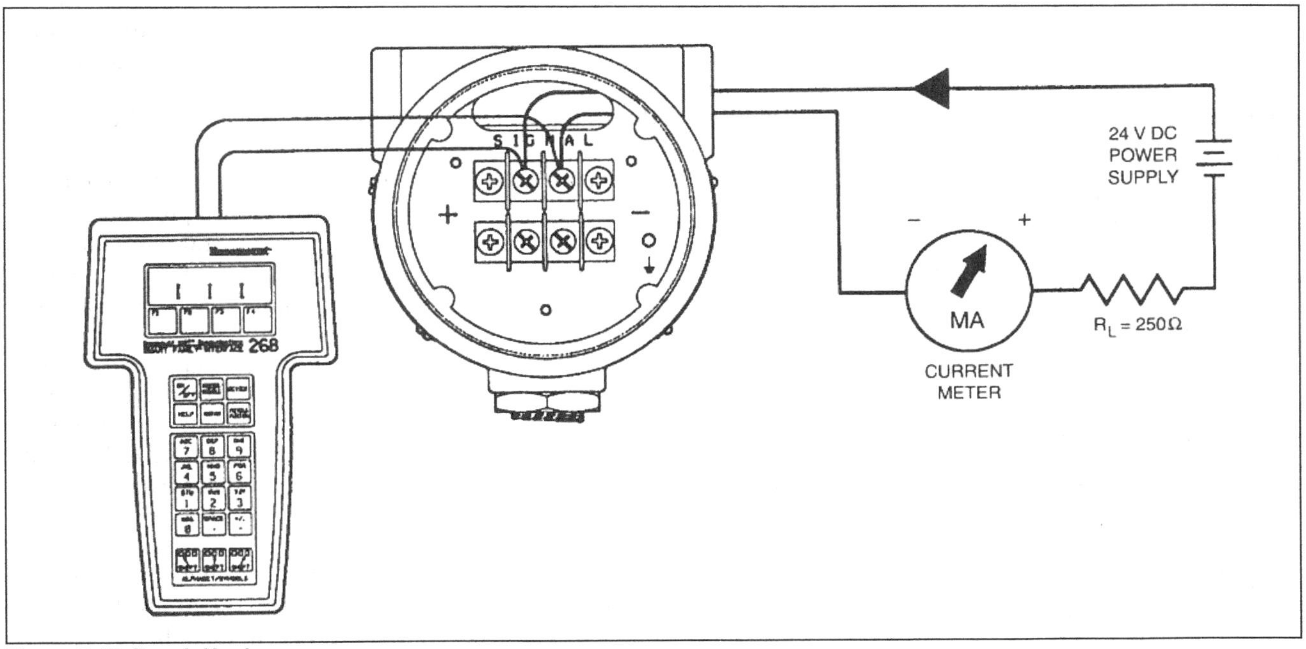

Figure 9-12. Bench Hookup.
(Courtesy of Rosemount, Inc.)

Before proceeding, some of the available test procedures may be performed. Testing is also recommended when component failure or a problem with loop performance is suspected. Smart devices may provide up to three separate types of tests: communicator test, transmitter test, or loop test. The communicator test tests the functionality of the hand-held communicator. The transmitter test will typically initiate a transmitter test more rigorous than the diagnostics run continuously in the device. The loop test allows for verification of transmitter output, loop integrity, and the operation of any recorders or similar devices. If commissioning is taking place on the bench, the loop test should be repeated after the transmitter has been installed in the field.

Each smart transmitter carries in its nonvolatile memory a database of information that is related to the operation and configuration of the transmitter. It is a good idea to review these parameters as set at the factory. This database may include but not be limited to the following:

The output current of the transmitter can be driven to predetermined levels in order to check the operation of all devices in the loop. Current levels on all devices can then be verified.

Transmitter model	Sensor fill fluid
Manufacturer	Isolator material
Transmitter type	Flange material
Tag	O-ring material

Transmitter range	Vent/drain material
Date	Integral meter
Descriptor	Remote seal type
Message	Remote seal fill fluid
Min. and max. sensor limits	Remote seal isolator material
4 and 20 mA point settings	Number of remote seals
Minimum span	Sensor serial number
Engineering units	Transmitter serial number
Configuration of output	Transmitter multi-drop address
Damping	Transmitter software revision level

As the last step in start-up and commissioning, the transmitter output should be checked. This can typically be done by using the process variable function on the hand-held communicator or the communicating host. This should display the process variable in engineering units. Any discrepancies from the desired configuration should be corrected at this point.

Any zero difference is usually due to a difference in transmitter orientation. Any span difference is most likely due to the fact that the pressure source does not match the one previously used to calibrate the transmitter.

Rearrange the transmitter to the desired range points using any of the three methods discussed previously.

After the transmitter has been rearranged to meet application requirements, one may wish to verify the digital calibration of the device using a pressure source at least three times as accurate as the transmitter. With the transmitter vented to atmosphere and in a stable, draft-free, vibration-free environment, observe the digital process variable reading. Next, apply the 20 mA set point pressure and again observe the process variable reading.

Since the analog signal is being used for control in this instance, the only benefit to performing the digital sensor trim would be that the digital process variable as viewed on the hand-held communicator would match plant standards.

At this point the transmitter may be left as is or corrections may be made using the sensor trim procedure described previously to optimize performance over a specific measuring range. However, if the 4 and 20 mA set points were set with a precision pressure input and the digital process variable is not going to be used for control purposes, it may not be necessary to perform the sensor trim function.

INSTALLATION

The complete calibration functions that a person performs on a smart transmitter have been demonstrated. The transmitter has been tested and meets the requirements of the application. The transmitter is now ready to install, and the rest of the commissioning can be performed on line.

The accuracy of a smart transmitter may depend a great deal on installation. For example, the accuracy of a pressure, DP flow, or level measurement depends, to a great extent, on proper installation of the transmitter and impulse piping. The piping between the process and the transmitter must accurately transmit process pressure to the transmitter.

The transmitter should be mounted close to the process with a minimum of piping to achieve best accuracy. Refer to the manufacturer's suggested installation for best performance.

All field transmitters should be installed to minimize vibration, shock, and temperature fluctuations where possible. The need for easy access, safety of personnel, and practical field calibration when installing field instrumentation must always be kept in mind.

Pressure transmitters are typically calibrated in an upright position at the factory. If this orientation is changed during mounting, the zero point will shift by an amount equal to the liquid head caused by the mounting position. This is corrected by performing a zero trim or re-zero.

Before installing a smart transmitter, remember there may be minimum loop resistance requirements to ensure that communications with the device will work. This may also increase the minimum voltage required at the transmitter terminals and have an effect on minimum power supply requirements. Check the manufacturer's power supply load limitations table to ensure minimum requirements are met.

The same general wiring practices should be followed with smart transmitters as with traditional analog transmitters. Do not run wiring in conduit or open trays with power wiring or near heavy electrical equipment. Signal wiring may be grounded at any one point on the signal loop, or it may be left ungrounded. The negative side of the power supply is a recommended grounding point. Signal wiring need not be shielded, but twisted pairs should be used for best results. Wiring of 24 AWG or larger should be used and should typically not exceed 5,000 feet. The transmitter case may typically be either grounded or ungrounded.

After the transmitter is installed, the last thing one may choose to do is to set the desired alarm level and the security feature to prevent accidental or deliberate rearranging.

Now that the transmitter is installed and corrections have been made for installation effects, the output can be verified at the loop control device. This task is greatly simplified by the use of the two-way communications in a smart transmitter.

Weight, Speed, and Vibration

WEIGHT MEASUREMENT

Weighing devices measure the net weight of products. The calibration is generally checked using certified known weights. Some scales utilize balanced arm against a known weight, measure pressure, utilize load cells, or strain gage.

MECHANICAL LEVER SCALES

Using the mechanical advantage between the load arm and the power arm (counterbalance arm), the scale mechanism is designed in such a way that without any weight or loads on either pan, both arms will be level. Known weights are placed on the power arm, and the load is placed on the load arm. The length of the power arm is directly proportional to the weight.

HYDRAULIC LOAD CELLS

Hydraulic load cells consist of a piston and cylinder, whereby the load exerts a counter force upon the piston which, in turn, is supported by hydraulic fluid confined within a sealed chamber. The pressure exerted in the hydraulic fluid chamber is proportional to the force exerted on the weight-supporting piston. The output of a hydraulic load cell may be connected to a pressure transmitter which will convert the pressure value to any type of signal for input to a controller, recorder, or computer.

PNEUMATIC LOAD CELLS

Pneumatic load cells are quite similar to hydraulic load cells. Pneumatic load cells employ diaphragms of a flexible material rather than pistons.

ELECTRIC LOAD CELLS

Electric load cells, the most widely used weight sensors today, are of two types: strain gage cell and induction cells. Cantilever load cells using a strain gage is a configuration designed to measure very small loads.

TENSILE LOAD CELLS

Tensile load cells are used for lifting machinery. Load is transmitted to the tensile member by two end fittings. An end-load measuring strain gage bridge is employed.

COMPRESSIVE LOAD CELLS

Compressive load cells are often used in weighing platforms. Compression load cells use a Wheatstone bridge circuit with strain gages as arms in the bridge.

CONVEYOR BELT SCALES

Belt conveyor weighing systems are utilized to measure the flow of solids. The weight of the belt and material it contains is transmitted to the load cell. If the belt is running at a continuous speed, the total amount of weight passing over the scale is easily weighed by integration.

NUCLEAR GAGE CONVEYOR BELT SCALE

This non-contact weight measurement uses gamma rays emitted to the target, where the rays are observed. An increase of material on the belt increases the amount of radiation absorption and decreases the input to the detector, which produces a low-level current output.

Speed Measurement

Speed can be measured using tachometers. Most tachometers count pulses from a proximity probe. Tachometers are critical instruments for use as an overspeed protection system. American Petroleum Institute's Standard 670 outlines the stringent requirements for rotating equipment.

Tachometers systems can not only measure rotation speed, but can measure rate-of-change, rotation direction, locked rotor, and zero speed. Stand alone speed indication generally uses a panel mounted meter directly wired to a suitable speed input, such as once-per-turn signal, and may provide power to the transducer. Accuracy is generally +/- 1 rpm.

Magnetic pulse transducers are generally mechanical in design and operation. Pulses are counted as the magnet passes a metallic object. Optical tachometers generally receive light pulses. The rotating device blocks the light being received by a photo diode. Eddy Current proximity probes are recommended for speed, overspeed, and phase reference applications. They are superior to magnetic pickups because they allow better circuit integrity checking, work frequencies as low as 0 Hz, have a linear output regardless of shaft rotative speed, and have excellent high frequency response characteristics. In many applications, a probe observing single-event-per-turn notch (keyway) or projection (key) is appropriate for both speed and phase measurement.

Vibration Measurement

Unlike proximity probes, which observe the machine's shaft directly, casing vibration transducers measure the vibration of the surface to which they are attached – usually a bearing housing. Devices in this category include both accelerometers and velocity transducers. Information is then sent to a control system, alarm, or a panel meter.

Proximity Measurement

A proximity detector is a sensor that produces an electric signal when the distance from the sensor to another object is less than a predetermined value. A proximity switch is a device that senses the presence or absence of an object without physical contact and activates an electrical circuit as a result.

Quantity Measurement

Quantity measurement can be just counting pulses as objects pass by a transducer or they may be level measurements to measure heights. An event can be classified as a quantity measurement which reports the on-off information. This can also be recorded against time to determine when events happen. These events may just show the presents or absence of an object.

Frequency Measurement

Frequency is the number of cycles a periodic variable passes through per unit time. A frequency meter is an instrument for determining the frequency of a cyclic signal, such as an alternating current or radio wave.

10

Analytical Instruments

Overview

Process Analyzers are a unique category of process instrumentation and require unique handling in their selection, design, and maintenance. Most of the other (not pneumatic) process monitoring instruments are monitoring a physical property (typically pressure, temperature, flow, etc.) of the process. They are electrophysical devices and consequently require electronic and physics understanding to implement and maintain properly. Process Analyzers are utilized to monitor a chemical property (typically composition related) with an electro physical (or sometimes chemical) device in the process. Consequently Process Analyzers also require an understanding of chemistry of the process and instruments measurement techniques. This chapter focuses on the issues that are somewhat unique, special, or different for process analyzers.

Field Analytical Instrument Systems

A field analytical instrument system can consists of the following elements:

(1) Sample Probe (a valved tube that inserts into the process through a packing gland or welded seal)

(2) Sample Transfer Line (tube or pipe often insulated and/or temperature controlled between the sample probe and sample conditioning system)

(3) Sample Conditioning System (A small process to convert the sample conditions to something that is acceptable to the analyzer. It often includes items like filters, demisters, heaters, coolers, temperature indication, flow splitters (by pass and analytical flows), pumps, aspirators, flow indicators, flow controllers, pressure indicators, and/or pressure controllers.)

(4) Field Analytical Instrument (see next section)

(5) Sample Return Line (tube or pipe to transfer the sample from the sampling system to a sample return point.)

(6) Sample Return (process, flare, tank, etc.)

(7) Calibration System (Known material used to calibrate and/or benchmark the current analyzer and/or analytical system (depending on point of introduction) to be calibrated.)

(8) Data Output (typically a digital, analog, and/or serial result scaled output of an electronic or pneumatic nature)

Field Analytical Instruments

Field monitoring instruments are normally used online to monitor a process or material being produced. Field analytical instruments monitor the chemical status of a sample/process on line. Many of these instruments are modified versions of laboratory analytical instruments. As modern process control techniques came into vogue, the need for fast process analytical results exceeded what could be done in the laboratory. After process versions of these analytical instruments became available, they not only generated faster results, but were also utilized in new and interesting applications. A couple of application examples where field analytical instruments are commonly applied today are for worker safety, process safety, and environmental. monitoring.

Field analytical instruments are often classified in two major categories. These are in situ and extractive. In situ analytical instruments are physically installed in the process and make a measurement under the physical process operating conditions (like most non-analytical process monitoring instruments). Extractive analyzers monitor a sample that has been extracted from the process and is usually physically conditioned (filtered, pressure, temperature, and/or flow conditioned) to make the sample more compatible with the analytical instrument and/or measurement technique being utilized. At least a majority of the sample is typically returned to the process unaltered. The sample conditioning system/process is usually the major source of analyzer maintenance and analyzer reliability issues. For this reason, analyzer maintenance technicians need to be well trained in the science/art of sampling technologies.

A few typical examples of common in situ analytical instruments would include pH, moisture, conductivity, viscosity, oxygen, combustibles, colorimeters, environmental, corrosion, refractometers, etc.

A few typical examples of common extractive analytical instruments would include chromatographs, spectrophotometers, titrators, sulfur, flash point, boiling point, viscosity, oxygen, environmental, moisture, etc.

Note that several of the techniques are listed as both in situ and extractive. That is because as much can depend on the process/monitoring sample conditions as on the analyzer type. For example, oxygen analysis can be monitored in situ at combustion stack conditions with a zirconium oxide sensor, for breathing/ambient air with a fuel cell, or extractively from a vacuum process with a fuel cell. Clearly process analyzer applications can be varied and unique, even for the same analyzer. Similarly, the maintenance requirements and needs of each of these applications can be quite different, even for the same brands and models of instrument (depending on the application).

Examples of Typical Field Analytical Systems Maintenance Issues

Typical questions to consider when reviewing maintenance issues for any field analytical system are given below.

SAMPLE PROBE

(1) Is the probe properly positioned? – Middle third of process pipe with opening oriented down stream.

(2) Is the process block valve open and probe inserted?

(3) Is the process sealing gland leaking?

(4) Is the instrument block valve open and clear?

SAMPLE TRANSFER LINE

(1) If heated is the tracing or heating fluid on?

(2) If temperature controlled does the insulation completely cover all components (any exposed elements or supports that are not properly insulated)?

SAMPLE CONDITIONING SYSTEM

(1) Are there sufficient sensors and controllers in place to properly monitor and control this sampling process (typically temperature, pressure and/or flow)?

(2) Are controllers properly sized to adequately control (e.g. multiturn valves that require multiturns to adjust the flow through a rotameter and heaters that control properly and don't often over shoot)?

FIELD ANALYTICAL INSTRUMENT

(1) This is very unique to each type of analyzer. A couple of general examples will be given. Your analyzer manual should indicate typical problems to look for with your system and how to identify and correct them.

(2) Are appropriate utilities and reagents available (e.g. helium for a GC, reagent chemicals for a titrator, etc.)?

(3) Is the analyzer detector working properly (e.g. FID for a GC, lamp and detector for a spectrophotometer, etc.)

(4) Are the sensor and solution ground appropriate for electrochemical measurements (e.g. pH, conductivity, etc.)?

SAMPLE RETURN:

(1) Is there an appropriate and stable pressure drop?

(2) Are there any leaks?

(3) Is the tank full?

CALIBRATION SYSTEM:

(1) Is the calibration sample introduction system tight and free of potential cross contamination issues?

(2) Is the sample stored and introduced under conditions that will keep it stable and accurate?

(3) Is there a chance that trace components can experience partial absorption before they are analyzed?

(4) For appropriate environmental measurements was the calibration sample introduced through the appropriate portions of the sampling system or just into the analytical instrument?

DATA OUTPUT:

(1) Is the wiring properly shielded?

(2) Is the signal properly scaled on both ends?

(3) Is the serial interface assigned the proper protocol?

Organization

There are many different successful analyzer maintenance organization styles; a few of the more successful styles are outlined below.

Large complexes tend to operate most successfully with a dedicated Process Analyzer Organization. These organizations utilize dedicated (full time Analyzer Technicians and Engineers). This seems to make the best use of individuals who have acquired special process analyzer skills and also provides them a close network of fellow workers with which to share their problems and learning. Many companies achieve optimum results when the Process Analyzer Engineers and the Process Analyzer Maintenance Technicians are under the same organization. This tends to simplify critical information exchange between these groups on installation, design, and maintenance issues. Other plants have an organization structure in which these groups are separated, with the Analyzer Engineers under the Engineering or QA department and the Technicians under the Maintenance Department. Even when these groups are not under the same management, they need to be encouraged to network very closely. When there are problems and they are under separate organizations there tends to be more finger pointing (bad design or bad maintenance). When the same group is responsible for both, the typical attitude is more often teamwork and lets get and keep it working.

Medium complexes typically do not have enough analyzers to justify a dedicated Process Analysis organization. Since many process analyzer skills are developed over time and experience, medium sized organizations tend to operate with dedicated or primary Process Analyzer Technicians within an instrument shop. These individuals typically have the most experience and training on these instruments. Similarly their Instrument Engineering organization may have an Analyzer Engineer or an Instrument Engineer who handles most of their analyzer projects. Plants where these analyzer oriented Engineers and Technicians network closely tend to have more successful process analyzer projects.

Small complexes typically have very few analyzers and their engineers and technicians tend to be very multi-skilled (do what needs to be done when it needs to be done). Even in this environment someone who is the primary analyzer person (and feels some ownership for both the instrument and the technology) is usually more successful than an organization where any available body is sent to do the job.

All of the previously mentioned organizational approaches will require close coordination with production, engineering, process control, maintenance, and the QA laboratory. Most other instrument maintenance personnel work with the first three to varying degrees, but usually it is only the process analyzer personnel and production who work closely with the laboratory personnel. It is critical that the process analyzer and laboratory groups work together. For production to make appropriate use of the numbers they are getting from the lab and the process analyzers they have to have confidence in them. For this to happen the numbers need to be the same and/or any difference needs to be rationally explainable.

Personnel

In part because of the unique nature of many process analyzer applications, maintenance issues can be very application specific. Factory manuals and/or maintenance support is often not aware of these special aspects. Learning gained through experience can be costly. Consequently experienced personnel can be worth their weight in gold. Union or nonunion, these technicians need to be trained and treated as the professionals they are. Because of the uniqueness of many applications and instrument technologies, it is too costly to retrain new people each time there is a problem or to transfer people from other areas of the plant to work on and learn all of the different applications. A career ladder and pay for performance are the best ways to achieve and maintain a skilled workforce within an operating plant.

Basic process understanding, chemistry, physics, electronics, data processing, and people skills are needed to be a successful analyzer technician. Some may ask why people skills are needed by an analyzer maintenance technician. Analyzer technicians, even more so than other maintenance technicians, are routinely working closely with process personnel, other maintenance personnel, QA personnel, engineers, and others. Process analyzer sampling systems are mini processes in and of themselves, consequently a process analyzer technician who must maintain them, must have a good understanding of the physical and chemical processes that operate within them. Appropriate testing and training programs need to be tailored to the needs issues present in each facility. Good personnel selection will help tremendously, but unlike most other instrument maintenance applications, prior experience on the same equipment in different processes and plants is only an indicator of potential and doesn't usually qualify one to do the job (This may be different if the experience was gained at an almost identical plant/process.).

Maintenance Approaches

Maintenance approaches are almost as dynamic and cliché based as management structures. Be it routes, scheduled, preventative, predictive, or reliability based; the goal is to do the most with the least (money and people). This is natural and correct. No processing plant wants a process analyzer for its own sake. They want to do something that makes/saves them money with the numbers it generates. Consequently the less often it is working the less money the firm can make or save.

Most reasonably successful process analyzer maintenance organizations are a blend of these types. This is true at least partially because of the nature of the beast. Many process analyzers, unlike most other process instruments (consume utilities and reagents on a somewhat predictable basis). These consumable items may include include standards, carrier gases, measurement reagents, etc. Because the rate of consumption is predictable these maintenance activities lend themselves to a scheduled approach. Process analyzer extractive sampling systems tend to be mini processes (filtering; flow, temperature, and pressure controlling; demisting, etc.). Fully automating them is quite expensive, so often they are monitored on a route and when a technician notices a change they adjust it. Most current generation process analyzers are microprocessor controlled and have constant internal diagnostics operating on the instrument itself. These diagnostics can be used to generate work requests and report instrument changes (sometimes before it has gone out of specification).

Service Factor

Service Factor is one of the terms used to describe the reliability/availability of an analyzer application. Since it is based on the time the analytical system is performing for the process it is monitoring the total application (analyzer and sampling system).

Two calculations that are often used to make this measurement are:

$$\text{Availability (\%)} = \frac{\text{time the analyzer is operating correctly}}{\text{time the process unit is operating}} \times 100$$

and

$$\text{Utilization (\%)} = \frac{\text{time the analyzer is used by the process operations}}{\text{time the process unit is operating}} \times 100$$

Availability and utilization percentages are key indicators directly tied to the idea that the plant wants to do something that makes/saves them money with the numbers the analyzer generates. It makes sense that the analyzer doesn't make/save them money when the plant isn't working and consequently the important time for it to be working is when the plant is operating. This encourages the maintenance group to schedule any down time or preventative maintenance during process down times and have the analyzer available when the process can make the best use of it.

There are additional maintenance performance monitoring approaches that are generic to all process instruments. Many or these like Reliability and Maintainability are based on the utilization of time based measurements like Mean Time Between Failure (MTBF) and Mean Time To Repair (MTTR). They are primarily used to compare instrument models and/or maintenance approaches.

Maintenance Work Load

What is an appropriate workload for a maintenance technician? This is a truly impossible answer to give. There is such a large variation in analyzer application maintenance requirements and analyzer technician capability, that providing a number would be a truly a useless exercise. How do you compare a simple pH analyzer application to a complex gas chromatograph or spectrophotometer? The numbers of potential things that can and do go wrong are decades apart. What about a pH application on clean water or steam condensate verses one on a process with coating issues? We know which is going to require the most maintenance and it isn't just dependent on the quality of the instrument or the maintenance it has received. It is heavily impacted by the application.

In spite of this statement, it is possible to make strides in monitoring a plant's technicians and applications to identify more reasonable workloads and to evaluate the performance of different technicians. Significant strides have even been made at trying to generate useful numbers on even a site basis for sites with similar applications (i.e. refineries). These numbers have even been used to try to evaluate overall site analyzer maintenance performances. One of the most successful approaches involves a calculation called Equivalent Analyzers per Technician (EQAT). This is a very complex series of calculations incorporating estimated adjustment factors that are based on a Base Equivalency Factor (BEF) for each type, brand, and application of analyzer. This factor is adjusted for application specific issues like detector type and number, number of streams, sampling issues, etc. The only generally publicized factor seems to be the BEF. A typical calculation would be:

EQA (Equivalent Analyzers) = Number of Analyzers X BEF + (the sum of correction factors for additional detectors, additional valves, additional streams, and additional components)

This may help evaluate the performance of individual technicians on similar processes and applications, but even this approach is questionable when one is trying to compare the performance of a site analyzer maintenance system with another site. In this case, the overall performance is not generally perceived to be accepted as the sum of the parts (technicians). A site evaluation also needs to incorporate a factor for servicing its analyzers by priority. For example Safety Protection, Environmental Protection, Asset Protection, and Profit Maximization factors are incorporated when the performance is compared on a site basis. This approach allows a maintenance team to give the most important analyzers the correct maintenance priority and service.

These concepts have been discussed in several ISA Analysis Division panels and papers. Another and a very complete documentation of this approach is in Engineering Equipment and Material Users Association (EEMUA) publication Number 187-1999. Firms that use the EEMUA approach should do a complete evaluation of their processes BEF and correction factors. The analyzers may be the same, but the sample and sampling system can make a huge difference.

Spare Parts

Vendors typically provide recommended spare parts lists with their instruments. These are a guide to users but unfortunately they are based on an average requirement of many applications and each application may require significantly different needs. For example a vendor may recommend that a pelistor type combustion sensor will typically last twelve months, but a plant may have silicones or sulfur compounds in its atmosphere and these compounds may poison the sensors every couple of months. This would also require more frequent calibrations than indicated for the average application. Similarly a vendor may indicate that a sample valve is typically good for a certain number of applications, but because of sample conditions (corrosivity, temperature, abrasion, etc.) or the application timing (allowing the valve to set between actuations), each plant may find significant different results on its application.

To conduct timely and efficient maintenance on process analyzers, appropriate utilities, reagents, calibration mixtures, and parts need to be readily available, when needed. Cost cutting/inventory reduction operations at most manufacturing facilities have increased concerns that maintenance materials may not be there when needed. Automated maintenance management and ordering systems and vendor supported twenty-four hour parts availability has allowed many users to safely reduce their parts inventory, but it isn't as reliable as having appropriate parts in hand. Each maintenance department has to identify their spare parts needs based on many factors, including maintenance experience and plans, known special factors, recommended spares, replacement spares availability, and other unique conditions (like planned shutdowns and maintenance). Once generated these numbers should not be cast in stone. These requirements need to be reviewed periodically and adjusted accordingly.

Vendor Support

Vendor support is an important maintenance consideration when selecting an analyzer supplier. What do they offer in the area? Do they offer training courses (what types – classroom, video, internet)? What type (telephone support, technician support, etc.) of back up service support do they offer and how much does it cost? What type of spare parts inventory do they carry and what type of delivery can be expected? Do you have to pay an annual fee to get their faster service? What is the quality of their maintenance manuals? The greatest and most cost effective instrument is not a good option if adequate local support is unavailable. For this reason the best choice in one part of the world may not be even a good choice in another.

Application Unique Issues

To some extent the maintenance performance of all instrument process monitoring applications can be impacted by variations between processes and are not solely based on the instrument capabilities. But because of the chemical nature of process analytical methods, this issue is even more prevalent in process analyzer applications than most other instrument monitoring applications. Each analyzer has a range of conditions that it was designed to operate in. Often this range is not compatible with the range of potential process conditions. When this happens a sampling system needs to be devised to introduce the sample from the plant process to a process (sampling system) that will prepare it for the conditions required by the analyzer and not allow the sample to change in any way that will adversely affect the accuracy of the final measurement. Often, even when the process is operating within an acceptable range for the analyzer, the process variation within that range can adversely affect the measurement repeatability to a degree that generates a need for a sampling system to reduce sample conditions variation.

Installation Issues

Many complex analyzers that are in open-air process environments often demonstrate poorer than average service factors. Obviously, accessible equipment housed in a protected and comfortable environment is likely to require less and to receive preferential maintenance attention. They have typically demonstrated superior service factors. Many plants utilize heated and air-conditioned process analyzer shelters for their complex process analyzers (gas chromatographs, spectrophotometers, boiling point, etc.) that utilize extractive sampling and are going into open air plants. There are at least two reasons typically mentioned for the improved performance of housed analyzer systems. The first is the fact that a technician is happy to give them their full service requirement in the foulest weather. The second and probably most influential is that most of these analyzers are based on heavily modified laboratory techniques that were developed to be conducted in a laboratory environment. Even the industrial hardened process versions of these laboratory analyzers have tended to perform better in the better environment. In some very moderate coastal climate areas, some plants achieving performance comparable to HVAC shelters with three sided freely ventilated shelters, but even these offered a canopy roof to minimize the daily solar temperature variations and direct rain contact.

11

Primary Elements and Final Control Devices

Overview

A primary element is a sensor or detector that responds quantitatively to the measured variable and performs the initial measurement operation. A primary element performs the initial conversion of measurement energy. The final control device is the element that directly controls the value of the manipulated variable of a control loop. Often the final control element is a control valve. It is an instrument that takes action to adjust the manipulated variable in a process. Final control elements can also be solenoids and servomotors.

Temperature

Temperature measuring instruments include liquid- or gas-filled thermal systems, electrical measurements using thermocouples (TCs), resistance bulbs (RTDs) and thermistors. Noncontacting sensors are not covered here as they are usually factory calibrated and need only an emissivity check or adjustment to read properly.

In today's technology, the thermocouple and the RTD are used extensively for industrial temperature measurement and will be discussed in more detail than other temperature-measuring devices. Many local field controllers, however, still exist and use the older filled-tube sensors that mechanically move a pointer in the local instrument. These devices require mechanical adjustments that increase or decrease the arc that the Bourdon spring takeoff moves in order to provide the correct match of the process pointer to the measured temperature. In the case of gas-filled temperature systems, the measurement is nonlinear. Calibration should be made using the scale that came with the instrument. The vendor's instrument instruction book will usually explain, in detail, how to make these adjustments.

The intent of this chapter is not to provide all the sensor calibration curves, but to provide only the practical information that is necessary for calibration. Most vendors have these tables in their catalogs or information sheets and can supply them if requested to do so. See the section on fluid bath calibration for more information on how wiring circuits affect the accuracy of TC and RTD temperature measurement.

Tables 11-1, 11-2, and 11-3 provide basic information regarding the types of thermocouples available, their ranges, accuracies, and upper limits when using a protection well or tube. If a thermocouple being used on a specific application is experiencing drift or early failure tendencies, a check of these tables may suggest a solution to the problem.

Table 11-1. Limits of Error for Thermocouples
(Courtesy of Transmation Inc.)

Type	Temperature Range		Limits of Error			
			Standard		Special	
	deg F	deg C	deg F	deg C	deg F	deg C
B	1600 to 3100	871 to 1705	±1/2%			
E	32 to 600 600 to 1600	0 to 316 316 to 871	±3 F ±1/2%		±2-1/4 F ±1/4%	
J	32 to 530 530 to 1400	0 to 277 277 to 760	±4 F ±1/4%	Note 2	±2 F ±1/8%	Note 2
K	32 to 530 530 to 2300	0 to 277 277 to 1260	±4 F ±1/4%		±2 F ±1/8%	
R or S	32 to 1000 1000 to 2700	0 to 538 538 to 1482	±2-1/2 F ±1/4%			
T	-300 to -75 -150 to -75 -75 to +200 200 to 700	-184 to -59 -101 to -59 -59 to +93 +93 to +371	 ±2% ±1-1/2 F ±1/4%		±1% ±1% ±1/4 F ±1/8%	

Table 11-2. Recommended Upper Temperature Limits for Protected Thermocouples
(Courtesy of Transmation Inc.)

Thermocouple Type	Upper Temperature Limit for Various Wire Sizes (Awg), deg F (deg C)				
	No. 8 Gage (0.128 in. (3.25 mm))	No. 14 Gage (0.064 in. (1.63 mm))	No. 20 Gage (0.032 in. (0.81 mm))	No. 24 Gage (0.020 in. (0.51 mm))	No. 28 Gage (0.013 in. (0.33 mm))
B				3100 (1705)	
E	1600 (871)	1200 (649)	1000 (538)	800 (427)	800 (427)
J	1400 (760)	1100 (593)	900 (482)	1700 (371)	700 (371)
K	2300 (1260)	2000 (1093)	1800 (982)	1600 (871)	1600 (871)
R and S				2700 (1482)	
T		700 (371)	500 (260)	400 (204)	400 (204)

Tables used for comparisons of RTDs resistance to temperature or thermocouple millivolts to temperature should be of the following vintage: thermocouple tables are referenced to International Practical Temperature Scale (IPTS) 1990 and RTD tables are referenced to the Deutsches Institut fur Normung e.V. (DIN) 43760 curve for platinum, dated October 1980. The slight change in ohms versus temperature appears to be to the third decimal place, and on higher temperatures there is a slight adjustment of the resistance versus temperature characteristics.

Table 11-3. Operating Range and Limits of Error of Common Thermocouple Types

Thermocouple Type	Temperature Range	Limits of Error	
		Standard*	Premium
T	0–350°C, 0–660°F –200° –0°C, –300° –32°F	±1°C or ±0.75% ±1°C or ±1.5%	±0.5°C or 0.4% —
J	0–750°C, 0–1400°F	±2.2°C or ±0.75%	±1.1°C or ±0.4%
E	0–900°C, 0–1650°F –200° –0°C, –300° –32°F	±1.7°C or ±0.5% ±1.7°C or ±1%	±1°C or ±0.4% —
K	0–1250°C, 0–2300°F –200° –0°C, –300° –32°F	±2.2°C or ±0.75% ±2.2°C or ±2%	±1.1% or ±0.4% —
R,S	0–1450°C, 32–2700°F	±1.5°C or ±0.25%	±0.6°C or ±0.1%
B	800–1700°C, 1500–3100°F	±0.5%	—
Tungsten-Rhenium 24 ga.**	0–2930°C, 32–4200°F	±13.3°C or 1%	—

*Whichever is greater.
**Smaller gages have lower temperature limits.

Most thermocouples in the field are calibrated using the IPTS 1990 thermocouple tables. For 100-ohm platinum, the most common RTD, the October 1980 resistance versus temperature table is the proper one to use for field calibration.

An explanation of the possible errors induced in direct wiring from the field sensor to the readout is necessary to explain why simply checking the TC or RTD sensor calibration will not always produce an accurate temperature reading. An unbalance in thermocouple lead wire occurs because the two lead wires have the same thermoelectric properties as the thermocouple itself; in effect, there are two different materials on the two equal length wires, and their resistance is not the same. For example, assume that a large motor in the plant starts up and lines of force cut across this thermocouple extension wire. Ohm's law states that when a line of force cuts a conductor, a current is induced in that conductor. Since both wires are subjected to the same lines of force but have a different resistance, the same current induced in them will produce voltage drop across that length of wire. At the other end, the receiver sees the difference in millivolts (or microvolts) due to those lines of force, decides that the temperature is related to the millivolt value at its input terminals, and may take action accordingly.

This phenomenon was probably always there but was not observed on older analog-type receivers or recorders, which had much longer time constants and slower speeds of response. However, today's new digital-type receivers with very quick response times will see these as sudden "spikes" and may make more 1s and 0s as a result of the sudden surge of more millivolts appearing at their input terminals.

If transmission accuracy is a problem, the answer may be to change to field-mounted transmitters mounted directly near or adjacent to the sensors. The signal from the temperature location of the sensor is now transmitted on two copper wires with a high-level, 4-20 milliamp current signal. If twisted, shielded pair cable is used and installed properly, the stray currents induced by motors starting and other "electrical noises" present in the field will cancel out.

Thermocouple and RTDs

Thermocouples (TCs) and RTDs require barriers when connected to receivers that could under fault conditions possibly release sufficient energy to enter the hazardous area and cause an explosion. The following sections on barriers for these services will help in understanding the wiring and use of barriers on thermocouples and RTDs.

THERMOCOUPLE CIRCUITS USING BARRIERS

The thermocouple, perhaps more than any other electrical sensor, is inherently a safe device. The voltage levels of which thermocouples are capable (less than 100 mV DC) are well below the voltage level required to ignite even the most susceptible air and gas mixture. This safety is recognized by unrestricted permission on the part of electrical authorities to use the thermocouple for temperature measurements in hazardous locations. Such freedom, however, does not apply to instruments that need to be connected to the thermocouple to amplify the measured signal and/or translate the same into a visible indication of temperature.

The amplifying, recording, and display instruments are often powered with 117 V AC or by built-in batteries that render them unsuitable for use in or direct connection to circuits in hazardous locations. The problem thus created is routinely solved by locating the amplifying equipment in a safe area and connecting the same to the thermocouple in the hazardous area via one or two intrinsic safe barriers. This practice, while it promotes safety, also creates some new measurement problems, some of which will be explored next. Figure 11-1 shows the typical circuit of a safety barrier.

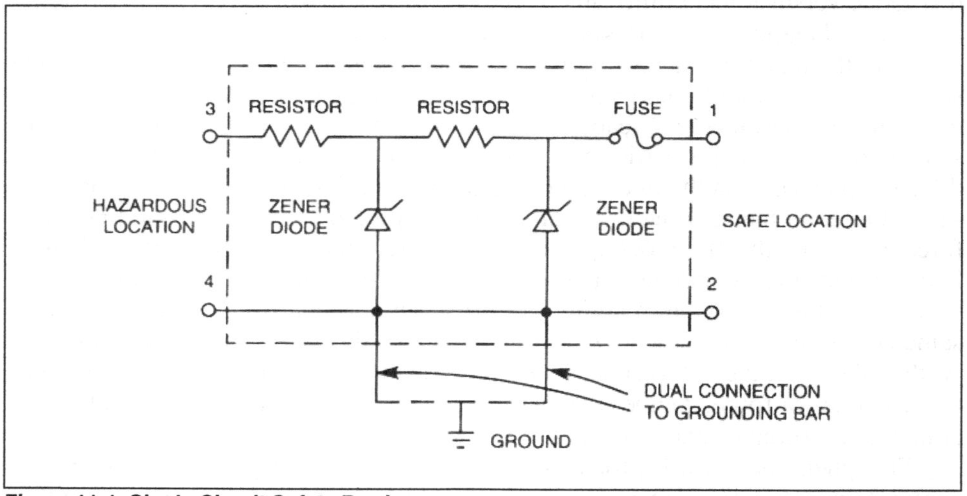

Figure 11-1. Single Circuit Safety Barrier.
(Courtesy of Transmation® Inc.)

The simplest thermocouple measuring arrangement that can be constructed with a safety barrier is shown in Figure 11-2. A single barrier carries one signal conductor, and the second conductor is grounded at the barrier grounding bar. The arrangement adds the least cost without compromise in safety. It requires that the measuring junction *not* be grounded. To do so would yield two groundings, one at the junction and one at the ground bar, with attendant indeterminate measurement results.

Different ground points are seldom at exactly the same AC and DC potential. The difference potential between the two ground points would cause a current in the grounded thermocouple conductor and a consequent voltage drop between the

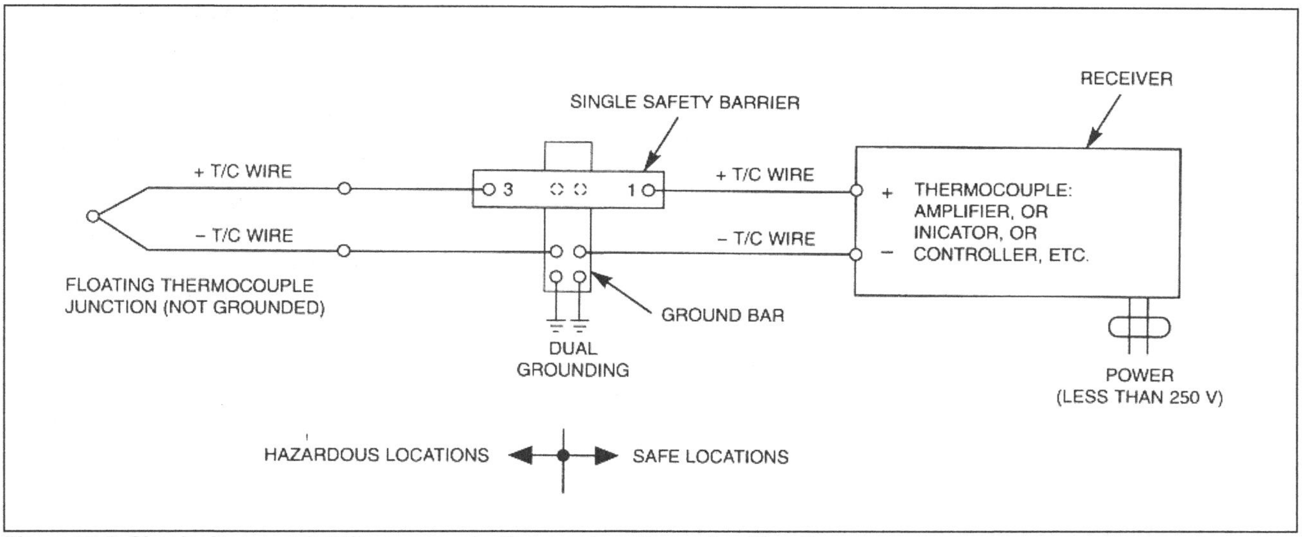

Figure 11-2. Simple, Ungrounded Thermocouple, Temperature Measuring Circuit.
(Courtesy of Transmation® Inc.)

two grounding points. That voltage drop would add to or subtract from the thermocouple voltage, leading to false results.

Thermocouple wire is attached to terminals 1 and 3 of the barrier. The barrier circuits are made of copper. Thus, a thermocouple wire to copper junction is formed at both terminals. The two junctions are of opposite polarity, and the terminal emfs will cancel provided they are at the same temperatures at these two locations. Points 1 and 3 are between four and six inches apart. By mounting the barrier in an enclosure, which is normally necessary to protect wiring anyway, the thermal gradient between terminals 1 and 3 can be kept to less than 0.3°F, normally an acceptable error. A TC to copper junction pair is also formed at the ground bar connections. These, however, are so close together that even without precautions little of a thermal gradient can exist and degrade the measurement.

The safety barrier introduces additional series resistance into the measuring circuit. This has two effects, both degrading. The barrier series resistance in conjunction with the input resistance of the signal receiver forms a resistive attenuator that decreases the thermocouple voltage available in the input terminals of the receiver. For example, a barrier with a series resistance of 1000 ohms, coupled to a receiver with an input resistance of 50 kilohms, will attenuate the signal by $(1/(50 + 1))$ times $100 = 1.96\%$ of value. The need for high input resistance receivers becomes clear.

Perhaps more important than high input resistance is low input offset current. The receiver may inject into the thermocouple circuit a small DC current. This current may arise from the nature of the amplifier employed, or it may be due to an open thermocouple measuring circuit. Both effects may be simultaneously present and be additive. The current so injected into the TC circuit causes a voltage drop across the resistance of the barrier. That voltage drop adds to or subtracts from the TC voltage and causes the measured value to be in error by this very amount. This type of error is independent of the measured value and is thus an "offset" error. Its value can be considerable.

Take a current of 100 nano-amperes $(100\ 10^9)$ and the 1-kilohm barrier. The resulting error is 100 microvolts. Given a type J thermocouple at 100°F with a sensitivity of 29 microvolts per °F, the error becomes 3.45°F.

Both the aforementioned errors, attenuation and offset, are barrier resistance-dependent. They can be reduced by reducing the barrier series, or end-to-end, resistance. That resistance decreases as the voltage rating of the barrier is reduced. The series resistance can be as little as two ohms for barriers rated at 1 volt. With thermocouples incapable of voltages greater than 0.1 volt, a 1-volt barrier is quite acceptable unless AC hum pickup on the thermocouple circuit is a significant problem. When that happens, rectification of the pickup by the barrier emerges as an additional and often unpredictable error contributor.

The typical polarized safety barrier (see Figure 11-1) is constructed of a string of series-connected resistors connected to the ground counter conductor with two zener diodes. The effect of the arrangement is that signals of the appropriate polarity but less than the zener voltage are readily conducted through the barrier. Signals greater than the zener voltage or of reversed polarity are shunted to ground via the zener diodes, thus limiting the voltage that may appear at the output terminals of the barrier. The arrangement is intended to keep excessive voltage from the hazardous location circuits, but it also has the potential of modifying the signals coming from the thermocouple circuit.

The problem of rectification is subject to two solutions:

(1) Reduce the hum pickup by more judicious wiring routing, additional shielding, and/or grounding of the circuit at or near the junction rather than at the barrier.

(2) Use nonpolarized or AC-type barriers that are less prone to DC voltage generation.

Figures 11-3 (a) and (b) and Figures 11-4 (a) and (b) show zener diode-based construction of a nonpolarized or AC type of barrier. The transfer function is also shown and will be noted by symmetrical positive and negative signals. The positive and negative halves of AC interference signals are rectified equally and cancel each other.

Figures 11-4 (a) and (b) also exhibit an AC-type barrier, constructed with diodes. This type of barrier is restricted to low level signals of less than 0.5 volts. It, too, is characterized by a symmetrical transfer curve and is particularly suitable for thermocouple signals by virtue of the characteristically low end-to-end or series resistance. Low end-to-end resistance is desirable to minimize the conversion of common mode signals (signals applied equally to ± TC conductors) to normal mode interference voltages. Readout equipment tends to be more sensitive to the presence of normal mode interference voltages.

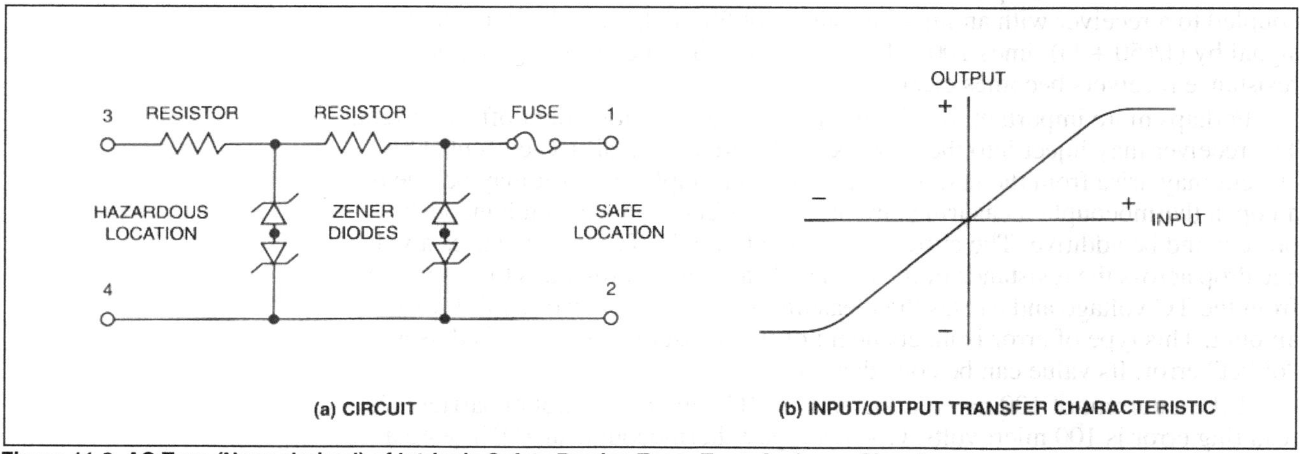

Figure 11-3. AC Type (Nonpolarized) of Intrinsic Safety Barrier, Zener Type, for Large Signals.
(Courtesy of Transmation® Inc.)

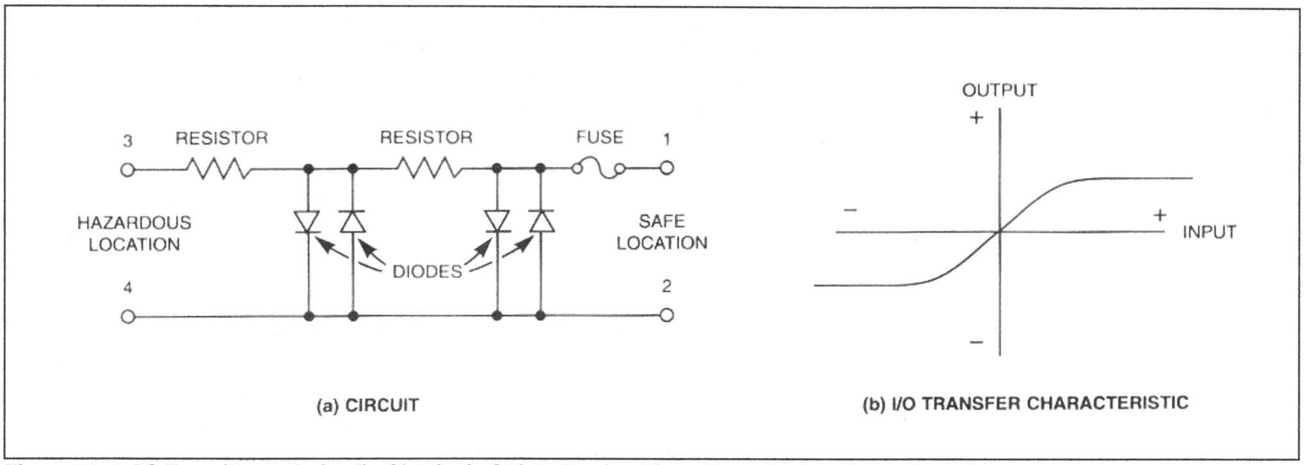

Figure 11-4. AC Type (Nonpolarized) of Intrinsic Safety Barrier, Diode Type, for Low Level Signals. *(Courtesy of Transmation® Inc.)*

Grounding the thermocouple circuit at or near the measuring junction requires the use of two barriers in order to avoid error-producing ground loop signals. The matter is illustrated in Figure 11-5.

In summary, accuracy of temperature measurement via safety barriers is promoted by choosing barriers with low series resistance, employing a receiver with low offset current and high input impedance, using nonpolarized barriers or using two barriers and grounding the thermocouple at the measuring junction, avoiding pickup of AC by judicious wire routing and shielding, and avoiding a temperature gradient across the safety barrier.

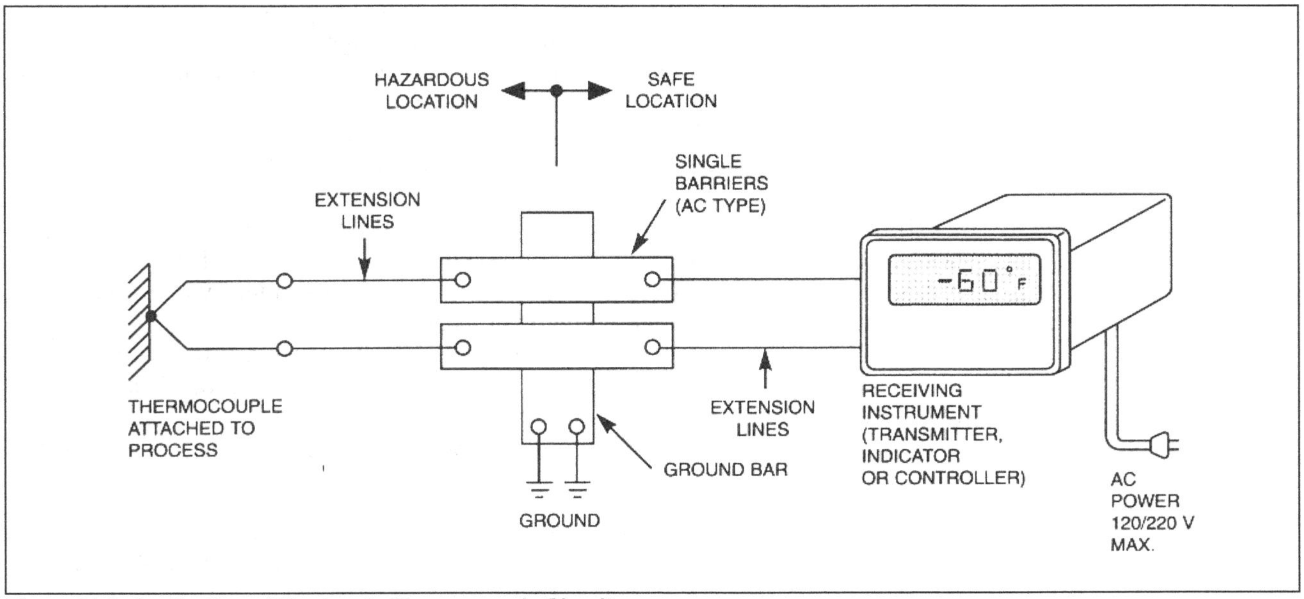

Figure 11-5. Application of Barriers to Thermocouple Circuit.

Primary Elements

Primary elements are covered to some extent in the chapter on calibration and also in portions of the transmitter section, but some maintenance and troubleshooting aspects will also be covered here.

Maintenance and troubleshooting are usually "after the fact" events. In order to reduce maintenance problems and make troubleshooting easier and more productive, the maintenance department should always be involved in the design and installation of any instrumentation. The placement of sensors and valves is critical to good measurement and control, but the maintenance aspect should also be considered. Sometimes a compromise must be made in order to allow proper maintenance of the devices, but without proper access, these devices will not be properly maintained. It is therefore essential that engineering work with maintenance when designing new systems or when modifying an older system. Every sensor, transmitter, converter, or valve should be installed with the capability of maintenance or removal without shutting down a major portion of the process or without the need for scaffolding when maintenance is required. Block valves, bypass valves, and insertion valves should be employed for ease of maintenance.

The following list of primary elements, with some maintenance and troubleshooting guides, may be helpful when dealing with an instrument problem. Proper selection of primary elements determine the accuracy and speed of response of the temperature measurement. The most common temperature transducers are:

- Noble metal thermocouples
- Base metal thermocouples
- Resistance temperature detectors
- Thermistors
- Integrated circuit sensors
- Infrared temperature transducers
- Optical fiber thermometry
- Thermometers

Temperature Sensors

Temperature measurement is not always what is expected. Although the sensing element can be accurate to within a tenth of a degree, the measurement of a gas or liquid temperature may be off by many degrees. The mounting and location of a temperature sensor is extremely important. If a temperature reading does not agree with a calculated temperature or simply does not agree with other temperature sensors in a system, the following should be considered: check the calibration of the transmitter.

If the transmitter uses a filled temperature bulb, check actual sensor calibration with a field "dry block" calibration device or a liquid bath. If the sensor is an RTD or thermocouple, first check the output of the transmitter by using a temperature simulator/calibrator, and remember to correct for the reference junction compensation if the device is not automatically compensated. If the loop calibration is still incorrect, check the actual sensor with a field "dry block" calibration device.

If the results of this calibration do not provide a correct reading, consider the following suggestions:

(1) Is the sensor compatible with the process? Some thermocouples deteriorate in oxidizing atmospheres but not in reducing atmospheres and vice versa.

(2) Is the sensor installed in a location that is conducive to temperature pickup? Any wall effects when used on heated or cooled tanks?

(3) Is there a thermowell effect? Lag due to a thick thermowell?

(4) Is the temperature sensor bottomed out properly against the end of the thermowell? Does the sensor fit too loosely in the well?

(5) Are the well and sensor inserted far enough into the line or tank for proper temperature pickup? Insertion length is very important. These dimensions are listed in most temperature books or in vendor's installation pamphlets.

(6) If the sensor is a filled-tube system, is there a crimp or leak in the capillary, causing errors or loss of signal? Check spring linkages in the transmitter or receiver.

(7) If the measurement is electrical, are all the connections tight? Is the correct thermocouple type being used and does the transmission wire match the thermocouple type? Check the color code of transmission wire and sensor wires. The red wire is always negative. Is 3-lead extension wire being used when long runs are used? Is the element grounded or ungrounded? Is the correct calibration curve being used for the type of T/C or RTD?

(8) Check for loose tubing fittings or electrical wiring problems. Follow normal transmission line troubleshooting routines.

(9) If a wire-wound ceramic tip type of RTD is used in an agitated vessel or where vibration is a problem, the wire-wound sensor tip can easily be broken. Try etched or ribbon-type sensor elements that can withstand severe vibration.

Special thermocouple calibrators have built-in reference junction compensation, but using a millivolt meter for calibrating thermocouples connected to ordinary receiving instruments having copper input terminals requires reference junction compensation. Add the equivalent millivolt value of the terminal temperature to the millivolt reading of the thermocouple at this connection. This value for the terminal temperature may be taken from the mV/temperature tables for the type of thermocouple being used. Conversely, when a nonthermocouple instrument is being used as a millivolt source, a value corresponding to the temperature of the instrument terminals must be subtracted from the setting of the millivolt source in order to arrive at the actual input to the thermocouple receiver. (A simple way to remember to add or subtract is to remember "subtract when sending" or "S = S.")

There are several types of noncontacting temperature measurement devices; infrared and pyrometers are the most common. The most common problems with these devices is with selection of the proper unit for the measurement being made. Emissivity, lens focus, and distance between the source and the measurement must be considered. The noncontacting temperature-sensing devices can be very handy for applications in which a probe is not practical. Temperature ranges can be quite high and accuracies are excellent when properly used.

Flow Sensors

Flow can be measured with direct or inferential devices. In most cases, flow is measured by the pressure drop developed across a flow device, such as an orifice plate, venturi, or any specific restriction in a process flow line. Newer flow devices work on various physical or magnetic principles. The ISA Practical Guides Series volume, *Flow Measurement*, provides a complete description of all the various flow devices available today. Although flow is measured in many different ways, the following checklist will assist you in troubleshooting flow metering problems.

Primary Element Location

The so-called "hi-tech" part of maintenance engineering has been discussed. The most return for good engineering practice comes from proper selection, location, and care of primary measurement elements and final control devices. This is where the marriage of practical process knowledge and computer skills becomes important. No computer in the world can totally compensate for a poor measurement. Someone has to know how and where to apply sensors in order to have a measurement foundation upon which to build control excellence.

Comforts of the modern control room tend to overshadow the need to pay attention to field conditions during engineering as well as the operating phase of a project. Not only must the proper sensing device be selected, but it must initially be, and forever remain, in calibration if it is to do the intended job.

A feeling sometimes prevails that if something is not new, it has no value. While technology has developed in orders of magnitude in recent years, the basic physics of processes remains fairly constant. It's just that now there are better tools to control the many simultaneous events in complicated processes.

The best resource for the determination of primary element location is the vendor from whom it was purchased. The vendor does not want the product to fail. It is important, however, to have open discussion so that the usage intent is well known.

Be wary of the supplier who is so eager to get the order that he or she is reluctant to obtain application specifics. Someone in the vendor organization knows exactly how their device will perform. Make sure access has been made deep enough in the organization to get to that knowledge.

It is also important to know the physics of the process to be controlled. The ultimate responsibility lies with the engineer who applies the sensor, so it is important to arm oneself with the knowledge and not just pass the application decision to someone else.

Primary elements face the worst characteristics a process has to offer. That which looks really good on a process flow diagram may work only for a short time before succumbing to the environment of the process. Common problems encountered are corrosion, coating, chemical reaction between the process and materials of construction, and misapplication. Other common problems relate to improper location of the sensor in the process stream, inaccessibility for maintenance while running, and physical mounting that allows process vibrations to mechanically affect, or even destroy, the device. It is imperative to know the characteristics of process devices such as pumps, fans, and unit operations modules.

How does one cope with these problems? The best defense is knowledge and training. Whether by self-study, peer assistance, on-the-job training, or trial and error, the lessons must be learned. It is most effective and least expensive to learn from the mistakes of others if at all possible.

Blindly ignoring the impact of the components that surround the primary element has spoiled many a measurement.

However, as careers develop it is possible to outdistance that which has been tried—requiring the need to be on the leading edge. Common sense and a strong knowledge base is all there is to rely on. It is good to have developed resources within research and development ranks to provide a sounding board for the untried.

Many new measurement devices are the product of a specific need for a process found by someone charged with maintenance duties who found a better way. The hot breath of a production superintendent breathing down one's neck can do wonders for innovative powers, especially at three in the morning.

Flow element locations are important. Certain measurement devices require straight pipe runs before and after the device. Sometimes these are not attainable after process installation. The time to establish flow element location is before the piping plans hit the field.

Other critical measurement locations belong with chemical measurements such as pH, conductivity, oxidation reduction potential (ORP), and air emission monitoring. If continued maintenance is to be the lot in life for an installed device, it makes sense to build in excess (or redundancy) to ensure minimum disruption of the measurement. Regulatory controls provide some impetus for doing that, but the well engineered installation will have necessary maintenance aids built in during the design phase.

It is disastrous to ignore seasonal environmental changes. Too hot, too dry, too wet, and too cold are potential problems for primary elements. Even if the process generates enough heat to prevent freezing and it runs 365 twenty-four hour days per year, someday it is going to be down when the weather is very cold. Many a southern mill has experienced freeze-ups while down for an unseasonably cold Christmas Eve. Whether anything is going to be done or not should be a conscious decision.

It is important to review installed life cycle costs up front. Often the original installation is minuscule compared to the maintenance effort required to keep the system running. A few extra dollars added to the initial design might have a multifold payback. Care must be taken to spend extra money only for critical items and not fall into the trap of overdesigning. A good tool for making that decision is historical data on existing system maintenance costs.

Don't fall into the trap of ignoring the unlikely.

The best guideline is to spend money as if it were one's own. Projects can be thrown into the disapproval basket because of design overkill. Conversely, some projects that made approval shouldn't have because of over-optimistic projected maintenance effectiveness.

Control Valves

Introduction

One of the most important parts of a control loop is the device located in the field whose function is to regulate the supply of mass or energy to the process. This role is most commonly given to the pneumatic control valve.

A control valve is the final control element in a process and is the muscle, or action part of the loop. Control valves start, stop, and manipulate flow to control such variables as pressure, temperature, and level.

The control valve is an integral part of the overall control system, and its selection and application have a dramatic impact on the resulting control loop's ability to perform correctly.

Here is an example of how a control valve functions in a typical process control loop (Figure 11-6). A sensor detects change in the value of a measured variable and sends a signal to a transmitter. The transmitter relays the signal to a controller, which compares the value of the measured variable to the setpoint. If there is a deviation, the controller changes its signal to the control valve. The valve responds to bring the value of the measured variable to setpoint.

Since a control valve is the final control element, it is often the last line of defense for safety or environmental hazards. Choosing the proper control valve for an application can help avoid situations that might compromise employee safety. In addition, choosing the correct control valve for a process will help to protect the environment by avoiding process fluid leakage.

Proper control valve selection will also have an impact on product quality. A control valve can affect product quality because it manipulates variables within a process. For example, if a control valve in a temperature dependent process is not capable of regulating flow in response to the controller, process temperature will not be properly maintained. This could result in poor product quality.

An improperly selected control valve may also require more maintenance and unscheduled downtime because of damage from cavitation or flashing.

To further complicate the problems associated with the selection of a control valve for a given process control application, ancillary devices are used in conjunction with control valves that are designed to overcome inherent problems that are common to this particular type of final control element. In order to troubleshoot control valves successfully, the technician must be familiar not only with

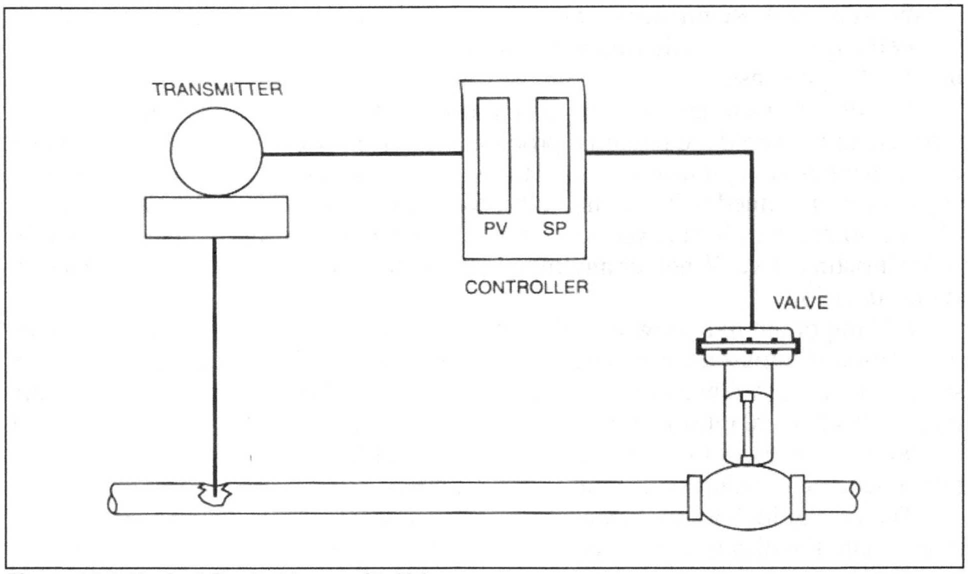

Figure 11-6. Typical Control Loop with Valve.

pneumatic control valves but also with the auxiliary devices used in conjunction with the control valves. Specifically, a working knowledge of valve positioners and volume booster relays may be beneficial when trying to troubleshoot pneumatic control valves.

Valve Selection

The basic criteria involved in the valve selection process involve a thorough understanding of the process itself and a knowledge of the control scheme being implemented. Some of the more important considerations for implementing the correct control valve might include the following:

FAIL-SAFE OPERATION

On a loss of the demand signal to the valve, the control valve will either be fully closed—no flow of fluid through the valve body (air to open)—or the control valve will be wide open—maximum fluid flow through the valve body (air to close). The inherent action of the valve is selected based on maintaining safe process conditions for the control loop in question.

VALVE CHARACTERISTIC

The ability of a control valve to regulate the flow of material passing through its body is a function of the valve size (Cv) and its characteristic. The valve characteristic is a function of the amount of fluid passing through the valve for a given valve stem position. For control purposes, valves usually exhibit either a linear or an equal percentage valve characteristic. The final choice of valve characteristic is dependent upon linearizing the overall control loop when the valve is installed in the field. No one specific valve characteristic suits all process requirements; therefore, the characteristic will change with the application. It is important to note that the valve characteristic, as purchased from the vendor (shelf characteristic), will change when installed in an actual process situation (installed characteristic). The amount of distortion that occurs for a given control valve characteristic can be calculated based on actual process conditions. One of the major factors involved in this situation is whether the control valve will be subjected to varying pressure drop as the valve is manipulated to maintain a viable supply of material to the process. This is usually the case. A control valve will be subject to maximum pressure drop as it throttles down to minimum valve stem position (closed) and minimum pressure drop when it is wide open (maximum flow). This pressure variation impacts the rated flow versus stem position and, therefore, changes its installed characteristic.

This point can be emphasized by a hypothetical situation. Flow control is a very common type of control loop in a production environment. Traditionally, flow rates are measured with a family of flow sensors defined as head-type sensors. Any device that generates a differential pressure as a function of flow rate falls in this category. These devices include orifice plates, venturi tubes, flow nozzles, etc. They all exhibit a common relationship of flow vs. pressure; namely, dP = flow rate squared. The differential pressure is measured with a dP cell, and this signal is fed to a flow controller as the measurement signal. The flow controller is designed to have a linear response. Its output signal is sent to the field to drive a pneumatic control valve. The correct valve characteristic for this loop is selected with the idea of linearizing the overall loop characteristic when the valve is installed in the field. At this time, the loop characteristic is nonlinear. The flow rate measurement was not linearized through the use of a square root extractor; therefore, the installed valve characteristic will have to exhibit a square root response curve. This is achieved by the selection of a control valve that has an inherent (shelf) characteristic that is linear. When installed in the process with a variable pressure drop across the valve, its installed characteristic will be changed to that of a quick-opening valve. By carefully sizing the valve, the flow rate measurement signal nonlinearity is offset by the nonlinear valve characteristic, and the loop is linearized without the use of another device (square root extractor). This proves to be a cost-effective way of linearizing the loop response and attaining effective control for the flow control loop.

Valve Sizing

Control valves are sized for a given process application, based on the process conditions such as line pressure, temperature, and fluid characteristics such as viscosity, specific gravity, etc. If a valve is undersized, the loop will not be able to deliver a sufficient amount of material to meet the process demands. If a valve is oversized, control is degraded. The valve usually is subjected to a reduced stroke and controls close to the valve seat. This means the fluid exhibits an unnecessarily high fluid velocity and excessive wear of the valve's wetted parts (valve trim); frequent maintenance may be required. In sizing control valves, other considerations may have an impact on determining the proper valve size. Flashing and cavitation may impact the choice of a control valve size where the fluid is a liquid. Sonic or choked flow may be a consideration for control valve sizing where the fluids involved are gases or vapors. All of these are taken into account when a valve is sized properly. Therefore, when replacing valves in the field, they should not arbitrarily be down-sized, nor should reduced trim be used.

Stroking

The pneumatic valve is a mechanical device, and its physical components should be in good working order. The procedure involved in checking out the operation of a control valve is called stroking the valve. This provides for an operational check of the valve components. A pneumatic signal that represents the throttling range of the valve signal is applied to the actuator or air motor. Typically, this is an air pressure signal of 3 to 15 psig. For a globe valve (a valve that works off vertical displacement of the valve stem), the mechanical movement (stroke) is adjusted based on the valve design. For example, with an air-to-open valve working with a 3 to 15 psig signal. 3 psig is applied to the valve actuator (the valve should be fully closed) (see Figure 11-7). If not, a mechanical adjustment to the valve is made until the valve is fully seated. This may involve an adjustment of the valve stem where it is mechanically attached to the valve bonnet assembly. Maximum signal is applied (15 psig) and the valve should be fully open. If not, the spring tension of the spring mounted internally in the actuator is reduced until the valve is fully open. This procedure is repeated until, with a change in air pressure of 3 to 15 psig, the valve travel is correct.

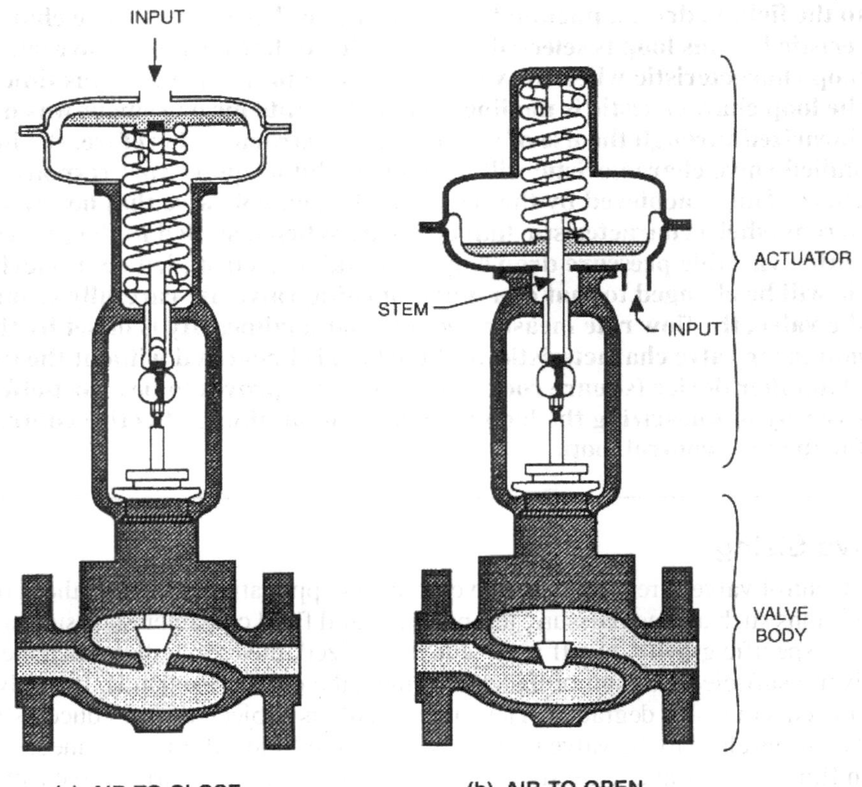

Figure 11-7. Globe Valve Operation.

In stroking a control valve, there may be some difference in the actual physical position of the valve system vs. demand signal, depending upon whether the signal was increasing from a lower to a higher signal or whether the signal was in a decreasing mode from a higher to a lower signal level. In other words, as the signal increases to the actuator from 3 to 15 psig, with a demand signal of 9 psig applied to the actuator the valve stem position may be 50% of valve stroke, while, with a decreasing signal from 15 to 3 psig, the same demand signal of 9 psig may result in a valve stem position of 55 or 60% of valve stroke. This is typical of control valves and is due to the fact that the valve is a mechanical device that is subject to friction and the forces of inertia. The word used to define this occurrence is hysteresis. Most control valves are subject to a hysteresis dead band of from 2% to 5% of their stroke. If a valve exhibits a large hysteresis dead band, the valve stroke vs. controller demand signal will not be accurate. The use of a valve positioner is indicated for this situation (see Figure 11-8). The valve positioner will correct for any inaccuracies with regard to the controller demand signal vs. valve stem position regardless of the direction of change.

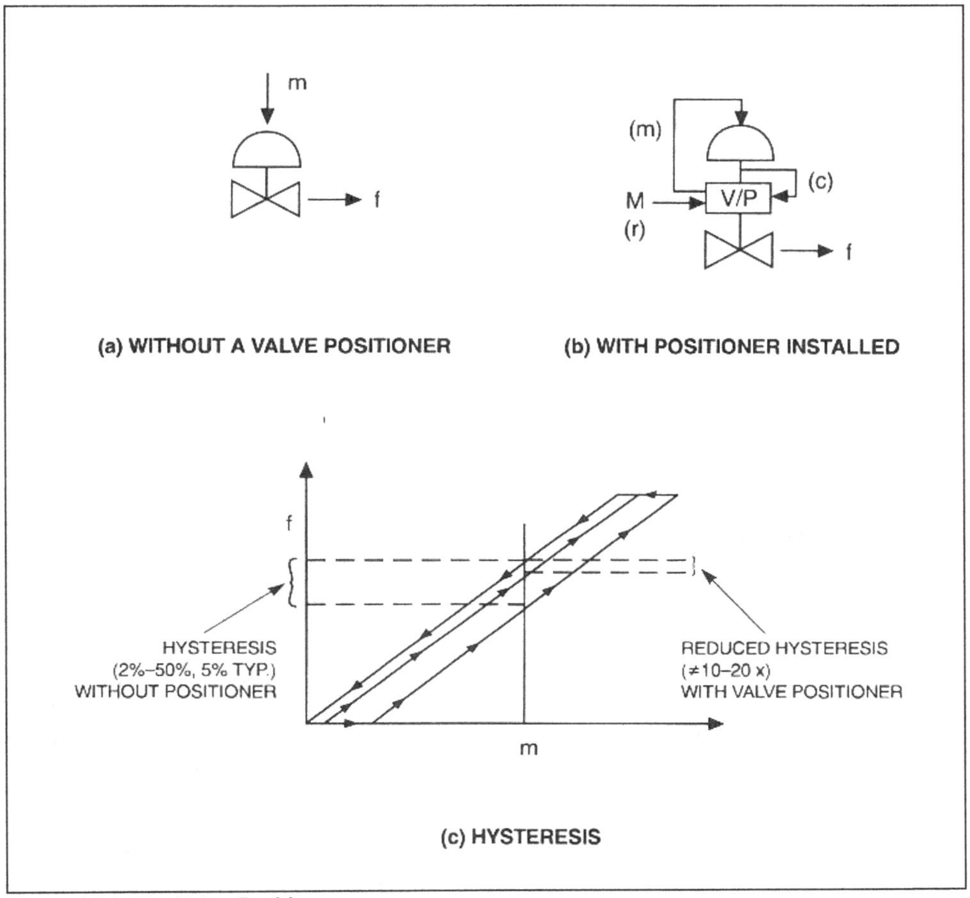

Figure 11-8. The Valve Positioner.

Another situation regarding valve operation may relate to the stroking time of the valve. How long does it take for the control valve to pass through its full travel or stroke? In a process in which the valve design incorporates the use of a large actuator or air motor, the capacity (large volume) of the actuator may result in a very long stroking time. This problem is addressed through the use of a volume booster relay (see Figure 11-9). The volume booster usually has its own air supply, and its output connection to the valve actuator is made through a larger diameter pneumatic line or pipe. The volume booster acts as a 1:1 repeater, as the controller demand signal changes from 3 to 15 psig, its output changes from 3 to 15 psig but at a much larger volume of air flow signal to the valve actuator or air motor. The increased volume of air flow at the required pressure value (3 to 15 psig) results in a faster valve stroking time and improves the response of the control loop.

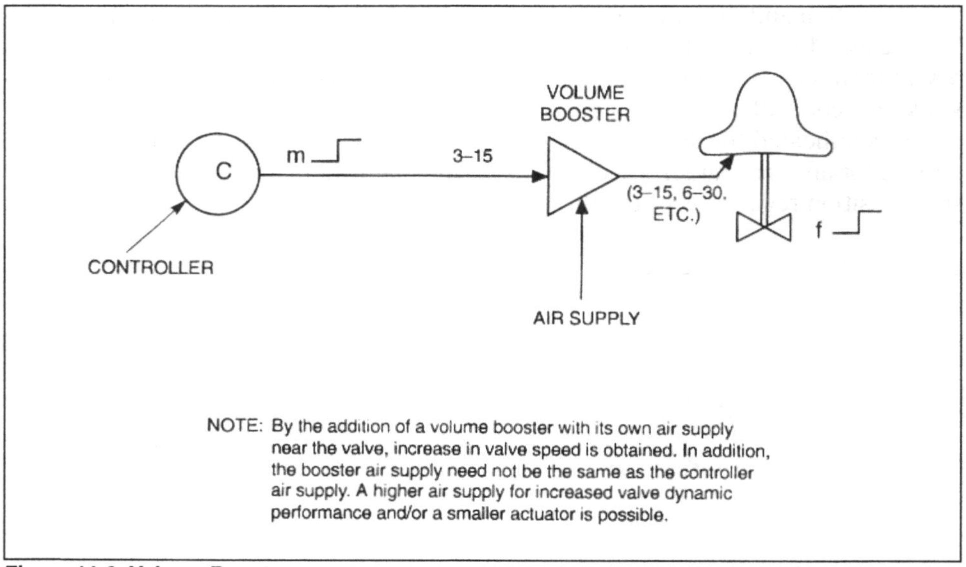

Figure 11-9. Volume Booster.

Valve Components

In the final analysis, all of the various components of the pneumatic control valve must be in good working order if the control valve is to perform as expected. These would include the following:

(1) *Actuator*. The actuator is comprised of the diaphragm assembly, the spring assembly that applies sufficient force to provide for fail safe operation, and the bonnet assembly that provides a mechanical connection to the valve stem.

(2) *Body*. The valve body houses the wetted parts of the valve and the valve plug, seats, and cage depending on the valve design. The valve body must also keep the process fluid contained within itself; therefore, valve packing around the valve stem as it attaches to the valve plug must also be considered.

(3) *Stem*. The valve stem is a connecting mechanism. It translates force generated in the air motor or actuator to physical displacement of the valve plug within the valve body. It is important that the valve stem be free to move and to be exposed to as little friction as possible. The valve stem is a highly machined piece of metal and its surface is highly polished to minimize frictional forces working against its movement. This condition should be maintained at all times. Over-tightening of the packing material at the gland of the valve body may have the effect of increasing friction on the valve stem and thereby reducing the effectiveness of the valve in its normal operation. Doing anything that mars the surface of the valve stem itself is to be avoided at all costs.

NOTE: Electronic instrumentation requires an electronic-to-pneumatic (I/P) transducer be used to interface the valve or pneumatic positioner with the 4-20 mA output signal from the electronic controller (see Figure 11-10).

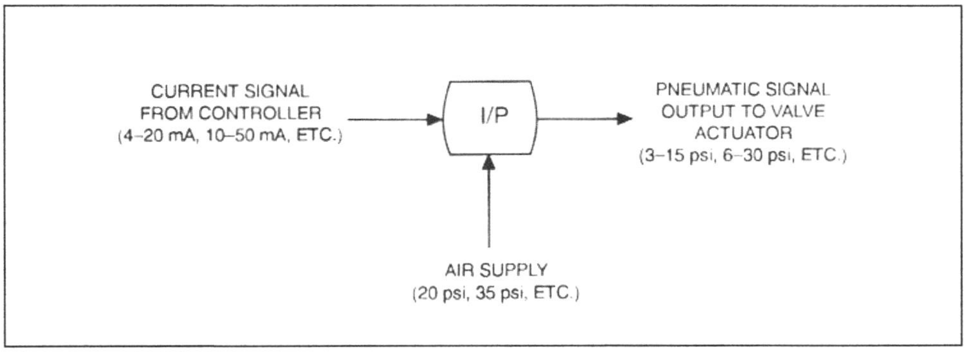

Figure 11-10. I/P Transducer.

Troubleshooting Guide

Improper sizing and installation, corrosion, cavitation, and misapplication of technology are problems most often encountered when troubleshooting control valves. The following troubleshooting guide (Table 11-4) coupled with previous knowledge of performance in a given service, manufacturer's instructions, and related facility procedures should be used with logical troubleshooting approach.

Table 11-4. Troubleshooting Guide.

Problem	Symptom	Probable Cause
Sizing		
Under	Insufficient supply to meet load.	Lack of process knowledge to calculate accurately. Incorrect valve installed due to lack of valve availability.
Over	Valve is almost closed, and possible oscillating during normal operating conditions.	Lack of process knowledge for correct calculation. Too many "just in case" factors included in calculations. Incorrect valve installed.
Corrosion	Trim parts consumed by process.	Trim/body material not matched to process medium under operating conditions, particularly temperature.
	Valve body attacked by process.	Foreign materials contaminating process medium.
		High velocity fluid removing protective barriers to process attack.
Misapplications		
Wrong valve characteristic	Difficult control.	Lack of process knowledge and of valve application experience.
Wrong valve type	Damaged valve parts, poor control.	
Wrong material for process conditions such as maximum temperature and pressure; entrained solids; high velocity flow	Damaged valve parts and valve body.	
Erosion		
	Abrasive wear of trim parts and of body.	Lack of information about process contents. High velocity flow. Incorrect material.
Cavitation		
	Gravel flow sound inside valve.	Too great a pressure drop being absorbed.
	Trim wear as though chiseled.	Frequently occurs on high pressure tight shutoff service.
Improper Installation		
In-line	Valve C_L offset from pipe C_L.	Insufficient experience. Lack of training.
	Wrong gasket.	Not reading instructions.
	Poor bolt torquing.	Gasket creep. Piping improperly supported.
Valve Assembly	Wrong or missing parts.	
	Incorrect stroke.	Wrong spring.

Table 11-4. Troubleshooting Guide (cont'd).

Problem	Symptom	Probable Cause
	Incorrect preload.	Preload not readjusted after change from ATO to ATC or vice versa. Positioner setting. Stops misadjusted. Age.
Leaks		
Seals	Wear.	Age.
	Misapplication.	
Packing	Wear and/or dirt.	Age/no wiper. Poor assembly.
	Loose.	
Actuator Diaphragm	Wear.	Age.
	Cracking.	
	Air contaminants causing deterioration.	Improper filtering and conditioning.
Galling	Trim parts (stainless steel) seized.	Insufficient process lubrication. Wrong trim hardness.
Friction and Hysteresis	Poor valve response. Poor control.	Overtightening of packing. Insufficient lubrication. Normal situation.
Choked Flow		
Gas and vapor	Flow will not increase with increase in *P* upstream.	Too much dP for valve size. Wrong valve location.
All fluids	Low flow.	Mechanical stops incorrectly set. Loose liner lodged in valve opening. Line blockage.
Linkage Wear	Lost motion.	
	Poor control.	Age. Poor lubrication. Contamination.
Fatigue	Valve stem fractured at stem guide.	High velocity flow. Wrong guiding.

12

Pneumatic Instruments

Overview

Pneumatic instruments and explosionproof electrical systems were used in hazardous locations in the early years. New ideas, such as intrinsic safety and sealed and purged electronic instruments, are now being used. Yet, many plants still use pneumatic instrumentation, mostly to move large control valves through an I/P transmitter.

Instrument Air Requirements

With all the electronic and electrical controls in place or on their way, one can get rid of the pesky high maintenance air compressors, right? Wrong! The prime movers for most control valves, damper drives, and fan louvers are still pneumatically operated. They are likely to be so for some time. Many current-to-pneumatic (I/P) and pneumatic-to-current (P/I) transducers are still in place, and many pneumatic control loops are still in service. Compressed air as a control necessity is going to be around for a while.

The two worst enemies of compressed air for control service are moisture and oil. Moisture is always a potential problem because the cooling of compressed air below its dew point causes moisture to condense. Oil can be dealt with more easily because of nonlubricated compressors. Precautions must still be taken to condition compressed air to be of instrument quality.

Types of compressors found in service for instrument air are rotary vane, rotary liquid ring, diaphragm, rotary screw, and centrifugal. Some reciprocating oiled piston compressors are still in service, but it would not be advisable to start a new installation with one because of the added concern for oil.

Compressor intakes should be well filtered and located away from contaminants. Each installation should be addressed for possible contaminants and proper scrubbers or absorbers installed to reduce ambient-imposed problems. Another source of internal corrosion is synthetic lubricants that attack seals and diaphragms.

The compression process causes heat. Inner coolers and after coolers are used to bring the temperature down, causing moisture to drop out as the temperature drops below the dew point. It is, therefore, necessary to provide air drying downstream of the compressor system. Types of air dryers are deliquescent, mechanical refrigeration, and regenerative desiccant.

The deliquescent type requires no utility but is limited in capacity (1 to 5000 at 100°F and 100 psig), and outlet dew point is 30°F or higher. Refrigeration types require electric power. They have a higher flow range of 0 to 10,000 at 100°F and 100 psig but are good only for dew points of only 35°F and higher.

Regenerative desiccant types come with heaters and without, and the motive utility is air; they can reach dew points of –40°F but can go lower if properly designed. Capacity is 0 to 10,000 at 100°F and 100 psig.

The dew point at line pressure should be at least 18°F (10°C) below the minimum temperature to which any part of the compressed air system will be exposed during any season of the year. In no case should the dew point at line pressure exceed approximately 35°F (2°C).

The selection of the proper type and size of air dryer must be based on the actual inlet and flow conditions under which the dryer is expected to operate. To assure continuous delivery of instrument air in compliance with the referenced standards, it is necessary to verify the quality of the system air. Tests would include oil content, dew point, particulates, and other contaminants. Tests should be conducted on initial start-up and retests routinely made at intervals depending upon local conditions.

Maximum total oil content, exclusive of noncondensibles, should be as close to zero as possible. Under no circumstances should it exceed one part per million on either a weight or volume basis.

Once a quality instrument air system has been established, it becomes the most expensive utility in the plant. It is, therefore, necessary to make sure the distribution system is leak-free. One report regarding leakage in industrial plants states that the average industrial plant has a 17% leak rate on compressed air fittings. There are two sides to the leakage problem: (1) piping supply headers and (2) tubing that is used to feed instruments off the header and signal tubing.

Care should be taken in the plumbed makeup of the air system. Screwed fitting technology has been around for over 150 years. It relies on the taper between the male and female threads, with a lubricant between them. Proper lubricant type and application determine success or failure of the joint.

Older tube fittings were of the flared type. Improvement of tube connections came with the development of compression tube fittings. They require proper tube dimensions and close tolerance on the fitting machining. The tube is inserted into the fitting and the compression nut is tightened, causing a ferrule to swage into the tube. Care must be taken to cut the tubing squarely and to remove any burrs caused by the cutting. It is also critical not to excessively tighten the compression nut. Compression fittings are available for stainless steel, copper, and plastic tubing. They are not interchangeable.

Pneumatic Field Instruments

In a discussion of process control in an industrial environment, it is necessary to be aware of the various devices used for the purposes of monitoring, controlling, or providing alarms in such an environment. The needs are many and varied depending upon the process and products produced.

Over the years, the process of automating various production facilities has taken on many different forms. Originally, manual operations and local indication and control approaches were all that were available to manufacturers in their attempts to produce a viable product. This was a relatively costly and inefficient approach and in some cases a very dangerous situation for plant operators as well. The idea of automation (having machines or equipment of a highly specialized design performing in these areas) led to a safer, more cost-effective method of producing a desired product. The progression of measurement and control automation from manual control to local automatic control, to centralized control, to distributed control systems has been based for the most part on the availability of hardware devices and technology. Automation at first consisted of using mechanical and pneumatic devices designed to certain performance criteria that made automation a practical reality.

The most simplistic form of continuous automatic control evolved over the years into the use of certain devices, working in conjunction with one another, to provide what is now commonly known as feedback control. The components of a feedback control loop may be very clearly defined by the devices being used in the control loop. A feedback control loop is comprised of the following:

(1) A field transmitter, whose function is to provide a continuous source of information about the control variable to the feedback controller.

(2) An automatic controller, whose function is twofold:

  (a) to determine if an error condition exists, where error is defined as the difference between the measurement signal as provided by the field transmitter and the set point (which is the reference point for controlling the loop) as adjusted by the operator; and

  (b) to provide a control response (a demand signal calculated by the control algorithm) as the controller output signal. The demand signal is sent remotely to a field device that will act to regulate the supply to the process to achieve a balance of supply and demand for the process control loop.

(3) A final control element, whose function is to regulate the supply of mass or energy to the process itself, based on receiving a remote demand signal. The most common type of final control element used in a process industry is the pneumatic control valve.

Pneumatic systems use a combination of mechanical components and an air pressure signal to provide useful information about process conditions or to provide automatic control capability. The most commonly used pneumatic systems found today operate on an air pressure signal of 3 to 15 psig to infer a change in a process variable of 0 to 100% of its measurement span (the algebraic difference between the lower range value or starting point of the measurement span and the upper range value or the end of the measurement span). The generation of this transmission signal is the main function of a pneumatic field transmitter. In general, all field transmitters have certain components that provide the device with the overall functionality of a field transmitter. These are:

(1) a detector mechanism (a device used to sense changes in a process variable),

(2) an amplifier (a device that magnifies or increases the detector signal to enable remote transmission of the information provided), and

(3) a feedback mechanism (a device that provides the stability required to provide accurate, repeatable information over the distance traversed by the transmitter's output signal).

All of these components may be found in any pneumatic field transmitter. The design and terminology may vary according to which vendor is supplying the device, but functionally they perform in a very similar manner.

The basic signal provided by the detector assembly is a back-pressure signal that is directly proportional to the displacement of a motion-type sensor. One term used to define this sensor system is flapper/nozzle detector. In a sensor system, a small piece of tubing is connected at one end to a stable air supply. The air supply should be at least 5 psig greater than the throttling range for the transmitter, which allows for the constant loss of air as the pneumatic system performs to indicate the value of the process variable being measured at any point in time. Because the pneumatic transmitter works in a relatively small area, the distances traversed by the detector sensor are very small, and the associated orifices used are of a very small diameter. The air supplied to the system passes through a fixed flow restrictor whose diameter typically falls within the 0.005 to 0.010 inch range. The other end of the tubing is connected to a specially designed orifice called the nozzle. The nozzle opening is typically on the order of 0.020 to 0.040 inch in diameter, and it vents the air present in the detector system to atmosphere.

The other component in the detector system is a thin piece of metal that is fixed at one end and allowed to move about a pivot point with regard to the opening provided by the nozzle assembly. The total amount of displacement of the flapper to the nozzle assembly may be on the order of 0.01 inch. The flapper nozzle assembly acts as a variable restrictor to air flow passing from the air supply, through the fixed restrictor, and out through the opening of the nozzle. As the flapper travels to the nozzle, it may, in fact, seal the nozzle opening, which results in the back pressure of the sensing system going eventually to the supply pressure value (i.e., 20 psig). Conversely, as the flapper travels away from the nozzle opening in the other extreme, maximum air is vented from the sensing system, which results in no back pressure being generated (i.e., 0 psig). Of course, neither condition is acceptable, and, to compound the problem of the sensing system itself, the displacement of the flapper to the nozzle and the resultant back-pressure signal is very nonlinear.

One approach to correcting these problems is to limit the flapper-to-nozzle travel distance. When this is accomplished, a flapper-to-nozzle displacement of 0.0006 inch will generate a linear back-pressure signal of 0.75 psig, which is proportional with flapper displacement or movement. All that remains is to find a means of connecting the flapper displacement to a changing process variable. For pneumatic devices, this is a function of design, the two most common methods relating either an increment of movement (motion balance) or a moment of force (force balance) to flapper travel.

The pneumatic amplifier is sometimes called a control relay.

The pneumatic signal generated at this time is accurate and repeatable, but its usefulness is limited. A three quarters of a pound per square inch gage air signal cannot effectively be transmitted over relatively long distances and function properly. To overcome problems associated with this condition, an amplifier assembly is required. A pneumatic amplifier is connected to the detector signal, and its output is used as a remote signal source to provide a transmission signal that is capable of traversing long distances without degrading signal integrity.

The pneumatic amplifier is a mechanical amplifier that has a fixed gain of 16. The detector back-pressure signal of 0.75 psig, when multiplied by this amplifier, provides an air signal on the order of 12 psig ($0.75 \times 16 = 12$). Thus, the instrument output span of 12 psig reflects a detector displacement of 0.0006 inch, which in turn reflects a process variable changing from 0 to 100% of its measurement span. The component parts used in the control relay are mechanical in nature and consist of the following parts:

(1) A metallic diaphragm of a fixed area. The detector signal impacts the area of the diaphragm and generates a sufficient amount of force to cause movement of a stem connected to the back part of the diaphragm.

(2) A ball valve. The position of the diaphragm stem, which seats the ball, allows the relay's independent air supply to access one of two ports in the control relay. The two ports are connected, respectively, to either the output of the relay or an exhaust port that allows for venting the pressure signal with a decreasing detector signal.

(3) A spring assembly. The spring assembly provides a counterforce that applies tension to the ball valve to allow for accurate and responsive positioning of the valve assembly.

All pneumatic devices that have transmission capability will have some type of amplifier assembly that performs in a similar manner. Design criteria may change slightly, depending upon equipment usage, but pneumatic devices such as field transmitters, controllers, converters, valve positioners, computational devices, etc., have this amplification capability.

So far, two of the three basic components found in a typical field transmitter have been explored. The mechanisms involved are able to detect and transmit information about a process variable; however, stability is still a concern. Since instrumentation works off a balance of forces as implied by the terms "force balance" and "motion balance," another mechanism is required to provide a useful field transmitter. At this point, if a slight increase in the process variable were being measured, the flapper/nozzle relationship would change; the flapper would travel to the nozzle, and, since there would be no opposing force to this movement, it would eventually seal the nozzle opening, resulting in the output signal going to a maximum value. If the process variable were to decrease slightly, the opposite condition would occur, and the transmitter output would go to a minimum output condition.

This problem is addressed through the use of a *negative feedback* mechanism. To achieve output stability for a given measurement value, a feedback bellows is applied at the detector mechanism. A portion of the transmitter output signal is fed to a bellows assembly; the bellows assembly applies a force that is essentially equal and opposite to the force or motion applied to the flapper portion of the detector assembly. Now there is a proportional relationship between the flapper position with regard to the nozzle assembly and the output signal of the field transmitter. This means that for every position of the flapper in relationship to the nozzle, a predictable output signal is generated.

The addition of a zero spring allows the generation of a live zero for the transmission signal. The lower range value of the output span is set to 3 psig instead of 0 psig; therefore, the transmission signal provided as the transmitter output changes from 3 psig to 15 psig with a measurement change of 0 to 100%.

In some cases an additional fulcrum point is added to the detector feedback system. This becomes a variable span adjustment and allows the same transmitter to work with different measurement spans.

Pneumatic instrumentation uses air pressure to generate signals that represent process conditions. Standard signals are usually between 3 and 15 psi, where 3 psi represents 0% of process variable (0% of chart) and 15 psi represents 100% of process variable (100% of chart). In order to develop this signal, a supply pressure (AS) of 20 psi is typically used and should be well filtered, oil-free, and dry.

To develop signals that are proportional to process conditions, a detector must be used. The concept can be demonstrated using a short section of tubing with two valves and a pressure gage (as shown in Figure 12-1) connected to the air supply and vented to atmosphere. V_1 will be referred to as the flow restrictor and V_2 as the pressure regulator.

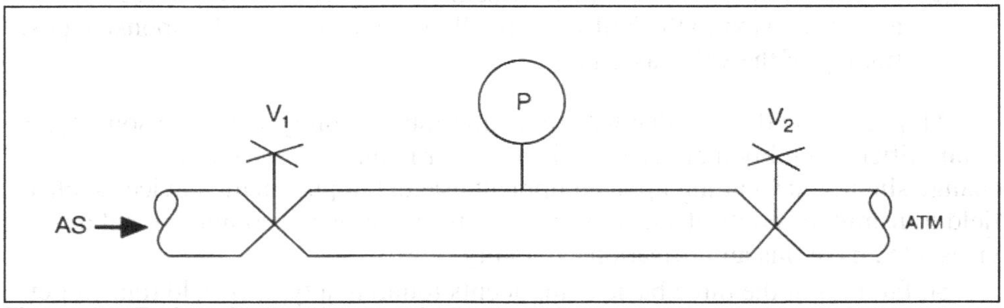

Figure 12-1. Tubing with Two Valves and a Pressure Gage.

Starting with both valves half open, a back pressure will be indicated on the gage (*P*).

If V_1 is gradually closed, P will decrease. V_1 will be closed until P reads zero. At this point, V_1 will be considered as a fixed restrictor (R).

The pressure *P* is now controlled by V_2. If V_2 is closed gradually, *P* will gradually increase to 20 psi. Every position of V_2 generates a unique, precise, and repeatable pressure signal. If the position of V_2 can be made to represent process conditions, *P* will be directly proportional to process conditions.

In Figure 12-2, V_1 has been replaced by a fixed restrictor and V_2 has been replaced by a flapper and nozzle detector assembly. The flapper is shown pivoted at its upper end. If we cause the flapper to move slightly left or right, we will find that each position of the flapper produces a unique back pressure (*P*). The flapper can now be coupled to the process so that *P* is proportional to process conditions. One method of doing this is shown in Figure 12-3 where a process pressure is sensed at the input bellows, which causes the flapper to cover the nozzle and increase *P* with increasing process pressure. A range and bias spring (*S*) is shown to limit the bellows travel and establish reference conditions of output (*P*) and input (process).

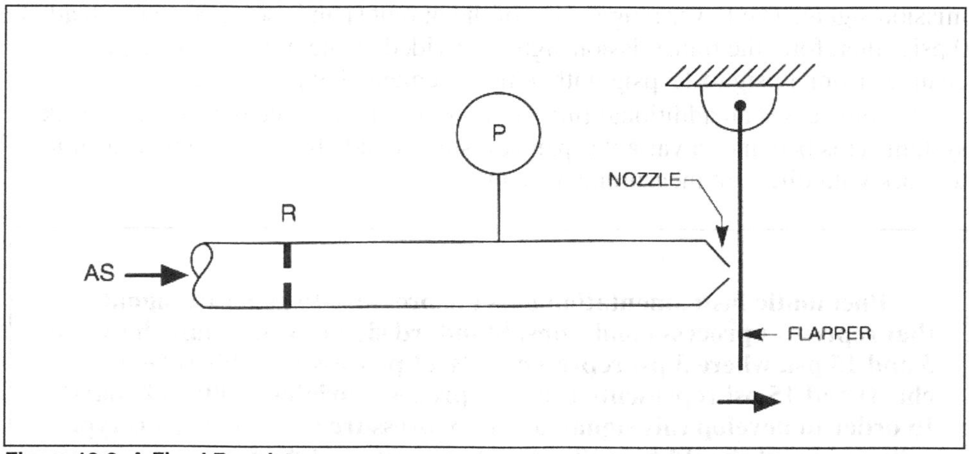

Figure 12-2. A Fixed Restrictor.

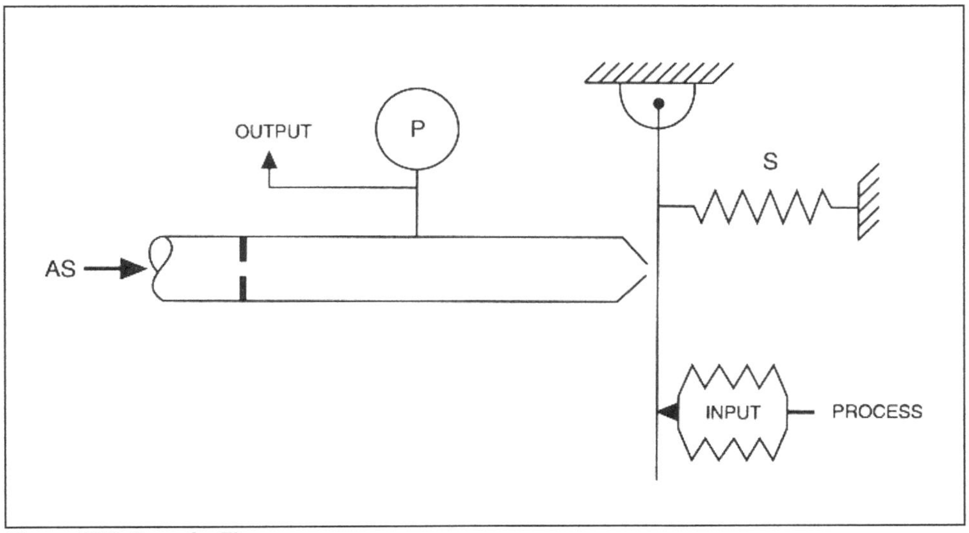

Figure 12-3. Use of a Flapper.

At this point, a brief discussion of detector sensitivity is necessary. The fixed restrictor is typically an orifice of between 0.005 and 0.010 inch. The nozzle opening is typically 0.020 to 0.040 inch. The travel of the flapper required to cause P to vary from 3–15 psi is only a few thousandths of an inch. The mechanism of Figure 12-2, while theoretically accurate, would in practice have many drawbacks if output is used to transmit information any significant distance. These drawbacks are the following:

(1) Poor repeatability. (The bellows and spring are not sensitive enough to position the flapper accurately or repeatedly. Friction and hysteresis are also excessive.)

(2) Poor response to change. (Any change in process conditions will cause P to change, but response will be slow. Resistance to flow is represented by R, while the gage and connecting tubing represents capacitance (C). The combination of RC represents a response delay.)

(3) Instability. (Any vibration will cause the flapper to move, and its high sensitivity to motion will cause P to vary.)

The problems listed above can be overcome by adding negative feedback. This is accomplished with the feedback amplifier (FB amp) and feedback bellows (FB B) shown in Figure 12-4.

The FB amp has its own air supply, so it can provide a powerful output signal (P_o) that is applied to the feedback bellows and is also the final output, which is now linearly proportional to input pressure.

The overall operation of the device is as follows:

(1) Initial conditions. Assume that there is a stable output between 3 and 15 psi and that the input increases slightly.

(2) Flapper moves toward nozzle.

(3) P increases.

(4) Amplifier responds with increased output (P_o).

(5) FB bellows generates higher force to balance input.

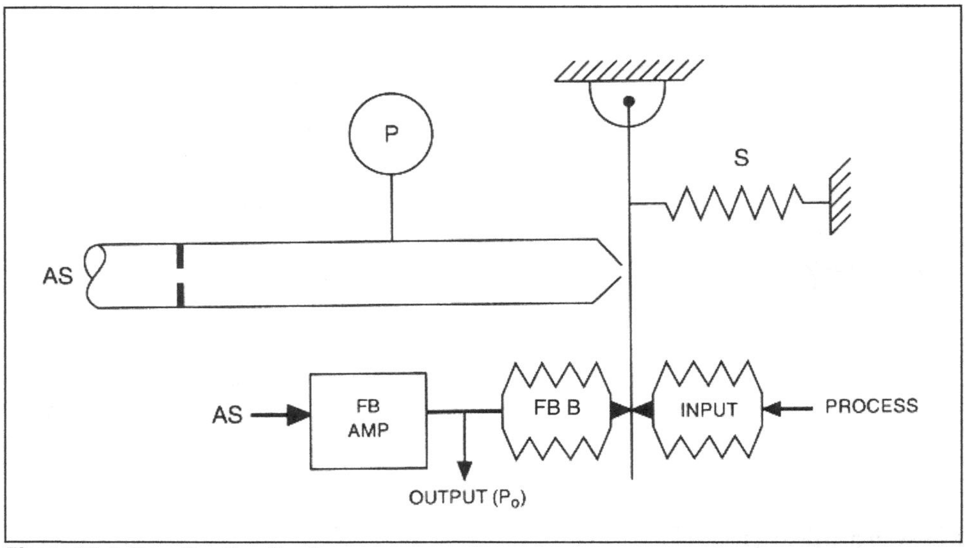

Figure 12-4. Negative Feedback.

(6) Flapper is stabilized at a new position when P_o balances input. Total flapper travel in the FB B system is less than 0.001 inch to generate 3–15 psi at P_o. Total change in P is less than 1 psi.

This basic concept of feedback stabilized pneumatic signal generator is used throughout the instrumentation industry in transmitters, controllers, timers, etc.

This type of assembly provides a dependable, repeatable, and powerful signal that is capable of being transmitted long distances (up to 1500 feet depending on the application) to drive recorders, valves, or other pneumatic instruments. The feedback balances output against input so that any minor changes in performance of the detector or amplifier will not affect the accuracy or repeatability of the output signal.

There are many variations on this basic mechanism to achieve a wide range of application requirements. As shown, the mechanism performs as a fixed gain amplifier, as used in some process transmitters.

By altering the relative size of input and feedback bellows, number and location of additional inputs and feedback elements, relative mechanical advantage of the input and feedback elements, and the type of components in the feedback loop, many other functions can be developed. Some of these, shown in Figure 12-5 through 12-11, are:

(1) differential amplifiers (Figure 12-5)

(2) variable gain amplifiers (Figure 12-6)

(3) summing amplifiers (Figure 12-7)

(4) comparators (logic amplifiers) (Figure 12-8)

(5) integrating and differentiating amplifiers (Figure 12-9)

(6) derivative amplifier (Figure 12-10)

(7) position (motion) amplifier (Figure 12-11)

In combination, these devices provide the instrument and control system designer with the equipment necessary for measurement and control throughout the process industry.

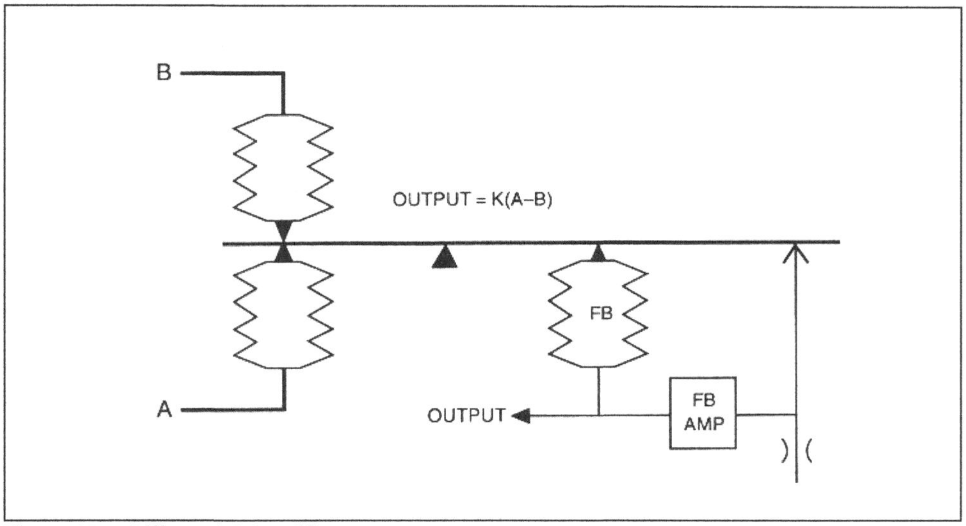

Figure 12-5. Differential Amplifier.

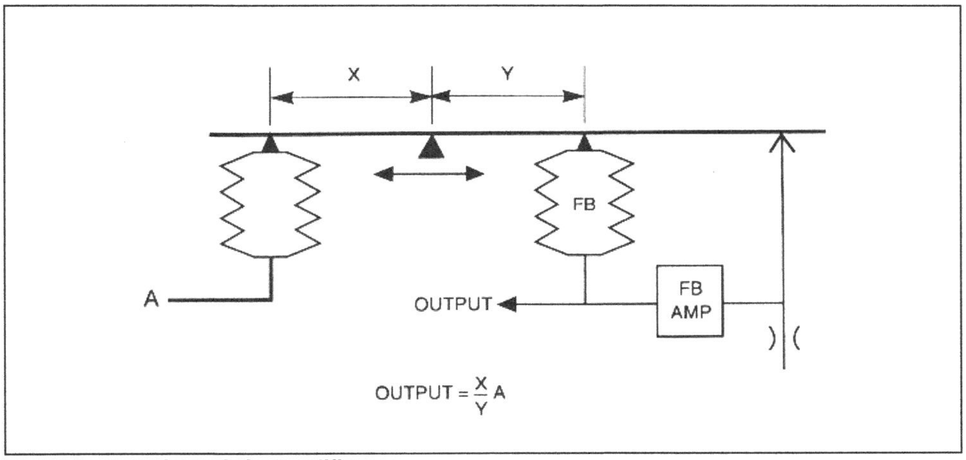

Figure 12-6. Variable Gain Amplifier.

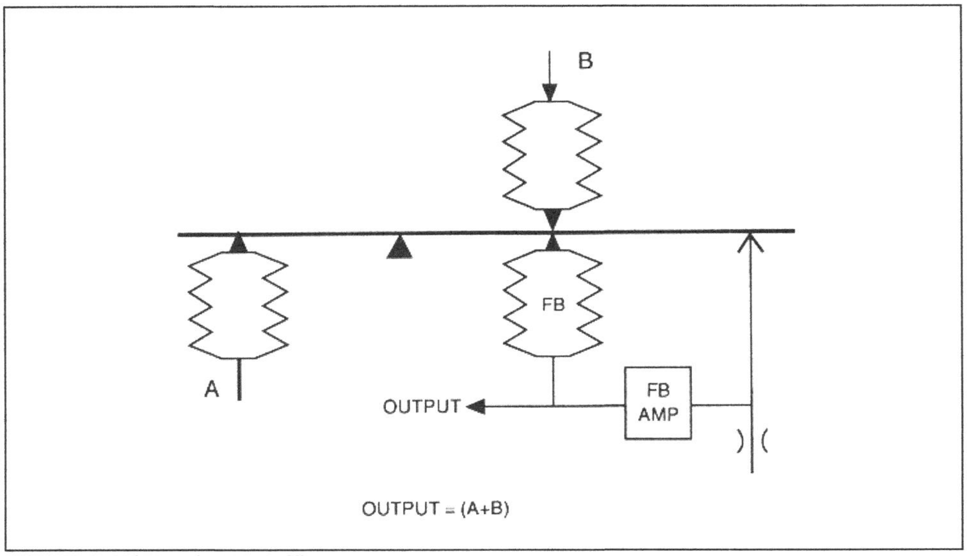

Figure 12-7. Summing Amplifier.

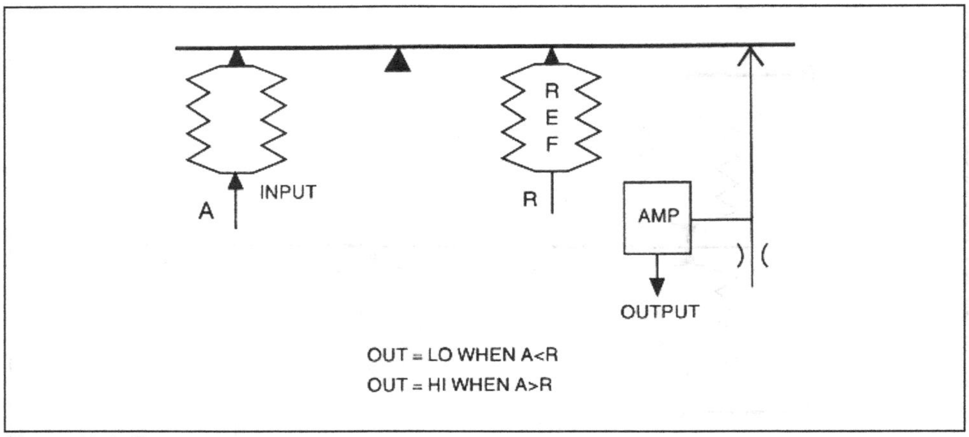

Figure 12-8. Comparator.

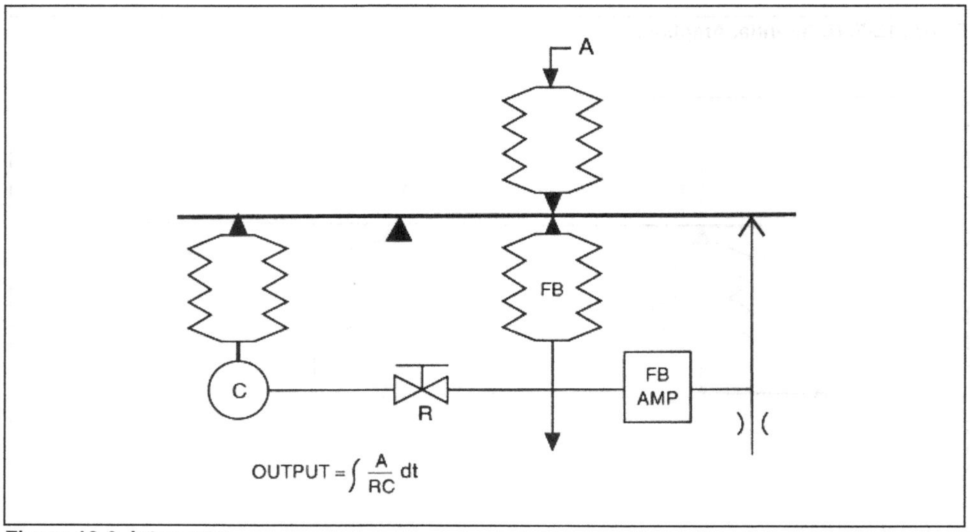

Figure 12-9. Integrator.

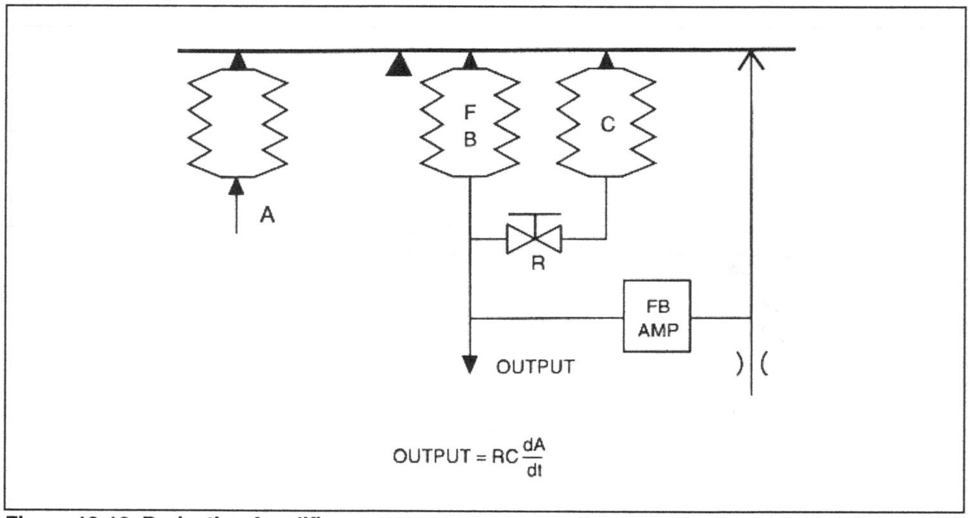

Figure 12-10. Derivative Amplifier.

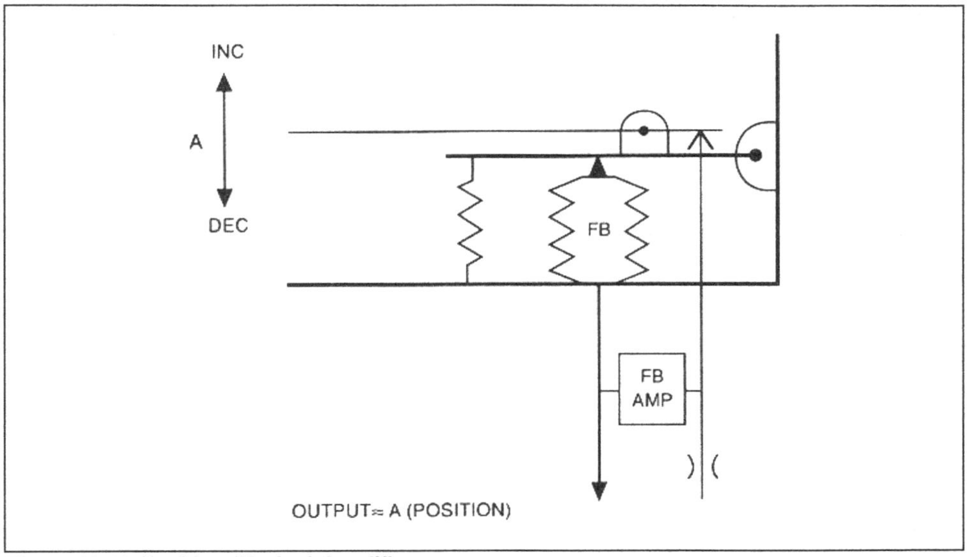

Figure 12-11. Position (Motion) Amplifier.

An indicator, to assist in analyzing problems, is beneficial in troubleshooting a pneumatic field transmitter. The transmitter's output signal is an ideal indicator of the operating condition of the transmitter. Table 12-1 shows potential problems that may be associated with the use of pneumatic field transmitters by analyzing the transmitter's output indicator.

Table 12-1. Potential Problems of Pneumatic Field Transmitters

Symptom	Probable Cause
No output	Bent flapper. No air supply; plugged restrictor (very common). Corroded control relay or components. Dirty control relay seats. Flapper is away from the nozzle due to freezing, improper adjustment, bent "C" lexure, or transmitter has been dropped. Leak in the feedback bellows. Leak in the nozzle circuit. Leak in the sensor pressure circuit. Disconnected or broken links in a motion balance transmitter.
Partial output	Plugged low-pressure leg on a dP cell. Worn control relay parts. Partially plugged supply screen or filter. Burr on the flapper assembly. Hole in the flapper assembly. Damaged feedback bellows. Worn capsule diaphragms. Warped or distorted "C" flexure or "A" flexure on a dP cell. Wrong range-sensing unit. Pin hole leaks in the control relay diaphragm.
Full output	Plugged nozzle. Ballooned capsule diaphragm. Loose nozzle lock nut. Blocked control relay vent. Sensing capsule impacted with process solids. Flapper assembly distorted or bent.
Zero shift diaphragms	Dirty flapper assembly set point capsule problems: coating, fatigue, warped. Temperature changes: either ambient or process temperatures. Process static pressure changes. Worn zero or span adjustments. Flapper is "dimpled" on the surface. Pin hole leak in the flapper. Flashing and/or condensate on either leg of a dP cell installation.
Output oscillates	Liquid in the feedback bellows (water or oil, etc.). "C" flexure lock nut loose. Close-coupled pneumatic system. Loss of capsule fill fluid. Hole in the feedback bellows. Loose bleed/vent valves. Flashing due to pressure variations.

13

Calibration

Overview

Calibration and Reasons to Calibrate

When one buys gasoline for the car, one expects to get exactly one gallon of gas when paying for one gallon. When one takes a child's temperature, one needs to be assured that the reading is correct. When the remote level indicator on a tank indicates that it is 90% full, can one be sure that the tank is not running over? Can the manufacturer be sure the process control system is providing the proper mix for the product? When a combustible mixture is being heated to just under the flash point, can one be sure that it won't ignite?

Continuing calibration assures the equipment continually meets the specifications required at installation, and it should be checked regularly thereafter. Calibration after any maintenance ensures the equipment still conforms to the required calibration data. Customers require letters of conformance or certificates of calibration.

Calibration is a test during which known measurement values are applied to a device and a corresponding instrument reading is compared to a standard series of values the device was to measure, receive, or transmit. These known measurement values are usually obtained by using test equipment that has traceability to the National Institute of Standards and Technology (NIST), formerly the National Bureau of Standards (NBS). Data so obtained are used to determine the locations at which scale graduations are to be placed, to adjust the output to bring it to the desired value within a specified tolerance, and to ascertain the error by comparing the device output reading against a standard.

Chapter 13 gives particulars on procedures, standards, records, traceability, and calibration of specific types of equipment.

Calibration may be defined as the process of checking the zero, span, and range of a device and its linearity or fit to a known calibration curve. Simply, this means how well the device duplicates the actual variable it is measuring, or how well the device duplicates the output it is being asked to follow. The device can be either a field device, such as a pressure transmitter, a thermocouple or a valve positioner, to mention a few, or it can be a panel or control room device, such as an indicator, a recorder, or a controller, and so on. The device used as a reference for the calibration should be traceable to the National Institute of Standards & Technology (NIST). That does not mean that the calibrator itself, must be calibrated at the national labs or some certified calibration lab. It only means that shop or field calibration devices should be checked against a certified standard (which could be the shop instrument NIST-certified gage), and any deviations from the test standard should be listed as a curve or correction factor to be added or subtracted from the field tester readings. All instrument shops should have at least one set of basic calibration standards that have NIST-traceable certificates.

Calibration may be thought of as the simple adjustment of the zero, span, and linearity of a transmitter, but, in reality, calibration should include the transmitter, the interconnecting cable, and the receiver, since what is wanted is to have the receiver read exactly what the transmitter sees. In some cases that may be pressure in pounds per square inch (psig) or flow in gallons per minute (gpm). In the case of flow, the transmitter may measure the pressure drop across an orifice plate or the voltage from a magnetic flowmeter tube, but the operator is interested only in the gallons per minute reading.

If a pressure, as indicated by a panel reading, is suspected to be in error, both the transmitter and the panel meter should be checked independently. If they are both in calibration, the transmission (cable or tubing) should be checked, as well as the transmitter installation, to determine if some outside influence might be affecting the signal.

From this short description of calibration, one can easily see that calibration isn't always a simple adjustment of a single device. Calibration can involve troubleshooting to discover and correct what was thought to be a simple calibration adjustment. The comments herein will, therefore, include some useful troubleshooting techniques that might help resolve a calibration problem.

The calibration techniques for process control instrumentation have changed greatly in the past few years because of the following advancements:

(1) High resolution transmitters

(2) Digital-type receivers

(3) Longer transmission lines

The advent of newer digital-type receivers with their improved resolution plus longer transmission distances has required higher-accuracy calibrators and more loop consciousness. One no longer reads a process signal on an 8-inch strip chart with "fat lines" and "thick needles," but on a digital control system, programmable controller, or digital readout that has more resolution than the analog world ever dreamed possible. As a result, many times calibration is performed in the field, in order to include the actual process variables of temperature, pressure, and electrical environment that may affect the loop accuracy and calibration.

Instruments may be checked in the shop, but the final check of calibration may be more meaningful if done in the field to include ambient effects, transmission cable and rack or control room readouts, and devices. Stray pickup, improper grounding, too much resistance in the loop, etc., can become a significant input to loop calibration.

A creditable calibration should include five points of reference. Points at 0%, 25%, 50%, 75%, and 100% are usually sufficient. Zero % is important, because in most cases, for a quick check, the transmitter can be valved off from the process and vented to atmosphere. If only the zero seems to be off, many times a complete calibration is not needed.

A complete calibration should include both an up and a down calibration in order to pick up any hysteresis or any lack of repeatability. Most modern transmitters have noninteracting zero and span adjustments, thereby reducing the number of calibration adjustments. Field calibrators available today are rugged, have the accuracy to provide laboratory-grade calibration, are battery powered, and are very portable.

Field Calibration

When adjusting span, always go back and verify zero.

Field calibration saves time since it does not usually require the removal of the instrument from the process or from its mounting bracket. In many cases it also allows the field device to be tested or calibrated at the true process and ambient conditions, which may be considerably different from instrument shop conditions.

A proper field installation should allow easy disconnect of the field device from the process. In most cases the transmitter or device can be disconnected from the process by simply closing isolating valves. The device is then vented to atmosphere and the test or calibration signal, whether an imposed pressure or simply an electronic signal, can be connected to the device. This procedure is usually performed as a first step. If the transmitter is found to be out of calibration, it is simply field calibrated. If the transmitter has other problems, the rest of the loop is checked before the transmitter is replaced. This maintenance concept makes sense because it includes field effects, saves considerable time, and includes troubleshooting when necessary.

Today's field calibration devices can be purchased with calibration accuracies equal to those of shop-type calibrators. They are made to withstand field abuse and have digital displays that maintain their accuracy and are easy to read. A discussion of the various types of calibrators is included under the section on basic variables.

Before disconnecting any pressure device attached to the process, determine what chemical or gas is being measured and at what pressure. Never simply loosen a process fitting or vent a transmitter before you have confirmed it is safe.

Pressure Calibration

Pressure calibration equipment must include a device to produce the calibration pressure, a device to accurately indicate the pressure being produced, and a readout of the transmitter output, either pneumatic or electronic. Some typical calibration equipment is described below.

HAND-OPERATED PORTABLE PUMPS AND COMPARATOR TESTERS

These devices use the same pump arrangement as do the shop-type dead weight testers, but they substitute a test gage in place of the standard weights. This device is somewhat more portable than a dead weight tester and can tolerate more field abuse. The tester uses a standard-type pressure gage for readout and has a connection to attach the device being calibrated. The tester gage should be periodically checked against a certified standard gage; a correction curve kept with the field tester (see Figure 13-1).

HAND-HELD, COMPARATOR TESTERS

These testers are relatively small, light in weight, and portable. They must be used in conjunction with an accurate test gage. Air-operated pumps can be hand-pressured to about 200 psig, but hydraulic hand pumps can calibrate to as high as 5000 psig (see Figure 13-2).

Small hand-held pneumatic calibrators are available that are calibrated in inches of water ranges for differential pressure (dP) transmitters or draft gage calibration. They are also made to cover a range of 0-18 psig for troubleshooting 3-15 psig instrument loops. These are very portable and excellent devices for troubleshooting as well as field calibration of valve positioners and I/P transducers. (see Figure 13-3).

Any portable pressure tester receives hard use and must have its readout periodically checked against a dead weight tester or other traceable calibration device in order to guarantee its accuracy. When in doubt, always check the field calibrator against an NIST-traceable standard instrument.

There are many more versions of these types of testers. Some can generate a vacuum, some use standard gages, and some use electronic pressure sensors and digital readouts. Still others come in larger containers and include several gages, switching valves, and so on, to cover a wide range of pressures or higher accuracies than the small hand-held types. The instrument tech needs to try several types and choose one that fits his or her needs of portability, ruggedness, flexibility, and desired accuracy.

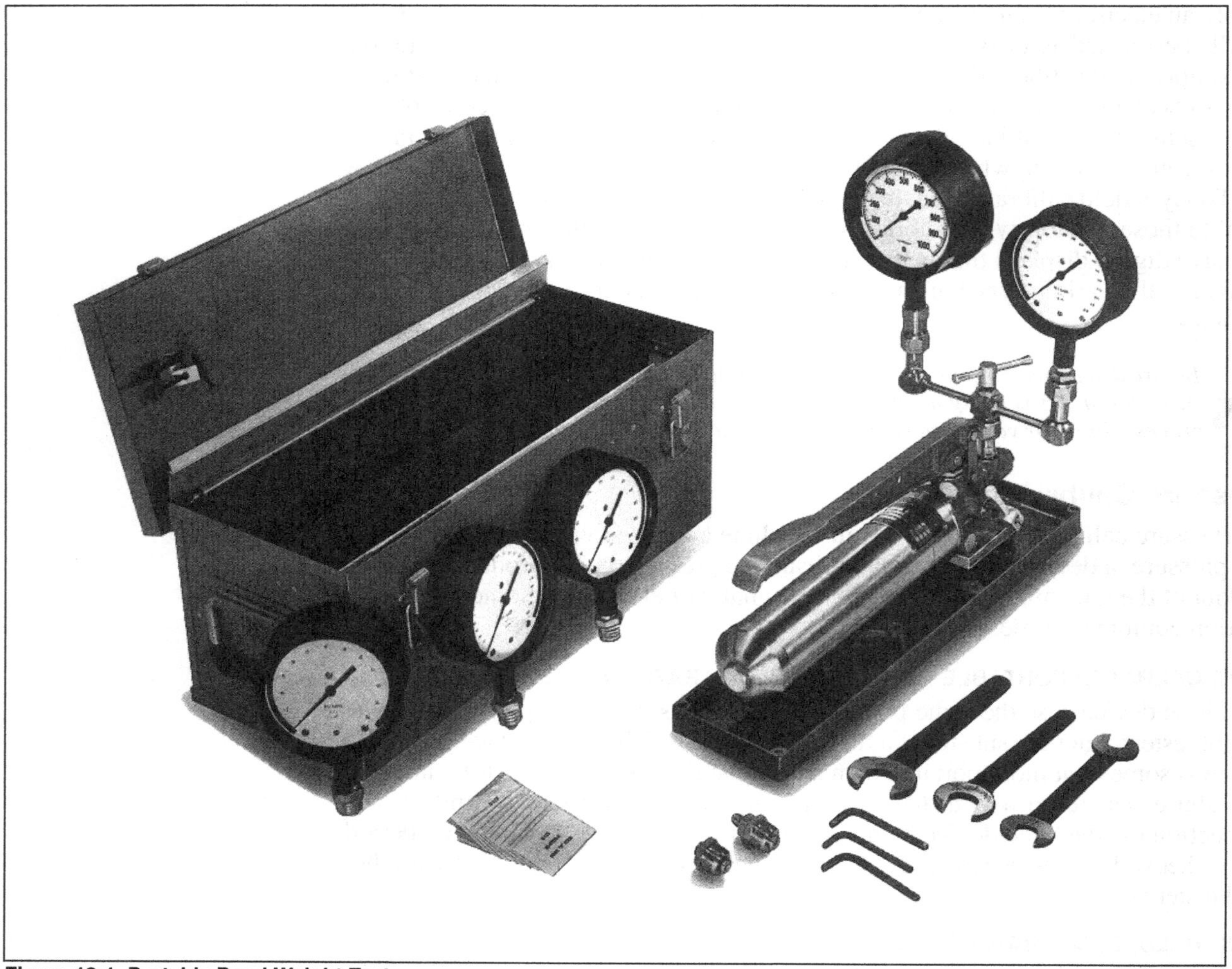

Figure 13-1. Portable Dead Weight Tester.
(Courtesy of Ametek)

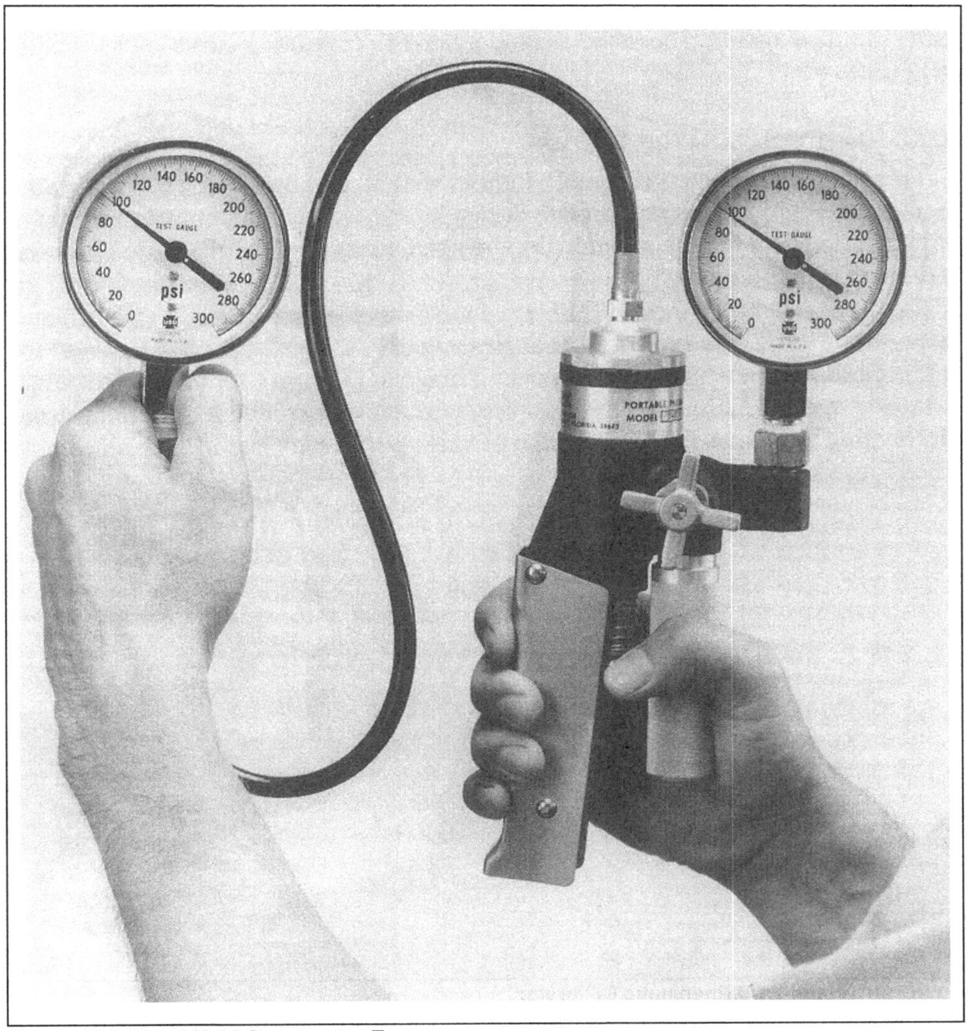

Figure 13-2. Hand-Held Comparator Tester.
(Courtesy of Ametek)

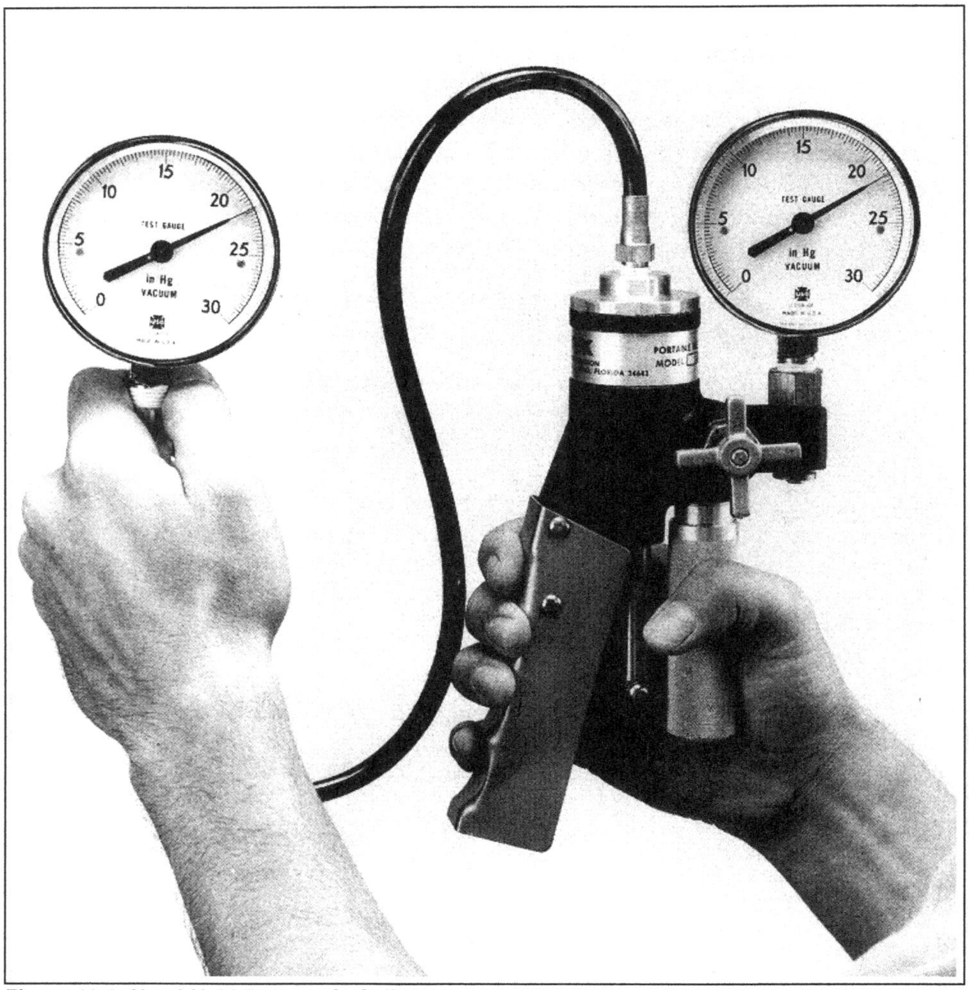

Figure 13-3. Hand-Held Pneumatic Calibrator.
(Courtesy of Transcat)

A major problem with pressure calibration is maintaining a constant calibrator pressure while making adjustments. To prevent leaks from the calibration source or any of the tubing, be sure fittings and any O-rings are clean and in good shape. Teflon™ tape is recommended on any screwed fittings. Tighten or adjust fittings until a fixed calibration pressure can be maintained long enough to make the adjustments.

When dealing with high pressures, care must be taken when bleeding the transmitter or tester to atmospheric pressure. Always bleed the tester before disconnecting the tubing and vent away from face or body. High pressure applications usually have a block-and-bleed manifold to allow controlled relief before disconnecting the process connection. Use the bleed valve before loosening any process fittings.

A technician brought a transmitter to the instrument shop, but nothing could be found wrong with it. The instrument was declared to be operational and was re-installed in the field. The output immediately went full scale. The cabling was checked and a ground loop was found. Full output when the transmitter is not at its upper limit usually indicates a ground loop.

If the calibration equipment has the ability to supply 24-volt DC power, one can then power the instrument in the field, with the independent and isolated power supply inside the calibrator, and find out if the loop is performing properly. If the loop performs properly using the isolated 24-volt DC power from the calibrator but does not perform when it is back on the main power supply, this is almost sure evidence that a ground loop exists, because there are two instruments using the same supply at different ground potentials, and a current is flowing between them.

Figure 13-4 is a troubleshooting guide developed for a hand-held field calibrator instruction manual. This particular case illustrates a field-mounted pressure transmitter and the step-by-step procedures to follow in order to check calibration and transmitter operability in the field. The use of a guide helps newer technicians develop the thought processes necessary to become proficient in the field of instrument maintenance.

HOW TO MEASURE RESISTANCE WITH A CURRENT/VOLTAGE CALIBRATOR

By making use of a calibrator's ability to measure an unknown signal and to output a test signal simultaneously, it is possible to measure an unknown resistance. Measuring an unknown resistance with the calibrator involves driving a known current through the unknown resistance and measuring the voltage generated across the resistance. The unknown resistance can then be calculated using the formula $R = V/I$, where R = *resistance*, V = *voltage*, and I = *current*.

To minimize the required arithmetic, use one of the following values for the known current signal: 0.01 mA, 0.1 mA, 1 mA, or 10 mA. The voltage value read on the digital display must be multiplied by the proper factor (see below) to obtain the unknown resistance value in ohms. For instance, assume that the current output is 1 mA. A reading of 1 volt = 1000 ohms, and a reading of 0.5 volts = 500 ohms.

Known Current Signal	Multiplication Factor
0.01 mA	100,000
0.1 mA	10,000
1 mA	1,000
10 mA	100

To measure resistance, follow the instructions listed below:

(1) Connect a test lead from the black (–) mA OUT terminal on the calibrator to the resistance.

(2) Connect a test lead from the red (+) mA OUT terminal on the calibrator to the resistance.

(3) Connect a test lead from the black (–) V IN terminal on the calibrator to the resistance.

(4) Connect a test lead from the red (+) V IN terminal on the calibrator to the resistance.

(5) Set the Display Mode Switch to OUT.

(6) Set the Output Function Switch to mA.

(7) Turn the Output Adjust Controls counterclockwise three full turns.

(8) Lift the Power Switch to ON.

(9) Adjust the Course and Fine Output Adjust Controls until the readout displays the desired signal input to the resistance.

(10) To monitor the voltage drop across the resistance, set the Display Mode Switch to IN and the Input Function Switch to V. Alternate the display mode using the Display Mode Switch as required.

(11) Perform the necessary calculations to obtain the resistance in ohms.

(This illustrated procedure uses a Transmation® Calibrator, but any calibrator with a dual input and output can be used.)

Excessive loop resistance usually shows up as a drop in the receiver reading as it approaches its maximum reading or in its inability to go to the higher portion of the scale. This is not a device calibration problem (which can be determined by

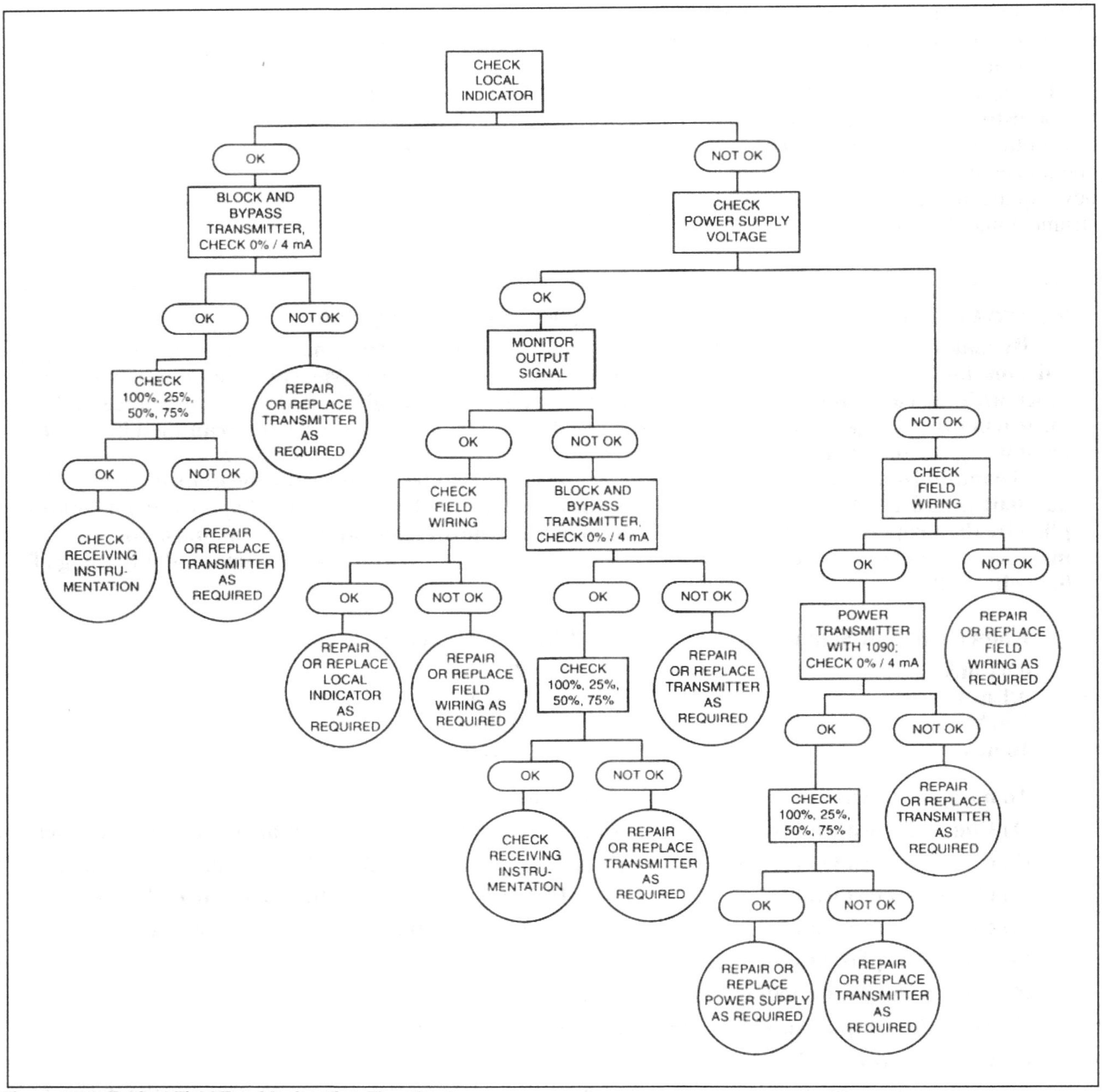

Figure 13-4. Typical Field Troubleshooting Flowchart.
(Courtesy of Transmation® Inc.)

testing only the transmitter); it is simply too much resistance in the loop. Someone probably has added another device. The loop must be reduced to a lower resistance by removing a device from the loop, or a 1:1 repeater or signal conditioner can be inserted somewhere in the loop to handle the added resistance.

Instruments often are added in series to a current loop, and the total resistance of all the devices in the loop exceeds the ability of the power source for the two-wire instrument loop to drive the signal full scale. The following is a method for using a field calibrator to measure loop resistance.

Differential Pressure Calibration

Calibration pressure is generated in the same manner as for pressure, but in most cases the differential pressure is quite small. The low side of the dP device is usually vented, and the test pressure applied to the high side of the device. The low side of the body of the dP device should be vented or drained to remove any source of potential pressure.

Three-valve manifolds allow equalizing the dP transmitter at line pressure or valving off the transmitter from the process for calibration or removal of the transmitter.

Five-valve manifolds allow draining the sensing legs to the transmitter without disconnecting the transmitter process side tubing.

When valving off a transmitter, always open the equalizing valve first to prevent possible overranging of the sensing cell.

Although the differential pressure may be small (in only inches of water), the line pressure may be at hundreds and even thousands of pounds pressure, or the liquid seal, if vented, may quickly become high temperature steam. Be certain of what is in the transmitter and any connecting tubing before disconnecting a transmitter from the process.

When calibrating a field dP flow transmitter, check that both taps come off the orifice or flow device at the same height or are piped to the same height before dropping to the transmitter. This will establish a solid zero when there is no flow through the meter. When the down legs (tubing to the transmitter) need to be refilled with water after a calibration, it is a good practice to provide filling tees at the high point of the process connection tubing.

When differential pressure transmitters are used for flow measurement transmitters, their calibration is in the inches of water range. Most orifice plate measurements are 0 to 100 inches of water but can range from as low as 50 inches to a high of 250 inches of water or higher. Pitot or averaging Pitot tube flow measurement ranges are usually just a few inches of water and require careful calibration and special low range calibrators.

See the section on flow calibration for more information on the use of differential pressure to determine flow.

When calibrating differential pressure transmitters for orifice plate or other square root signal sensing devices, finalize the calibration by checking zero. This is the most sensitive portion of the scale and, if off by even a small amount, can contribute to flow indications or totalizers counting when there is no flow. If the transmitter has a low signal cutoff or the system has a software scheme to allow for a stable zero display and use, the absolute zero is not as important as when there is no method to take care of this sensitivity.

Level Calibration

When a standard differential pressure transmitter is used to measure level by sensing liquid back pressure, calibration is the same as for pressure or differential pressure, with one exception. When a pressurized tank level is measured and a wet leg is used, the differential pressure transmitter is reverse calibrated so that, with no pressure on the level-measuring side (high side), the back pressure from the filled reference leg (low side) will cause the transmitter output to be 3 psi or 4 mA (zero level), and the output will be 15 psig or 20 mA (100% level) when the high side pressure equals the low side pressure, i.e., the differential is zero.

Bubble-tube level measurements use differential pressure or low pressure transmitters to sense bubble-tube back pressure and are calibrated the same as any pressure transmitter application.

When a diaphragm-type level transmitter is to be calibrated, a special fitting must be used that will allow the level transmitter sensing head to sense the pressure-calibrating medium. Various special fittings can be shop-built or purchased to handle any type of pressure-sensing head transmitter that needs calibration.

For level applications, most of the pressures do not exceed 30 to 40 psig, and instrument air is used as the pressure medium. The three- or four-inch flanged connection on the level transmitter is connected to the special calibration fitting, and instrument air to the fitting is adjusted to the desired pressure for calibration. A test pressure gage on the calibration fitting indicates the set calibration pressure.

Diaphragm-type level transmitters must be removed from the tank, chest, vessel, etc., for calibration when the vessel level cannot be adjusted for a wet calibration. If the vessel cannot be emptied easily, a special flanged gate valve must be initially installed between the vessel flange and the diaphragm level transmitter so the vessel opening can be closed before removing the transmitter. These gate valves must have special flanges to mate with both the tank and the transmitter. They also have a ratchet-type actuating handle, since the transmitters are so close to the tank that a standard circular handle will not fit. Several valve manufacturers can supply this special isolating valve.

Capacitance, resistance, infrared (IR), sonar, radar, and other special level transmitters do not rely on the standard testers. Most of these transmitters have built-in calibration procedures that allow one to calibrate against the actual tank level. These procedures are spelled out in the vendor instruction books.

Many older, special types of level transmitters, especially the capacitance type, had built-in calibration adjustments, but some adjustments had to be made by nonmetallic screwdrivers or special tools supplied with and sometimes attached within the housing of the transmitter. Most modern transmitters are now digital and require simple push button programming. Always replace, correctly, any special covers or shielding material used to protect the transmitter from radio frequency interference (rfi).

Flow Calibration

There are about a dozen basic flow-measuring instruments. Some work on similar and generally accepted principles, while others use very different and more complex principles.

THE ORIFICE PLATE

The orifice plate is still probably the most used flow device, but magnetic flowmeters, vortex shedding meters, and Coriolis meters are rapidly replacing the orifice plate. Accuracy, linearity, no flow restrictions, mass flow capability, and actual lower cost of ownership has made these newer flow-measuring devices very popular.

Flow measurement using an orifice plate relies on a differential pressure device to sense the difference between the upstream tap pressure and the downstream tap pressures. With clean fluids or gases, the measurement is relatively easy. Location of the dP transmitter is important in order to either keep the transmitter body full of the flowing medium or free of liquids when a gas flow is being measured. Locating the transmitter above or below the measuring point with the proper process line slope to the transmitter and special blowdown pots, when required, is essential and will affect the calibration of the transmitter once it is valved into the process.

Proper manipulation of the three valves on the 3-valve manifold that connects the dP transmitter to the process is essential to preventing the accidental overpressuring of the dP cell.

To valve-out a transmitter, always open the equalizing valve first to prevent possible overranging of the sensing cell.

To valve-in a transmitter, with the high and low side valves closed and the equalizing valve open, slowly open the low side valve (to apply equal pressure to the high and low sides). Close the equalizing valve, and then slowly open the high side valve (this prevents sudden pressure surges that can cause a calibration shift in some older instruments).

On steam service or very hot liquids, prefill the high and low side process legs with water or cool product before opening any of the process connections. Always check that the downcomers are the same length to produce zero differential when there is no flow. Steam flow dP transmitter lines should always be filled with water during instrument commissioning and after recalibration. This prevents hot steam or condensate from filling the body of the dP transmitter and damaging the instrument when it is valved back into the line.

The differential developed across an orifice plate produces a square root signal. This signal can be square root extracted in the dP transmitter if the transmitter is purchased with this feature. In most DCS systems, the square root signal is transduced in the DCS, using software. Sometimes the signal is used in its original square root form, and a special square root chart or scale is used. The only advantage in using the square root signal occurs if one is not controlling and usually operates above 50% flow. The wider scale divisions provide better visual readout. However, most flow signals are linearized and digitally converted for ease of use.

Calibration of the dP transmitter in the shop does not guarantee that the transmitter will produce an accurate flow measurement. An accurate flow reading depends heavily on the physical installation of the flowmeter and transmitter. If the flow transmitter calibration does not produce accurate flow readings, proceed as follows:

Before calibrating any flow transmitter, look at the piping (or process) and instrumentation drawings and any available logic and loop drawings to determine if and where the square root conversion is being done. This prevents confusion or possible double conversion during calibration and determines what hardware or software is required.

(1) Check for the type of tap connections being used, then check the orifice sizing sheet for use of the correct sizing equation.

(2) Check for proper upstream and downstream distances of straight pipe before and after the orifice as well as for correct distances of tap locations for vena contracta or pipe taps, if either of these methods of installation is used, rather than flange taps.

(3) Check that the sharp edge of the orifice is facing the flow.

(4) Check density, temperature, pressure, etc., of the medium and compare with the orifice specifications. Perhaps some variable has changed.

(5) Check that the orifice edge is not worn, the orifice plate is correctly installed in the line, and the process line is clean of trash, etc.

(6) Check all process tubing. Are transmitter and process valves open, as required?

(7) Is transmitter, controller, or DCS damping adjusted too high?

(8) Check for transmitter vibration, high ambient temperature, and the physical condition of the transmitter.

(9) Is the line kept full of fluid with no air in the line, or kept dry if a gas is being measured?

These are a few of the more obvious things to check when troubleshooting an orifice plate flow reading or a flow calibration problem, but if the problem still exists, do not limit the search only to these suggestions.

Other "pressure drop" devices develop three halves or five halves flow signals. If nonlinear outputs are developed, a multisegment linearizer will have to be programmed into the calibration.

VORTEX SHEDDING FLOWMETERS, CORIOLIS FLOWMETERS, AND MAGNETIC FLOWMETERS

These flowmeters produce linear signals, which is the reason they are so popular. Calibration of the newer versions of these types of transmitters are usually integral in the transmitter and do not require an outside source calibrator. Most require only simple key strokes at the transmitter. These devices are precalibrated at the factory, but if an actual product flow calibration is necessary, a volume check must be made using a bucket, barrel, or container that can be measured to hold a fixed volume or can be easily weighed. Accurate cutoff of the flow is required, and the calibration runs should be repeated until the numbers are consistent with the standards desired. This method works only on small flows. For large flowmeters, a certified flow lab would have to be contracted to do the calibration. See the individual instruction books for details about specific transmitters.

With these flowmeters, correct installation is important to obtain the accuracies stated by the vendor. The installation manuals should be carefully read when the calibration does not produce an accurate flow measurement.

Some magnetic flowmeter vendors use a standard Schedule 40 pipe flow table as a calibration reference, and the ft/sec velocity relating to the flow, as read off the pipe tables, is used to establish the calibration. Others use a meter calibration factor that is specific to each meter. This factor must be used when performing a field calibration. The factor is stamped on the flowmeter identification tag affixed to the meter. These meters require a special field calibrator that is specific to the brand of magnetic flowmeter being calibrated. The newer digital transmitters can usually be programmed in the field, on a digital pad that is integral within the transmitter, without a calibrator, or they can be remotely calibrated using a handheld programmer. Check the instrument manual for the exact procedure to use.

POSITIVE DISPLACEMENT FLOWMETERS AND TURBINE FLOWMETERS

These devices can be checked only by confirming that each revolution produces a pulse output. The meter must always be kept full with no air in the line. The instrument does not know if the cavities are full or half full. The installation must be designed to keep the meter full when product is flowing through it, otherwise there will be large errors. The turbine meter has a minimum flow cutoff or inaccurate flow area that usually covers the first 5% to 10% of the range. Check the instruction book for this break point when doing a calibration.

SOLIDS FLOW MEASUREMENT

If solids flow is being measured inferentially by counting the revolutions of a flight conveyer or the speed of a pulley, etc., check that there are sufficient pulses per revolution to obtain good flow resolution and a positive zero when the device is stopped. Also check that the speed pickup is mounted the proper distance from the notch, bump, magnetic material, or gear tooth to obtain a positive signal 100% of the time. Use a frequency generator/counter to check calibration and be certain of any pulse-to-rpm ratios that are part of the gearing or sensing.

Some solids flowmeters use capacitance, radiation, or other principles. Refer to the vendor's instruction or maintenance book for details of calibration.

When calibration does not solve an accuracy problem, remember that all meters rely on the fact that the line is full and there is no trapped air in the line. For liquid flows, a vertical line with the flow flowing up is the best installation. A low spot in the line, where the pipe is kept full, is also a good location when a liquid is flowing to an open vessel. Liquid flows should not be measured at the high point of a line, as air bubbles form in these areas and cause errors in measurement. Gas flowmeters are best placed in a vertical line. If steam is being measured by an orifice plate in a horizontal line, provide a bleed hole in the bottom of the orifice to allow trapped condensate to pass through the orifice plate. Sometimes calibration involves troubleshooting the installation or checking software.

> It is usually preferable to use TC and RTD transmitters when the signal is used in a control loop or a high degree of accuracy is required, with minimum field interference.

HOT OIL, WATER, AND TEST THERMOMETER CALIBRATION

Before the advent of portable temperature calibration devices, the technician had to make his or her own field calibration device. Depending on the temperature range required, the technician would heat water or oil, using any available bucket and a steam hose or blow torch. Obviously, this method is not highly accurate, and there are definitely safety hazards when calibrating to the higher temperatures. This method is not recommended unless there are no other choices.

Safety should be the highest priority, and extreme care must be taken when using hot oil. Since one cannot maintain a steady temperature, the calibration is poor at best, but it is better than no calibration. If this method is used, a test thermometer must be used as a reference. When using a test thermometer, be sure to emerse the thermometer to the correct depth. Most glass test thermometers have a marked point on the stem, called the emersion point. Emerse the thermometer only to this point for an accurate reading. This method of field temperature calibration is not recommended, but since it may still be used, it is mentioned only to cover all the possible methods used for temperature calibration.

DRY BLOCK CALIBRATOR

The modern way to calibrate a temperature probe is to use a "dry block calibrator." These devices are portable, have accuracies to within 0.3°C (0.5°F) with resolution to 0.1 degree and stability to within 0.1°C (0.2°F). They are safe since they do not use liquids as heat transfer medium and, therefore, do not have flash points to consider or noxious fumes and do not require an air supply. Ranges are from –25°C (–35°F) to 650°C (1200°F). Readouts are selectable between °C and °F. Ranges depend on the manufacturer and may require two different models of the calibrator. The units can be purchased with multiple-sensor insert holes for various diameter bulbs or TC or RTD probes and can have fast heat and cool cycles plus other automatic features.

Dry block calibrators are very simple to use. Just place the sensor in the appropriate insertion tube, switch the unit on, select the readout in degrees, and dial in the temperature required. Wait for the temperature to stabilize (some have alarms that sound when the desired temperature is reached), and then calibrate the device to this point. Continue to select calibration points that provide a good sampling of the span of the sensor being calibrated. Some sophisticated types of dry block calibrators have an RS-232C interface to a personal computer.

THERMOCOUPLE OR RTD SIMULATORS

Calibration of the probe is sometimes not required if a well or protecting tube protects the sensing element and the element is new. The easiest check is to calibrate the transmitter using a hand-held TC or RTD simulator-type tester. The simulator produces an electromagnetic force (emf) or resistance that duplicates the output of these two types of sensors. If this check verifies a zero or span correction is needed, recalibrate the transmitter, using the simulator. Reconnect the sensor to the sensing head and check the output reading against any local temperature reading that is available and is accurate. For a complete calibration, remove the TC or RTD from the well, line, or piece of equipment, and, using a dry block sensor calibrator, calibrate the sensor. If these two calibrations do not correct the temperature readings, the fault could be in the receiving device or in any software associated with this loop; or it could be a grounding or induced voltage problem or some other problem that must be corrected before the temperature loop reads correctly.

ISOLATION AND GROUND LOOP

Figure 13-5 illustrates a ground loop caused by buildup of condensation and corrosion inside a thermowell that results in a grounded thermocouple. Always specify input-to-output isolation on 2-wire, field-mounted thermocouples, RTDs, and low-level millivolt transmitters.

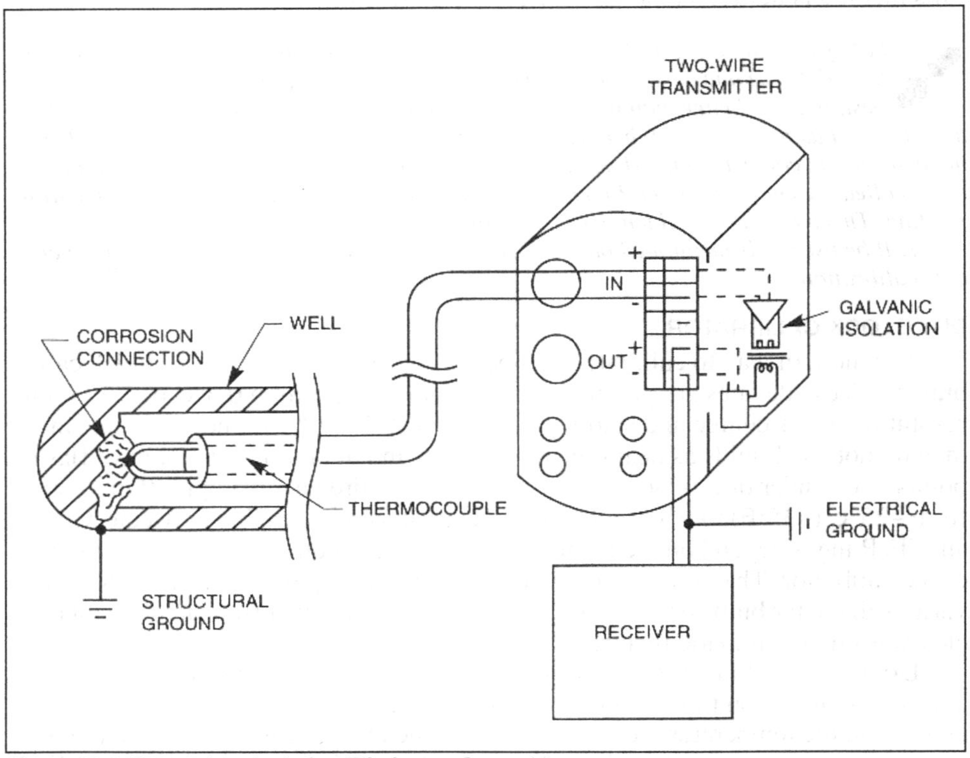

Figure 13-5. Transmitter Isolation Eliminates Ground Loops.

Note also that an RTD is not grounded, but some problems appear on the signal due to the lines of force created by large electric motors and other high-voltage electronic equipment that may exhibit a surge on start-up. The natural resistance and capacitance in the isolation circuit acts as an R/C filter, which helps to attenuate these stray interfering electrical noises. Thus, isolation does more than merely isolate input-to-output, it is also an electronic filter.

RADIO FREQUENCY INTERFERENCE

Figure 13-6 illustrates radio frequency interference (rfi) from a hand-held FM transceiver ("handi-talkie") commonly used for plant communication. Sometimes a poorly functioning temperature or other type of transmitter is the result of rfi interference. If the terminal block of the temperature transmitter is not protected electronically via feed-through capacitors and ferrite beads, which forms an L/C low pass filter, or if the case is not made of a conductive material with no cracks or openings, radio frequency energy can be picked up from "handi-talkie" transmitters. Test field transmitters for radio frequency pickup by triggering a "handi-talkie" held 1 to 3 feet from the transmitter that is suspect. If the loop is a control loop or has an alarm trip as part of the loop, be sure to have the operator place the loop on manual before doing the test.

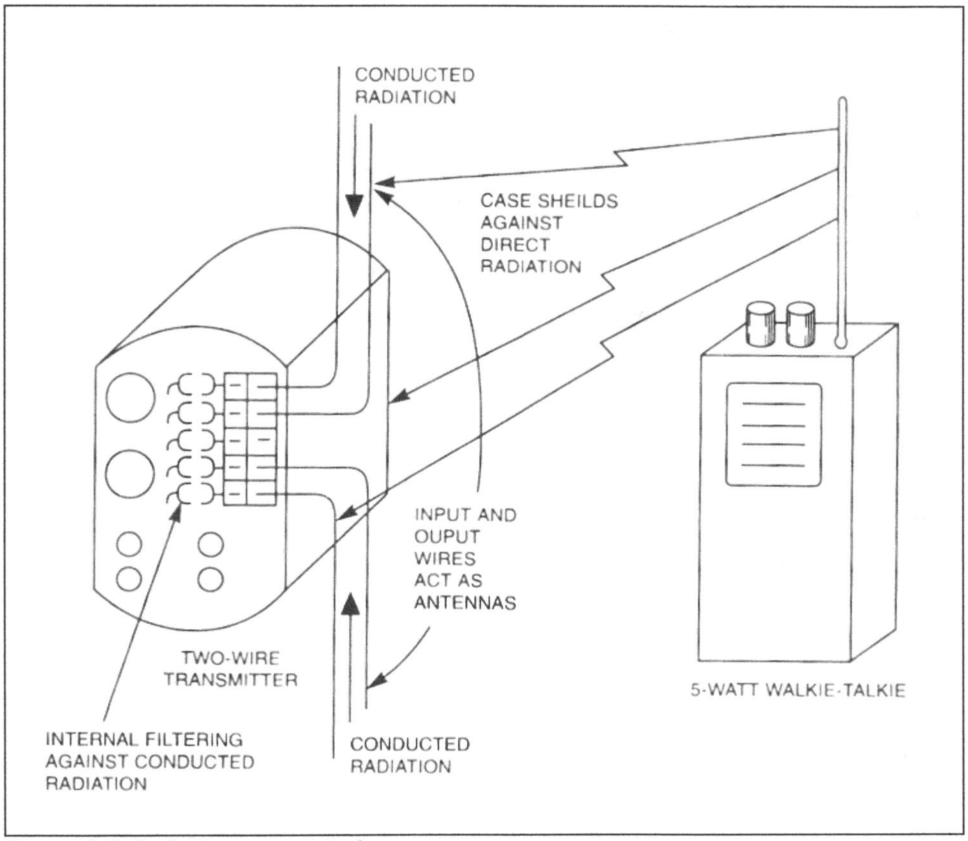

Figure 13-6. Radio Frequency Interference.
(Courtesy of Transmation® Inc.)

Calibrating in Hazardous Locations

The use of electrical instruments and control devices in areas where explosion hazards are present carries with it the potential for disaster unless specific preventive measures are taken. Hazards exist in the form of escaped flammable gases such as acetylene, hydrogen, propane, and others. Metal, coal, flour, and other dusts as well as some fibers suspended in air are capable of being ignited, with destructive consequences.

When calibrating any temperature system, check the device to determine if the measurement is linear. Check that the scale or readout is compatible with the sensing device. Sometimes the differences are small, but this may be the reason that it is difficult to read correctly through the entire span of the device.

In order that the seriousness of calibration in hazardous areas be understood, the following information on hazardous locations is presented.

Plant locations within which specific hazards may be present because of the handling of flammable materials must be classified as outlined in Article 500 of the National Electrical Code. Table 13-1 indicates the classification scheme in abbreviated form. Attention is drawn particularly to Division 1 locations. There, flammables may be handled openly and the risk of fire and/or explosion is ever present. Such locations require the ultimate in safety measures and practices.

The recognition of safety hazards is not limited to the U.S. Other countries throughout the world have either originated or adopted location and hazard classifications. Table 13-2 compares the location classifications of some countries.

Many countries distinguish between locations that are continuously or predominantly hazardous and those that are intermittently hazardous. Table 13-3 gives an approximate comparison of the hazard groupings. The test gas groups are made up of a number of gases with the most easily ignited of any given group furnishing the group name. The groupings are not quite the same for the different safety codes. Comparison is further made difficult by multiple codes that exist in

Table 13-1. Area and Materials Hazardous Classifications (NEC).

Classes

Class I — Flammable gases and vapors
Class II — Combustible dusts
Class III —Flyings

Groups

Group A Atmospheres

Chemical
acetylene

Group B Atmospheres

acrolein (inhibited)[2]
arsine
butadiene[1]
ethylene oxide[2]
hydrogen
manufactured gases containing more than 30% hydrogen (by volume)
propylene oxide[2]
propylnitrate

Group C Atmospheres

acetaldehyde
allyl alcohol
n-butyraldehyde
carbon monoxide
crotonaldehyde
cyclopropane
diethyl ether
diethylamine
epichlorohydrin
ethylene
ethylenimine
ethyl mercaptan
ethyl sulfide
hydrogen cyanide
hydrogen sulfide
morpholine
2-nitropropane
tetrahydrofuran
unsymmetrical dimethyl hydrazine (UDMH 1, 1-dimethyl hydrazine)

Group D Atmospheres

acetic acid (glacial)
acetone
acrylonitrile
ammonia[3]
benzene
butane

Chemical
1-butanol (butyl alcohol)
2-butanol (secondary butyl alcohol)
n-butyl acetate
isobutyl acetate
di-isobutylene
ethane
ethanol (ethyl alcohol)
ethyl acetate
ethyl acrylate (inhibited)
ethylene diamine (anhydrous)
ethylene dichloride
ethylene glycol monomethyl ether
gasoline
heptanes
hexanes
isoprene
isopropyl ether
mesityl oxide
methane (natural gas)
methanol (methyl alcohol)
3-methyl-1-butanol (isoamyl alcohol)
methyl ethyl ketone
methyl isobutyl ketone
2-methyl-1-propanol (isobutyl alcohol)
2-methyl-2-propanol (tertiary butyl alcohol)
petroleum naphtha
pyridine
octanes
pentanes
1-pentanol (amyl alcohol)
propane
1-propanol (propyl alcohol)
2-propanol (isopropyl alcohol)
propylene
styrene
toluene
vinyl acetate
vinyl chloride
xylenes

Divisions

Division I — A location where the hazard is expected to be present in the normal operation of the plant

Division II —A location where the hazard exists only in the event of failure in the processing equipment or plant

Notes:

[1]Group D equipment shall be permitted to be used for atmospheres containing butadiene, provided all conduit runs into explosionproof equipment are provided with explosionproof seals installed within 450 mm (18 in.) of the enclosure.

[2]Group C equipment shall be permitted to be used for atmospheres containing allyl glycidyl ether, n-butyl glycidyl ether, ethylene oxide, propylene oxide, and acrolein, provided all conduit runs into explosionproof equipment are provided with explosionproof seals installed within 450 mm (18 in.) of the enclosure.

[3]For classification of areas involving ammonia atmosphere, see Safety Code for Mechanical Refrigeration (ANSI/ASHRAE 15-1994) and Safety Requirements for the Storage and Handling of Anhydrous Ammonia (ANSI/CGA G2.1-1989).

[4]A saturated hydrocarbon mixture boiling in the range 20–135°C (68–275°F). Also known by the synonyms benzine, ligroin, petroleum ether, or naphtha.

[5]Flyings are materials not normally in suspension in air, i.e., they are of larger particle size than dusts. Flyings include materials such as cotton linters, sawdust, textile fibers, and other large pargicles that are usually more a fire hazard than an explosion hazard.

Europe and are in various stages of being coordinated. Those who must deal professionally with classifications and groupings are well advised to resort to the code or standard documents applicable to their tasks.

Electricity in the form of a spark is an ideal agent to commence the ignition process. Sparks might be generated in the process of opening switches, at loose wires on terminal blocks, and within any number of faulty electrical components. Neither the avoidance of hazardous situations brought on by the open handling of flammables nor the use of electrical equipment is always practical or even desirable. The taking of precautionary measures, even though they may be expensive, is then indicated.

Purging, explosionproofing, and intrinsic safety are the three dominant techniques used to guard against the electrically induced explosion hazard.

PURGING

Purging is the act of surrounding the potential ignition-causing electrical device within a housing with a stream of uncontaminated air or inert gas such as nitrogen, at some pressure slightly above atmospheric, in such a way that the clean air or inert gas leaks away from the electrical device. By this act the dust or vapors are prevented from getting to the potential ignition mechanisms and safety is promoted. The purging method requires interlocks between clean air or inert gas pressure and electrical power. In the event of failure of the former, the latter is turned off. This safety measure, while effective when working as intended, can be easily defeated by unauthorized modification of the interlock circuit, and it is not always deemed acceptable. ISA–12.04.01–2003 (IEC 60079–2 MOD)–Electrical Apparatus for Explosive Gas Atmospheres–Part 2 Pressurized Enclosures gives further details and requirements.

Table 13-2. Comparison of Hazardous Location Classifications

	USA and Canada	IEC* and Europe
Location Continuously Hazardous	Division 1	Zone 0
Location Intermittently Hazardous	Division 1	Zone 1
Location Hazardous under ABNORMAL Conditions	Division 2	Zone 2

**International Electrotechnical Commission*

Table 13-3. Comparison of Hazardous Materials Classification

Test Gas Groups	USA and Canada Group	Group per IEC and EN 60079-10
Acetylene	A	IIC
Hydrogen	B	IIC
Ethylene	C	IIB
Propane	D	IIA
Methane	D	I

When required to calibrate an instrument or system located in a purged enclosure, the enclosure must be opened. Before opening the enclosure, call the safety inspector or engineer and have a sniff or explosive limit test made in the surrounding area. If the area is safe, then and only then may the enclosure be opened to the atmosphere. The atmosphere must be monitored while the enclosure is open. When the work is complete, button up the enclosure and test the purge system before leaving the area.

EXPLOSIONPROOFING

Explosionproofing requires that the electrical equipment and wiring be contained within substantial boxes, containers, and pipes in such a manner that an explosion within any enclosure is contained within the enclosure and none of the resulting hot gases leaks out at a temperature that would promote flame or explosion propagation. The method is effective but also very expensive. It requires the rigid adherence of codified installation and maintenance procedures to function as intended i.e., to relieve pressure but still retard flame propagation. It is the only method by which high-power electrical and lighting equipment can be rendered safe in hazardous locations.

When opening an explosionproof enclosure, the same sniff or explosive limit test must be made and the area declared safe by the safety person. If the enclosure uses a bolted cover, carefully remove the bolts, place the cover with the mating side up.

Never place the cover face down in the dirt, gravel, or cement; the surface of the joint may be scratched or pick up some material that will prevent a proper seal once the cover is re-bolted to the enclosure.

Always inspect both the face of the cover and the face of the enclosure before reassembling the enclosure, and always replace and tighten all bolts to maintain the explosionproof integrity of the enclosure. See Chapter 17 for more safety-related information.

INTRINSIC SAFETY

Intrinsic safety (IS) can be applied to the characteristically low power needs of instrumentation and can gain significant economic benefits for the user. Intrinsically safe equipment and wiring is designed and constructed to be incapable of releasing sufficient electrical or thermal energy under normal or abnormal conditions to cause ignition of a hazardous atmospheric mixture in its most readily ignitable concentration. The concept of intrinsic safety recognizes that an energy discharge limit exists below which ignition cannot be brought about. The concept also recognizes the possibility of failure within the electrical equipment and requires that safety not be impaired by the occurrence of faults within the equipment or protective mechanism. ANSI/ISA–RP12.06.01-2003, Recommended Practice for Wiring Methods for Hazardous (Classified) Locations Instrumentation Part 1: Intrinsic Safety, gives further details.

SAFETY BARRIER

The energy flow into the hazardous location is limited with the safety barrier. Safety barriers may be of the positive, negative, or AC type and admit that form of energy with reference to ground. Safety barriers may be of single-, dual-, or multiple-circuit construction. A safety barrier typically contains resistors to limit the current through the barrier to a safe value, and zener diodes, or TRIACs plus zener diodes, to limit the voltage that may appear at the terminals in the hazardous area to an acceptable value. At least two of each of these components are used in such a way that the failure of one does not render the arrangement unsafe. Finally, a fuse is used to open the circuit when excessive current threatens the circuit. All components are encapsulated to prevent tampering, and the failure of any component, particularly the fuse, renders the device useless and in need of replacement. No repairs to a barrier may be made; however, some recent types support replaceable fuses.

Figure 13-7 shows the circuit of a typical single-circuit safety barrier. The barrier is grounded by a dual connection that carries any excess or fault currents to ground, thus preventing its entry into the hazardous location.

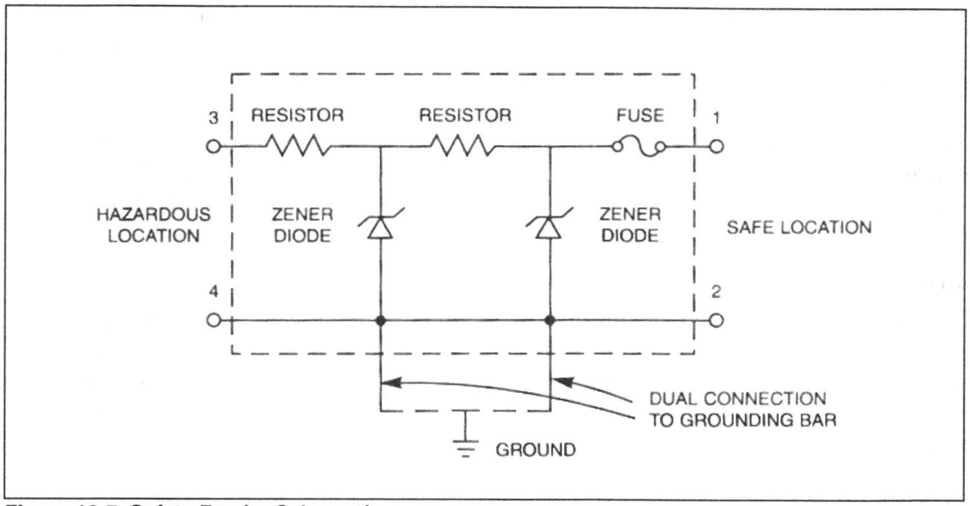

Figure 13-7. Safety Barrier Schematic.
(Courtesy of Transmation® Inc.)

Figure 13-8 shows an example of how barriers might be used with a grounded thermocouple in the hazardous location. The barriers in this setting prevent the entry into the hazardous location of the power that is used to operate the receiving instrumentation.

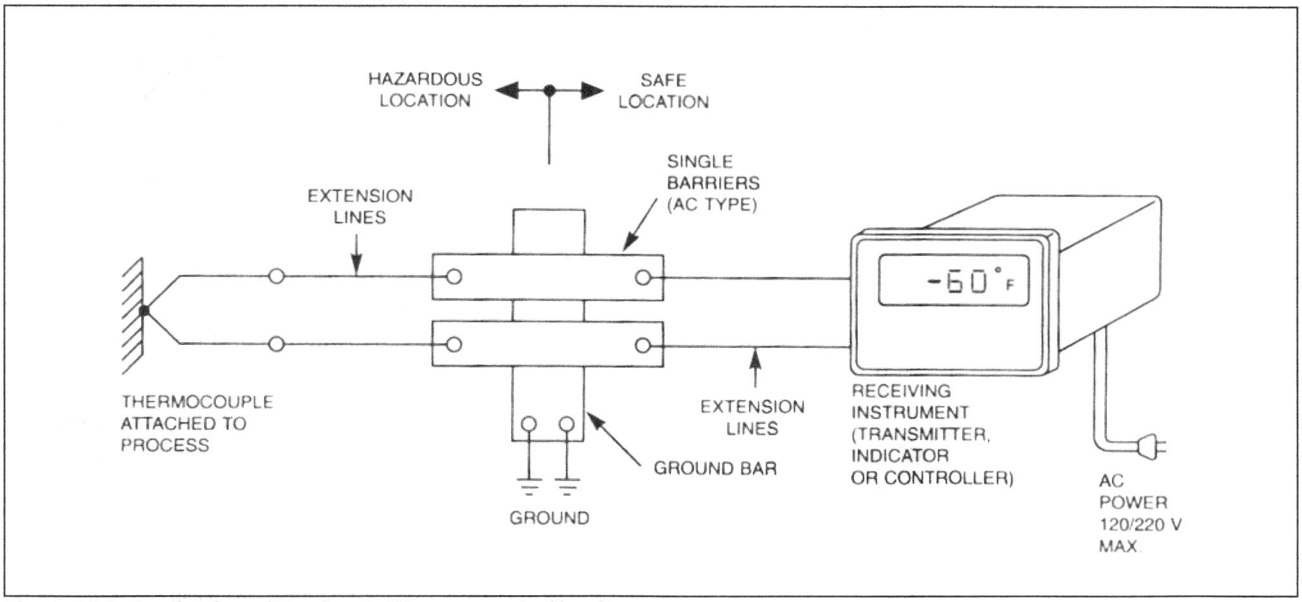

Figure 13-8. Application of Barriers to Thermocouple Circuit.
(Courtesy of Transmation® Inc.)

Barriers are available in a broad range of voltage and current ratings. High voltage ratings are associated with low current ratings and vice versa. Barriers always carry a limiting specification to the maximum voltage that may exist within the safe area equipment to which they are connected. For many barriers this is 240-V AC or DC.

The various combinations of voltage and current ratings of barriers are dependent upon the hazard classification assigned to the barrier by design-approving and design-certifying agencies.

When applying barriers, the user must be sure that the barrier selected is compatible with equipment in the hazardous as well as the safe locations and that its hazard group rating is equal to or better than that which applies to the hazardous location. Installations that may use traditional wiring techniques generally require approval from local safety authorities and/or underwriters.

BUILT-IN BARRIERS

Barriers are the instrumentality by which nonintrinsically safe equipment in a nonhazardous location may be connected to approved and intrinsically safe equipment in hazardous locations. Barriers as such are the add-on equipment through which the connection between safe and hazardous locations become permissible. It is, however, not the only method. Manufacturers who supply equipment for use in hazardous as well as nonhazardous setting have found it expeditious to incorporate the components of safety barriers directly into their safe location equipment. This done and following its approval, that equipment, while located in a safe area, may then be connected to approved equipment items in hazardous locations of suitable classification without the further addition of safety barriers. The user is spared the cost of the barriers and their installation but has to pay a lesser premium for the intrinsically safe approved nonhazardous location instrument.

Test Equipment

Portable test equipment may be transported only into classified locations that have an equal or lower hazard rating than that which is plainly noted on the face of the equipment. Also, test equipment may be connected to intrinsically safe circuits only when the instruments and/or barriers in these circuits are especially approved for connection with the test equipment. Exceptions to this are thermocouples and resistance sensors not otherwise powered; these may be connected to approved test equipment without the sensors having specific combination approval with the test equipment.

This information is not intended to be a substitute for the applicable codes and legislations that apply to hazardous location classification. Neither is this writing intended to substitute for the instructions and admonitions that must be supplied by equipment manufacturers in connection with instruments and testers that may be used in classified locations.

Classifications and restrictions of use specifically enumerated in codes and instructions and on the labels of instruments or required by local safety authorities must be adhered to at all times.

In many cases a field-mounted transmitter is installed in a hazardous location in a process plant or refinery. Calibration devices are available that are certified to work in the field under hazardous conditions. The two classifications are intrinsically safe and nonincendive. Many times a field-mounted transmitter, due to its location, is in an out-of-doors location and, therefore, will be considered to be Class I, Division 2 (see Table 13-1).

Certain warnings must be observed when using an instrument in a hazardous location in order to meet the nonincendive classification. Groups A, B, C, and D refer to certain types of gases that could be present in a Division 2 hazardous location.

In nonincendive Class I, II, III, Division 2, Groups A, B, C, D, or G locations, the following precautions should be taken, even when using a qualified calibrator:

(1) Do not make any connections to circuitry and/or components inside an instrument in hazardous locations. Connect only to nonincendive field circuit wiring terminals in hazardous locations.

(2) Do not connect to any circuit containing more than 50 V in hazardous locations.

(3) Do not open the case or charge batteries in a hazardous location.

(4) Maximum input parameters allowable are: $V_{max} = 50$ V, $I_{max} = 50$ mA, $C_i = 0.8$ µF, $L_a = 90$ mH

(5) Always follow any warning attached to the test gear being used and follow safe operating procedures. When in doubt, call the plant safety officer or responsible person who issues working permits for hazardous areas.

Note that in a Division 1 hazardous location, gases are usually present. In a Division 2 area, gases may be present under a fault condition. Since an outdoor location makes it more difficult to concentrate the gases, most outdoor locations are Division 2.

Without a calibrator approved by a recognized testing agency for operation in such hazardous locations, one is forced to cut power, remove the instrument, and return it to the instrument shop for calibration. Thus, having a calibrator with Factory Mutual or another recognized testing agency approval can be a great time saver as well as provide a safe means of field calibration in a hazardous area.

The concept of intrinsic safety recognizes that an energy discharge limit exists below which ignition cannot occur. Thus, intrinsically safe equipment and wiring are designed and constructed to be incapable of releasing sufficient electrical or thermal energy under normal or abnormal conditions to cause ignition of a hazardous atmospheric mixture in its most readily ignitable concentration. This includes the possibility of failure within the electrical equipment and requires that safety not be impaired by the occurrence of faults within the equipment or protective mechanism.

In areas designed to be intrinsically safe, any electrical calibrator or piece of test gear must also be certified to be intrinsically safe. The possibility of an explosion still exists and the test equipment must also be incapable of causing a spark or generating sufficient energy to cause ignition of the hazardous atmosphere.

In practice, intrinsic safety is realized by limiting the amount of energy brought into a hazardous location and limiting the effect to which energy may be stored in operating equipment. One common method by which nonintrinsically safe equipment can be used in a hazardous location is via safety barriers. Barriers are add-on devices through which the connection between safe and hazardous locations becomes permissible. The other popular method is the incorporation of the safety barrier components into the actual equipment. Such equipment, after being tested and approved, may be connected directly to equipment in hazardous locations without further safety barriers. The user is spared the cost of the barriers, but normally pays a premium for the intrinsically safe approved equipment.

Two-wire transmitters, operating at the low energy level of a 4-20 milliamp signal, can potentially be approved as intrinsically safe devices and may subsequently be connected to the safe location via barriers or approved receiver/power supply devices. In the case of nonisolated transmitters, the power supply may not be grounded, and two barriers are required — one for each output current conductor. When calibrating intrinsically safe loops, check that the correct type and number of barriers are being used, use an intrinsically safe calibrator, and follow the instructions carefully.

RTD BASIC WIRING

Resistance temperature detectors (RTDs) are favored in many applications for the measurement of temperature. Their outstanding stability, good predictability and high accuracy, among other factors, contribute to this favoritism. It cannot be taken as unusual then that users may wish to employ the RTD for making measurements in hazardous locations.

Figures 13-9(a), 13-9(b), and 13-9(c) illustrate how RTDs are typically connected to the excitation and sensing circuits. The two-wire connection shown in Figure 13-9(a) is recommended when the connecting wires a-b and c-d combined have a value of resistance that is insignificantly small compared to the value of the resistance of the RTD. In this event, neither the connecting wire resistance nor its change with ambient temperature will affect the sensing circuit, which is composed of balancing resistors R_1, R_2, and R_b, excitation source E, and readout instrument G. When resistance of the connecting wires becomes a significant fraction, such as 1% or more, of the value of the RTD resistance, the simple two-wire circuit becomes sensitive to connecting wire length and connecting wire ambient temperature. The sensing circuitry cannot distinguish between resistance changes due to the different causes. All are equated equal and interpreted to be temperature changes at the RTD. Going to the three-wire connection arrangement shown in Figure 13-9(b) solves the problem.

The effectiveness of the three-wire arrangement rests on the replacement of two of the connecting wires into opposite bridge legs such that they mutually compensate for both total resistance and like resistance changes due to ambient temperature. The two wires in question are a-b and c-e. They must be equal in gage, length, and location to achieve the desired result. Third wire c-d is part of the excitation branch and needs only to be adequate to supply the required current. Though as a matter of practicality all three wires are normally of the same length, gage, and location, they often constitute a cable. The three-wire connection scheme is enhanced in the perfection of the wire compensation when resistances R_1 and R_2 are made many (10-1000) times larger than the value of the RTD resistance and its balancing resistor R_b. It may be noted that the current through the compensating leads a-b and c-e can be equal only to one measured temperature, being slightly different at all others. Large values of R_1 and R_2 make the currents less RTD- sensitive and, hence, promote quality of compensation. Current tracking circuits and other techniques are also used to improve the performance of the basic bridge circuit.

The four-wire connection shown in Figure 13-9(c) is a classic in that the value of the RTD is determined by knowing the current through it and sensing the voltage across it. The results obtained are independent of the connecting wires and dependent only on the accuracy of the current source and the voltmeter. While this is the best solution as far as independence from connecting wire effects is concerned, it is not a perfect method. It, like the others, suffers due to the cable leakage resistance and thermal emfs at connection points. Care must always be taken to avoid these by using high quality insulated cables and by using only copper wire.

BARRIER CONNECTED RTDS

Safety barriers are electrical protection networks that prevent the flow of excessive electrical energy from the safe location into a hazardous location. A barrier is required to be inserted in each and every wire going from the safe to the hazardous location, except the grounded return wire if such is used. That wire may be grounded at the grounding bar of the safety barrier (see Figure 13-10).

The end-to-end resistance of barriers increases with voltage rating and is at its lowest value, about 10 ohms, for a nominal 1-volt barrier. At 10 volts the resist-

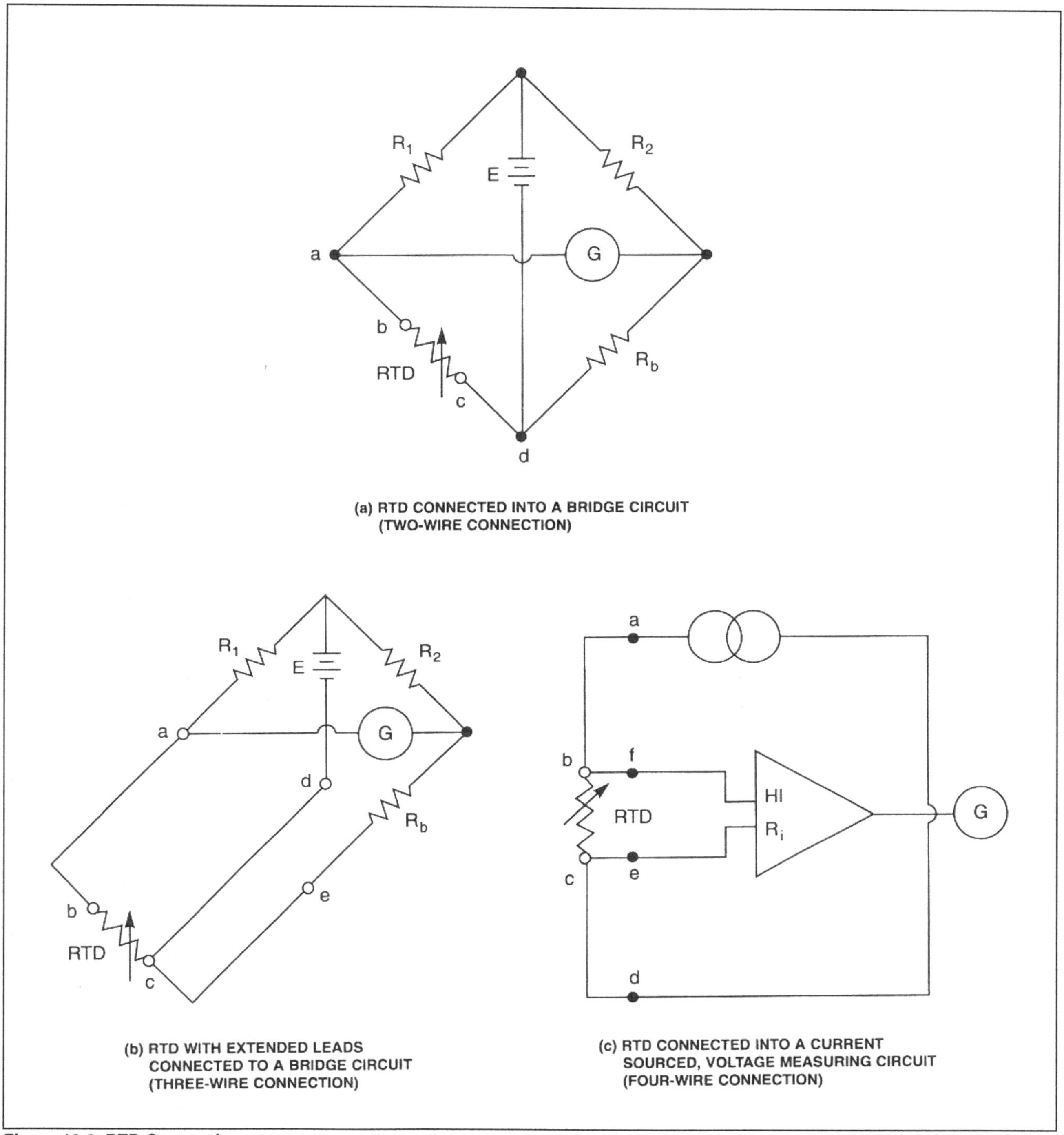

Figure 13-9. RTD Connections.
(Courtesy of Transmation® Inc.)

ance is 80 ohms. Clearly, the resistance introduced by barriers into a circuit is not trivial.

The protection diodes, which are internal to the barriers, contribute a leakage current that is also significant. The current may be held to be an acceptable value such as 1 micro-amp by operating the barrier well below its rated voltage. Thus, a 1-volt barrier might be operated at 0.1 volt, a 5-volt barrier at 0.5 volt, a 15-volt barrier at 11 volts, and a 28-volt barrier at 24 volts.

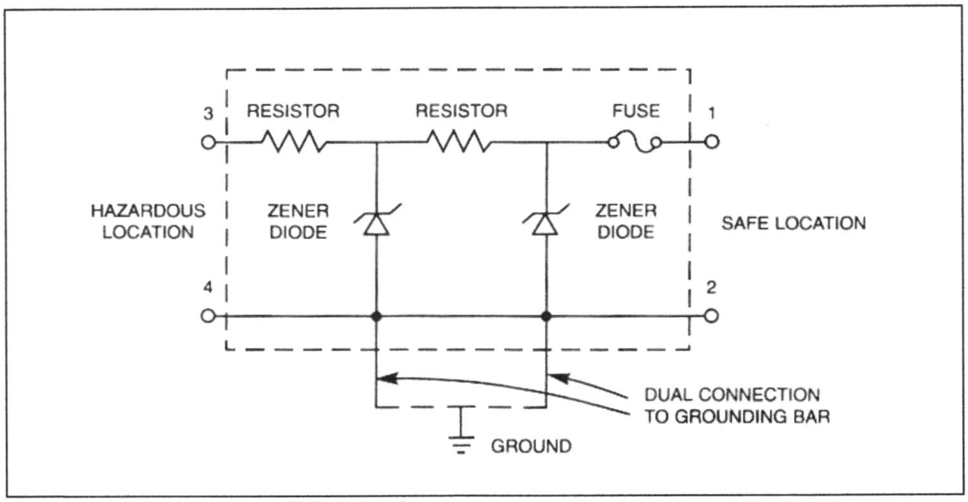

Figure 13-10. Single Circuit Safety Barrier.
(Courtesy of Transmation® Inc.)

Given these factors, some general remarks about RTDs and barriers can be made. First, low resistance RTDs such as 10-ohm elements should be avoided. Three-wire circuits with 100 to 500 ohm RTDs are satisfactory when special barriers with tightly controlled and low temperature coefficient end-to-end resistances are used. The four-wire arrangement can handle all the exceptions and does not require special barriers. It is, however, the most expensive circuit to implement since at least three barrier circuits are required.

The grounding of one conductor, shown in both Figures 13-11 and 13-12 is a common practice and saves a barrier circuit. RTD elements are normally well insulated from their protective tube, and this grounding introduces no special burden. It does, however, establish the grounding point for the readout equipment, which must not be grounded a second time.

The choice of barrier in any given instance must take into account the voltage level of operation as established by the RTD and its associated current and the effect of any leakage current on the accuracy of the results. When freedom of choice exists, the use of a 100-ohm platinum RTD, operating at 3 mA or less, in association with 5-volt barriers and transmission circuits of AWG 20 for 500 feet or less, AWG 16 for 1000 feet, and AWG 14 for 1000 to 2000 feet will generally be found satisfactory. What is and is not satisfactory, of course, depends on the accuracy required. When the matter is critical, a detailed evaluation is necessary. While such an evaluation is beyond the scope of this writing, a few observations may be helpful. RTDs are not perfect. Their standard resistance accuracy is on the order of 0.1% at 0°C increasing to larger values at higher and lower temperature. An optimum design tends to be one that no more than doubles the inherent inaccuracy of the sensor. Asking for better results is costly; asking for less saves little. Accuracy is degraded by barrier leakage. Its value at the maximum working voltage and maximum ambient temperature of the barrier should not be more than 1/1000 of the excitation current. "Balanced" conductors can be expected to differ in resistance by up to 5%. This is no problem in most cases and can be zeroed out. However, the 5% potential imbalance is subject to the cable ambient temperature variations. In outdoor service the temperature of the cable must be expected to range from –40 to 55°C. Thus, putting all factors together, the cable resistance must be expected to change 2% through the season. With a 10-ohm conductor, this is 0.2 ohm.

The barrier end-to-end resistance also has a temperature coefficient. Depending on manufacturer and type of barrier, this may range from 500 to 2000 ppm/°C.

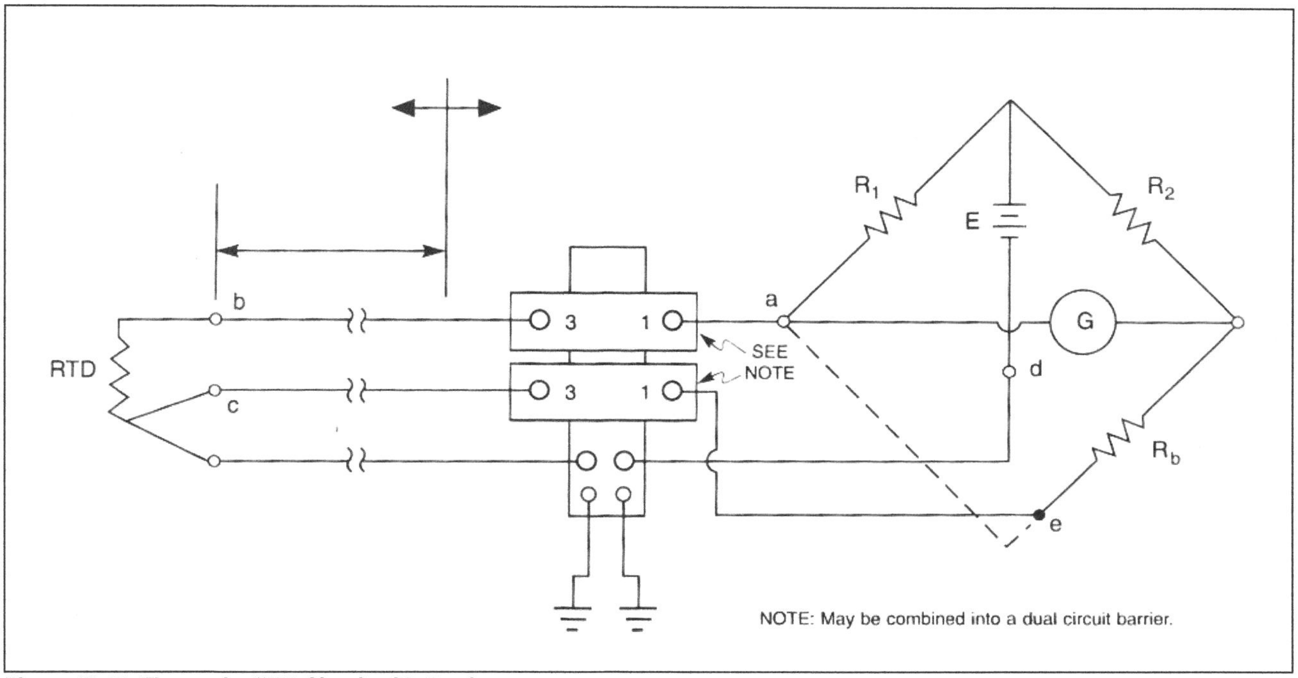

Figure 13-11. Three-wire RTD Circuit with Barriers.
(Courtesy of Transmation® Inc.)

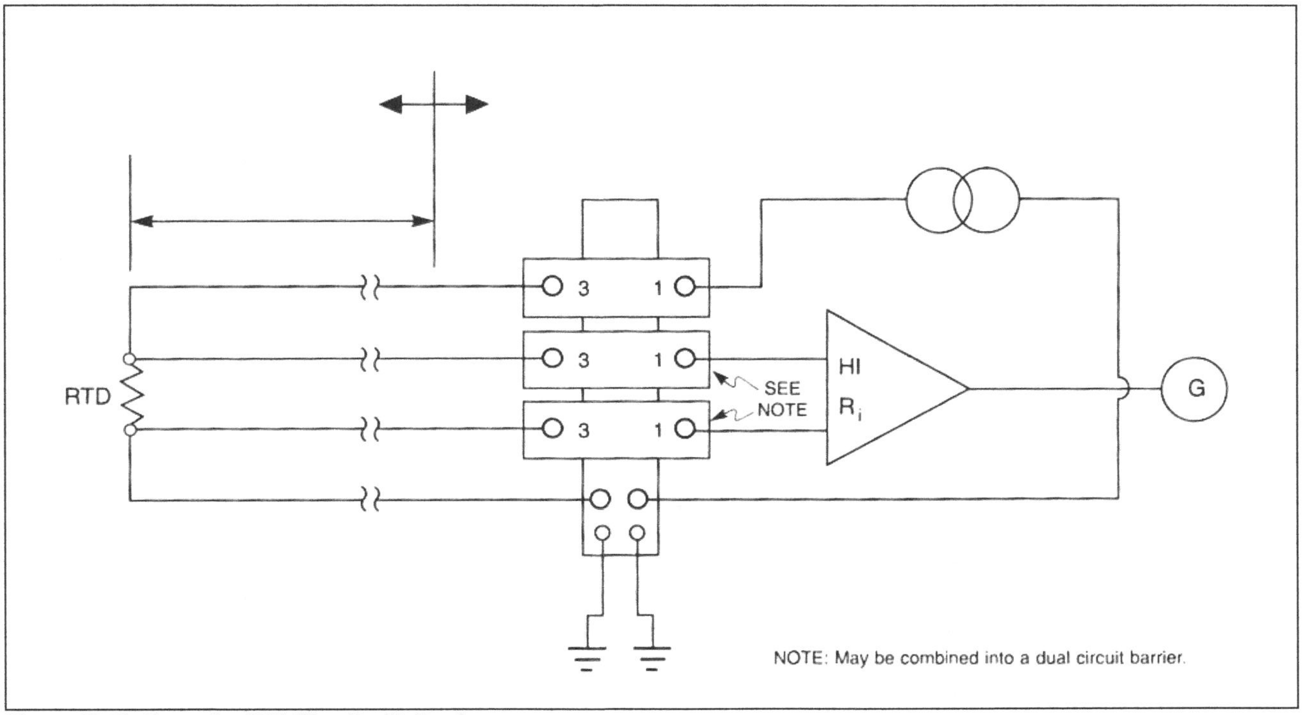

Figure 13-12. Four-wire RTD Circuit with Barriers.
(Courtesy of Transmation® Inc.)

Barriers, however, are often mounted indoors and are subject to a narrower environmental temperature range. Fifteen to 35°C might be typical. Thus, the contribution of a low-temperature coefficient barrier might be 1% of its value, or 0.2 ohm. With two barriers in the sensing circuit and compensating each other to within 1/20, the error from this source becomes about 1/20 of the 0.2 ohm or 0.01 ohm, a negligible value, but one that can become very significant when barrier resistance and/or temperature coefficient are large.

Summarizing the cautions and conditions that lead to successful RTD and barrier combinations:

(1) Keep RTD resistance high and connecting lead resistance low.

(2) Operate RTD at a current that is as high as possible without introducing a self-heating effect and thereby minimize barrier leakage effects.

(3) Maintain balance and symmetry in all circuits.

(4) Use only copper wire.

(5) Lay out wiring and equipment to minimize ambient temperature changes.

(6) For difficult applications, go to four-wire circuitry.

(7) Make an expected error estimate and be satisfied with it before going ahead with purchases and equipment installation.

Thermocouples and RTDs require barriers when connected to equipment located outside the hazardous area. Thermocouples require two barriers; RTDs, depending on their wiring, may require up to three barriers. When calibrating these sensors be sure to connect the lead wires back to the proper barrier.

There is a lot to know when choosing and working with intrinsically safe barriers. The information given is minimal. One should be well versed in intrinsic safety wiring and testing before working on such loops.

In-Shop Calibration

In-shop calibration is done in essentially the same fashion as field calibration, with the exception that the device or transmitter is disconnected from the process, cleaned as required, and taken to the shop, where it is visually given a good going over and then is mounted on a test stand at a calibration bench. The test bench provides all the necessary input and output indications to perform the calibration. The test equipment is certified, and any corrections to the test instrument are usually graphed, labeled, dated, and displayed next to the test instrument.

A new instrument should have a calibration certification. A typical certification might state: "The calibration of DC voltage, current, and resistance products is directly traceable to the National Institute of Standards and Technology (formally National Bureau of Standards) via our calibration standards, which have been certified by NIST and are subject to a program of periodic rectification." Each unit should carry a label that indicates the date of calibration and the date of future recalibration or maintenance required.

The convenience of shop calibration is that the test area is clean, quiet, and well lit and has all the necessary tools within easy reach of the technician. Bench testing, however, will not find any problems or calibration inaccuracies due to installation, cable pickup, or ambient conditions. Portable test gear should be periodically checked against certified shop test equipment.

Shop test instruments, particularly pressure gages, tend to be large 12-inch to 16-inch precision gages with mirrored scale surfaces to eliminate parallax. However, today's technology is driving towards precision digital test pressure indication that is NIST traceable. Whichever the case, shop test instruments do not require the ruggedness of field test instruments. Digital indicators usually cannot be overranged or damaged as easily as pointer movement indicators and are a good choice for field test equipment.

Bench Test

Bench testers generally consist of a "standard" gage and a means of producing a test pressure. The gage is usually 6 inches to 16 inches in diameter and is a highly accurate "standard" type of gage. Digital pressure displays are becoming quite common, but the old reliable bench gage is still seen in most instrument shops. Instrument air is the medium used for testing to about 100 psig. A high precision air-reducing valve is used for the pressure adjustment. Either water or oil is used as a medium for the higher pressures. When a liquid is used, it is usually compressed, using a threaded plunger with a convenient handle that can be smoothly and accurately adjusted. When required, a bleed system is incorporated to allow lowering the calibration pressure, as needed. Most bench testers are constructed on the job site by the technicians; some are ordered from vendors as complete systems. The benches have the various hoses and pump devices labeled and arranged so that technicians can easily and quickly switch from low to higher pressure testing.

Deadweight Tester

Every instrument shop should have a deadweight tester. They come in many types and ranges but consist basically of a hand pump to increase the pressure, through a liquid medium or air, to a piston that multiplies the pressure to a floating cylinder that supports a series of standard weights and also has a connection for the device being calibrated (see Figure 13-13).

Although they can be purchased as portable testers, most are bought as shop testers and remain in the shop where there is less chance of the test weights being nicked or damaged in a way that will reduce the accuracy of the tester.

Deadweight testers are usually used to calibrate the portable test gages or test equipment, which is then used for field calibration work.

The deadweight tester uses a standard weight or series of weights that are placed on a weight platform or pressure plate; the pressure is then pumped up until the weight platform begins to float freely on the fluid or air pad. The weight platform is generally spun slowly to make certain that the platform is not binding in any way. Test ranges start at about 10 lbs and go as high as 15,000 lbs. Most deadweight testers use oil as the pressure medium, but some models use water or air.

Fluidized Bath Calibrator

Most instrument shops have temperature baths for calibration of TCs, RTDs, and filled-tube systems. These can be oil-filled or sand-filled. When using an oil-filled calibration bath, always know the flash point of the oil used, and do not adjust the calibration too close to this point. Keep the exhaust fan and ducting clean of oily deposits, since the ducts can also catch fire. Fluidized hot sand is safer because it does not have a flash point, but one cannot tell the temperature of sand simply by looking at it. Be careful when working near or using a calibration bath, as they take some time to cool down and serious burns can occur if the temperature transfer medium is touched. Leave the exhaust fan on until the bath cools down to near ambient.

A fluidized bath works on the principle that small particles (typically sand) can be "fluidized" by means of a suitable gas or air stream at a constant pressure circulating through the bath.

Air supplied to the chamber is controlled by a valve into a diffuser. The diffuser ensures a uniform flow of air across a full section of the container and also acts as a support plate for the sand inside the unit.

The control valve should be opened slowly to make sure that the solid bed of sand remains undisturbed and the air finds its way between the particles; therefore, the bed will behave as a fluid and is said to be "fluidized." The air stream to accomplish this is normally 3 pounds per square inch, which can come from a pump or an air line that is usually contained inside the fluid bath calibrator. The fluidized bath is much cleaner than an oil bath, and the sensors do not have to be

Figure 13-13. Deadweight Tester.
(Courtesy of Ametek)

cleaned after they are removed from the bath. A light dusting is all that is required. The thermowell and temperature sensor must be clamped in the unit because the air rushing through the fluidized bath makes it quite buoyant. The air pump should include a safety valve and filtering to be sure that it is clean, dry air that is being passed through the bath at all times.

Typical fluid baths are capable of being operated from –100°C to +1100°C, although a single bath cannot cover this complete range. Although some baths may have fast heating and cooling, it is practical to turn on a bath at least 1-1/2 hours before it will be used. Cooling time can be as long as 2 hours after power is removed.

A fluidized bath can be a source of toxic fumes. It is important that an adequate air extraction system is installed and that one takes precautions when breathing air directly over the bath.

When the bed does not fluidize, the most common problem is a blocked air line or incorrect air supply. Check the filters often since they may become blocked, and check for air leaks. Also, be sure the control valve is working correctly. The sand bed should be dry before being used. If the bath does not work properly and appears to be boiling in one spot, check the diffuser plate at the bottom of the bed for cracks. The porous plate, which is the diffuser, is usually a readily available spare part and can be replaced.

In operation, the RTD or thermocouple should be inserted in the bath and allowed to stabilize for at least one hour. Set the temperature at the desired point on the fluidized bed bath, then check the output in millivolts on the thermocouple either directly with the millivolt meter or with one of the newer digital readout thermometers. Check that the millivolt output of the thermocouple or the resistance of the RTD matches the temperature curve table within the limits of error, which is typically specified by the American Society for Testing and Materials (ASTM.) Thermocouples are typically within 2 or 3 degrees, while RTDs can be within 0.1 degree depending on span and type. Consult the manufacturer's specification sheet for the accuracy specifications on any resistance temperature detectors.

Most glass thermometers used with liquid or sand baths have a fixed insertion length (the point at which it was emersed when initially calibrated). When using a fixed emersion length test thermometer, emerse it to the emersion mark to obtain the specified reading accuracy.

Agitation of oil and sand is essential for good heat conduction and transfer. Be sure to allow sufficient time between changing the bath temperature and using the reading for calibration. Most well equipped shops will have two baths so the device being calibrated can be quickly cycled between two temperatures without having to reset the bath temperature each time it is tested between, say, high and low.

When doing an initial calibration of a thermocouple or RTD sensor and transmitter, it is best to calibrate the sensor with the transmitter. This way, the proper range, span, calibration curve, and linearization are automatically checked.

If using a millivolt-only meter for thermocouples, remember to make the proper arithmetic calculations for the cold junction reference. Remember: you are measuring the millivolt output at room or ambient temperature, but the table is based on zero millivolts at 0°C (32°F); therefore, you must compensate for the difference between ambient and the freezing point of water. Carefully measure ambient temperature. Look up the millivolt value for that type of thermocouple in °C or °F and add this to the reading that you see on the millivolt meter when it is connected to the thermocouple.

When calibrating TCs or RTDs, always confirm that the sensor is matched to the correct transmitter type. If you are using a Type J thermocouple, be sure you have the transmitter set up for the Type J input.

Shop calibration does not take into account any misapplication or improper installation of the device. The device might pass shop calibration with flying colors but fail field calibration. Shop calibration is necessary, at times, to eliminate outside effects in order to help determine if a problem is in the sensor or the transmitter or is the result of some field effect; i.e., calibrating a valve without a positioner in the shop without the process pressure across the valve will produce a zero and span that are different from those produced under field conditions.

The red lead wire on thermocouples is always negative.

Testing of electronic parts, disassembly, and replacement of electronic instrument parts and cleaning of electronic components are best done in the instrument shop under clean conditions.

All electronic parts should be kept in sealed plastic bags and stored in a clean atmosphere. Precautions should be taken to ground oneself when working on sensitive electronic components. Usually the vendor uses special packaging and warning stickers to alert of the possibility of damaging the hardware. Follow vendor rules under these circumstances, in order not to void any warranties.

Other Aspects of Calibration

Device Calibration

Device calibration infers the calibration of only one device. This could be a magnetic flowmeter transmitter, a single pressure transmitter, an I/P transducer, or simply a solenoid valve or limit switch. Each component of a system or loop is a single device, with the exception that a few devices may be coupled together such that both devices must be adjusted or manipulated in order to check out either of the devices. Examples are the stroking of a valve to check its position switch, and the stroking of an I/P and checking the valve position rather than the output of the I/P. These are slight deviations from the device concept, but conditions sometimes require deviations. Device calibration can refer to shop or field calibration.

Loop Calibration

Loop calibration involves more than one device. A simple loop could be pressure energizing a local alarm. Simply checking the pressure switch high alarm setting is not a loop calibration. The alarm light and horn must come on at the high setting and go off at some pressure below the setting. The pressure switch dead band, if adjustable, should be set to prevent chatter or continued cycling. A control loop calibration should include the calibration of the transmitter, the controller, any intermediate readouts, the I/P, the positioner, and the valve and a check of any alarms or interlocks associated with the loop. Transmitter failure modes should also be checked. Sometimes forgotten is a check that the transmitter range and the controller and associated readouts all have the identical ranges, scales, and curves. Sometimes a square root device is read out on a linear scale or may be extracted in two places. Loop calibration should involve a look at the P&IDs, loop drawings, and any DCS or other software involved with the loop.

Test Equipment Calibration and Traceability

Every instrument shop should have at least one "standard" test instrument for each variable requiring calibration that is traceable to the National Institute of Standards and Technology (NIST). The calibrating devices or readouts that are used as shop standards should be sent back to NIST on a yearly basis for recalibration and certification.

NIST calibration correction data should be prominently displayed next to the standard readout so that any deviations from the readings can be corrected by reference to the correction table.

The standard test device should be at least one order of magnitude more accurate than the device being calibrated; i.e., calibrating a 0–100 psig 1% gage would require a test gage with an accuracy of at least 0.1%. Thermocouples and RTDs have published standard accuracies and curves plus millivolt and resistance tables that can be used for calibration. Most modern temperature calibrators contain these curves, which are based on the type of sensor selected on the calibrator. These curves and accuracy comparisons can be obtained from most temperature sensor manufacturers.

Elevated and Suppressed Ranges

Elevated and suppressed ranges are usually associated with level measurements but apply equally to temperature, pressure, and other variables.

ELEVATED RANGES

Range elevation is required when a transmitter is located below the bottom of a tank and the high side of the transmitter sees a pressure even when the tank is empty. In this case, the elevation can be adjusted until the output of the transmitter reads zero when the tank is empty. The elevation equals the distance the transmitter is below the zero level of the tank times the density of the tank liquid.

SUPPRESSED RANGES

To suppress the range of a transmitter is to "start the sensing at some point above zero"; i.e., a suppression of 10 psig requires 10 psig to be applied to the transmitter before the output begins to move upscale. As an example, a pressure is critical to a process at around 250 psig and must to be read within ± 0.5 psig. The existing transmitter has a range of 0–300 psig with an accuracy of 1/2% of span. In order to read ± 0.5 psig, the span cannot exceed 100 psig. The only way to do this is to calibrate the transmitter for a span of 100 psig and suppress the zero 200 psig. This will produce a range of 200 to 300 psig, with an accuracy of 1/2 psig and the critical sensing point at midscale.

When suppressing a transmitter range, always check the maximum upper range limit and the available turndown ratio of the transmitter in order not to exceed the upper range limit of the transmitter or to find that the transmitter cannot be turned down to the span desired.

In order for this particular transmitter to be able to be recalibrated to the new suppressed range, the transmitter span must be capable of adjustment to a span of 1/3 the full range. Every transmitter has a minimum span and a maximum upper limit. The elevation plus the span of the transmitter cannot exceed the full range, or the upper limit, of the transmitter cell.

Accuracy versus Resolution

A very important step in the calibration of any instrument is the understanding of accuracy versus resolution specifications.

ACCURACY

Accuracy indicates the limit that errors will not exceed when the instrument is operated under the specified conditions.

Accuracy normally includes conformity, linearity, hysteresis, and dead band, expressed as % of full scale, % of span, and % of Reading. The tightest method is percent of reading, the most common is percent of full scale.

***Accuracy as % of Full Scale Example*:**
Range equals 0–200 psig.
Accuracy equals 1% of full scale or ± 2.0 psig at any reading.

At 200 psig, accuracy is ± 2.0 psig, which is 1% of reading.
At 100 psig, accuracy is ± 2.0 psig, which is 2% of reading.
At 50 psig, accuracy is ± 2.0 psig, which is 4% of reading.
At 10 psig, accuracy is ± 2.0 psig, which is 20% of reading.

The accuracy is poorest at the low end of the scale.

***Accuracy as % of span* (using the same example)**

Range equals 0–200 psig. Span equals 100 psig.

If the span is set at 50–150 psig for a span of 100 rather than 200, the accuracy equals 1% of span or ± 1.0 psig at any reading. By reducing the span to 100 psig, a 1% accuracy gives a ± 1.0 psig reading rather than the ± 2.0 psig reading, when the span was 200 psig

At 150 psig, 1% of span is ± 1.0 psig, which is 0.66% of reading.
At 100 psig, 1% of span is ± 1.0 psig, which is 1% of reading.

Although the accuracy is still 1%, since the span is half the original range, or 100 instead of 200, the absolute reading is ± 1.0 psig rather than ± 2.0 psig. By decreasing the span of a transmitter that is accurate to percent of span, the accuracy of the indication or reading can be improved.

Accuracy as % of reading **(using the same example)**

Range equals 0–200 psig, Span equals 0–200 psig.
Accuracy equals 1% of reading.

At 200 psig, accuracy is 1% of reading or ± 2.0 psig.
At 100 psig, accuracy is 1% of reading or ± 1.0 psig.
At 50 psig, accuracy is 1% of reading or ± 0.5 psig.
At 10 psig, accuracy is 1% of reading or ± 0.1 psig.

Obviously the percent of reading cannot apply as you approach a zero reading but accuracy as a percent of reading is certainly much better than as a percent of full scale accuracy or even percent of span accuracy. It is important that these three types of accuracy statements are understood when referring to the accuracy of a device.

RESOLUTION

Resolution is defined as the smallest increment that can be distinguished or displayed. The resolution of a pressure gage is usually considered to be half a graduation. If a 0–100 psi gage has 100 graduations, the resolution would usually be considered as half pound or 1/2 a graduation. Depending on the size of the scale, if the scale were large enough to recognize 1/4 of a scale division, one could possibly read 1/4-pound increments. This is a judgment call and would depend on the lighting, condition of the pointer and scale, and the eyesight of the individual doing the calibration.

The resolution of a digital calibrator depends on the number of digits in the display module. The accuracy of the calibrator is stated separately from the resolution of the readout. Most calibrators have a 0.1% or better accuracy, yet have readouts with 3 to 5 digits and resolutions as high as 1 part in 30,000.

ACCURACY VERSUS RESOLUTION

Resolution can exceed accuracy, but accuracy cannot exceed resolution.

A 3-1/2 digit indicator display divides the input signal into 1999 parts. Therefore, the smallest increment is 1 part in 1999 or 0.05%. This means the resolution is 0.05%, which limits the accuracy of the instrument to 0.05%. Decreasing the resolution will decrease the accuracy, but increasing the resolution will *not* increase the accuracy.

Assume a multimeter with 0.05% accuracy full scale, 0.005% resolution full scale, and a 4-1/2 digit display. The accuracy does not change the additional digit that is provided by the resolution. The least significant digit (LSD) is meaningless. Do not assume all digits are accurate; compare the accuracy to the resolution.

Maintenance and Calibration Records

Proper maintenance of process control equipment must always consider records and documentation of repairs and recalibration information. Each field device should be tagged when recalibrated. A simple stick-on tag should be affixed to the device where the technician can easily see it when checking the device. Preferably, the tag should be affixed to the inside cover of the device or any other area inside the device where it can be seen. It should not become unreadable with time. A typical stick-on tag is shown in Figure 13-14. After any recalibration, the tag should be removed and replaced with a new tag, using current dates and calibration data.

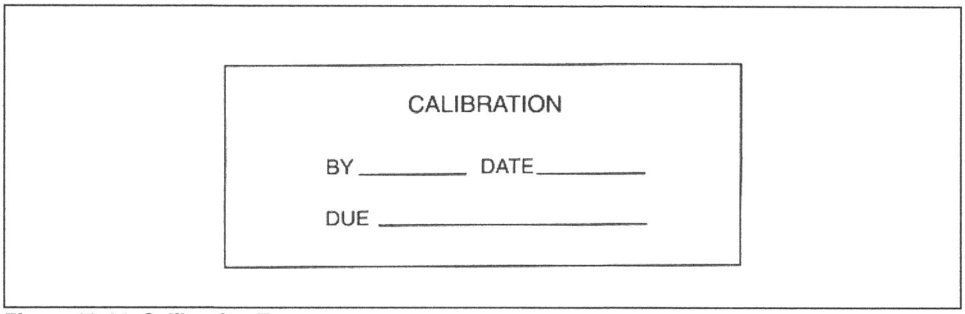

Figure 13-14. Calibration Tag.

As a general rule, instrument data points should include readings made at 0, 25, 50, 75, and 100% of the range of the instrument. Calibration should include both upscale and downscale testing to determine repeatability and hysteresis.

Calibration and testing records of an instrument should always show the "as received" data, new calibration date, the final calibration data, initials of the technician who did the calibration, and the date when the next calibration is due.

Figures 13-15 and 13-16 show two typical calibration report sheets. These are only two of many types used and are shown only to illustrate some of the items covered in a report.

The newest types of calibration equipment are capable of storing calibration protocol for a group of instruments, performing the calibration, then downloading this data into a personal computer for record keeping. The record can include a linearity and an accuracy curve and a digital calibration record of each point calibrated. The information can be used as an electronic reminder of the required date of recalibration and checkout. (The frequency of calibration is to be determined based upon the condition of the equipment at the various times it is checked out and calibrated.)

Test or recalibration intervals should be set up for each type of instrument and should be chosen to be 3 months, 6 months, or 1 year, depending upon the results of previous calibrations. Secondary standards used in the instrument shop should be rechecked for NIST certification at least once each year. Most manufacturers and several instrument repair companies can offer full traceability to NIST and certify that the equipment meets or exceeds manufacturer's specifications. These companies should have test equipment that is certified directly by the National Institute of Standards and Technology in Washington, DC.

By periodically reviewing the maintenance and calibration records, one can then determine the frequency of calibration and assign work orders at the proper 3- 6- 9- or 12-month intervals for rechecking the instruments in the field.

Preventive Maintenance Form No. 856			
TRANSMITTER, DIFFERENTIAL PRESSURE (ELECTRONIC)			
Plant No.	Date	Week/Schedule	Revision 2-5-82
Equipment No.		Supervisor's Initials	
Service:			
		Checked By	

Frequency	Action Code	Test Number	Perform Procedures Below Reference: Rosemount 1151 Manual	Record	Checked O.K.	Needs Repair	Repaired
SA		1	Note excessive corrosion or coating deterioration.				
SA		2	Repair conduit, flex, seals, vents, drains, covers, and wiring.				
SA		3	Repair process leaks.				
SA		4	Zero check local indicators. Repair or replace, if necessary.				
SA		5	Inspect internal parts and clean interior.				
SA		6	Lubricate cover gaskets and cover screws.				
			Replace defective or missing gaskets, O rings, or screws.				
SA		7	Check ZERO. Note change: Zero:				
SA		8	Check purge rotameters or filled legs and suppression.				
SA		9	Check winterization. Perform minor repairs. Note other repairs				
			needed.				
YR		10	Record calibration data below:				
			Zero: PSID = ma				
			Full Scale: PSID = ma				
YR		11	Check ZERO and SPAN in place. If either has shifted, take to				
			shop and inspect capsule with particular attention to corrosion				
			of welded disc in center of each diaphragm.				
YR		12	If span shift cannot be accounted for by range change requests				
			perform static alignment test and check for ZERO shift at				
			pressure equivalent to process.				
NS		13	If high accuracy is desired refer to instrument specifications				
			regarding ambient temperature effect and static pressure effect.				

Figure 13-15. Typical Calibration Report.
(Courtesy of Georgia-Pacific Corp.)

Preventive Maintenance Form No. 909			
POSITIONER, ELECTRO-PNEUMATIC			
Plant No.	Date	Week/Schedule	Revision 7-20-81
Equipment No.		Supervisor's Initials	
Service:			
MANUFACTURER		MODEL:	
SERIAL NO:		Checked By	

Frequency	Action Code	Test Number	Perform Procedures Below	Checked O.K.	Needs Repair	Repaired
SA		1	Note Excessive Corrosion or Coating Deterioration.			
SA		2	Repair Air Leaks and Blowdown Filter Regulators.			
			Replace all Gauges as Required			
SA		3	Inspect & Clean Torsion Rod, Armature, Feed Back Spring and			
			Nozzle.			
SA		4	Record Name Plate Input Signal and Output Pressure Ranges Below:			
			Input Output			
SA		5	Inspect, Clean & Lubricate Linkage.			
SA		6	Disconnect Linkage, Induce Input Signal to Positioner, Inspect			
			Action of Torque Motor and Pneumatic Relay.			
SA		7	Calibrate, Reconnect Linkage & Stroke Valve with Induced Input			
			Signal.			
SA		8	Clean & Lubricate Cover Threads and O-Ring.			
N/S		9	Note Reason For Shop Repair Below:			

Figure 13-16. Typical Calibration Report.
(Courtesy of Georgia-Pacific Corp.)

Smart Device Calibration

The first "smart" transmitters were introduced in about 1983 by Honeywell to measure pressure and differential pressure. Many other companies have since brought out "smart" transmitters, some of which are designed to communicate only with the distributed control system of that particular company.

Smart transmitters are becoming more intelligent with each new release. Their long-term drift and ambient temperature effects have less and less effect on their absolute accuracy. This does not mean, however, that calibration should not be checked on a periodic basis. Certainly, if all one needs is 1% accuracy and one has a 0.1% transmitter, calibration need not be checked every 3 months. Calibration must be checked when a plant operator complains about the readings or when there is a problem with the process readings.

A smart transmitter can report certain problems, such as loss of signal, low power voltage, high resistance readings, etc., but the device does not know if it is off calibration by, say, 1%. They depend on high accuracy, improved compensation, and fewer parts but still require calibration at some point during their lifetime.

A smart transmitter actually is digital from the sensor back to the distributed control system or programmable logic controller. As such, one must have more than just a serial interface; the interface must also know the protocol, the baud rate, and the other information that is required in order to communicate with the host computer. However, the digital signal of zeros and ones is not affected by stray pickup, minor power fluctuations, or the cabling effects of analog transmission.

Another advantage of the smart transmitter is that most can be remotely addressed, ranged, zeroed, suppressed, elevated, recalibrated, configured, etc. For nonaccessable or difficult locations, remote addressing is an important maintenance feature.

It is good practice to ask the operator to go to manual control before you enter the loop and "talk" to a smart transmitter. Some smart transmitters should not be calibrated on-line, as programming may affect the controller output.

In some cases, a hand-held communicator made by the company that designed the smart transmitter is the only means of interfacing the transmitter. In other cases the DCS system or a standard lap top or small portable computer may be used to "talk" to the transmitter from the control room. Some calibrators affect the signal during programming, while others do not. If working on a controlled loop, it makes good sense to have the operator place the controller on manual before doing any remote configuration.

It is much more convenient to be able to calibrate or interrogate a transmitter from the ground or the control room area. In addition, the smart electronics can continuously monitor ambient temperature and correct for any ambient temperature fluctuations that would cause a change in, say, the cold junction reference on a thermocouple transmitter.

A big disadvantage with digital transmitters is that they are not universal at the present time. Until a universal standard is available (exact digital signal configuration (fieldbus)) each smart transmitter will have to be configured and used with its host distributed control system, which has the exact same transmission language. This interface information is often proprietary.

Smart transmitters differ from standard transmitters in that they can be shipped with a specific tag name, calibration data, etc., embedded in the electronics, allowing them to be installed and checked out without an initial calibration, if one accepts the vendor's factory stated accuracy.

Smart Calibrators

Some smart calibrators have the capability to simulate and/or read virtually every process variable in use today. The list of variables depends on which calibrator is discussed.

The database field software package is menu-driven and operates on any MS DOS™-compatible system. The database fields include all pertinent process data, instrument data, maintenance plan data, and calibration data. The calibration data includes the instrument setup, ideal input versus output, test points, error tolerances, and all historical calibration data. Up to 20 test points can be specified per instrument. Capacity is limited only by available hard drive storage. The software downloads to the calibrator information, which consists of tag number, test points, ideal input versus output, error tolerances, and technician ID.

Through an RS-232 port, calibration parameters for many instruments can be downloaded into the calibrator storage. A tag number is selected from the display, and the calibrator sets the appropriate input and output configuration. The calibration test can then be done automatically, with the calibrator stepping through the prescribed test points, recording the actual input versus output, and the calibration status of each point (pass, fail, or alert). In the case of a pressure calibration, the pressure must be adjusted and the reading logged manually with an ENTER key. The calibrator prompts the technician for each test point in this case.

When work in completed, the software will retrieve data from the calibrator, including the as found and as left data, the number of calibration attempts, the maximum error found, an analog trace, and a digital display of the calibration points, date and time of the test, calibrator ID number, and mode of data entry.

14 Tuning

Overview

The word loop is often used in discussion about process control. A loop, the basic unit of process control, typically consists of one process measurement, one set point, and one output.

A process usually has more than one loop, and many processes used in industry have thousands of loops. Some of these loops are independent of other loops; some loops are interconnected with other loops.

A variable that is measured by the process instrumentation is usually known as the measured variable or process variable. Examples of measured variables are liquid and gas flows, levels, temperatures, and pressures.

Some variables are not directly measured from the process but are calculated by the control system (usually using measured variables). Examples of calculated variables are differential temperature (calculated by subtracting two measured temperatures), and a flow ratio (calculated by dividing two measured flows).

Loop Classification by Control Function

Each loop has a measured (or calculated) variable. These variables are usually measurements made from the process using sensors or transmitters. Variables can be classified based on how they are used in the control system.

Controlled Variable

A controlled variable is held by the control system at either a target or set point. The primary function of a control system is to control these variables.

Manipulated Variable

For every controlled variable, at least one must be manipulated, that is, directly adjusted by the controller or the operator. The manipulated variable is usually a flow that passes through a control valve or some other final control element. It may or may not be measured.

Indicated Variable

Many variables are measured but are not controlled or manipulated. These are indicated only for operator information or for record keeping, but no effort is made either by the control system or by the operator to keep that variable at a particular value.

In Figure 14-1 the level of the tank is a controlled variable; it is desired to hold this variable near a set point to prevent the tank from emptying or overflowing. To *control* the level, the flow out of the tank is *manipulated*. Other than for the control of the level, there is reason for requiring a particular discharge flow rate. It may be measured for indication or recording, but this measurement is not needed for control.

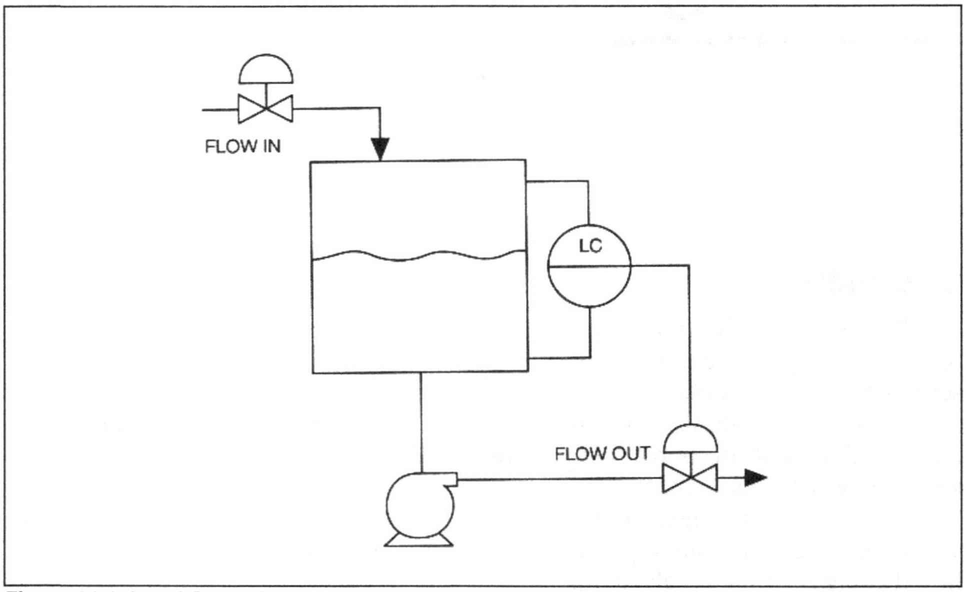

Figure 14-1. Level Control.

Flow loops usually contain both controlled and manipulated variables. In a few cases the flow that is manipulated is not the same as the controlled flow. For example, the total flow in Figure 14-2 is the controlled variable, while flow A is the manipulated variable.

The terms manipulated, indicated, and controlled variables include not only variables that are directly measured but also those that are inferred: variables that are calculated based on other inputs. In Figure 14-3 a ratio between two flows is manipulated to control the composition of the blended flow.

In Figure 14-4 a differential temperature within a process is to be maintained. The controlled variable is the differential temperature; that is, it is inferred from two individual temperatures.

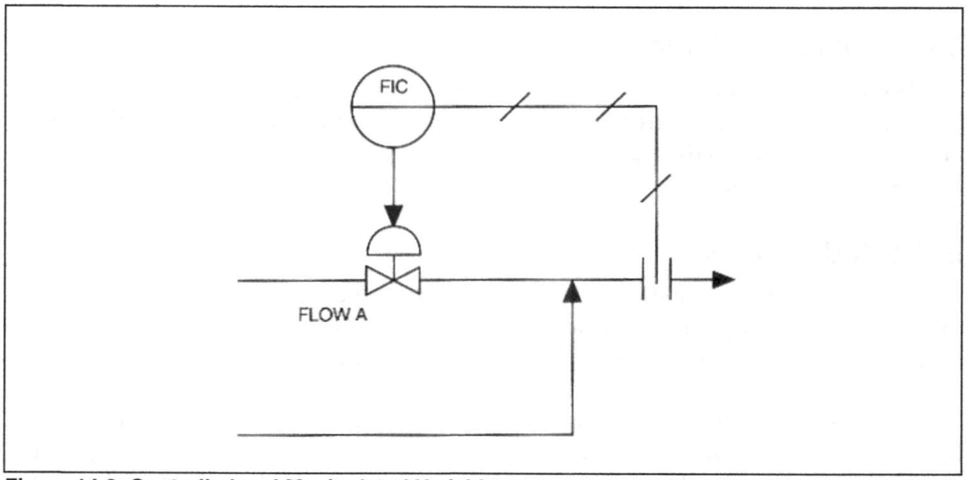

Figure 14-2. Controlled and Manipulated Variables.

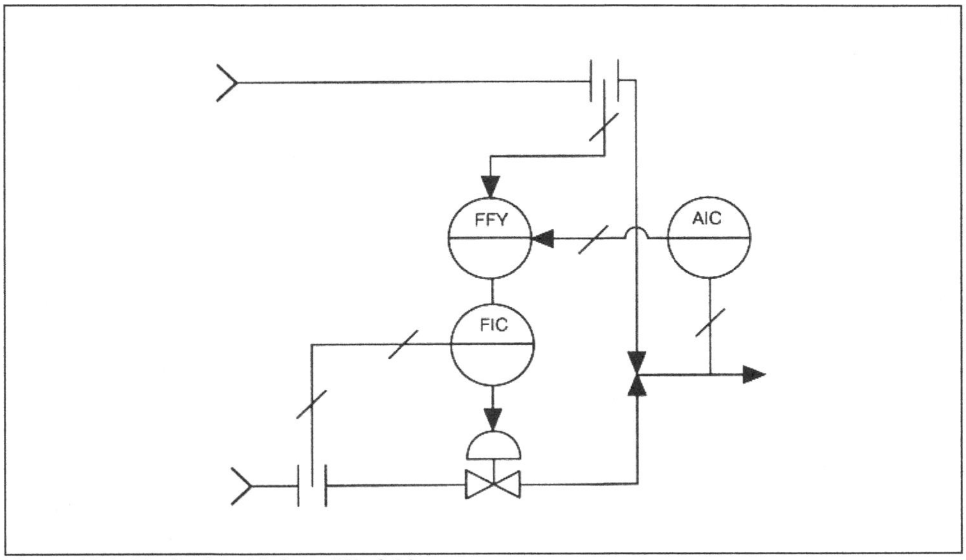

Figure 14-3. Manipulated Inferred Variable.

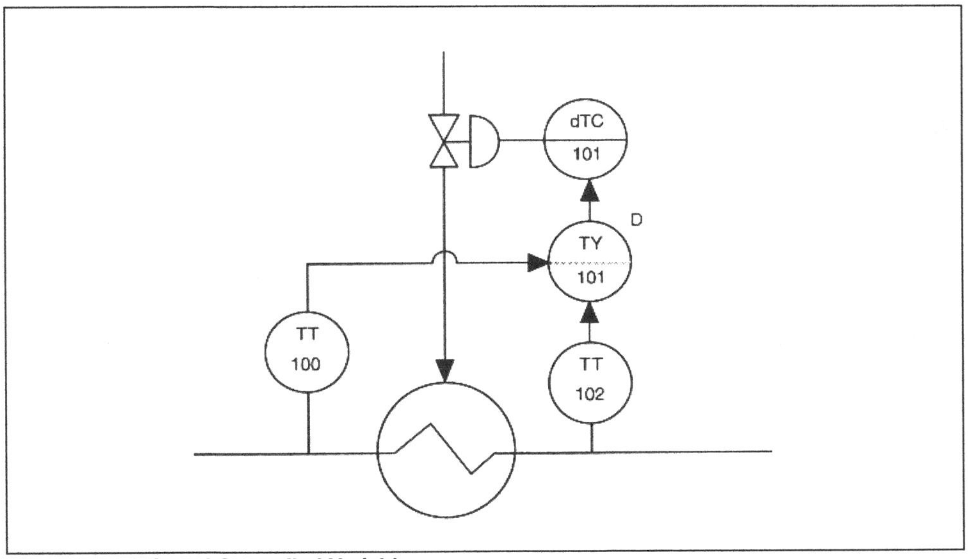

Figure 14-4. Inferred Controlled Variable.

Set Point

Each control loop has a set point—the desired value for the controlled variable. For example, the set point for the temperature of a room, as set on the thermostat, may be 70°F. The set point may remain constant or may be often adjusted by the operator, by other control loops, or by a computer program.

Control Algorithms

The controller compares the measured variable to the set point and determines an output to the manipulated variable. To make this determination, an "algorithm" is used, which is a step-by-step procedure for solving a problem. Explanations of some of the simple basic algorithms follow; then the PID algorithm used in industrial control is examined.

Basic Algorithms

The most basic form of control algorithm is a simple on/off control. A switch detects whether the process variable is above or below the set point. The state (on or off) of the manipulated variable is changed when the controlled variable crosses the set point.

This type of control is used successfully in many situations. Most home appliances use on/off control to switch the power to the heating element on and off to maintain the desired temperature.

Because on/off control often tends to cause the manipulated variable to be turned on and off too often, a variation of the on/off control adds a gap or dead band in the comparison of the controlled variable with its set point. This causes the manipulated variable to be changed less often but also increases the amount of variation in the controlled variable.

The PID Algorithm

The most common algorithm used in industrial process control (almost the only algorithm used) is the time-proven proportional-plus-integral-plus-derivative (PID) algorithm.

The PID control algorithm does not "know" the correct output to bring the process to the set point. It merely moves the output in the direction that should move the process toward the set point. The algorithm must have feedback (process measurement) to perform.

The PID algorithm must be tuned for the particular process loop. Without such tuning, it will not function. To properly tune a PID loop, each of the terms of the PID equation must be understood. The tuning is based on the dynamics of the process response.

Most control systems in industrial process control allow the operator to place a loop into automatic or manual control. Manual mode allows the operator to manipulate the output to bring the measured variable to the desired value. This mode is often used during the start-up of the process (see Figure 14-5). In automatic mode the control loop manipulates the output to hold the process measurements at their set points. This should be the most common mode for normal operation. See Figure 14-6.

The most important configuration parameter of the PID algorithm is the action. This determines the relationship between the direction of a change in the input versus the resulting change in the output. Direct action means that an increase in input results in an increase in the output. Reverse action means that an increase in the input results in a decrease in the output.

The PID controller, as the name implies, is made of three basic responses: proportional (or gain), integral (or reset), and derivative. This section presents the individual responses that make up the PID controller.

The controller action is always the opposite of the process action.

The error is the difference between the process and the set point. If the controller is direct-acting, the set point is subtracted from the measurement; if it is reverse-acting, the measurement is subtracted from the set point. Error is always in percent.

Error = Measurement – Set point	(Direct action)
Error = Set point – Measurement	(Reverse action)

PROPORTIONAL

The most basic response is proportional, or gain, response. In its pure form, the output of the controller is the error times the gain added to a constant known as manual reset.

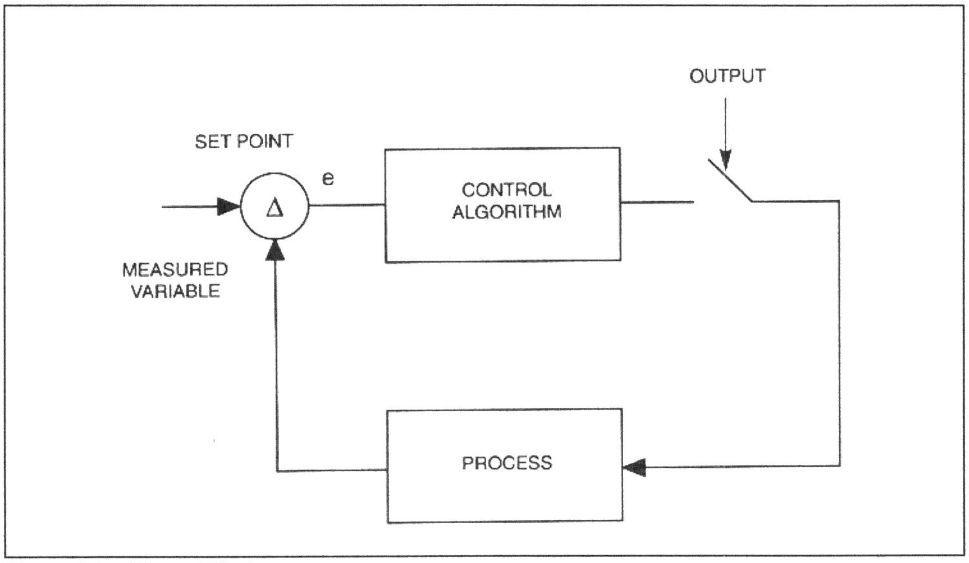

Figure 14-5. Manual Mode.

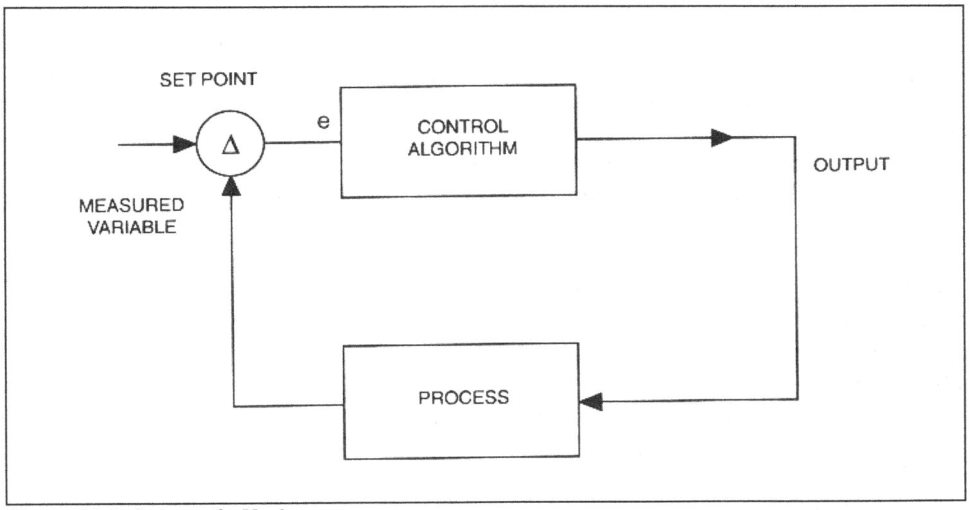

Figure 14-6. Automatic Mode.

$$\text{Output} = G \times e + K$$

where G is gain, e is error, and K is manual reset.

If the manual reset remains constant, there is a fixed relationship between the set point, the measurement, and the output.

The output of a proportional-only controller always follows the error, with no dynamic (time-based) difference.

When there is a disturbance (or load change), proportional-only control reduces but does not eliminate the error. The remaining error is known as offset.

The tank in Figure 14-7 has liquid flowing in and flowing out under control of the level controller. The flow into the tank is independent and can be considered a load by the level control. The flow out is driven by a pump and is proportional to the output of the controller.

Assume first that the level is at its set point of 50%, the output is 50%, and both the flow in and the flow out are 500 gpm. Then assume the flow in increases

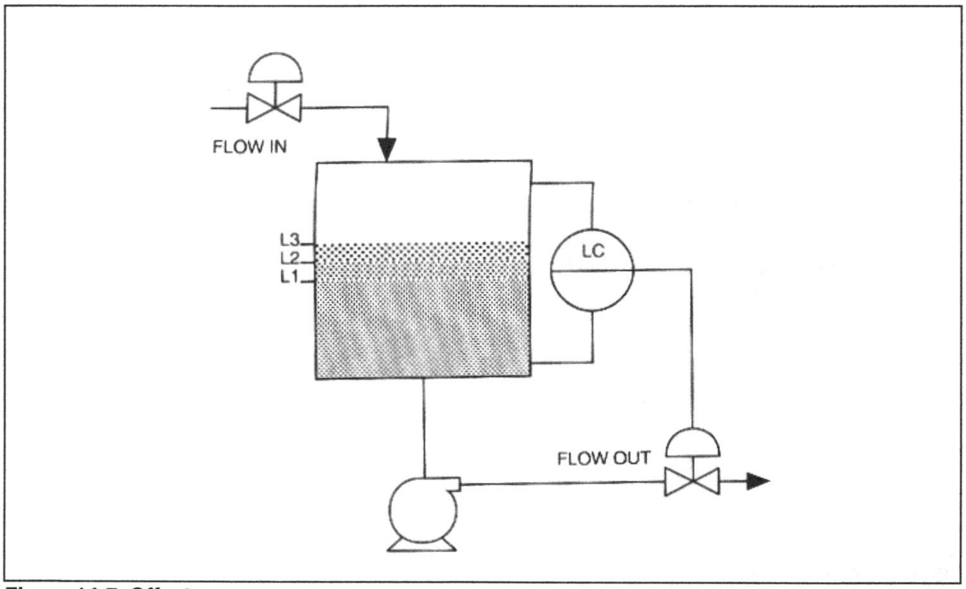

Figure 14-7. Offset.

to 600 gpm. The level will rise because more liquid is coming in than going out. As the level increases, the valve will open and more flow will leave. If the gain is 2 for every one percent increase in level, the valve will open 2% and the flow out will increase by 20 gpm. Therefore, by the time the level reaches 55% (5% error) the output will be at 60% and the flow out will be 600 gpm, the same as the flow in. The level will then be constant. This 5% error is known as the offset.

Offset can be reduced by increasing gain. Repeating the above experiment with a gain of 5, for every 1% increase in level the output will increase by 5% and the flow out by 50 gpm. The level will have to increase only to 52% to result in a flow out of 600 gpm and cause the level to be constant.

Increasing the gain will reduce the offset, but only an infinite gain will totally eliminate offset.

Figure 14-8 illustrates the relationship between measured variable and output. With no load, a line can be drawn from the vertical (output) axis to the line L1. A line from the intersect is then drawn down to the horizontal (measured variable) axis. Likewise, a line is drawn from the measured variable axis to the control line. From the intersect a line is drawn left to the output axis.

If we switch from the line L1 to line L3 (indicating a change in the load), the line drawn from the original output to L3 and down will indicate the change in measured variable if there is no control action. Compare the offset with low proportional gain and high proportional gain.

Offset can be eliminated by changing the manual reset every time the load changes. After the load has changed, the operator can increase the manual reset to open the valve further or decrease the manual reset to close the valve. The operator continues to adjust the manual reset until the offset is eliminated.

INTEGRAL

Whenever the load changes and the operator notices an offset in the control loop, the operator "resets" the controller (to move offset) by adjusting the manual reset.

If the load does not often change or the offset is too small to create a problem, proportional-only control is sufficient. If the offset and the need to frequently adjust the loop is a problem, the manual reset may be replaced by automatic reset. Automatic reset is a function that continues to move the output whenever there is

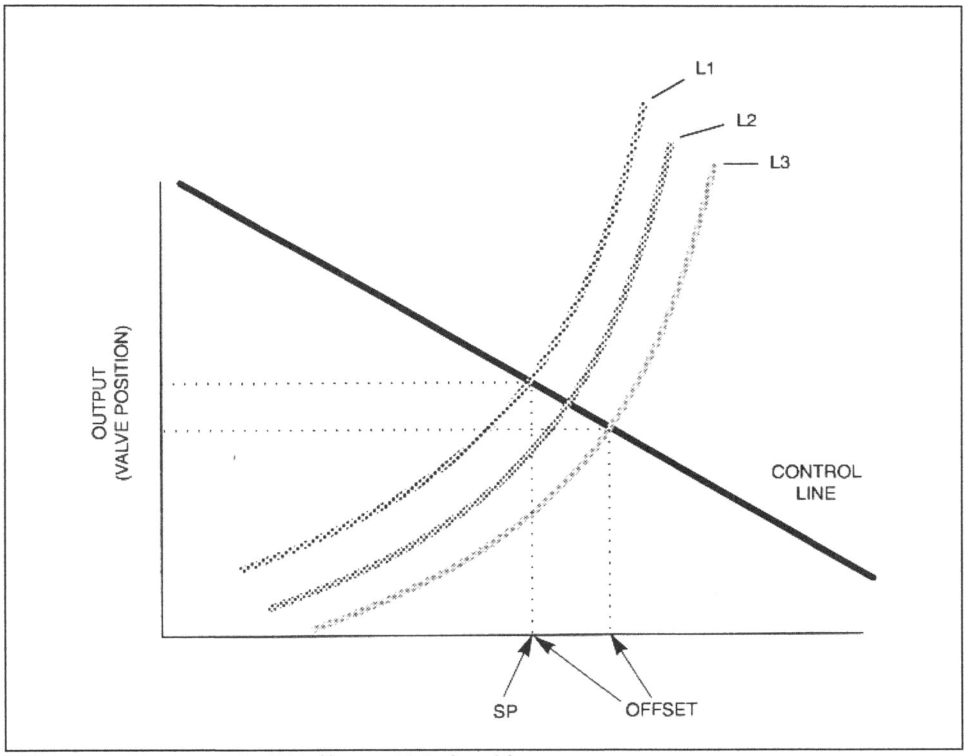

Figure 14-8. Measured Variable/Output Relationship.

any error. Automatic reset is usually called *reset* or *integral action.* One way to add reset action to a controller is to add a circuit or function that takes the controller output, performs a lag, and adds the result to the product of the gain times the error.

Note the use of the positive feedback loop to perform integration (see Figure 14-9). As long as the error is zero, the output will be held in place. However, if the error is not zero, the output will continue to change until it has reached a limit.

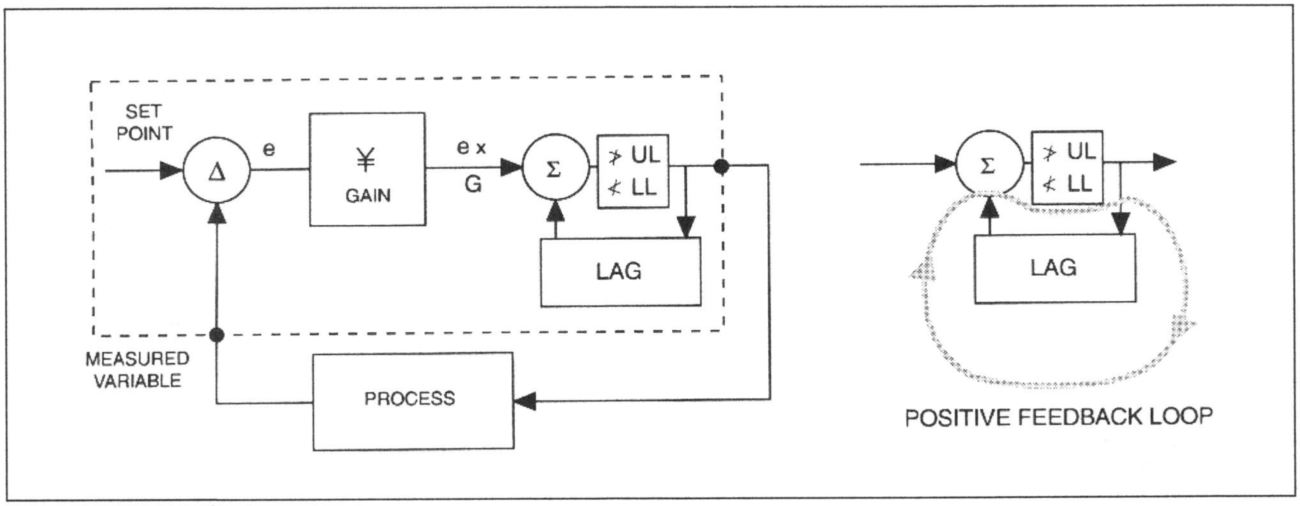

Figure 14-9. Integration.

From a more mathematical point of view, the reset contribution is:

$$\text{Output} = G \times K_i \times \int e \; dt$$

where:

G = gain
K_i = reset rate, repeats per min.
e = error, %

Figure 14-10 graphically represents pure reset.

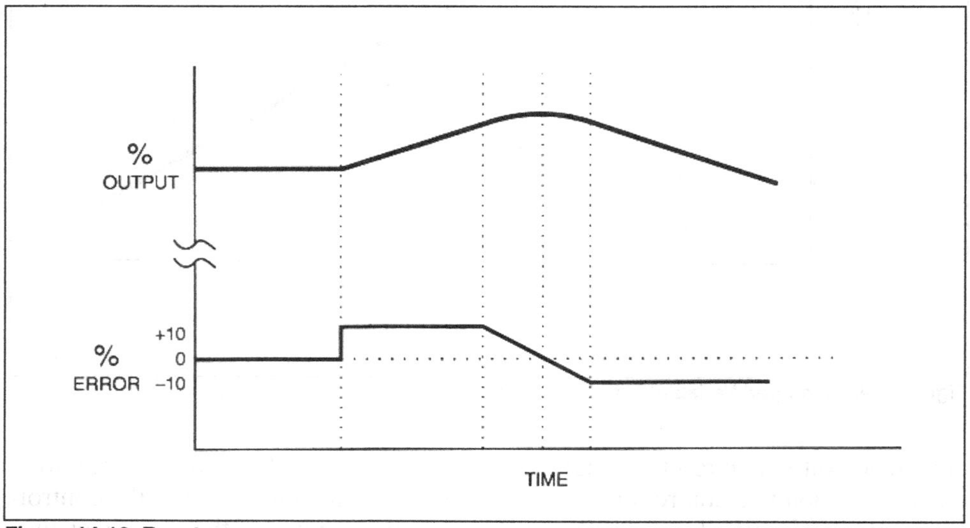

Figure 14-10. Reset.

If one looks at the output of a controller (not connected to the process) over time following a step change in the error, two distinct effects (see Figure 14-11) are obvious. Immediately the output changes by the amount of the change in the error (in percent) times the gain. This is the *gain effect.* Then the output continues to move as long as there is an error. This additional change in the output is known as the *reset effect.* The time between the change in the error and the point in time when the reset effect equals the gain effect is known as the repeat time or reset time.

One problem with the reset function is that I may "wind up." Because of the integration of the positive feedback loop, the output will continue to increase or decrease as long as there is an error (difference between set point and measurement) or until the output reaches a defined limit.

This normally is not a problem and is a normal feature of the loop. For example, a temperature control loop may require that the steam valve be held fully open until the measurement reaches the set point. At that point, the error will cross zero and change signs, the output will start decreasing, and the steam valve will begin to throttle back.

Sometimes, however, reset windup may cause a problem. Actually, the problem is not usually the windup but the "wind down" that would be required.

Some control vendors measure reset by *repeat time* in minutes. Others measure reset by *repeats per minute* the inverse of minutes of repeat.

Suppose the output of a controller is broken by a selector, with the output of another controller taking control of the valve. In Figure 14-12 the lesser of the two controller outputs is sent to the valve. The controller that has the lower output will control the valve. The other controller is, in effect, open loop. If its error would

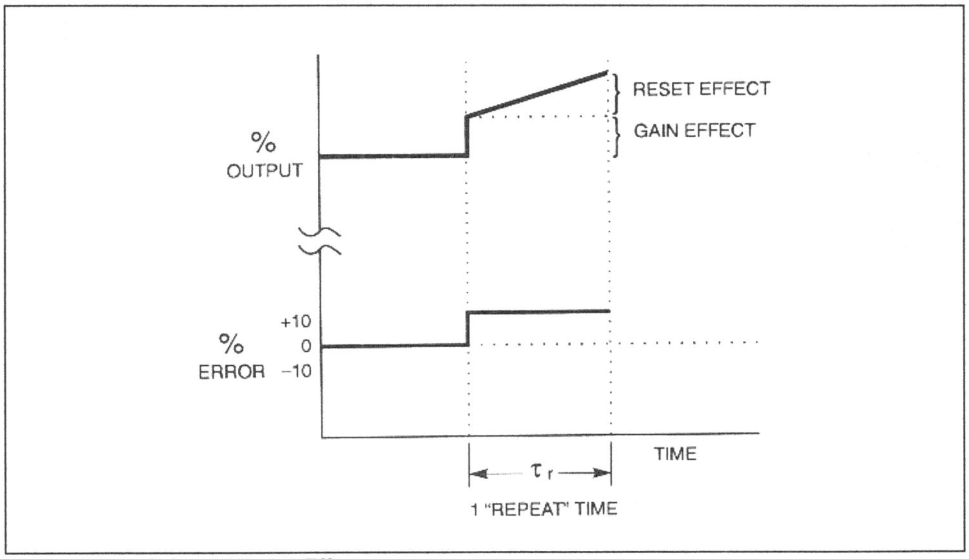

Figure 14-11. Gain and Reset Effects.

make its output increase, the reset term of the controller will cause the output to increase until it reaches its limit.

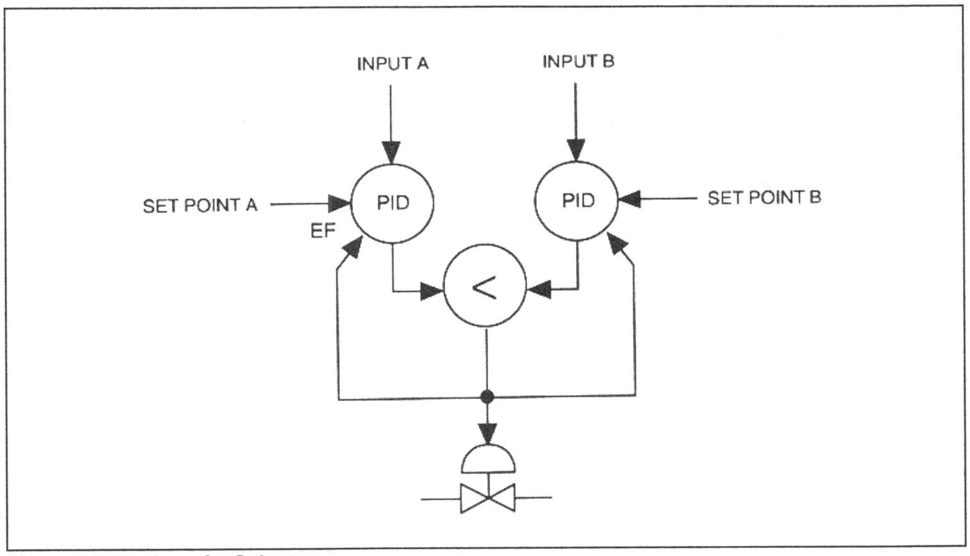

Figure 14-12. Use of a Selector.

The problem is that when conditions change and the override controller no longer needs to hold the valve closed, the primary controller's output will be very far above the override signal. Before the primary controller can have any effect on the valve, it will have to "wind down" until its output equals the override signal.

The positive feedback loop that is used to provide integration can be brought out of the controller. Then it is known as "external feedback."

If there is a selector between the output of the controller and the valve (used for override control), the output of the selector is connected to the external feedback of the controller. This puts the selector in the positive feedback loop.

If the output of the controller is overridden by another signal, the overriding signal is brought into the external feedback. After the lag, the output of the controller is equal to the override signal plus the error times gain. Therefore, when the error is zero, the controller output is equal to the override signal. If the error becomes negative, the controller output is less than the override signal, so the controller regains control of the valve.

DERIVATIVE

Derivative was first used as a part of temperature transmitter to overcome lag in transmitter measurement. It is also known as rate.

The derivative contribution can be expressed mathematically as:

$$\text{Output} = G \times K_d \times \frac{de}{dt}$$

where:

G = gain

K_d = the derivative setting, minutes

$\frac{de}{dt}$ = the rate of change of the error, %/min.

The response, over time, of a controller with gain and derivative is shown in Figure 14-13. While the error is changing, the output is given an additional boost ("derivative effect" in the figure). After the error has stopped changing and is constant at a new value, the derivative effect is eliminated and the only change in the output is the gain effect, just as if the controller had proportional-only control.

The output with the same gain setting but no derivative effect is shown as a broken line in the figure. If a horizontal line is drawn from any point on the broken line to any point on the solid line, it can be seen that the addition of derivative "advanced" the output by a certain time. That is, the output reaches a given value earlier if derivative is used than if proportional-only is used. The amount of time that the derivative action advances the output is known as the "derivative time," which is measured in minutes.

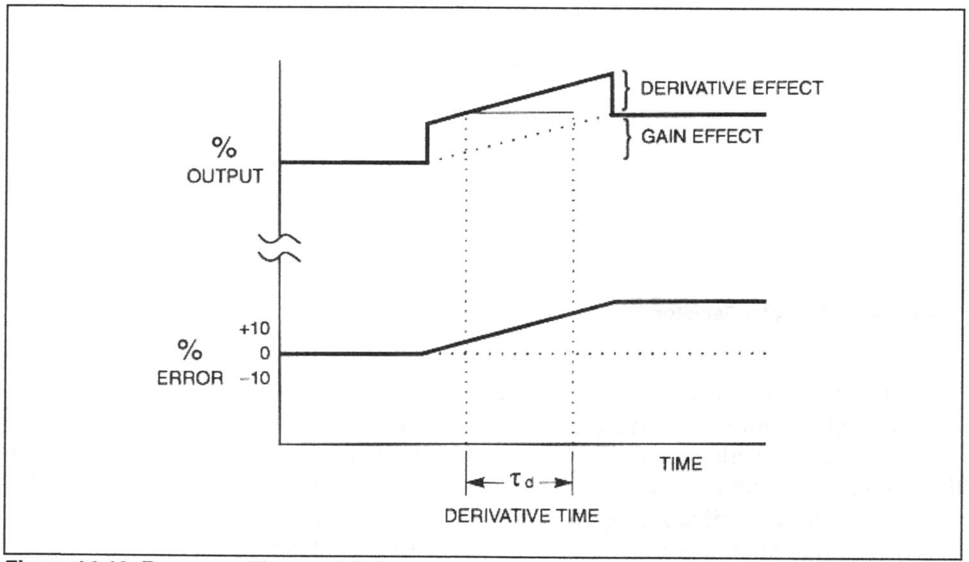

Figure 14-13. Response Time and Gain.

COMPLETE PID RESPONSE

All major vendors measure derivative (rate) in minutes.

Combining the gain, integral, and derivative elements of the PID equation we have:

$$\text{Output} = G\left(e + K_i \int e\,dt + K_d\right)\frac{de}{dt}$$

where:

G = gain
K_i = reset (repeats per minute)
K_d = derivative (minutes)
e = error, %

While any combination of the three modes (P, PD, ID, etc.) is possible, the combinations of P-only, PI, and PID are the most commonly used. At least 75% of all control loops use PI control.

Loop Tuning

Once a loop is configured and started up, someone has to adjust the gain, reset, and derivative values in the controller. Unless these adjustments are made, the controller will not properly control the process. At one extreme, the control loop will be unresponsive, just as if the control loop were in manual. At the other extreme, the control loop will be unstable, with the manipulated variable swinging between its low and its high limits. The adjustment of these parameters is known as loop tuning.

Tuning Criteria or "How Do We Know When It's Tuned?"

One often-overlooked aspect of control loop tuning is the determination of when the loop is correctly tuned. Of course, if the loop is unresponsive to set point changes or if it is wildly swinging, it is not correctly tuned. But between these extremes it is more difficult to determine how well the loop is tuned.

Informal Methods

These methods involve observation and measurement of the measured variable after the set point has been changed or the loop has been disturbed. The choice of methods depends upon the loop's place in the process and its relationship with other loops.

OPTIMUM DECAY RATIO (QUARTER WAVE DECAY)

This common method of judging tuning is known as the "quarter wave decay" method (see Figure 14-14). It has been shown that if a loop is tuned so that the oscillation decays with each wave being one quarter of the previous wave, it produces satisfactory, if not optimum, set point response and disturbance rejection.

MINIMUM OVERSHOOT

For some loops the most important criterion is that the loop be able to respond to a set point change with as little overshoot as possible. For these loops the tuning will be adjusted to minimize the overshoot, even at the expense of reduced disturbance rejection (see Figure 14-15).

MAXIMUM DISTURBANCE REJECTION

In other cases the goal is to reduce the peak response to a disturbance even if this causes the loop to take longer to stabilize at the set point or causes an overshoot following a set point change.

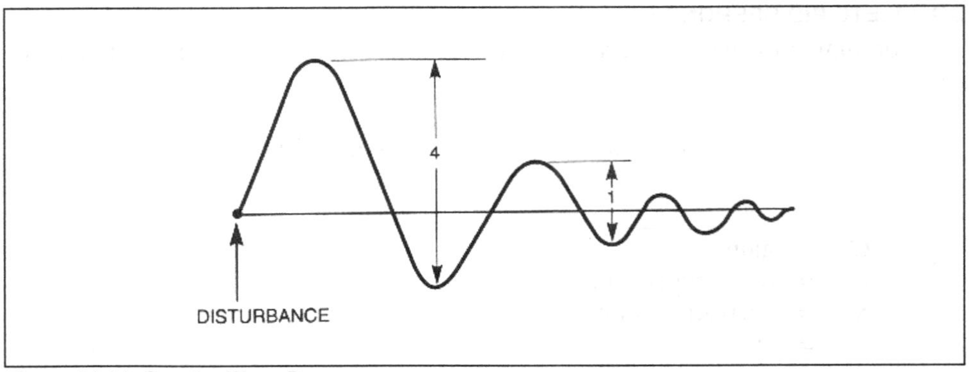

Figure 14-14. Quarter Wave Decay.

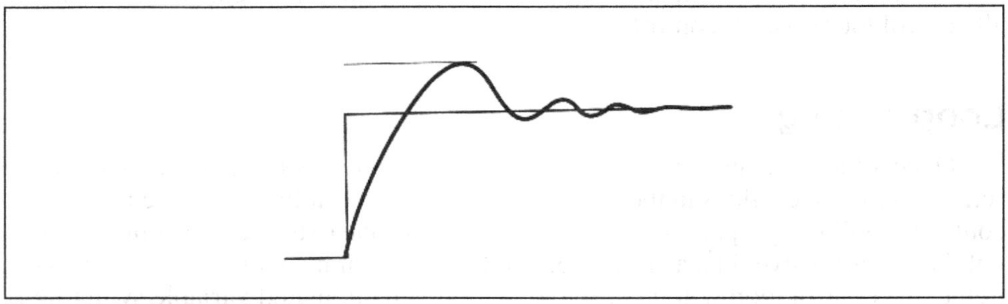

Figure 14-15. Minimal Overshoot.

Mathematical Criteria—Minimization of Index

Certain mathematical criteria numerically compare the results of a loop following a disturbance or set point change. After either a set point change or a disturbance, the error is measured and integrated using the methods listed below. After the process value has settled at the set point, the result of the integration is noted.

IAE—Integral of absolute value of error:

$$\int |e|\, dt$$

ISE—Integral of error squared:

$$\int e^2 dt$$

ITAE—Integral of time times absolute value of error:

$$\int t|e|\, dt$$

ITSE—Integral of time times error squared:

$$\int te^2 dt$$

These mathematical methods are used primarily for academic purposes, together with process simulations, in the study of control algorithms.

What's Really Important

The real key to judging the tuning of a loop is knowledge of "what's really important" to the operation of the process. It may be important that the process be held extremely close to the set point. It may be important that the process be held in the general vicinity of the set point but with a limit of the effect of a disturbance. In other cases it may be most important that the process rapidly follow changes in the set point.

Too often loop tuning is tested by checking the set point response. It is an easy test to do: one simply makes a small change in the set point and then observes the response of the process. However, the response to set point changes is an overrated measure of loop tuning. While some loops are subject to set point changes and must respond quickly, the set points of many loops typically do not change. For these loops the set point response is irrelevant.

Experience-Based Tuning

The most common form of tuning, used by experienced instrument personnel, is performed as follows:

(1) Enter an initial set of tuning constants from experience. A conservative setting would be a gain of 1 or less and a reset of less than 0.1 repeats per minute.

(2) Put the loop in automatic with the process lined out.

(3) Make step changes (about 5%) in set point.

(4) Observe the response. Based on experience, determine the change required (if any) and adjust if further tuning is needed.

Of course, this method requires experience. Other loop tuning methods do not require an extensive knowledge base, although they are often used by experienced personnel.

Ziegler-Nichols Tuning Method, Open-Loop (Reaction Rate)

Ziegler and Nichols suggested two methods of loop tuning that are still used today. For most processes these methods will result in quarter wave decay tuning (describe above). This tuning is usually a good compromise between tuning that optimizes set point response and tuning that optimizes disturbance rejection.

The first of the Ziegler-Nichols methods is also known as the "reaction curve" or "open loop reaction rate" method. This method cannot be used with integrating processes that will not level off at some acceptable value when a small change is made in the output.

The process must be lined out, that is, stable and not changing. With the controller in manual, the output is changed by a small amount (5% is typical). The process is recorded as it reaches its new level (see Figure 14-16).

The following measurements are made from the reaction curve:

ΔX	%	Change of output
R	%/min.	Rate of change at the point of inflection (POI)
L	min.	Time until the intercept of tangent line and original process value agree (lag time)

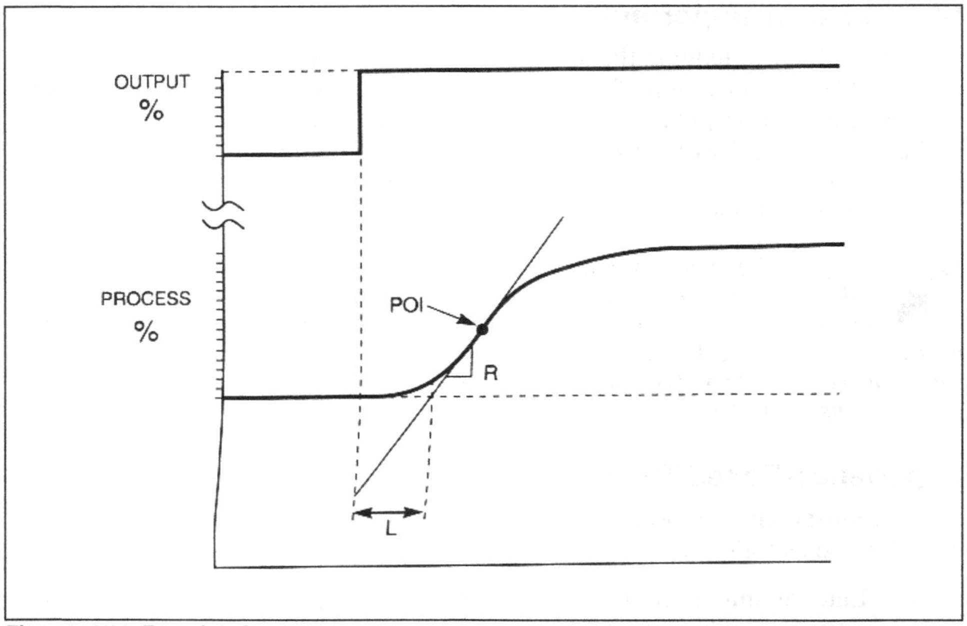

Figure 14-16. Reaction Curve.

Ziegler-Nichols Tuning Method, Closed Loop (Ultimate Period)

Another method developed by Ziegler and Nichols is the closed-loop method. In this method the loop is kept in automatic with the rest of the process in a stable condition. Unlike the open-loop method, this method can be used with integrating processes.

Steps:

(1) Place the controller into automatic with low gain and no reset or derivative.

(2) In steps, increase the gain and then make a small change in the set point. Watch for oscillations to start.

(3) Adjust the gain to make the oscillations continue with a constant amplitude. If the oscillations tend to grow, slightly decrease the gain. If the oscillations tend to decay, increase the gain. The gain that results in constant oscillations is known as the ultimate gain (G_μ). The period, in minutes, is called the ultimate period (P_μ).

The gain, reset, and derivative are calculated using the following:

	Gain	Reset	Derivative
P	$\frac{\Delta X}{LR}$	—	—
PI	$0.9\frac{\Delta X}{LR}$	$\frac{0.3}{L}$	—
PID	$1.2\frac{\Delta X}{LR}$	$\frac{0.5}{L}$	$0.5L$

Controllability of Processes

The "controllability" of a process depends upon the gain that can be used. As a general rule, the higher the gain, the greater rejection of disturbance and the greater the response to set point changes. Therefore, with most loops it is preferable to use the highest gain that does not result in undue oscillations.

In looking again at the response of a loop to a step change in the output, dynamic lag can be calculated. The *predominate lag* (*L* in Figure 14-16) is based on the largest lag in the system. The maximum gain that can be used depends upon the ratio *LR*. Notice the calculation for the gain in the Ziegler-Nichols open-loop method above.

From this we can draw two conclusions: (1) decreasing the dead time increases the maximum gain and the controllability, and (2) increasing the ratio of the longest lag to the second longest lag also increases the controllability.

Flow Loops

Flow loops are too fast to use the standard methods of analysis and tuning.

Some flow loops that use analog controllers are tuned with high gain. This will not work with digital control. With an analog controller, the flow loop has a predominate lag of a few seconds and no subordinate lag. The scan rate of a digital controller can be considered dead time. Although this dead time is small, it is large enough when compared to the lag to require that a low gain be used.

For most flow loops, good control will be obtained with a gain from 0.5 to 0.7, a reset from 15 to 20 repeats/min., and no derivative.

15

Distributed Control Systems

Overview

This overview of distributed control system (DCS) maintenance consists of a definition of maintenance goals and objectives, comparison of alternative maintenance strategies, an introduction to the maintenance planning process, and implications of distributed control system design characteristics on maintenance. It concludes with a discussion of maintenance program implementation and management.

Distributed Control System Maintenance

A distributed control system typically consists of a control room network connecting multiple control, display, and communication devices. These devices include operator and engineer workstations, continuous and discrete process controllers, field network interfaces, and computing resources, as well as other node types. The distributed control system architecture allows these devices to be distributed by either function or geographic area. The individual devices are optimized for the specific function they serve. However, this geographical and functional distribution and the quantity and complexity of the individual devices and their interconnections make the development of an effective maintenance program a significant challenge.

Figure 15-1 shows the configuration of process-connected and control room equipment as well as their interconnections at field, control room, and plant networks within a representative distributed control system architecture.

Maintenance Goals and Objectives

The successful development of a distributed control system maintenance program requires a long-term commitment to life cycle planning, budgeting, and training. The planning process should begin before selecting the control system vendor and continue throughout the life of the system. The plan should have the agreement of plant maintenance, operations, and management personnel. Roles and responsibilities must be clearly defined and understood. A quality improvement process should be in place to periodically monitor and evaluate the effectiveness of the maintenance program.

The primary goals of a distributed control system maintenance program are to:

(1) minimize the number of component failures in the system,

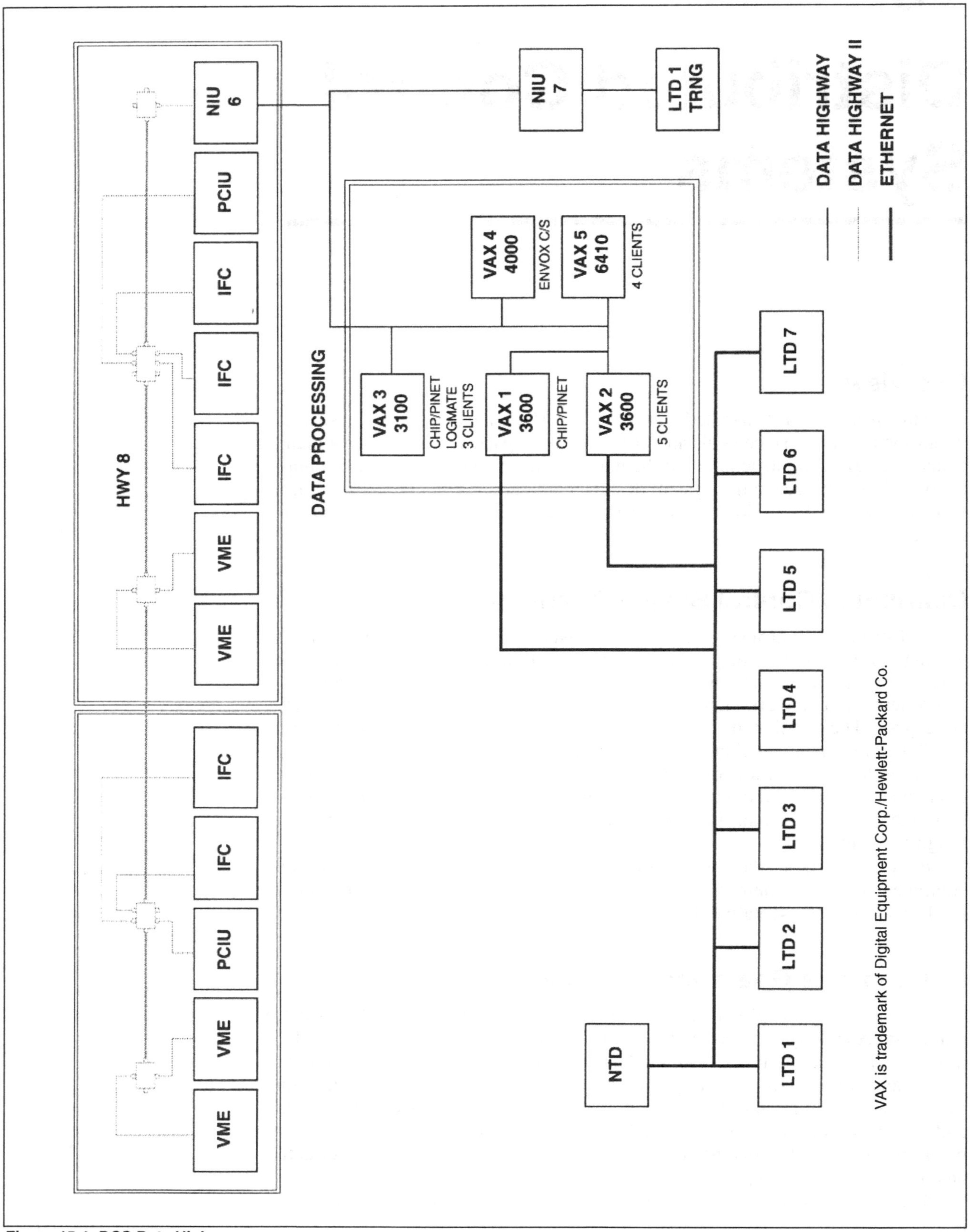

Figure 15-1. DCS Data Highway.
(Courtesy of Fisher-Rosemount)

(2) minimize the process impact of failures that do occur, and

(3) effectively diagnose the cause of failures and take corrective actions as quickly as possible.

Maintenance Strategies

REACTIVE MAINTENANCE

Historically, maintenance has too often been associated with breakdowns, the philosophy of running equipment until it failed then sending in an emergency repair crew is a waste of resources, both human and material. The reactive or "run-to-failure" approach is a costly and difficult strategy to manage. Maintenance costs are high, and don't account for production losses, damage to machinery, high inventory, poor morale, and the long range effect on product quality, unit cost, and overall competitive position.

PREVENTATIVE MAINTENANCE

Preventative maintenance (PM) planning allows plant and maintenance engineers to implement maintenance activities more effectively by scheduling periodic shutdown for inspection and parts replacement based on equipment-repair histories. Advance preparation is the first step in the transformation towards a proactive maintenance methodology. Since unscheduled maintenance work costs two to three times as much as work planned in advance, this alone can produce significant savings in the maintenance budget.

Benefits of a well-implemented preventative maintenance program include the following:

(1) documented repair history

(2) better planning and scheduling

(3) more accurate inventory control

(4) better control of costs

(5) increased labor utilization

(6) access to work and cost repairs

(7) work order tracking

(8) asset-management capability

(9) better communications and decision making

(10) better inter-departmental communications and responsiveness

(11) conformance to regulatory and quality control requirements

Although preventive maintenance offers clear benefits, there is a downside. Because preventive maintenance activities are performed on a set schedule, there is a risk that functioning equipment may be shut down or replaced unnecessarily, each of which carries its own cost in lost production and wasted capital. Experienced maintenance professionals know that the process of shutdown, inspection, and re-assembly itself can introduce a new set of problems. In addition, preventive maintenance cannot entirely eliminate emergency repairs.

PREDICTIVE MAINTENANCE

These operating realities have given rise to the relatively new concept of predictive maintenance. Predictive maintenance is based on accurate measurements of real-time operating conditions. This information is readily available and routinely collected automatically by the distributed control system. These parameters are used to generate a prediction based on certain criteria, for example, number of days until an alarm condition. Work orders can be scheduled according to demand and required parts ordered — "just-in-time" maintenance.

Because it is parameter-driven rather than schedule-bound, predictive maintenance eliminates the expense of unnecessary preventive maintenance on functioning equipment, making resources available for production or other important areas. Predictive maintenance can also prolong the intervals between preventive maintenance activities, an additional source of savings in labor and parts. And because problems are identified early, predictive maintenance prevents catastrophic failure. The clear benefit: predictive maintenance delivers availability (see Figure 15-2).

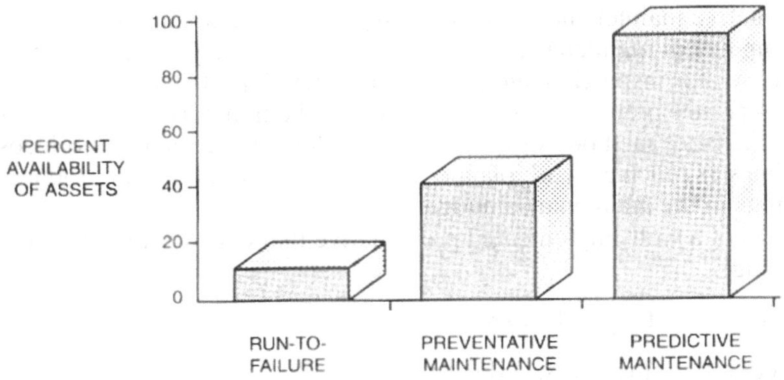

Figure 15-2. Comparison of Maintenance Strategies.

Maintenance Planning

The development of an effective maintenance plan involves a careful analysis of several factors. A thorough assessment of external and internal requirements and capabilities must be completed before implementing the plan. The sections that follow address some key factors to be considered in the development of a distributed control system maintenance program.

PROCESS CONSIDERATIONS

The first factor to be considered in the development of a distributed control system maintenance program is the process under control. The process should be evaluated to determine the control requirements and the safety and economic impact of component failures within the control system. Some degree of redundancy is required in almost every distributed control system. The redundancy requirements must be determined before purchasing the system. The level of redundancy required is a function of the process control requirements.

Listed below are some factors to consider in selecting the proper level of control system redundancy:

(1) Economic impact of component failures that result in a loss of control

(2) The safety impact of component failures that result in loss of control

(3) The expected reliability of components within the system

(4) The availability of spare parts from either in-house or vendor stocks

(5) The response time of maintenance personnel during both normal and off-hours

These factors should be used to evaluate each critical component in the system to determine the appropriate level of redundancy. Keep in mind that even the best maintenance program cannot overcome problems resulting from an insufficient level of redundancy in the control system.

IN-HOUSE EXPERTISE AND TRAINING

In the development of a distributed control system maintenance plan, the experience and expertise of in-house maintenance personnel must be carefully evaluated. Many organizations have experienced maintenance personnel; however, most of their experience may be with electronic, analog or even pneumatic control systems. This experience cannot be directly translated to the maintenance of distributed control systems because they require a totally different set of skills.

When maintaining analog control systems, technicians had to be proficient in analyzing and troubleshooting analog circuits. The use of vendor-supplier test equipment was common, and repairs to the analog circuits were often made on-site. Since analog control systems were implemented by physically connecting various hardware components, troubleshooting consisted of locating the failed hardware component. Once identified, the hardware was replaced or repaired.

With distributed control systems, control strategies are implemented with a combination of hardware and software, with software playing an ever increasing role. Often, it is difficult to distinguish between hardware and software problems. Even if distributed control system maintenance technicians are not responsible for software, they are usually first on the scene and need the skills necessary to make an initial evaluation.

The maintenance technicians must be familiar with the function of each component in the distributed control system installation. The determination of hardware problems with distributed control systems will usually involve the use of computer-based diagnostic software. The maintenance technicians must be proficient in the use of this software and the procedures associated with removal and replacement or repair of the actual hardware.

The level of both initial training and continuing education required for distributed control system maintenance is usually more significant than initially thought. Ever-increasing distributed control system reliability increases the difficulty of maintaining a satisfactory degree of proficiency to resolve problems when they occur. A realistic estimate of the training requirements should be considered in the development of the overall maintenance program. The effectiveness of the program is only as good as the quality of the personnel who will be responsible for its implementation.

VENDOR INVOLVEMENT

Some degree of vendor involvement is required for any successful distributed control system maintenance program. The role played by the vendor will vary from site to site and vendor to vendor.

In deciding the vendor's proper role, several issues must be addressed. Does the vendor maintenance support originate at the factory or at a local service office? Factory support personnel are usually dedicated to maintenance activities and may be better trained, have more experience, and have more up-to-date information. However, because you get a different person on each service call, the individual may not be familiar with your particular installation. This includes the distributed control system equipment, the site environment, and the relevant history of both. This makes developing an effective working relationship more difficult and may lead to a loss of continuity on long-term maintenance issues. The response time of factory personnel also is a factor. Typically, the support person must travel by air. This will greatly increase response time and cost of support.

The quality of maintenance support from a local sales office can vary dramatically. Staffs will be smaller, and the experience and expertise of support personnel may vary a great deal. However, the personnel probably will become more familiar with your installation and be better able to establish a good working relationship with your in-house maintenance staff. This can be of great value if support personnel remain properly trained. Response times from local offices are usually better, as travel is normally by automobile. However, because of small staffs, the risk exists that support personnel may not always be available.

In developing the distributed control system maintenance plan, a realistic evaluation of vendor support capabilities must be made. The system must be supported and any deficiencies in vendor support capabilities must be made up by in-house personnel.

SPARE PART INVENTORY CONTROL

An effective inventory control system is an important part of a distributed control system maintenance program. A variety of spare parts are required to maintain a distributed control system. These parts may range from entire replacement computers to electronic circuit boards to commodity wiring items. In many distributed control systems a single circuit board may be used in several different devices. For example, the same board may provide field network communications for several control devices. In this situation a single spare may be sufficient for several devices. In other cases a circuit board may have a unique function and require a spare of its own.

Selecting the parts to stock and determining the quantities is a difficult but critical task. Insufficient spares may lead to increased downtimes or process shutdowns. Stocking too many spare parts may tie up capital that could be better utilized in other ways. Each component in the system should be analyzed to determine the optimal spares requirement. The following are some factors to consider in this evaluation:

(1) Quantity installed in the system

(2) Expected reliability or frequency of failure

(3) Level of installed redundancy

(4) Impact of component failure

(5) Allowable downtime

(6) Time required to repair or replace

(7) Cost of spares

The vendor can play an important role in this evaluation. The distributed control system vendors will make spare parts recommendations. Their experience with the hardware can provide valuable information to be used in the evaluation. However, vendors will sometimes overstate the reliability of their equipment. Vendor spare parts recommendations are often based only on installed counts and do not take all of the above factors into consideration.

An additional challenge associated with distributed control system spare parts is firmware revision control. Many vendor-supplied circuit boards have firmware installed on them. Over time this firmware will undergo revisions. Boards returned to the factory for repair may be returned with the latest firmware installed.

For example, a two-year old input-output processing board fails and is returned to the factory for repair. The board is returned with the latest firmware installed. When the board is placed back in service, it may fail if the firmware is incompatible with that of the process controller. When spare parts have been acquired over a period of time, it is quite common to find boards with different identification numbers, due to firmware and other changes, that may still be interchangeable. The result is needless overstocking of some spare boards.

Failure to track firmware revisions of both installed and spare parts could lead to system failures.

Design for Maintenance

The major implications of distributed control system architecture and design principles on maintenance strategy and tactics are the following:

(1) Faults are rare, and little preventive maintenance is required.

(2) The faults that do occur are tolerated with little or no degradation of system performance, and they affect only the module in which they occur.

(3) For most failures, the system itself determines the correct optimum replaceable unit (ORU) — a board, power supply, peripheral device, etc.

(4) The variety of optimum replaceable units is held to a minimum, so the spare parts inventory is minimized.

(5) Test equipment is seldom needed, and no special or complicated test equipment is required.

FAULT CONTAINMENT AND FAULT TOLERANCE

First, some terms must be defined (see Figure 15-3).

Fault—A temporary or permanent condition in a functional unit that makes it deviate from its expected sequence of operations. A fault may exist even though it cannot be seen outside the unit, because the fault may be corrected each time it occurs.

Error—An abnormal or undesired result of an operation caused by a fault. An error may or may not be overcome by retrying the operation.

Failure—The inability of a functional unit to perform its expected task because of a solid error or an error that recurs too frequently.

Figure 15-4 shows one strategy for fault containment. If a fault occurs in a module of the control room network, the following happens:

(1) If possible, the result of the fault is corrected.

(2) If the result is not corrected, the operation is retried one or more times, in the hope the error won't recur.

(3) If retries are not successful, or if too many errors recur in the module, the module has failed. The operator is notified through an alarm and the module's operation is stopped in a safe state.

Figure 15-5 shows how the distributed control system architecture contributes to fault containment and fault tolerance. If a module on the control room network fails and has a redundant module backing it up, the backup takes over in five seconds or less. If there is no backup, the system continues to operate without the failed module. Some capacity, throughout, or functions may be lost, but the system remains operational. Also, with two exceptions, a failure of one module has

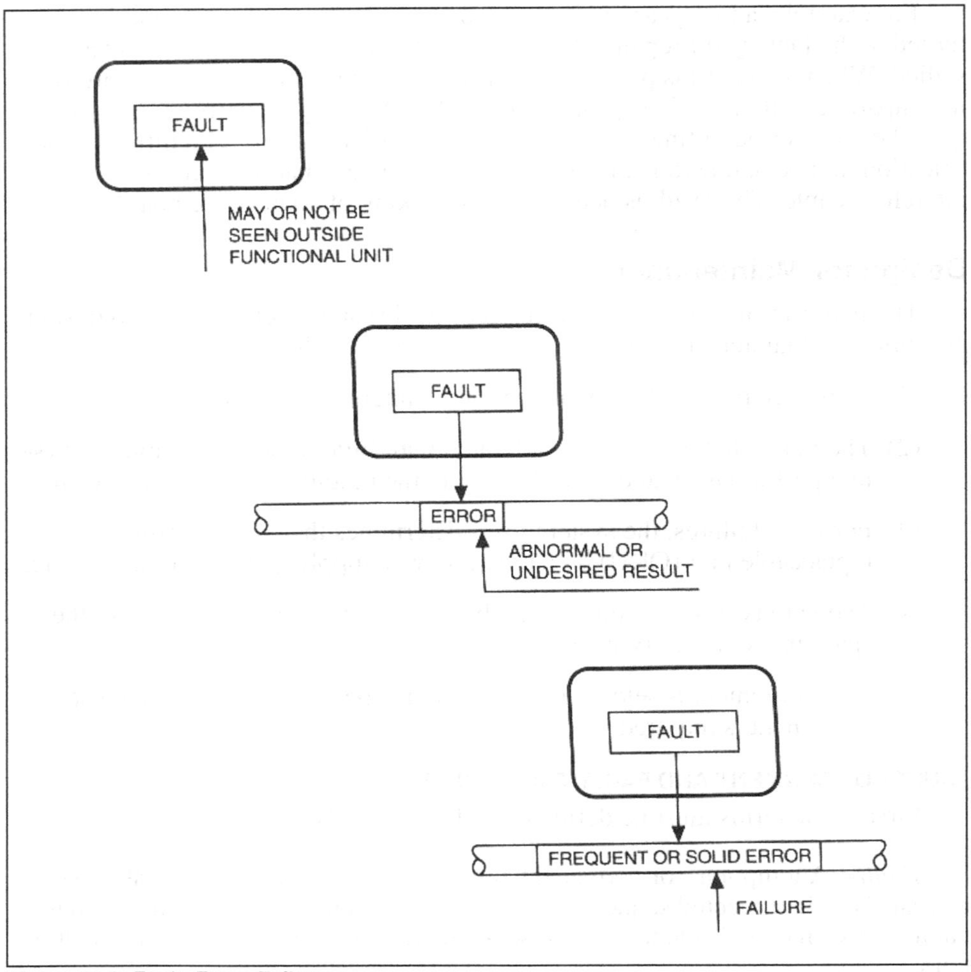

Figure 15-3. Fault, Error, Failure.

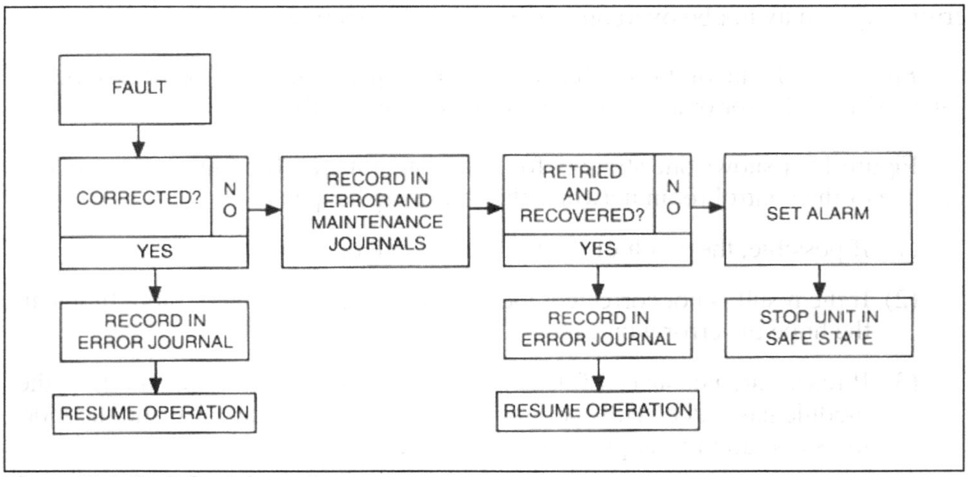

Figure 15-4. Fault Containment.

no effect on any other functional unit in the system. Most units in the system are independent of other units for their operation.

The exceptions are the field network interface or gateway. If a distributed control system had a single field network interface or gateway linking the control room network to the process-connected devices, the process connections would be lost if this module failed. This is why field network interface or gateway redundant pairs are used in many installations.

The same strategy applies both to control room equipment (modules on the control room network and devices connected to these modules) and to process connected devices. Process controllers on the field network are available with redundant control hardware and some redundant input-output processors. Redundant processors can also be specified for logic control devices.

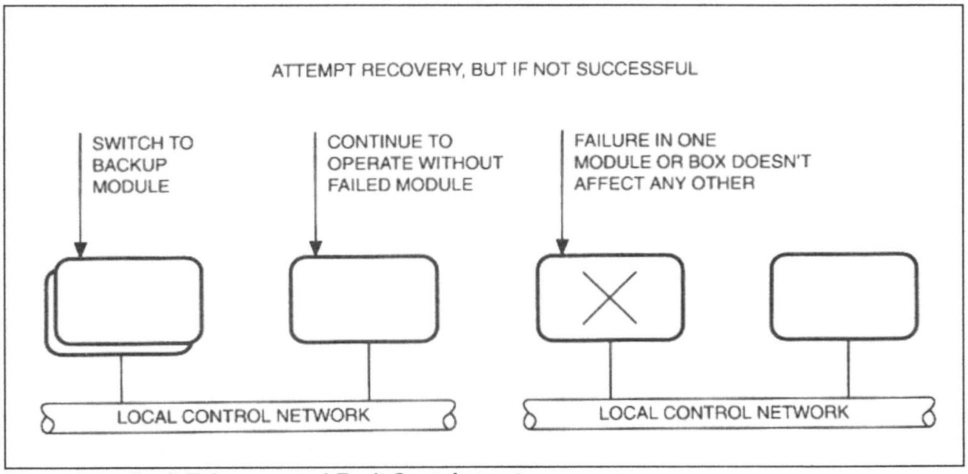

Figure 15-5. Fault Tolerance and Fault Containment.

SELF-DIAGNOSIS

The tests built into the control room equipment are designed to diagnose a high proportion of all failures and pinpoint them to the optimum replaceable unit without intervention by the operator or a maintenance technician while the system is on line and controlling the process.

The process-connected boxes on the field network all have built-in self tests, which either diagnose or help diagnose failures in those boxes (see Figure 15-6).

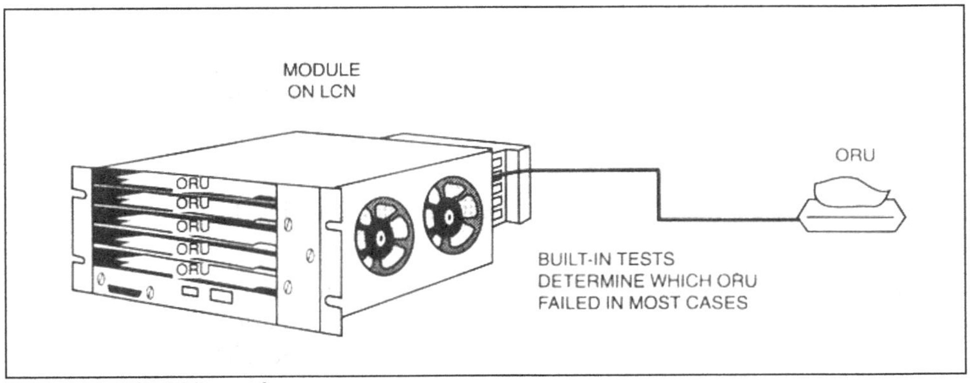

Figure 15-6. Self-Diagnosis.

MAINTENANCE SOFTWARE

An operator or engineer workstation typically has maintenance software that is used to diagnose failures in modules on the control room network that are not diagnosed by the built-in tests and to test the operation of the field network and their process-connected boxes. When an on-process work station is being used for maintenance, one or more of the remaining workstations in the console take over its on-process functions (see Figure 15-7).

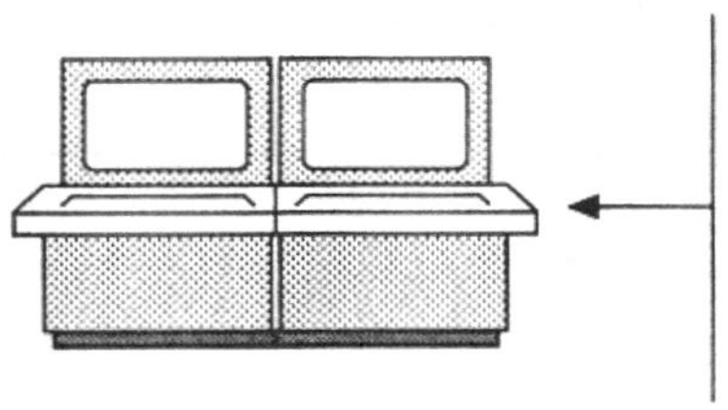

Figure 15-7. System Maintenance Activity.

SIMPLIFIED REPAIR PROCEDURES

The key to reduced repair times for distributed control systems is the replacement of easily isolated and removed parts and their return to the vendor for repair. The selected replacement parts are referred to as optimum replaceable units (ORU); they offer an effective tradeoff between the time-based costs of problem isolation and resolution, and the replacement costs of the parts themselves.

For some devices, such as printers and CRTs, two levels of optimum replaceable units are defined, thus expanding the range of time-versus-parts costs tradeoffs. The primary ORU level is the device itself. The secondary ORU level consists of lower-level service parts that require some additional skills to isolate and replace. The control room equipment is designed with the objective of minimizing the complement of optimum replaceable units. This is accomplished partly by using boards that are common throughout the control room network. The optimum replaceable unit concept also applies to the field network and process-connected devices.

Any module on the control room network, any peripheral device, and any process-connected device on the field network can be maintained and repaired while the remainder of the system is on-process. All on-site repairs are made by replacing optimum replaceable units. Optimum replaceable units themselves are repaired by the vendor.

BUILT-IN MAINTENANCE AIDS

Every module on the control room network has three levels of built-in self-test:

(1) *Start-Up Self-Tests*: These test programs are stored in read-only memory on each board that has a microprocessor. They run after power is applied, or when the module containing the board is restarted. These tests verify the correct operation of the basic logic on each board and of the module. The self-test on the module's microprocessor board tests the memory boards and the module bus. Any detected failure lights up an indicator on the failed board(s) and can also be detected over the control room network by other modules.

(2) *Quality Logic Tests*: Quality logic tests are automatically loaded and executed after the start-up self tests. Quality logic tests verify the correct operation of the module hardware and qualify it for loading of its on-process software.

(3) *On-Process Tests*: These tests are part of the on-process software of each module and execute whether a module is primary or backup. They run periodically to test the hardware and integrity features. After recovery from an error, these tests record the error in the system error journal for subsequent analysis. A nonrecoverable failure can cause the module to be shut down. The failure and any maintenance recommendation are reported to the operator and are recorded in the error and maintenance journals.

All errors and maintenance recommendations generated by the built-in tests are recorded on a system printer for later reference by maintenance personnel. Additionally, this information is accumulated by a storage device for use of the on-process analysis function. This function's primary role is the collation of error information by error type for each module, in order to perform predictive failure analysis and the issuing of maintenance recommendations.

On-process analysis also provides a printed error history that can be returned to the factory with the optimum replaceable unit being replaced.

An operator or engineer work station's maintenance capability is used when automatic tests cannot resolve a problem. The maintenance technician communicates with the station through interactive displays and the keyboard or pointing device.

The maintenance software of the work station provides these maintenance related functions:

(1) Interacts with the system maintenance journal: displays information in that journal, including active maintenance recommendations, and accepts maintenance activity entries.

(2) Reads and displays the hardware, firmware, and software revision status of the modules of the control room network.

(3) Displays a snapshot of system status, including the status of modules on the control room network, field network, and process-connected devices.

(4) Displays the content of memory locations in the modules on the control room network and process-connected devices.

(5) Displays error history from pools kept by the modules or from accumulations kept in an assigned storage device.

The workstation on-process operator software plays a significant role in maintenance by:

(1) Reporting entries in the system error journal to the operator,

(2) Inquiring into and displaying the status of modules, as requested,

(3) Stopping modules from communicating on the control room network when they have failed, and

(4) Displaying test results.

REMOTE DIAGNOSTIC SUPPORT

Improvements in the reliability and cost effectiveness of connectivity products have made remote access to distributed control systems practical. The many possible applications of remote user access to the distributed control system include, site support, remote diagnostics, central engineering, and management monitoring (see Figure 15-8).

This technology enables new approaches to distributed control system maintenance. From a remote location, you can:

(1) access host site file management,

(2) perform engineering functions, and

(3) view any host site display (group, schematic, trend).

Planning for secure communications is an important consideration in successfully applying this technology. Decisions must be made as to who will initiate the connection, who will pay for the cost of the connection, what level of access a remote user is allowed, and what capability for verbal communication between sites is required.

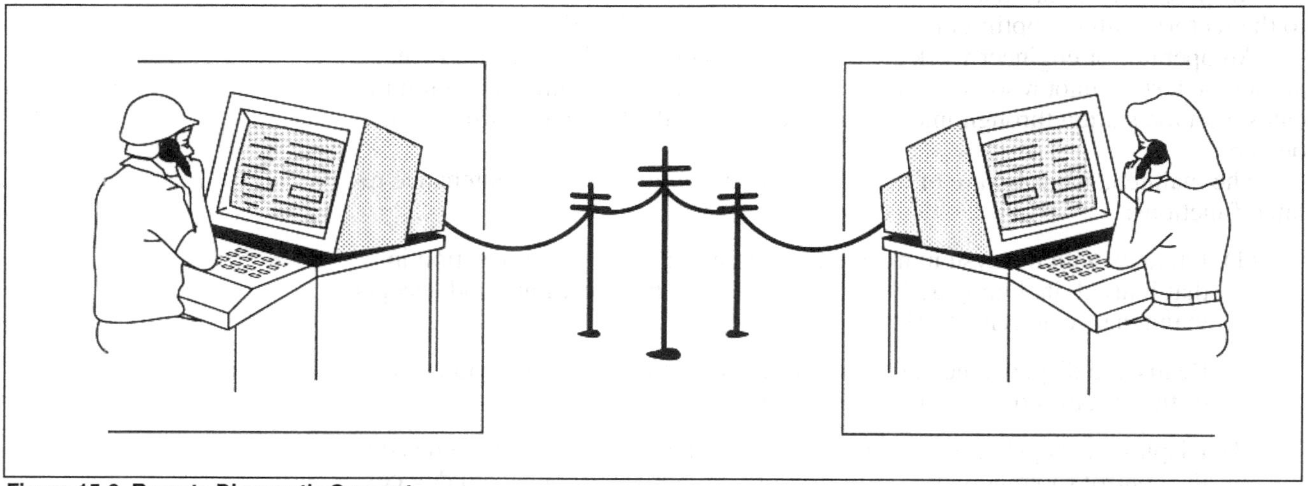

Figure 15-8. Remote Diagnostic Support.

Maintenance Program Implementation and Management

THE MAINTENANCE PROCESS

Maintenance of a distributed control system throughout its life cycle can be thought of as a sequential series of maintenance cycles. Each maintenance cycle starts with a request for maintenance to be carried out and ends with the issuing of a report giving the outcome of the work done.

Each maintenance cycle requires four key activities:

(1) Completing the requested maintenance action

(2) Re-testing the system

(3) Recording details of the activities carried out

(4) Noting any resulting changes made to the system configuration.

TYPES OF MAINTENANCE ACTIVITIES

There are basically three types of maintenance actions:

(1) Preventive maintenance, periodically scheduled routine actions

(2) Corrective maintenance, in response to faults and anomalies

(3) Perfective or adaptive maintenance, brought about by the need to simplify or improve the system or make system changes to reflect changes in the environment.

To ensure smooth maintenance operations, it is recommended that during the project stage a link into the engineering maintenance data base is established. If this is not possible, a paper or electronic system should be developed. The electronic system offers many advantages over the paper system but can often be no more effective or cost beneficial.

Preventive maintenance involves all actions that are intended to discover and clear faults before they affect service. It includes cleaning, routine testing and periodic replacement of parts that are at the end of their useful life (such as air filters, batteries, etc.) preventive maintenance is scheduled by the maintenance organization. The periodic replacement of parts demands a rigorous adherence to standard procedures. At the conclusion of this maintenance activity, the system should be tested before being put back into service and a report generated to confirm to the operations personnel that all maintenance actions have been successfully completed and tested.

Corrective maintenance involves all actions that are intended to diagnose and clear faults or anomalous behavior (nonconformance). The diagnosis should identify the cause of, and the circumstances surrounding the problem. If the corrective action required is the simple replacement of a faulty part without the basic configuration of the software or hardware of the system, an unscheduled part replacement is performed. It is important that during part replacement the configuration database is checked to ensure that switch settings and options are correctly set, that the card information for the failed and the replacement card are recorded on a failure report, the successful completion of the operation is reported and revalidation testing takes place. On completion of this stage the equipment is put back into service.

The use of perfective or adaptive maintenance ensures that any proposed change to a system baseline definition or build level is prepared, accepted and controlled in accordance with set procedures. Changes to be controlled include corrections, modifications, and enhancements brought about by the need to simplify or improve the system. Plant management will compare the estimated cost of modification with the benefit to the business of doing the modification. Some modifications may not be cost effective but may be essential to the safety or operation of the system. It is important that the modification is planned like any other activity. The configuration change needs to be approved by plant management who must also agree to the implementation plan.

SAMPLE PREVENTIVE MAINTENANCE ACTIVITIES

As noted earlier, one of the main goals of a distributed control system maintenance program is to minimize the number of failures that take place within a system. The two most important factors in achieving this objective are the quality of the initial installation and an effective preventive maintenance (PM) program. Preventive maintenance should be part of any distributed control system maintenance program. The following list may serve as a useful starting point for developing an effective preventive maintenance program:

(1) Check for proper power and grounding on a regular basis.

(2) Replace on-board batteries per manufacturer's recommendations.

(3) Test switch over to redundant devices on a regular basis.

(4) Keep the installation as clean and dust-free as possible. Change cabinet and air conditioning filters frequently. Keep the distributed control system equipment room doors and cabinets closed.

(5) Check field wiring connections for loose or corroded connections.

(6) Install and maintain an air filtration and treatment system in accordance with vendor recommendations. Where corrosive elements are inherent in the process, monitor for their presence in the distributed control system equipment room. Special reactivity coupons are available for this purpose at minor expense.

(7) Periodically review distributed control system performance with control room operators to bring forward any otherwise unreported incidents or concerns. Coordination with operations is essential in planning for new system installations, major system upgrades, and minor system updates. Periodic operations reviews can also be used to assess the need for additional or refresher training.

(8) Finally, consider integrating the distributed control system preventive maintenance program with your field instrument calibration program. The distributed control system will function only as well as the date it receives from the field.

VENDOR PARTICIPATION

Every distributed control system maintenance program will include at least some vendor participation. The degree of vendor participation should be determined after evaluating the factors discussed earlier. Three typical distributed control system maintenance scenarios are discussed below. Each scenario requires a different degree of vendor and owner participation.

Scenario 1: Owner Provides Labor and Stocks All Spare Parts

In this scenario, spare parts are purchased and stocked by the owner. The owner provides all diagnostic and maintenance expertise. Failed components are returned to the vendor for repair or replacement. The vendor may also be used for assistance on unusual or difficult problems.

This scenario requires the greatest commitment to training. The availability of spare parts must be adequate for all failures that could occur. A firmware revision control system must be in place to insure that replaced components are compatible with the rest of the system. This scenario has the advantage that repairs can generally be quicker because spare parts are on-site and manpower is dedicated. However, the capital cost of spare parts will be greater, as will the cost of training.

Programs of this type are found in organizations with larger installed distributed control systems and dedicated maintenance technicians. It may also be required in situations where timely vendor response is unavailable.

Scenario 2: Owner and Vendor Share Spare Parts and Labor

In this scenario, the vendor and the owner share the responsibility for spare parts and labor. The owner would stock the most frequently used spare parts and those of a critical nature. The vendor could be used to stock the more expensive and less frequently used parts, or to reduce the quantity of spares you need to stock by assuring rapid replacement. The vendor must be capable of rapidly delivering the spare parts. The vendor and owner must work together to insure that all parts necessary for maintenance of the system are available at either the vendor's or owner's location.

A significant commitment to training of maintenance technicians is still required. Primary diagnostic and troubleshooting responsibilities still rest with the owner. Care must be taken to insure that sufficient redundancy is in place or downtime is acceptable to allow for vendor-stocked parts to arrive. A firmware revision control system must still be in place. The system may be handled by the owner, the vendor, or both, depending on the circumstances.

This system may be appropriate for small organizations or those with a small installed base of distributed control system hardware. The vendor must have an adequate response time and have the necessary local expertise to assist in troubleshooting. Communication between the owner and the vendor is important to insure that parts and service are available when required.

Scenario 3: Vendor Provides Spare Parts and Service

In this scenario the vendor is totally responsible for the maintenance of the distributed control system. In-house maintenance technicians may provide initial diagnostic efforts but will rely on the vendor for both spare parts and labor. This type of arrangement is usually handled under a maintenance contract that should specify the response time. This approach minimizes the capital cost of spares and reduces the training of in-house personnel. The cost of the vendor support services will be significant.

The owner is totally dependent on the vendor for support. The vendor must have adequate facilities and personnel to support the maintenance effort. This arrangement may lead to increased downtime. Take care to insure that sufficient redundancy is installed unless the downtime is acceptable. Another approach to improve availability is to contract for vendor supplied on-site support for distributed control system software, hardware, parts and labor. This scenario is appropriate for small organizations with minimal in-house capabilities, or for organizations that prefer to focus on their core competencies by out-sourcing responsibility for distributed control system maintenance. The system requires strong local vendor support and a good working relationship between owner and vendor.

Recommendations

An effective maintenance program is critical to the success of a distributed control system installation. A successful program requires a long-term commitment to planning, budgeting and training. The following recommendations should be considered when developing a distributed control system maintenance program:

(1) Begin planning your maintenance program before purchasing the distributed control system. Vendor support capabilities should be a key factor in the selection process.

(2) The level of installed redundancy has a significant impact on the requirements of a distributed control system maintenance program. When evaluating the necessary level of redundancy, consider the impact on the maintenance effort.

(3) The level of training required to maintain a distributed control system is almost always underestimated. Make sure that sufficient funds are budgeted for training. Work closely with the vendor to define the initial and on-going training requirements and to make sure that the initial training is complete before startup of the system.

(4) Some degree of vendor involvement is required in almost any distributed control system maintenance program. Work closely with the vendor to determine the most appropriate role for them to perform.

(5) Spend adequate time to develop an effective spare part inventory control system. Consider more than the installed count where determining the quantity of spares to stock. Ensure that the system includes provisions for handling revisions to installed firmware.

(6) Make a preventive maintenance program an integral part of your overall maintenance effort. Work closely with the vendor to develop the preventive maintenance program and make sure that it is implemented as designed.

Programmable Logic Controllers

Originally introduced as a relay replacement for the automobile manufacturing industry, programmable logic controllers (PLCs) facilitate the replacement in software of many discrete electromechanical components including timers, counters, and various sequencing devices. The absence of moving parts and the reliability of today's microelectronics have contributed to high maintainability and acceptance by many other industries.

PLC Advantages: Replace many discrete components; No moving parts; High reliability.

Most programmable logic controllers share similar functionality. They require regulated power at several voltage levels free from the irregularities of the plant power system, a central processing unit to process the stored logic, memory to store the logic and I/O status, and I/O modules to allow the central processor real-time access to plant sensors. A rich variety of I/O modules allows the central processor access to almost any variable of interest. Remote I/O processors are often utilized to manage the communication of I/O status between the central processing unit and remotely located clusters of I/O.

Modules are packaged by function and usually mounted in a card cage, providing plug-in power distribution and central processing unit communications. Certain locations are generally reserved for power supplies and central/remote processing units. The bulk of the module space required will consist of I/O modules, which generally group 8 to 16 channels of I/O per module. Communications between local racks is generally carried over special cables, while communications with remote locations is usually routed over coaxial cable.

PLC modules and power supplies are designed to withstand the rigors of many plant floor environments, facilitating their use as a plant floor controller. Temperature extremes range from 32° to 140°F and humidity extremes from 0-95% non-condensing. Exposure to conducting dusts or powders and even trace amounts of gases corrosive to copper, such as chlorine dioxide, hydrogen sulphide, or chlorine, must be avoided. Metal cable fittings should also be protected from such environments. Power supplies will generally tolerate voltage supply fluctuations of ±15%.

From a maintenance perspective, the PLC is simply a highly specialized digital computer that requires full maintenance and a technician trained in digital computer maintenance. The complexity, diversity, and high density of microelectronic chips used in PLC modules further complicate module repair at a component level by plant technicians. Functional modularity and plug-in mounting promote fast, simple replacement in the field by plant electrical/instrumentation technicians, making repair-by-replacement a viable maintenance option. Since logic system downtime usually translates to production downtime, the ability to replace an entire module or processor in minimum time more than offsets the cost of extra spares in most installations. This approach is enhanced by repair or replacement programs from most of the major manufacturers. To realize the principle advantages afforded by current technology and repair-by-replacement requires restricting the number of manufacturers and models installed within the plant. Only then will maximum advantage in training and spares be realized. A common

I/O system that is compatible with all models further enhances the advantages of repair-by-replacement.

Successful repair-by-replacement requires isolation of the malfunctioning module. A number of techniques are available (depending on the manufacturer) to facilitate identification of a faulty module:

(1) Built-in diagnostics

(2) Diagnostic flow charts

(3) Diagnostic PLC language instructions

The last of these techniques allows the PLC to monitor I/O module status and detect malfunctions. The disposition of this information is left to the programmer. I/O malfunctions at the channel level can be verified by disconnecting field wiring and using the programming unit to force the status of an output or verify the status of a known input condition. Intermittent malfunctions are much more difficult to diagnose and usually require a trial replacement of the suspected unit.

Many maintenance problems are rooted in poor design or construction practices. During these phases of system installation, every module, rack, and cable connector must be readily accessible and removable. After all, that's how the unit is going to be maintained. After the unit is in operation, the usual rules for good housekeeping should be followed:

(1) Keep the enclosure clean and free of dust.

(2) Keep the wiring in the raceways and as neat as possible.

(3) Keep the covers for the raceways in place.

(4) Keep an up-to-date set of the wiring and other applicable documentation in the enclosure.

(5) Keep the enclosure cover or doors closed.

When suitable equipment is available and a data highway is in use, measure and record line losses on the data highway after installation and at available opportunities (during shutdowns or turnarounds). The resulting record should allow early detection of deteriorating communications on the highway. Monitor the age of memory backup batteries and schedule their replacement during planned outages.

Always make a fresh backup of logic and data prior to changing these batteries, since the logic and data will be lost during the battery change.

Invest in a software maintenance package. Keep the software up to date. Keep multiple copies of the software in different locations.

Regardless of how well a system is physically maintained, poor software maintenance can seriously degrade the effectiveness of an installation. Do, by all means, invest in one of the many excellent software documentation packages available from either the manufacturer or a third party vendor. Keep the software documentation up to date at all times. Last, but certainly not least, keep two identical magnetic media copies of the installed software in two separate locations to avoid loss by fire or other disaster. The ultimate nightmare is to lose the loaded version of the software by whatever means and discover the backup copy is on unreadable media.

Troubleshooting Communication Circuits

Communication circuits are many times difficult to troubleshoot because they represent not only voltage or current levels but they also transfer coded information. In addition, the transmitter and receiver also interact via control signals. The basic tools for troubleshooting are a line analyzer, which can be used to monitor and interact with the transfer of information, and the break-out box, which monitors the voltages on the communication line. Line analyzers for RS-232 serial links are readily available both in hardware and in software on a PC platform. Line analyzers for RS-422, RS-485, Mil-188, RS-530, LANs, WANs, and European standards such as X.25 are also available. Proprietary communication protocols generally require the vendor to supply the analyzer. The general procedure for troubleshooting communication circuits is as follows:

(1) Determine that the communication signal is present. Light-emitting diodes (LED) are many times provided, but one must be careful about what they indicate. Many times they indicate only highway traffic, not that a particular device is communicating. The presence of the signal must be verified at all points along the communication path.

(2) Determine whether the proper control signals are present. The proper control signals must be present at each device in the communication path.

(3) Use a line analyzer to decode the transmitted and received signals. Determine if any error codes or statuses are being transmitted.

PC software-based line analyzers provide a cost-effective line analyzer if you already have a PC. These packages start at about $250 and have a wide range of features.

For RS-232 circuits, the most common problems upon start-up are (1) the transmit and receive lines are crossed; (2) the control lines 4 and 5 and/or 6, 8, 20 are not correct; (3) baud rate is wrong; (4) parity is set wrong; or (5) the stop/start bits are wrong. Documentation of the wiring and identification of device type (i.e., data terminal equipment (DTE) or data communication equipment (DCE)) is very important because while RS-232C is a standard, it does not define specifically what all the pins are, and some vendors have used some of the pins in nonstandard connections. After start-up, the problems are typically hardware-based or possibly due to change in port communication parameters because of resets/reboots or perhaps outside interference. Records should be kept of all communication parameters for each device in the communication loop. Noise also can be a problem, and slowing down the baud rate may help.

Multidrop communication systems can present more difficult troubleshooting problems. Modules that are talkers or load down the system, ground loops, and module addressing are some of the problems seen in this type of system. Removal or isolation of modules is one of the troubleshooting methods used to isolate these problems.

Ground loops are a common problem. In general, it is good practice to use isolating modems or converters. This is critically important if the ends of the circuits are in different buildings. The use of fiber optic links can greatly reduce this problem.

Personal Computer Maintenance

Driven by the development of low-cost software and greater capability at ever-decreasing hardware costs, the application of personal computer systems in the process industries continues to expand at a rapid pace. Systems have evolved far beyond typical transaction-driven, commercially oriented software applications. Many applications now place these systems in on-line, real-time control applications. The development of light-weight, portable lap top units holds promise that these units will become a valuable maintenance tool for the instrument technician. Current applications include their substitution for manufacturer-specific PLC programming units.

Like PLCs, most desktop systems share common functionality and architecture. A typical system consists of a central processing unit, a keyboard, monitor, and a printer. Central processing units are typically packaged with a hard disk drive and one or two floppy disk drives. Keyboards are typically 101-key enhanced units providing 10 to 12 special-purpose function keys. Monitors are available in a wide variety of sizes, resolutions, and color capabilities. Laser and ink-jet printers are the predominant choice for most systems.

The internal layouts for most desktop (or tower design) central processors are functionally similar. The typical unit contains a power supply, a mounting rack for the disk drives, and a motherboard that contains the microprocessor and system control chips, a socket for a coprocessor or overdrive chip, the base memory, and a bus structure with connectors for several expansion boards. Units are typically furnished with the disk controller and a parallel/serial interface card installed. The expansion slots are used for additional parallel/serial interfaces, extended memory, bus mouse, and special application interfaces. Repair by replacement is recommended for failed power supplies, disk controllers and drives, expansion slot interface boards, and chips provided with sockets. Outright replacement or repair by a factory authorized service center is recommended for monitors.

Since keyboards involve moving parts, they frequently require minor maintenance. Construction is generally quite simple, allowing removal of the rear cover and the printed circuit board. A soft cloth or small vacuum should be used to pick up the dust and lint. Solvents, available from several manufacturers, can be used to clean the keys and contact surfaces. Care must be used in replacing the circuit board to avoid distortion and check to make sure the keys operate smoothly without sticking. The low cost of these units encourages some users to replace rather than repair them.

Repair by replacement is the only viable maintenance option for a failed floppy or a hard disk drive. Preventive maintenance for these units will extend their useful service life, however. Minute amounts of dust, lint, or magnetic oxide particles can cause a number of problems with floppy disk drives. Head cleaning kits are available and will keep the drive heads free of such contaminants. These kits consist of a diskette with a soft absorbent surface and a cleaning fluid. Dispense enough of the cleaning fluid to saturate the exposed surface of the cleaning diskette and insert the diskette in the drive.

This procedure should normally be used once a week, but more often if usage is heavy or if dust is a problem.

Preventive maintenance can significantly extend the life of hard disk drives. Alignment drift and repeated start-up of the read-write heads over data areas eventually lead to deterioration of the drive's low level format. Utility software allows the user to refresh the low-level format before failures occur. Such software also verifies the operation of the disk controller, disk caching, and the main memory. A low-level scrub of the disk surface can take several hours, so it is best to run this type of software overnight. A "park" utility provided with the software can be used to ensure that the read-write heads do not begin operation over an area that contains data. If an apparent failure is experienced, try cleaning and reseating the ribbon cable that connects the drive and the controller board. Many apparent drive failures are due to faulty connections. If repeated failures of this type occur, replace the cable and make sure the unit is not exposed to vibration.

16

Software and Network Maintenance

Overview

In today's computer-oriented society, most control modernization projects take advantage of the new and expanding capabilities of microprocessor-based digital control systems [DiDCs, digital distributed control systems, also called distributed control systems (DCSs) and programmable logic controllers (PLCs)]. Because these systems are capable of performing both regulatory and supervisory controls, plant personnel have had to learn how to maintain computer-controlled systems. Because these systems are more software-oriented than hardware-oriented, maintenance of these systems requires a knowledge of software configuration and programming.

Configuration and programming tell a computer how it should sequentially process data to accomplish a given goal. Configuration and programming also define the resultant list of instructions, which a computer interprets to guide its processing actions.

A computer term to remember is GIGO, which stands for Garbage In, Garbage Out. This is to remind computer users that the machines are only as useful as the programs people write and the information used in the computers.

Remember from basic computer courses: The processor reads an instruction from memory. The instruction describes an operation (command) and usually an address to find or put the data used. The set or list of instructions that are created to accomplish some specific objective is called a program. The act of coding, entering, keying, or punching the program into the computer is called *programming*. The program and, in fact, all of the instructions that command the processor operation are part of the system's software.

To most instrument and control systems engineers, configuration and programming have additional meanings. These definitions have grown out of the evolution of industrial control systems over the years. Programming generally refers to the development of custom code in which the engineer builds a program, utilizing a computer programming language. Programming is also a term used to describe the engineer's work with PLCs to create the logic flow sequence. The language used is commonly called ladder logic.

Configuration, on the other hand, implies the engineer is working with a DCS or some other system in which the functional operation is described by a linking of preprogrammed tasks that have been specifically developed for industrial control applications. Configuration implies a programming exercise in which the user is the subject matter expert (process control engineer), not a computer scientist or programmer analyst.

The application of computers to the industrial control industries stemmed from the need to advance control capabilities. This need was defined by economic, quality, and safety justifications for better control and automation and the benefit of less costly installation.

The application of computers also means that the scope of the maintenance skills required to keep these systems running must include the ability to understand and work with the system's software configuration and program. The definition and functionality of the configuration or program is contained in the documents that are produced for the design, implementation, and maintenance of the system.

Operating System Software

At the base of the software hierarchy is the computer's operating system. The operating system (OS) is the software that allows the user to manage the computer and its resources. There are many operating systems, most of which were developed by the hardware manufacturers. By far, the most widely used is microsoft's Windows. Another operating system, UNIX, has been a favorite with software developers for a number of years. A UNIX variation, Linux, is also gaining popularity.

Computer Operating Environment

The electronic components of digital computers are designed to be operated by electrical current that meets certain specifications and under certain conditions of temperature, humidity, and cleanliness. Failure to respect these conditions may cause components to fail or function erratically.

General-purpose computers are intended for use in homes and offices, where the environment offers few threats to electronic components. Their design, therefore, provides minimal protection against adverse conditions. One way to increase the level of protection is to change the design of the computer to incorporate more rugged components. An extreme example of this is the "battle hardening" of computers in military equipment. Water- and dust-tight cases, shock-absorbent mountings, high-temperature electronics, and other such measures can improve the odds of the computer's survival in a hostile environment.

This approach, however, entails a substantial increase in cost, especially since solving some environmental problems may create others. For example, sealing up the outer case of the computer to exclude water, dirt, and harmful vapors will also cut off air circulation, resulting in excessively high temperatures inside unless large heat sinks and cooling systems are added.

For this reason, many industrial computer installations use the approach of enclosing unmodified general-purpose computers in a protective room where an environment suitable for the computers can be maintained. However, the latest distributed control systems are designed for use in most industrial environments.

It should be noted that for programmable logic controllers (PLCs) the situation is somewhat different. PLCs were designed from the outset to function on the plant floor, and their environment requirements are, therefore, less stringent. This does not mean, however, that PLCs do not need any protection. They are vulnerable to the same environment threats as general-purpose computers, only somewhat less so.

This section discusses specific environmental threats to computers and how to protect against them.

Dust and Dirt

Dust and dirt will quickly form a blanket over all accessible surfaces in a computer. The layer of dirt may prevent physical contacts from closing (as in keyboards) or may clog ventilation and other openings. If the dust is conductive, it may cause short circuits; if insulating, it may lead to heat buildup and component failures.

The obvious preventive for dust infiltration is to minimize the number of openings through which dust can enter. This approach is taken in many PLCs and industrially hardened computers, although it may create heat dissipation problems. The other approach is to filter the air, either at the inlet to the computer room or as it goes in through particular openings in the computer case.

This method works well if filter media are appropriate to the contaminants to be guarded against and are changed regularly. Another simple though possibly unpopular measure is to ban smoking in the vicinity of computers.

Chemical Vapors

Chemical vapors, if corrosive, can eat away at contacts and conductive traces inside a computer, eventually breaking their electrical continuity. They can also damage data recording surfaces on disks.

The usual defense against chemical vapors is either to make the computer or the control room tight against the vapors in question or to install appropriate filters. It may be necessary to set up the computer room at a distance from sources of chemical vapors to ensure they remain at a sufficiently low concentration.

Some graphics software will fail to work with a display adapter card that does not include a certain minimum amount of onboard memory. Try adding more memory to a card if problems arise.

Water

Water can get into a computer in two ways: by direct splash (installation near process equipment using water or accidental spillage) and by condensation. Obviously, water can cause catastrophic short-circuiting inside a computer, and a chronic water problem can also lead to corrosion and failure of components. The operating specifications of most computers include a humidity range within which the computer is designed to function.

Designing an entire computer to be watertight is difficult because of heat dissipation requirements, but watertight keyboards or key pads are frequently used on the plant floor and are connected to a computer nearby in a protected room. Keeping coffee cups and other sources of non-process liquids at a safe distance from computer is a sensible preventive measure.

Condensation is usually the result of low temperature or extremely high humidity in the ambient air. If the problem is local, a heater may be used to raise the temperature, which lowers the relative humidity for a given air moisture content per unit volume. In an environmentally controlled room, a humidifier or dehumidifier may also be used. An air conditioning system will control both temperature and humidity.

Vibration

The vibration that often accompanies operating process equipment can break contacts and solder joints inside computers and cause malfunctions and data loss in disk drives. The simplest remedy is to isolate the computer from the vibration source, either by moving it away from the vibrating equipment or by mounting it on a shock-absorbent base.

Electrical Disturbances

Electronic computer components require a relatively unvarying supply of low-voltage, low-current electrical power. Standard distribution voltages (110-120 volts or greater) must be stepped down to supply these components, usually by a stepdown transformer (often called the computer power supply), which outputs the voltage used on the system bus (typically 5 volts). Further voltage reductions may be required for specific components.

One common source of voltage and current transients is a "dirty" plant electrical supply. Isolation transformers and other power conditioning devices will prevent most power line instabilities from reaching your computer.

VOLTAGE AND CURRENT TRANSIENTS

An overvoltage or overcurrent condition, if severe or prolonged, will simply burn out electronic components. Less severe transients may be interpreted as a spurious change of state, causing erratic program function or corruption of data. With too low an initial voltage, the computer will simply fail to function, because the voltage will be insufficient to raise electronic devices above their threshold values. A severe voltage drop while the computer is in operation may trigger an unplanned restart (warm boot). The most serious condition is fluctuating low voltages, which may cause repeated partial restarts, leaving memory and data in unpredictable states and possibly hanging up the computer altogether. The effects of a complete power loss are actually less drastic, except that the contents of volatile RAM are lost and the contents of a file may be corrupted if a write operation is in progress when the power loss occurs.

INTERFERENCE

Noise in the form of radio frequency interference (RFI) or electromagnetic interference (EMI) can be harder to track down and correct. Radio frequency interference arises most often from lengths of cable acting as antennae or from mobile communications sources in the plant such as walkie-talkies and lift truck radios. Electromagnetic interference can come from any piece of equipment that generates an electrical or magnetic field (most commonly, motors). Do not neglect to check possible temporary sources of interference; arc welding is a notorious culprit. Note also that computer components such as cables, long communication lines, or disk drive motors may themselves generate interference. Normally, however, computers are required to meet government specifications for RFI and EMI emissions so that they do not interfere with the operation of other electronic equipment.

The effect of electrical interference is similar to that of fast transients from the power line: the generation of spurious signals that the computer mistakes for data. These may cause erratic operation, or in extreme cases, total computer failure.

The fundamental remedy for interference is to move the computer away from the source. This may not be easy, especially if the source is the machine the computer is controlling or if the source is difficult to identify. If either is the case, careful shielding and grounding of cables and cabinetry is the best defense. Pay special attention to cable penetrations where ambient electrical noise can leak in. Use of noise-resistant technology such as fiber optic cables also make for a more robust computer system.

POWER FAILURE

If the continued operation of a computer is critically important, an uninterruptible power supply (UPS) should be added to the incoming power line. During normal operation, the line power charges a battery, which, in turn, feeds the power supply and the computer. In case of a blackout, the power stored in the battery can maintain the computer in operation for a limited time, typically one to several hours. If all that is needed is time to copy the contents of memory to disk and shut down the computer in an orderly fashion, twenty minutes to half an hour should be enough; otherwise the cost must be balanced against the importance of the process and the normal length of power outages. The wattage rating of the UPS should be sufficient to support the memory, display, and disk drives (typically the biggest power consumer).

STATIC ELECTRICITY

When working on electrical or electronic components of a computer, take care to avoid connecting devices to inappropriate power sources. A serious but not as obvious threat is static electricity.

A static discharge is typically low current which can reach extremely high voltages that can damage computer equipment. Components such as memory chips and integrated circuit boards are typically shipped in static-proof bags to protect them from accidental discharges during shipment. Once these components are installed, they are protected by the computer's grounding subsystem. During installation, however, they are vulnerable, and personnel should take appropriate precautions, such as wearing ground straps and static-free clothing and using anti-static mats. Maintaining adequate humidity levels helps control static electricity.

Heat

Electronic components are designed to work within an operating temperature range that is stated in their specifications. Outside this range, their characteristics change beyond design limits, their behavior becomes erratic, and their life is shortened. In extreme cases, a fire hazard may be created.

The usual method of cooling computer components is by convention to the surrounding air, often assisted by a fan that draws air over the components. Other devices such as monitors (screens) have no fan and cool themselves only by convection. For this method to work, the surrounding air has to be substantially cooler than the components, which is one reason computer rooms are often air-conditioned. Computers have vents to draw cooling air in and out, which must be left unblocked even if they seem to provide a nice shelf for manuals, drawings, and bag lunches.

If convection alone cannot maintain the components within their specified operating range, components with a higher temperature tolerance must be substituted, or a refrigeration system must be added to the computer. Both of these solutions entail substantial extra cost.

Network Operations

Networks by their nature enable computing systems to reach out to the wider world to obtain and provide data. However, this very accessibility carries with it risks and vulnerabilities. File corruption, intruders, viruses, and unauthorized use of computing resources are only some of the threats a network must deal with. In addition, networks are large and complex systems, which require ongoing troubleshooting and maintenance.

SECURITY

As corporate networks become more and more connected to the outside world, they provide companies not only with a way to get information out to their employees, customers, and suppliers, but an open door that allows others to get in. through this open door, competitors can steal proprietary information, corrupt vital databases, and cripple computer facilities, seriously damaging a company's operations, competitive position, and reputation. This threat is added to the existing challenge companies face to protect internal networks against unauthorized access and use.

ISOLATION

For critically important operations, the safest solution may be physical isolation, that is, disconnecting a particular computer from the network and/or physically preventing certain operations (e.g. a computer without a keyboard, screen, or diskette drive). However, this approach also makes data acquisition, maintenance, and troubleshooting more difficult.

FILE SECURITY

As soon as two or more people can use the same file or device, the possibility arises that they may try to use it at the same time. In the case of disk files, a common solution is to "lock out" one user until the other has finished. This locking may be done at the volume, file, or record level. With volume locking, no other user can get at any file on the same disk volume until the first user has finished. File locking prevents other users from getting at the file that the first user is accessing; while record locking prevents access only to the record the first user is working on. Clearly, the smaller the unit that is locked, the less the users' free access to data is compromised. Managing the locking and unlocking, however, becomes more complicated.

With several users, it becomes more important to control which files each can use and how. It may not be desirable for everyone to have access to payroll or personnel files. On the other hand, it may be desirable for them to be able to look at, say, process data histories or download them to their spreadsheets without being able to change the master copy. Most operating systems include some capability for controlling access to files (setting a read-only bit in the operating system file description, for example). More sophisticated systems can restrict access by individuals, groups, or job functions and allow administrators to decide whether each user may look at the file, make a copy of it, write to it, or delete it.

FIREWALLS

A *firewall* scrutinizes all transmissions going into or out of a corporate network and blocks or quarantines any questionable ones. The most common kind of firewall is a *filtering router*, which looks at data packets as they go by and passes or blocks them depending on the address or information in the packet header. This is not as simple as it sounds, in that the filtering software has to deal with a variety of messaging protocols, including FTP (used for Internet file transfers), DNS (Domain Name System), and X11 (Unix windowing protocol). Your firewall may leave you unable to use one or more of these. Anecdotal evidence suggests that files that are attached to electronic mail messages often get corrupted or disappear as they pass through firewalls.

A *host-based firewall* or *bastion host* is a computer that is dedicated to firewall functions, which provides more sophisticated capabilities than a filtering router. A bastion host normally contains information about the applications in use on the system and the nature of the files they generate. This permits more sophisticated screening than a system that is one based just on addresses and headers. Since general-purpose application programs are not designed with security as a high priority, firewalls often use their own secure versions with stripped-down functions, known as *service proxies.* Of course, this bastion host computer is a major point of vulnerability in your network and must itself be well protected.

Firewalls slow down traffic into or out of your network because of the extra processing they do on each and every data packet. If these delays become too severe or the other restrictions on traffic too burdensome, users may search for alternative means such as modem connections that bypass the firewall. This, of course, defeats the original security objectives.

VIRUSES

A computer virus is a malicious piece of executable code that is intentionally imported into a computer or a network and usually damages some aspect of the computer's or the network's operation. The virus may damage files, directories, other software, system configuration information, or any or all of the above. It may also make unauthorized use of information such as password files or email address books in order to gain access to other computers. In extreme cases, a computer system may need to be shut down and reconfigured from scratch. One of the major tasks of firewalls and other network security measures is to protect against viruses.

Most virus code is written by eager, ethically deficient programmers for whom breaking into other people's computer systems is an intellectual challenge. Fortunately, a virus can usually be recognized by characteristic patterns in its code or its behavior. Commercial software is available that detects and neutralizes incoming viruses and can be regularly updated as new viruses are developed.

Virus detection has often depended on the fact that a virus had to be an executable file with an *.exe, *.vbs, *.pif, or *.com extension. More recently, however, macro viruses have taken advantage of the ability of certain programs (mainly Microsoft Word and Excel) to include small sets of executable instructions (macro viruses) in a document or spreadsheet. A virus masked as one of these instructions sets as opposed to a separate file is much more difficult to detect and remove.

One of the most effective types of virus protection is user education. In a plant environment, the consequences of a virus infection are not only personal inconvenience and possible data loss, but also potential damage to the computer systems that run an operating plant. The safety and productivity issues this raises should give anyone pause.

NETWORK MAINTENANCE

Every network needs a person—a network administrator—whose primary responsibility is to keep the network up and running, even if that role is less than full-time for a smaller network. Needless to day, a sufficient portion of this individual's available working time must be allocated to network-related duties. Networks are sufficiently demanding of attention that no one should be expected to "fit them in" around and above other job demands.

Like all large complex systems, local area networks (LANs) have their share of problems, and it is vital that someone be able to solve them quickly. Once you have a network, you quickly become dependent on it. After all, much of your software and data as well as such services as printers and electronic mail will be accessible only via the network. When an industrial process is involved, the efficient operation of the process or the accuracy of the process information may be compromised if the network goes down. In most of the process control and factory automation systems installed in the past few years, the automation network is vital to the plant's operation. While a good control system design allows a plant to operate through a brief network outage, during prolonged network outages continued operation must not be left unsupervised by human operators. Clear installation and troubleshooting instructions and good vendor support should be high priorities.

Simple overloading is a prime suspect in frequently occurring network breakdowns or poor performance. Storage gets crowded and fragmented, buffers overflow, or frequent collisions and repeated transmissions clog the network. The preventive or cure for this is to select the right network, install it with guidance from the supplier, get plenty of capacity for present needs, and leave an easy expansion path open for future needs.

Diagnosing network problems is often difficult since so many components interact. Networks' error-recovery capabilities actually make diagnosis harder because in most cases transmissions still get through, although more slowly, so there is no clean break to locate and fix. In fact, today's networks can be so fast that the slowdown caused by a chronic problem may go unnoticed, especially when the network is lightly loaded—only to become problematic during peak demands.

Open Systems

Early computer installations were self-contained. If computers were networked at all, it was with other equipment from the same manufacturer, via proprietary interfaces. In the plant, isolated, single-purpose computers monitored or controlled "dumb" equipment. The option of communicating with other devices or computers was not an issue. Early networks encompassing a process area or department were usually purchased from a single supplier and were operated separate from other plant computer systems. The smooth functioning of the network was the responsibility of the supplier, and again connectivity with outside devices or computer systems was not a requirement.

However, networks soon grew to attain plant or corporate scale. Inexpensive computer hardware (PCs) meant a computer on every desk and in every corner of the plant, from various manufacturers, each with its own inventory of add-on hardware and software. And programmable logic controllers (PLCs) and "smart" instruments became more and more like specialized computers with their own interfacing needs.

In this situation, incompatibilities among different manufacturers' hardware and software started to become troublesome. Companies wanted to take advantage of new technology, yet, could not be sure it would work with their present systems. They could not meet their expanding and changing needs within existing proprietary systems, but also could not afford to have a custom solution built from scratch or spend the time and money needed to resolve interfacing and communication problems.

The open systems concept was an attempt to solve these problems by defining a system architecture that could seamlessly incorporate products from different manufacturers. The concept was initially used by Sun Microsystems in the mid 1980s as a marketing strategy to differentiate Sun from its mainframe and workstation competitors, which allegedly sold "closed" systems. The strategy proved so successful that other computer manufacturers quickly moved in the same direction. Large computer users in business, government, and defense quickly realized how much money they could save and flexibility they could gain by moving to open systems, so they started to specify interface standards to which their computer suppliers had to adhere.

The essence of "open systems" is expressed in this definition from the Software Engineering Institute of Carnegie-Mellon University:

"An open system is a collection of interacting software, hardware, and human components

- Designed to satisfy stated needs
- With interface specifications of its components that are
  - Fully defined
  - Available to the public
  - Maintained according to group consensus
- In which the implementations of the components conform to the interface specifications."

The important features of open systems are the interface (should be well-defined, widely used, nonproprietary), the role of standards and the systems' conformance to them, and the provision made for upgrading and expanding with minimal impact on the existing system. These translate into the anticipated business benefits of compatibility, flexibility, and cost reduction. A 1996 survey identified the following benefits as being important to a group of computer system buyers in the United States, United Kingdom, and Europe (Table 16-1) (The Open Group, 1998):.

Table 16-1. Open Systems Benefits Cited by Computer System Buyers

Benefits of Open Systems	% Citing
Flexibility	70
Freedom to choose technology from different vendors	67
Products from different vendors work together	66
Access across multi-vendor environments	65
Protect investment in existing computer system	61
Ability to use/share information anywhere in the world	59
Cost savings	55
Interoperability/portability across various platforms	54
Organizational change no constrained by IT system	49
Cost of ownership	49

One way that companies often achieve these benefits is by making greater use of COTS (commercial off-the-shelf) products rather than proprietary system components. Experience has shown that in many cases the commercial product is "good enough," if not perfect, and that it offers the advantages of reduced cost, widespread product support, and quick implementation when new technology and upgrades are released. However, COTS purchases impose the additional work of exploring the marketplace and evaluating the options available before each component is purchased

Interface and protocol standards are central to the success of open systems. There are a number of applicable standards, often either operating-system specific or emanating from different standards bodies. Among the most important are Motif (the Unix graphical user interface), IEEE POSIX (Portable Operating System Interface), IEEE 802 LAN, and Java (the programming language from Sun Microsystems).

As important as de jure (legal) standards are, de facto (in fact) standards based on common widespread use are even more important. Microsoft has established many of the de facto standards of the computer and network industries. The Windows operating system and its supporting MFC (Microsoft Foundation Class) libraries are used universally in the PC industry. This includes Microsoft's basic COM (Common Object Model) and DCOM (Distributed COM) technologies, which are widely used in industrial communications applications. Both are at the core of OPC (Object linking and embedding for Process Control), which is the most widely used specification for industrial automation communications and enables systems built by a wide variety of suppliers to seamlessly interoperate.

One of the challenges in achieving true "openness" is ensuring that all manufacturers interpret specifications in the same way so all their products do in fact work together. This implies some kind of conformance testing by a neutral third party that can provide believable certification that a product adheres to the standard. For open business systems, this role has been played by a consortium called The Open Group (www.opengroup.org). This consortium licenses and promotes The Open Brand (formerly X/Open), which is essentially a guarantee offered by the vendor to the customer that the product conforms to open-systems specifications and that the vendor will correct any nonconformities.

Within industrial automation, the conformance testing for OPC is left to the suppliers and a few independent testing labs. However, OPC is so broadly accepted that it is found on practically all systems whose components are supplied by different vendors.

Industrial Networks

The primary function of an industrial local area network (ILAN) is to serve as a data highway that connects sensors, controllers, computers, and other electronic devices in an industrial plant. ILANs also support the movement of data that allows human operators and plant engineers both to query and change the operation of these plant devices and to retrieve data for use in business or engineering applications.

High-level ILANs are high-speed networks because of the data transfer rates required. They must also provide a high degree of reliability and equipment redundancy. Lower-level networks, with a minimum of extra features, are also available that are specially designed for the efficient movement of real-time sensor and actuator data.

EVOLUTION OF INDUSTRIAL NETWORKS

Process control was initially conceived as an automation technique to relieve human process operators from constantly having to adjust control valve position in response to changing process conditions. The first of these control devices were mechanical. A transition gradually occurred from mechanical control to pneumatic control to electronic analog control to digital continuous control and finally to distributed control systems (DCS). DCS represented the transition from very early computer control or DDC (Direct Digital Control), which grouped the control of many loops together so as to save money and add control flexibility. As microprocessors became smaller, more capable, and more robust, they were incorporated into process devices to perform functions such as temperature compensation, calibration checks, and single-loop control.

The DCS achieved economy by sharing its computations across many control loops. However, the capacity of the early microprocessor limited this sharing to a relatively small number of loops (typically, 16 to 64). As a result, most real processes had to use many controllers. Uniting these controllers and the CRT-based graphic operator stations was the task of the ILAN, which for early DCSs was usually a proprietary communications link.

Initially, DCSs were entirely proprietary, a complete package of hardware, software, and networks. This led to problems when process-area DCSs in different parts of the plant—often manufactured by different suppliers and installed at different times—tried to exchange data and coordinate their activities. With the increasing power and diversity of the devices available, engineering also found that the limitations of the proprietary DCS network prevented them from making effective use of new technology. Today, control systems must be built for change, allowing control loops to be easily rearranged and devices to be easily replaced and reconfigured. The closed systems of the past, locked into a particular manufacturer's products, are a liability in this new age.

These developments have led to a trend in industrial networks that is analogous to the open-systems movement. In any case, as industrial networks move toward digital instrumentation and control, they are becoming more and more like computer networks anyway. A number of industrial network protocols have been developed in recent years, each of which represents its own tradeoff between technical sophistication, robustness, flexibility, "openness," and cost.

21st Century Maintenance Technology

To meet global competition, industrial management is now forced to continuously reassess manufacturing processes to meet the objectives of round-the-clock production, high quality products and high levels of manufacturing flexibility. Additional major requirements include ensuring plant safety and environmental protection. These goals must be achieved while at the same time reducing manufacturing costs. Each of these objectives is directly affected by the maintenance strategies applied.

The blend of skilled personnel and the application of maintenance technology continues to evolve. This evolution has moved significantly beyond on-site analysis and correction of process system errors toward the practical application of the concepts inherent in the science of maintenance, including maintainability as measured by MTTR (Mean Time To Repair) and most importantly, reliability associated with MTBF (Mean Time Between Failure). The latter is associated with predictive maintenance and root cause analysis, i.e., identifying potential problems before failure and if possible, establishing root cause to ensure complete elimination. Beyond the use of these as maintenance technologies is the movement toward applying the same techniques to enhance process performance.

To meet these objectives, maintenance technology has made significant advancements during the last decade. Advances in secured remote support capability allows for more effective use of immediate diagnostic support and predictive analysis, pro-active self-monitoring and alarm analysis with dial-back capability, and remote assistance (the vendor has the ability to see and use the customer's unique system displays) including web-enabled information distribution. The application of these technologies facilitates the partnership of responsible plant personnel with correct consultative resources regardless of subject and resource location.

In addition to the adoption of these tools to maintain system availability, the same technologies can be applied to enhance actual process performance and optimization. This is accomplished using the remote connection with virtual workgroups (experts connected electronically). The process system is the gateway to more extensive plant-wide evaluation. Finally, these tools could then be adopted to allow plant-wide asset availability management.

Given the objectives stated, an integrated support strategy must be developed. Since industrial customers are often located in remote or isolated areas, providing effective support in a timely manner has always presented a major challenge. This requires the design of a very structured support system that allows for the rapid deployment of support resources to:

- Improve response time and quality of interactions.
- Reduce travel to customer sites.
- Cooperate and share knowledge in a highly cohesive manner.
- Apply the most qualified resource, regardless of geographic location through virtual workgroups.
- Ensure project schedules are met through timely assistance.

Structured Global Connection

Progressive vendors have developed state-of-the-art technology to remotely connect the experts to customers located around the globe.

The use of TCP/IP network strategy takes advantage of the off-the-shelf routers and sophisticated firewall technologies. This innovative approach provides the ability to create a virtual support organization that securely links multiple performance centers together. This enables delivery of uninterrupted, consistent quality service to all customers around the world.

Given the flexibility that this technology allows, the very best people anywhere in the world working in maintenance support, advanced application, or economic performance consultation can be connected in seconds. Working in virtual workgroups, efficiency is gained and costs controlled by not having to fly people from one location to another. The result is an enabling vehicle allowing the leveraging of global expertise to access remote client systems anywhere in the world through secure network connection.

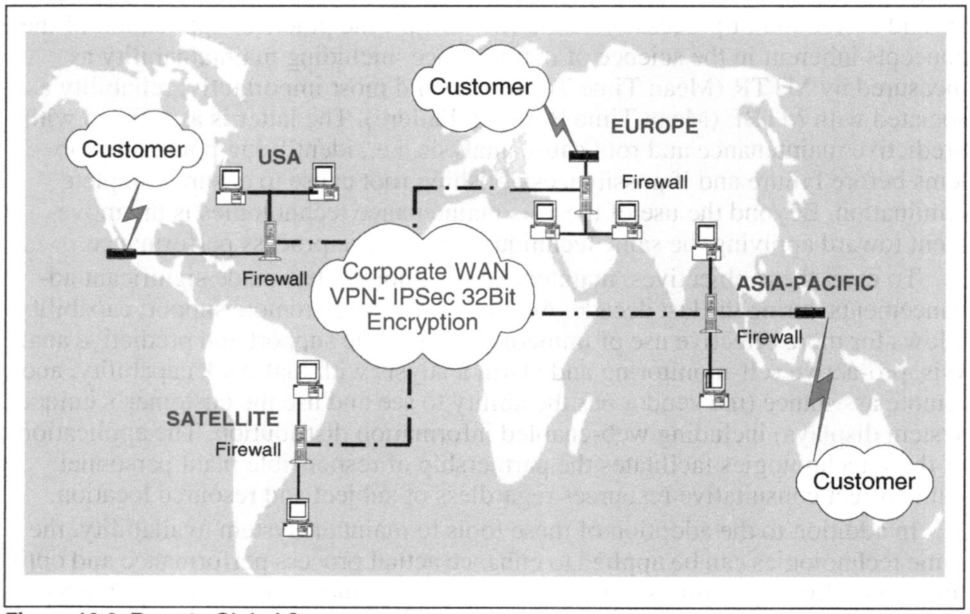

Figure 16-2. Remote Global Support.

Security is a major key to success. Customers require safe, secure and reliable operation of their process control systems during the remote session. The guidelines should include:

- Customer authorization required for all connections. All remote log-on sessions follow strict procedures to ensure no actions are taken that might affect the ongoing operation of the connected system.
- The vendor's remote center is a locked facility accessible only to authorized personnel who remotely access the customer's system.
- The remote support engineers should undergo specialized training in remote analysis techniques. In addition, they should be knowledgeable in general plant operations to help ensure plant safety procedures are followed. Advanced application experts can use the same techniques. This greatly expands the breadth of support capability that can be applied.
- All communications between the customer system and remote center pass through a secure, virtual private network (VPN) environment, using the highest exportable encryption algorithms.
- Access to the server database containing sensitive information such as passwords and system configurations is restricted to workstations vali-

dated as members of the domain structure. The database is password-protected and all passwords meet strict criteria.

- Each remote center operates with a firewall configured to allow only predefined networks access to customer sites. This firewall filters traffic to allow only designated protocols and applications to be communicated among its interfaces.
- Video displays are recorded and all keystrokes are logged and archived. A record of all actions taken can be provided upon request.
- Once a log-on session is mutually agreed upon between vendor and site personnel, a call-back connection is initiated as follows:

  (1) The remote center phones the customer's on-site system to give the correct password authorization.

  (2) The connectivity software loaded on the customer's system receives the call, breaks the connection and reconnects to the vendor's remote center system.

  (3) Only then does the flow of data actually begin between the customer's system and the vendor's remote center system.

Remote Diagnosis

Once the connection is established, software-embedded diagnostic tools resident in a dedicated maintenance module can move maintenance effectiveness to a very high level.

When some users encounter a problem, they first try to resolve it on their own. If they fail to fix the problem, they call the vendor of the process control system and try to solve the problem by discussing it over the telephone. The use of remote monitoring, diagnostics and maintenance of the process control system is a faster and better way of accomplishing the same process.

Typically, a fully featured process control system consists of multiple-bus structures and many components to support the distributed system architecture. Various computers, controllers and system stations or other modules with specific responsibilities communicate with each other via different types of networks.

Making sure all these modules work in harmony is "timing logic" that controls both the overall system's own operations—such as sending messages, accessing hard drives — and the control strategies. Network-related problems can be diagnosed by allowing information packets to be filtered and captured for analysis. As a result, the entire process — from failure identification to problem correction — takes on a new perspective. At the heart of remote support are high-level technical specialists who provide realtime analysis of developing problems.

In effect, remote support actually gives plant personnel greater control over how and when equipment service is scheduled and performed. For instance, when plant personnel first encounter a problem, the remote support function acts as an online diagnostic tool to provide immediate analysis. It logs information relevant to the problem and retains that information for problem resolution. Remote support personnel at the support facility can immediately use this information and initiate appropriate tests.

Once the problem is identified and severity assessed, a resolution plan can be developed with activates scheduled as appropriate. These activities may include actual remote resolution assistance.

Predictive Analysis Using Expert System Technology

Remote diagnostics can be enhanced with a second service, one that can actually predict and circumvent a potential problem before it interferes with the system's availability. This capability gives plant management the option of deferring a repair action until it can be scheduled at an appropriate time. This is a powerful concept that enhances the reliability of the system.

The need to administer a system takes on a special significance if the equipment is to work at peak efficiency throughout its product lifecycle. Over the years, the individuals working on and maintaining the equipment may also change.

Further, as the size of a network-based system grows over time, it becomes increasingly important to monitor the performance of all the various pieces of equipment. This is because degradation in the performance of one component can affect other components on the network.

With predictive analysis, however, users can monitor the overall system and gradually build up a performance profile that can serve as a relative indicator of its "health."

For example, each system component may be monitored to ensure that it is running at the correct revision level and to check that the error levels are within acceptable limits. Each bus is analyzed to ensure that it is operating within specifications, and various critical files are diagnosed to detect corrupted, duplicated or misplaced data.

The idle time and memory availability of the central processing unit (CPU) that runs the various applications, and the work load of the control processor, are among the key capacity measurements that are tracked. Such measurements can quickly become complex, since a single process controller can consist of several different subsystems. Also, the devices that are most likely to cause problems are identified and monitored closely to keep the overall system's performance from degrading.

In fact, plant personnel can remain in control of the total system only when they are able to factually pinpoint the source of the trouble. The ability to identify and eliminate potential problems obviously enhances plant safety and reduces environmental upsets.

Event-Activated Response

The reliability of the process control system can be increased via continuous self-monitoring for designated alarms. If a critical system event triggers an alarm, an automatic event-activated response function simultaneously notifies designated customer operations personnel and the vendor remote support center. Once connected, the system uploads all the data related to the alarm condition for analysis by remote support engineers.

Process systems and their associated problems are too complex to handle individually and manually. The task of predicting and managing failures become more manageable when the system itself begins the diagnostic process. Alert notification begins with vendor-supplied software that continuously listens to the system and initiates an "event-activated response" when a process abnormality appears.

When such problems occur, the plant's technical personnel are alerted via the system's alarm flashing on their workstations. A message describing the alarm type will usually accompany the alarm. At the same time, a program in the remote-monitoring module initiates remote connection to the vendor's technical service center and uploads the particular alarm sequence. At the vendor site, service engineers begin analyzing the problem based on the incoming data. If appropriate an engineer will initiate a phone call to the plant site to confirm that there is a problem and initiate a process-monitoring session. All the information generated is stored for future reference.

This capability can also extend to performance monitoring. This response kicks into effect when certain types of process alarms are encountered. Immediately, all information connected with the system's performance is recorded, followed by data analysis to reveal the source of the problem.

Knowledge Base Web Access

The ability to research an extensive library of support documentation is invaluable because it enables the user to answer questions quickly to determine if an issue has been previously resolved. The Knowledge Base website delivers up-to-the minute technical reference information, news and on-line communication 24 hours a day, 365 days per year.

Topic categories may include software correction text files, helpful hints, Product Specification Sheets and special notices, among others. One asset of a technical knowledge base website may be access to downloadable software. Another function enabled by website technology is the ability to "push" pertinent news bulletins to registered website users.

Key word search methodology facilitates users' abilities to locate extensive technical support information on vendor products. Additionally, the website provides an efficient means to communicate with technical specialists as well as company staff.

If a product issue is experienced repeatedly or there is a desire to formally request an additional feature or enhancement to a product, a customer request mechanism can be instituted as the preferred channel of communication and documentation, again via the website.

The integrity of website documentation must be well-maintained by technical personnel. The day-to-day activities of support personnel can generate much of the technical information posted.

On-line Interactive Assistance

User assistance is a critical requirement to customers seeking help in all aspects of equipment utilization — software, hard drive and memory allocations, how to configure programs, initial setup and usage, and system and application engineering assistance. This new remote service capability revolutionizes customer support in this increasingly important area. The ability to view exactly what the user sees locally is a far superior support technique than blind telephone discussion. The sharing of remote display screens allows both user and vendor personnel to easily guide each other through complex multi-step procedures, leading to faster problem resolution.

Remote Performance Support

In today's competitive global marketplace, the appropriate utilization of sophisticated advanced process control technology can often provide a significant economic advantage. Advanced remote application support capabilities enable geographically distributed, virtual work teams to effectively support customers' advanced control applications to maximize lifecycle benefit.

Thus far, we have concentrated on using support technology to ensure the integration of system availability and accuracy of defined functions. Only after this is accomplished can the enhanced utilization of the system application be assured.

The same tools are again applied to the next level of remote performance support. After economic studies are completed and baselines established, remote monitoring is critical to ensure that the on-going benefits are maintained.

Since plant facilities are often located in remote areas, it has always been a major challenge for vendors to provide effective support in a timely manner to plant facilities that are often located in remote areas. This requires vendors to design a very structured support system using technology that rapidly deploys resources to:

- Improve response time and quality of interactions.
- Reduce travel to customer sites.
- Cooperate and share knowledge in a highly cohesive manner.
- Apply the most qualified resource regardless of geographic location through virtual workgroups.

In the future, the same technologies applied to support the process control system can also be applied through the same system to effect the optimization of other critical plant assets. The control system is connected to thousands of different pieces of equipment, ranging from motors, pumps, transformers, valves, piping, vessels and heat exchangers to more complex systems such as turbines.

While not every one agrees on a definition of "asset optimization," most everyone — users and vendors alike — seem to agree that optimizing production assets is key to enabling a company to meet its business objectives. Asset utilization enables a company to:

(1) Ensure that production equipment is available when needed.

(2) Balance existing product assets against demand for product(s).

(3) Reduce maintenance costs without negatively impacting equipment availability.

(4) Meet all regulatory requirements.

17

Safety

Overview

Safety is the most important thing to consider when doing any job. Failure to perform any job safely is a no-win situation for the technician and the company. While many safety practices are dependent on the hazards involved, certain general practices apply to all jobs. Eleven general safety practices are listed below.

(1) The job should be studied carefully to distinguish all the hazards present and assure that all the safeguards and safety equipment are provided for the protection of the technicians, other persons, and the equipment during the performance of the job.

(2) All safety devices, tools, gloves, and test equipment should be examined before they are used to make sure that they are in good shape and that they are adequate for the intended work. Appropriate clothes and safety apparel, such as safety glasses, face shields, gloves, etc., should be worn.

(3) The applicable safety regulations, standard operating procedures, standard maintenance procedures, facility practices, etc., should be known before the work is begun.

(4) Safeguards, such as danger signs, roped off adequate space, etc., should be provided as necessary to protect persons close to, but not engaged in, the particular work being done.

(5) Safety should be planned into each job. Orderliness, organization, and good housekeeping are essential to personnel safety.

(6) The result of each act must be considered. No maintenance person should perform an unsafe act to accomplish the work or do anything that would endanger any person.

(7) Adequate lighting and work space must be provided for the intended work.

(8) Each person assigned to a job should be satisfied that the working conditions are safe. The care exercised by others cannot always be relied on.

(9) Safety must be designed into an installation from the very beginning.

(10) The equipment waiting to be worked on should be double-checked to be sure it is the equipment that is supposed to be worked on.

(11) Safety must never be taken for granted.

Remember, safety is a state of mind coupled with conscious action and maintained by continuous vigilance.

Electrical Hazards

Electrical hazards to personnel come in two forms: electrical shock and physical damage resulting from the release of electrical energy.

An electrical shock is defined as a flow of electrical current from an external source through a living organism. When a person touches an electrically energized surface, an electrical shock occurs. The physical effect of the electrical shock depends on the physical flow of the electricity, the magnitude, and the duration of exposure. For example, if the physical flow of electricity is from the right index finger to the right elbow, the effect for the same magnitude of electrical current would be less than if the flow were from the right index finger to the left index finger, i.e., across the heart; or, for example, the effects of 1 mA (perception) and 80 mA (in the range that can cause ventricular fibrillation); or, for example, a current of 20 mA for 1 second may be safe whereas 20 mA for 60 seconds may be hazardous.

The physical path of electricity is determined by the point of contact with the energized circuit and the point of contact with the electrical return path, typically a contact with a ground.

The magnitude of the electrical shock current is determined by Ohm's law:

$$\text{Shock current} = \frac{\text{Contact voltage}}{\text{Body impedance}}$$

At power line frequencies (50–60 Hz), the body's resistance is primarily a function of surface contact resistance, i.e., skin resistance. The surface contact resistance is affected by the skin moisture, the skin contact, (direct skin) penetration, and the applied voltage. At power line frequencies, the body is essentially resistive, while at frequencies of 300 Hz and above, skin effect comes into play and the body AC impedance increases. In most industrial countries, the "safe" voltage is generally considered to be 30 volts rms. The National Electrical Code (NEC) 110.27 requires that guarding be used at 50 volts and above. The effects of electrical current in the human body are given in Table 17-1.

Table 17-1. The Effects of Electrical Current

Current	Effects
0–1 mA	Imperceptible
1–3 mA	Perceptible
3–5 mA	Annoyance
6–9 mA	"No-let-go"
30 mA	Asphyxiation
80 mA	Ventricular Fibrillation
5–up A	Cardiac arrest, burns

The duration of exposure also determines the effects of the electrical shock. The longer the exposure, the greater the risk of damage.

In dealing with electrical equipment, the following safety practices are suggested:

(1) All work should be done according to National Fire Protection Association codes (NFPA-70, The National Electrical Code (NEC); NFPA-70E – Standard for Electrical Safety Requirements for Employee Workplaces; NFPA-101, Life Safety Code; and the OSHA Code of Federal Regulations, Title 29, Chapter XVII, Part 1910, Subpart S, Electrical.

(2) The electrical equipment or circuit to be worked on should be double-checked to be sure it is the correct one.

(3) All electrical equipment and circuits, unless the equipment or circuit is to be worked on energized, shall be locked out and tagged prior to starting work. The equipment or circuit shall be verified to be de-energized prior to starting work. Lockout tags and locks shall remain in place on the isolation device until removed by the technician whose name appears on the tag. In the event that the technician is unavailable, it may be removed after a thorough investigation by the technician's supervisor. When it is necessary for the technician to transfer the job, the replacement's name shall be added to the tag and the replacement's lock put on the lockout. NFPA-70E, Standard for Electrical Safety Requirements for Employee Workplaces, details the proper lock procedure. Equipment or circuits that are to be worked on energized shall fall under formal facility practices for working on energized equipment.

The facility must have a formal lockout procedure per OSHA requirements. This type of procedure is detailed in NFPA-70E and CFR 29, Part 1910.

(4) For work other than testing, work on energized circuits below 150 volts to ground should be done by a minimum of two people. Work on energized circuits 150 volts and above should be done on a permit basis and in the presence of a supervisor. All work other than testing that is done on energized circuits above 50 volts should be on a permit basis. Work near energized circuits that represent a hazard should be considered work on energized circuits.

Two contract electricians were working without a permit on a 480-V AC motor control center installing a conduit. They had drilled a pilot hole for a hole punch in the end of the MCC. One electrician was on the end of the MCC, and the other electrician reached up into the MCC to attach the backside of the punch. This electrician contacted the 480-V AC bus. This electrician suffered extensive physical damage and burns; the other electrician was blown off the end of the MCC and was killed. If a proper permit had been applied for and proper supervision of contractors had occurred, this accident would have never happened. The moral of this story is that stupidity and incompetence can be terminal!

(5) If it is necessary to interrupt the work in progress, upon returning to work, all circuits shall be verified as being de-energized before work can proceed.

(6) After the work is complete, a complete check of the circuits and equipment must be made to verify that the work has been completed, that the connec-

tions are tight, all guarding devices are in place, and the wiring is properly installed before re-energizing the equipment or circuits. After the power has been applied, proper operation of the equipment should be verified.

(7) One should never stand in front of equipment being energized, but to the side.

Because instrument people typically work at low voltages, they sometimes are not aware of the tremendous amount of energy that can be released from an electrical circuit. However, considerable energy can be released from even a 120-V AC circuit. A tech was working on a pressure switch live that had 120-V AC on it. He inadvertently shorted the 120-V AC to ground and was surprised to find that the screwdriver he was using now had a melted tip. To melt steel, it is usually necessary to have temperatures in the 2000 degree range. What does 2000 degrees do to one's skin? The moral of the story is to have a healthy respect for all electricity for a longer, safer life.

Hazardous Areas

Electrical Hazardous Area Classification

In order to have a fire or explosion, the three components (the fire triangle, Figure 17-1) must be present: air, a flammable material in an ignitable mixture, and a source of ignition. As air is readily available (one leg of the triangle) and facilities that handle flammable or combustible materials can provide another leg of the fire triangle, the remaining leg (an ignition source) is all that is required to cause an explosion or fire. Electrical equipment could provide the required source of ignition to complete the triangle.

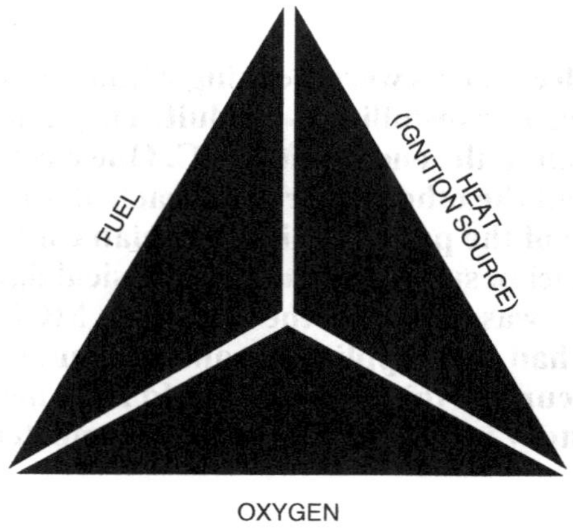

Figure 17-1. Fire and Explosion Components.

Since electrical equipment could be a source of ignition, methods that prevent the electrical equipment from serving as a source of ignition must be used to allow the safe placing of the equipment in areas where a flammable material could be present. The first requirement is to determine the nature, potential, and extent of the flammable material hazard.

Electrical hazardous area classification is a method used in the National Electrical Code (NEC) to classify area hazards in regard to using electrical equipment in the classified area. Article 500, "Hazardous (Classified) Locations," covers what the NEC defines as area classification. Articles 501-555 further explain the use of electrical equipment in hazardous areas.

Area classification is divided into three designators: Class, Group, and Division; therefore, each hazardous area is also classified according to Class, Group, and Division. Areas that are not classified are considered unclassified or nonhazardous.

Class identifies the physical nature of the hazard. The Class designators are as follows:

Class I — A class I location is a location in which flammable gases or vapors are or may be present in the air in quantities sufficient to produce explosive or ignitable mixture.

Class II — Where combustible dusts are processed, handled, or stored.

Class III — Where easily ignitable fibers or flyings are processed, handled, or stored.

The Group designator identifies the chemical nature of the hazard. Some of the chemical properties of interest are maximum explosion pressure, rate of rise of explosion pressure, and maximum experimental explosion gap (MSEG — the maximum gap where explosive gases can be vented and not set off the outside world). The Group designators with their representative chemicals are as follows:

A — Acetylene

B — Hydrogen, ethylene oxide, propylene oxide

C — Ethylene, acetaldehyde, carbon dioxide, methyl ether

D — Gasoline, methane, ethane, propane, propylene

E — Combustible metal dusts with a resistivity of less than 100 ohm-centimeter.

F — Combustible dusts: carbon black, charcoal, coal coke dust with a resistivity of 1,000–100,000,000 ohm-centimeter

G — Combustible dusts with a resistivity of greater than 100,000,000 ohm-centimeter, corn, wheat, polypropylene, polyethylene

Note that Groups A–D are Class I chemicals, while Groups E–G are Class II chemicals or materials. Class III does not have any groups.

The Division designators identify the probability and extent that the flammable or combustible mixture will exist in the area at any given time. The Division designators are as follows:

Division 1 — The flammable or combustible mixture exists under normal conditions. Normal conditions can include during regular maintenance activities, chemical releases that happen regularly, etc.

Division 2 — The flammable or combustible mixture exists under abnormal conditions. Abnormal conditions include malfunctions, pipe rupture, equipment leaks that do not normally occur, etc.

Another Division under consideration is Division 0, which would be similar to Zone 0 in Europe. Division 0 would include highly dangerous areas that are currently part of Division 1 and require extra means to ensure safe operation of the electrical equipment.

As part of the Division designator, the physical extent of the classified area must be determined. The extent is the physical boundaries, both horizontal and vertical, where the flammable or combustible mixture can exist. In general, there must also be transitional areas, i.e., a Division 1 area cannot transition into non-hazardous without there being an intervening Division 2 area.

The hazardous area classification hierarchy is illustrated in Figure 17-2.

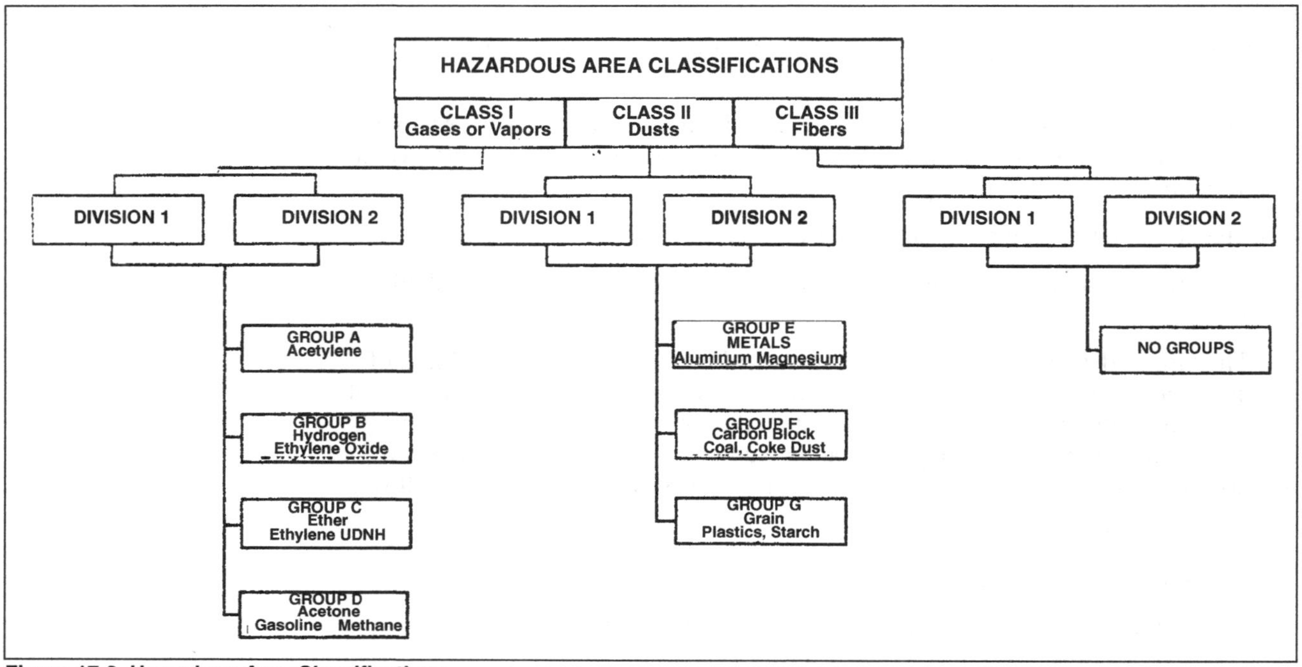

Figure 17-2. Hazardous Area Classifications.

A plant makes polypropylene in two areas: the process area and finishing area. In the process area propylene gas is considered hazardous material, and, in the finishing area the hazardous material is polypropylene dust. In the process area, propylene in a flammable concentration is considered to be there all the time, while in the finishing area, the dust is airborne only under abnormal conditions. What is the area classification of this plant? Figure 17-2 shows that since propylene is a gas at ambient conditions, it is a Class I material; also note that propylene is a Group D chemical. Since propylene is considered to be there all the time, the process area classification is Class I, Group D, Division 1.

In the finishing building, polypropylene dust is a Class II, Group G material. Since it is airborne only under abnormal conditions, the finishing area classification is Class II, Group G, Division 2.

DETERMINING AN AREA CLASSIFICATION

In determining an area classification, the normal procedure would be to form a review committee of people from engineering, maintenance, operations, and any other interested parties to review the area. In making its decision this committee should determine what materials are processed; the amounts processed; the pressures, temperatures, and specific gravity of the materials involved; the location, extent, and amount of any potential material releases; weather conditions; plant construction and layout; past history, and any other pertinent information for both normal and abnormal operations. NPFA 497 and RP500 should be used to provide area classification guidelines. Once the area classification has been determined, an area classification plot plan should be drawn.

It is a requirement to know the area classification before installing electrical equipment and wiring in a hazardous area. It is NOT sufficient just to know it is an "explosionproof" area or a Division 2 area. One must know the full area classification to determine the means to install electrical equipment and wiring in hazardous areas.

Installation of Electrical Instrumentation in Hazardous Areas

In order to put electrical equipment in hazardous areas, one must first determine the area classification and then determine the means to do so based on the NEC Articles 501–503 covering Classes I, II, and III, and Articles 510–555 covering specific locations such as gasoline dispensing and service stations, bulk storage plants, health care facilities, etc.

Means used to install electrical equipment in a hazardous area must be approved for that specific area classification. Classifications are not cumulative, i.e., equipment rated for Class I, Group C, Division 1 are not rated for Class I, Group D, Division 1. Approved means acceptable to the authority having jurisdiction for the facility (see NEC Article 90-4). In addition, approved equipment and enclosures must have a label that identifies the areas for which the equipment is rated.

DIVISION 1 AREAS

There are four basic means for installing equipment in Class I and II areas: explosionproof/dust ignitionproof, intrinsic safety, purging, and approved equipment. In Class III areas methods vary and include ventilation, Class III-approved equipment, and dusttight enclosures. For Class III installation, refer to NEC Article 503. Explosionproof (in Europe called flameproof) equipment used in Class I areas are so constructed that they will contain an explosion, vent the hot explosive gases in such a manner that they will not serve as an ignition source, and limit the surface temperature of the equipment to less than 80% of the auto-ignition temperature (AIT) in °C of the material in the area classification with the lowest AIT (AIT being defined as the temperature at which the material will self-ignite). Venting is typically done in two ways: flange and thread. In flange venting, the hot gases are vented between machined flanges that allow the gases to cool before reaching the atmosphere. In thread venting, the gases are vented through threads (minimum of 5).

Dust-ignitionproof equipment is similar to explosionproof equipment except that it is designed for dust areas and is used in Class II areas.

Note that an enclosure or equipment can be rated for more than one Group and/or Class. For example, it is common to have an explosionproof enclosure rated as Class I, Group D, or Class II, Groups E, F, G.

The effectiveness of explosionproof/dust-ignitionproof enclosures will be compromised if all the bolts are not secure and the flange surface integrity is not maintained for a flanged enclosure. For a threaded enclosure, all the required threads must be engaged.

Explosionproof enclosures should never be sealed.

Inspections of many facilities have shown on numerous occasions that most of the bolts have been removed from explosionproof enclosures because there were "too many bolts" for the maintenance personnel to be bothered with. If just one bolt is loose on an enclosure, the integrity of the enclosure is compromised. On one occasion, it was observed that a maintenance man removed all the bolts on an explosionproof enclosure, carefully removed the cover, and then placed it face down on gravel. The gravel scratched the machined flange, thus compromising the enclosure. The moral of the story is that while the explosionproof enclosure is probably the closest to an "install and forget" equipment, it does require some care and feeding to ensure that its integrity is maintained.

If explosionproof/dust-ignitionproof equipment and enclosures are not properly installed and maintained, they will not function properly. This defeats their purpose and may lead to a fire or an explosion!

Intrinsically safe systems are systems in which the hazardous areas cannot under normal or abnormal conditions release sufficient energy to ignite the most easily ignited mixture for the rated area. This means that if there is a short, open, grounding of conductors, or contact with other nonintrinsically safe systems, there will not be a spark or hot temperature sufficient to serve as an ignition source.

Intrinsically safe wiring systems allow wire to be run in hazardous areas using nonhazardous means, but there are separation requirements between intrinsically safe systems and nonintrinsically safe systems as well other wiring requirements unique to intrinsically safe systems. In addition, maintenance of intrinsically safe systems must be done in a manner that ensures the integrity of the intrinsically safe system is maintained. This normally requires special training for the maintenance technicians and additional administrative controls on maintenance activities that involve intrinsically safe systems.

Two general ratings are given to intrinsically safe apparatus: system and component. In a system-rated apparatus, the whole system has been tested and approved or listed by a testing agency. In a component-rated apparatus, only the component has been tested and the apparatus intrinsically safe parameters determined; it is up to the system designer to integrate the system components together to make an intrinsically safe system.

The intrinsically safe system must be rated or designed for the specific area classification in question.

Another method for placing electrical equipment in a Division 1 area includes purging and pressurization. Purging and pressurizing are used to reduce the area classification in enclosures. It does so by providing a positive pressure inside the enclosure, allowing no temperature greater than 80% of the AIT, in °C, and providing adequate safeguards to ensure the integrity of the purging system. Purging is the common term that refers to both purging and pressurization and will be used as such in this discussion.

The basic principle of purging/pressurizing is the reduction of the interior of the enclosure area classification to a classification suitable for the equipment in the enclosure. This is done by the maintenance of a positive pressure in enclosures to preclude the entrance of flammable gases, vapors, and dusts and not allowing any surface temperature both inside and outside the enclosure to exceed 80% of the AIT in °C or a temperature determined safe by test. In addition, ventilation of the enclosure is provided at equipment start-up to sweep any residual gases or vapors before power is restored and is generally provided (though not required) during normal operating conditions. Purges come in three types:

(1) Type "X" — Reduction from Division 1 to Nonhazardous

(2) Type "Y" — Reduction from Division 1 to Division 2

(3) Type "Z" — Reduction from Division 2 to Nonhazardous

Note that Types "X" and "Y" apply to Division 1 areas.

General requirements that apply to enclosure purges are (1) sufficient enclosure mechanical strength, (2) positive pressure, (3) identification labels, (4) visual indication or alarm of failure to maintain positive pressure, and (5) purging of the enclosure for hazardous gases before energization of equipment. Each type of purge will have its individual requirements. Purge requirements are also determined by the volume to be purged, whether hazardous gases are brought into the enclosure, and the area classification in which the enclosure will be located.

Purges have many times been the "install and forget" type of system with only a local indication of purge. This is certainly not a good practice, since anything can deteriorate over a period of time. For example, during an inspection of an old plant, a purged enclosure that had no label was inspected, and a rotameter was installed on the purge flow into the enclosure. The rotameter indicated that there was a good flow of purge gas entering the enclosure. Unfortunately, about 80% of the bottom of the enclosure had corroded away. A number of things are obviously wrong with this scenario. First, it was obvious that no one had inspected this enclosure nor verified that the purge was working in a long time. Second, as this enclosure purge was not monitored on a regular basis, the visual indication and/or alarm requirements were not met. Third, the rotameter installation was an inappropriate indication of the purge as it indicated only that purge gas was entering the enclosure and not that a positive pressure was being maintained. Fourth, there was not proper labeling of the purged enclosure. And last, the integrity of the purged enclosure was not maintained. This enclosure was potentially a match looking for a bomb to set off!

A rotameter on the inlet to a purged enclosure does not meet the visual indication requirements of NFPA 496 because it is not a positive indication that positive pressure is being maintained on the enclosure.

Approved equipment may include explosionproof/dust-ignitionproof and intrinsically safe equipment or other equipment that has been tested and determined acceptable for the area.

INCREASED HAZARD AREAS

There may be areas in a Division 1 area that are particularly hazardous and will require additional safeguards. These areas are equivalent to the European Zone 0 or the proposed NEC Division 0. Installation of electrical equipment in these areas is accomplished typically by applying more than one Division 1 means or a Division 1 means and a Division 2 means. Just as there are requirements for equipment in a Division 1 area, there are also requirements for wiring means, raceways, grounding, etc. Refer to the NEC for these requirements.

DIVISION 2 AREAS

In Division 2 areas, the basic requirement is that the equipment not provide an ignition source under normal conditions. Five basic means can be used to locate electrical equipment in a Division 2 area: Division 1 means, nonincendive systems, no source of ignition rules, purging, and approved equipment. All the Division 1 means, with the exception of the Type "Y" purge, may be used. Nonincendive equipment in a Division 2 area is similar in concept to intrinsically safe systems in a Division 1 area. Where intrinsically safe systems require that the system in the hazardous area under normal and abnormal conditions could not provide sufficient energy to serve as a source of ignition, nonincendive systems require that the system in the hazardous area under normal conditions cannot serve as a source of ignition. The Type "Z" purge can be used in a Division 2 area to reduce the area classification in the enclosure from Division 2 to nonhazardous. The "no source of ignition rules" in the NEC allow the placing of electrical equipment in a Division 2 area if no source of ignition is available. This can be accomplished by hermetical seals or immersing in oil the current interrupting contacts. Equipment with no current interrupting contacts may also qualify. Exposed surface temperature must be considered and may not exceed 80% of the AIT in °C or be tested to be safe. Approved equipment may include explosionproof/dust-ignitionproof and nonincendive equipment or other equipment that has been tested and determined acceptable for the area. Just as there are requirements for equipment in a Division 2 area, there are also requirements for wiring means, raceways, grounding, etc. Refer to the NEC for these.

NONHAZARDOUS AREAS

While an area may be classified as nonhazardous within an industrial facility, the area may still require more than the ordinary means allowed for nonhazardous areas. This can include the use of Division 2 means for installing electrical equipment and wiring.

Contamination

The National Fire Codes, particularly NFPA 70, and the NEC, are the "bible" for placing electrical equipment and wiring in hazardous areas. Don't leave home without them!

All field instruments that are connected to a process or are in environments that are exposed to process chemicals should be considered contaminated with the chemicals to which they are exposed. The danger involved is dependent upon the nature of the chemicals involved. When these instruments are removed from service, they must be decontaminated prior to repair. When an instrument is removed from the field, it should be tagged with its identity or tag, its service, who removed it, the date, and who it is to be returned to. A tracking identifier can be used to provide a consistent tracking means, and a charge identifier can be used to allocate charges for maintenance. Sufficient information on instrument service should be provided to allow maintenance personnel to determine the method and extent of the decontamination process. Once the instrument has been decontaminated, the instrument ID tag should be signed off, indicating the instrument has been cleaned. The use of colored stick-on circles on the ID tag can be used to easily identify the cleaned instrument. This method or equivalent should be used even if the same maintenance man is doing all the work, because work can be interrupted, delayed, or shifted, and, without some method of tracking, the status of the instrument could be lost. Failure to do this might lead to exposure of maintenance personnel to hazardous chemicals.

Maintenance procedures should be in place covering the decontamination of instruments. The decontamination process should clean all exposed surfaces as well as hidden pockets. The cleaning should be done with appropriate cleaning materials. Sometimes it is not possible to clean all areas of contamination without disassembly (gaskets, O-rings, diaphragms, packing, etc.), which would normally occur during the maintenance of the instrument. If the chemical involved is considered dangerous, the instrument should be tagged as a hazard so that downstream maintenance personnel will be aware of the danger.

Personnel who disassemble instruments should always take care even if the instrument has been "decontaminated." Others may not have been as careful. When in doubt, assume that the instrument has NOT been decontaminated.

It is good practice NOT to reuse sealing materials such as packing, O-rings, gaskets, etc.; they may be contaminated.

A 6-inch control valve was sent to the central maintenance facility for repair. The facility had no procedure for tagging instruments, and the valve appeared to be clean from the outside. An apprentice instrument tech was given the task of disassembling the valve for maintenance. When the tech disassembled the valve, a small pocket of sulfuric acid was inside the valve, some of which got on the tech and caused a small chemical burn. The moral of the story is, "When in doubt, decontaminate!"

Decontamination can work in another way. Sometimes an instrument is removed from service; when repaired, it is put in a general instrument replacement pool for later use rather than returned to the service from which it came. Since it is not known what service the instrument will go into, the instrument should be completely decontaminated, including replacement of any gasket, O-ring, etc., that are exposed to the process. In addition, the instrument should be marked, identifying its wetted parts.

Some special services, such as oxygen, require special cleaning. Maintenance procedures should be in place that control these special services. Instruments for special services should be identified as such, and maintenance personnel should be certified through training and testing to perform these types of cleaning and the installation of these types of instruments.

Working on field instruments can lead to contamination of the environment. Care should be taken to minimize the release of chemicals into the environment. Areas that can be problems are improper fitting makeup, loose connections, uncapped lines, leaking seals, open vents, etc. All process connections should be checked for proper makeup before being pressurized.

In a research pilot plant, a modification was made that had compression-type fittings. Upon applying pressure and after a short operational period, the plant released a flammable process fluid. The release occurred because the ferrule material was different from the nut material. In another case, when a fitting blew out, it was found that the fitting had not been tightened properly. Another fitting broke off because the compression fitting was tightened too tight. In these cases, the plants were pressure tested prior to start-up. During these types of tests, however, the plant is seldom at operating temperature and is not typically subject to pressure cycling.

Pressures and Vacuums

Pressures and vacuums can represent hazards to maintenance personnel. Most processes operate at pressures different from atmospheric pressure. When the process pressure is exposed to atmosphere, process material is released to the atmosphere or air is sucked into the process. In the case of pressure above ambient, a material could be released to the atmosphere, which can result in represent a health hazard to humans or to the environment. The material may also be flammable, thus representing a fire or explosion hazard. In the case of vacuum, air can be sucked into the process, which may contaminate the process or may form a flammable mixture. This can be subtle in that vacuum leaks are harder to detect and locate.

Instruments must be selected to meet the pressure specification. Instruments that may also be exposed to vacuum must also be rated for vacuum service.

Overpressurization can represent a physical hazard through rupture of the pressure-containing enclosure. Liquid pressure presents little hazard since liquid pressure has little stored (potential) energy. A large release of material can result, however, if the pressure is backed by a strong source of pressure (a pump, for example). Gas pressure represents a much larger hazard because of the stored energy in the compressed gas. An analogy demonstrating this is filling a balloon with water until it bursts: it makes a mess but there is not much violence (energy). If a similar balloon is filled with air until it bursts, the "explosion" of the balloon is much more violent. Extra care should always be taken in dealing with pressurized gas systems. This is particularly true when air is used to pressure test a system in lieu of hydrotesting.

It should be remembered that all pressure ratings have an associated temperature specification that must also be met.

A maintenance man was removing a control valve motor and mistakenly removed the bonnet housing bolts. This released pressurized propylene gas, which ignited, causing a large explosion. This type of valve was banned from the facility after the explosion.

High Voltage

High voltages are not commonly found in process instrumentation.Voltages are generally in the 120–208V AC and below range. However, high voltages are found in the CRTs of DCS systems and in human-machine interfaces. Maintenance on equipment that contains high voltages requires additional precautions, some of which are listed as follows:

(1) First and foremost, know what you are dealing with, at what voltage levels, and their locations.

(2) Do not work on high voltage equipment energized unless there is no other option.

(3) Remember that de-energized equipment can still "bite." Many high voltage systems have capacitors that will maintain a charge for a period of time after being de-energized. These capacitors must be discharged before work can proceed.

(4) Separation is your friend. The further away from the voltage, the safer you are. Safe separation for 120V AC is not the same as for 10,000 volts! Separation by barriers that are permanently installed or temporarily installed for maintenance will also provide additional protection.

(5) Never work alone on high voltage.

(6) Never leave guards off for convenience.

Care should be exercised when replacing high voltage parts and wiring. Any substituted parts must meet the original manufacturer's specifications, particularly the voltage level.

Moving and Rotating Machinery

Instrumentation on rotating and moving equipment has become more prevalent and more sophisticated. Many times maintenance personnel are required to troubleshoot or perform other maintenance activities on equipment that can move, is moving, or is in the neighborhood of moving equipment. Great care must be taken to ensure that this is done safely. A "ready to work" permit should be obtained from operations before any work should proceed. There should be standard maintenance instructions (SMI) or procedures (SMP) to cover working on moving equipment.

If the work is to be done while the equipment or portions of the equipment are de-energized, the equipment must be locked and tagged out and verified as completely de-energized.

Lockouts are covered in NFPA 70E— Standard for the Electrical Safety Requirements for Employee Workplaces, and OSHA CFR 29, Part 1910.

A maintenance man was killed in an accident in which he locked out a 480V AC pump motor and began work without verifying that the equipment was de-energized. He had locked out the wrong pump! Another incident occurred when instrument technicians were working on equipment they thought they had de-energized, when in fact there was more than one power feed, one of which they had not de-energized. While working on the equipment, they caused a fault that shut down a section of the plant.

If the work is to be done while the equipment is not moving but still energized (for example, troubleshooting), care must be taken that the equipment does not inadvertently begin moving. Generally, the maintenance person should not be in contact or near the moving portions of the equipment during these activities. If it is necessary to be near the moving parts, barriers should be in place to prevent contact with the moving parts if they should begin to move or the equipment should be mechanically blocked against moving.

If the equipment is moving, contact or near contact to moving or rotating parts should be prohibited. Barriers should be used if necessary to prevent this contact and allow safe maintenance of the equipment. Under no circumstances should safety interlocks or guards be bypassed to allow the work near the moving parts. In addition, administrative controls (permits, SMP, SMI, SOPs, etc.) should not be bypassed for expediency.

Care should always be taken that one does not inadvertently start or energize equipment one is working on. As the moving equipment many times can be remote from the part of the system being worked on, care must be taken not to injure someone who is around the remote equipment.

High and Low Temperatures

All high (above 113°F) and low temperature (below 0°F) surfaces should be identified. Where possible, guarding should be provide for these surfaces. It should be remembered that the removal of temperature motive force does not remove the hazard immediately. Proper time should be allowed for the equipment to reach ambient before work can proceed. If work must be done on equipment that has a temperature hazard, suitable separation or barriers should be used to ensure safe maintenance. The opening of process lines that represent temperature hazards should be done with care. Care should be exercised around steam and electrical heaters used in heating instrument lines and instruments.

An inexperienced instrument engineer was tracing analyzer sample tubing through a pipe rack. The engineer inadvertently put his hand on a vaporizing regulator and received a minor burn. There was no identification of the hazard.

Gases and Chemicals

All primary instrumentation and final control elements interact with the process that they are measuring or controlling as do some secondary elements. All instrumentation materials must be compatible with the process. This compatibility must provide satisfactory chemical resistance at the pressures and temperatures involved for all wetted parts. Wetted parts include all parts or surfaces that are exposed to the process such as diaphragms, O-rings, packing, pressure containing surfaces, etc. The chemical resistance must meet both normal and abnormal conditions as well as ambient conditions.

In a large petrochemical plant on the Gulf coast where one of the chemicals made was chlorine, it was discovered that the 316 SS dP cell bolts were deteriorating; in fact, in some cases they were crumbling. It was determined that there was a high level of chlorine in the air. The high stress in the bolts, high ambient temperatures, and the high humidity combined with the chlorine was causing chloride stress cracking. The bolts were replaced with a high strength steel bolt and the problem went away.

Maintenance on instruments that are used in hazardous chemical service (and in these days, most chemicals are considered hazardous) involves two areas: field maintenance and shop maintenance.

In the field, the following must be adhered to:

(1) The nature of the chemicals being dealt with must be known.

(2) When the instrument is to be serviced, removed, or installed where there is interaction with the process, suitable protection should be worn. At a minimum, safety glasses and leather shoes should be worn. Chemical goggles, face shield, slicker suit, respirator, fresh air mask, or SCUBA should be used as required. Convenience should never be a consideration in the use of protective gear.

(3) Once the instrument has been removed, it must be properly cleaned before maintenance is performed. If it is to be sent to a central shop for repair, it

must be properly cleaned and the process chemicals involved should be tagged on the instrument; if it is to be cleaned at the central shop, the instrument must tagged with the chemicals involved and be transported in a manner suitable for the hazards involved.

(4) When installing the instrument, all connections, bolts and nuts, seals, etc., must be verified before pressure is applied. Any substitute instruments should be verified as suitable for the process involved.

Two instrument techs were installing a dP cell. When pressure was applied, a stream of 10% caustic spurted out of the cell and struck one of the techs in the chest. The tech who was installing the cell had not ensured that one of the side vents on the cell was closed. The moral of story is be make sure that all connections are tight and be careful where you stand when the pressure is applied.

Maintenance in the shop may prove particularly hazardous because instruments are coming in from different places and the link between the process materials and the instrument can be lost. It is vitally important that this link be maintained. All instruments that are exposed to the process or adverse ambient conditions should be tagged as to the materials involved and whether the instrument has been cleaned. It is generally better for the source of the instrument to clean it because they should be more knowledgeable in cleaning the instrument than the shop. Tagging of the materials involved also helps ensure that any replacement parts are of the proper compatibility with the process.

A 6-inch control valve came into the shop for repair. An inexperienced technician was given the job of dismantling the valve. Some of the process material was trapped in the body of the valve, which the technician got on himself. Fortunately, the material was not very hazardous and timely washing was all that was required. This should, however, be considered a near miss in that the material could have been much more hazardous.

One other aspect is ensuring that the repair is proper for the process materials involved and that the instruments go back to the proper location. All instruments should be identified as to the wetted materials used and the original instrument tag.

ISA's 92 series of standards and recommended practices give more detailed information on safety factors using gases and vapors.

Heights and Confined Spaces

Tie-offs

OSHA requires those climbing to tie-off to prevent falling. Proper training is required to utilize the harness and properly tie-off as you climb.

OSHA also requires that you wear a harness properly tied-off when you enter an enclosed vessel. Before entering any closed vessel, check for proper oxygen level and the absence of dangerous gases.

Program Changes, Software Control

Making software changes on operating equipment is fraught with dangers. These changes can be made on-line or off-line. The basic problem is that these changes are typically made to solve a problem or provide additional operational capability.

If the change is made through an engineering design change, proper time is generally allotted in the design cycle to evaluate the effects of the change. If it is done "on the spot," this consideration is not always given and the change may be viewed from a too narrow perspective. All changes must be viewed from the perspective of whether it does the job intended and that is does not affect the system in an adverse way. Operational changes should be reviewed with operations and other knowledgeable persons before implementation.

Changes that affect the safety of the system should have a safety review. If the change has been done off-line, i.e., when the equipment is down, the change must be completely tested before being released to operations. If a safety system is involved, the extent of retesting of the whole system must be evaluated.

On-line operational changes should be approached with great trepidation. Once made and if in error, they may not be able to be revoked before damage or operational losses occur. Again, a review of the change from a functional perspective and a system perspective must be done. No safety system changes should be done unless failure to do the change is more hazardous than the change. Making a change on a safety system may compromise it because the safety system is not completely retested to ensure that the whole system is capable of performing its appointed safety function.

All changes must be noted (red-lined) on the field documents, and the changes must be returned to the organization responsible for the original documents for updating and re-issue. Changes should go through administrative controls to ensure that the changes are documented. This is particularly important for software changes, as these are essentially invisible compared to hardware changes. Nothing is more useless to a maintenance technician than out-of-date documents. Job security "rat-holing" of information and "documentation in the desk drawer" should not be allowed. The maintenance department can be a major contributor to poor documentation if changes are not documented because operational or maintenance priorities override documentation needs. Failure to maintain documentation is a time bomb waiting to go off. OSHA requires up-to-date documentation. Remember, the job is not done until the paperwork is complete.

Failure to maintain the documentation loads and wiring on several large 100-kVA UPS systems caused so much uncertainty that the people were afraid to add additional loads. The company had to do an expensive study of the system to bring documentation up to date. The expense was considerably more than the cost of keeping documentation up to date to begin with, and this does not even consider the associated problems they had with maintaining the system.

Personal Dehydration

Working in hot temperatures near boilers, furnaces and outside in the very hot weather, will cause dehydration if you do not drink lots of water. In these areas, the plants should have cold water sources available nearby.

Noise

Noise can be dangerous in many ways. One is that it can affect your hearing and it also can overshadow other potentially dangerous things that are happening around you. You may not be able to hear those calling you in case of an emergency.

Proper Tools and Test Equipment

Proper tools and test equipment must be provided to ensure that maintenance work gets done efficiently. Many times tools and test equipment are considered secondary to getting the job done. It is assumed that good technicians will make do. The real measure is what the "time to repair" is worth to the company. Given the proper tools, testing equipment, and the training to use them, the facility mean time to repair (MTTR) will be reduced, the availability (up time) increased, and maintenance costs reduced.

What is the cost of a particular system to be down? $100, $1000, $10,000, per hour? It does not seem reasonable that, if it costs $5,000 per hour for a system to be down and a $500 piece of test equipment would save half an hour in the maintenance, the test equipment not be provided. This also can apply to simple tools. What is it worth to the company for a technician to have to go back to the shop and look for or borrow a hand tool to get a five-minute job done?

Safety also enters into the question of proper tools and test equipment. If technicians do not have the proper equipment to do the job, they may improvise an unsafe substitute.

Work Area Cleanliness

Management emphasizes keeping the work area clean, many times evoking complaints from maintenance personnel because management seems to put more importance on the cleanliness than on the actual work. Cleanliness does, in fact, have a solid relationship with a good job. Cleanliness can remove potential safety hazards such as trip hazards, slipping hazards, fire hazards, hazard material exposure, and so on. Cleanliness can also improve the organization of a job: tools replaced in their proper and known location will make tools more readily available, and construction materials will be easier to find.

People who observe a messy work area many times will assume sloppy work, which may unnecessarily reflect adversely on the maintenance group's reputation.

Observation indicates that when high quality work is done, a clean work area generally follows. Care should be taken not to extend this to the theory that a clean work area leads to quality work. Clean work areas are an indication of a solid work ethic, and the company should create a work culture to encourage the work ethic and to ensure a safe and clean work area.

Safety Meetings

Safety must be a prime consideration in any work. The company must create an atmosphere that encourages and helps ensure a safe work environment. Safety meetings are one method to educate and train workers on safety issues as well as to encourage a safe environment.

Regular safety meetings can help reinforce the safety culture. They can cover a wide variety of subjects ranging from home safety to personal job safety to maintenance procedures that cover safety practices. General safety meetings should be held monthly as a minimum. Work group meetings may be on a weekly basis, and job specific meetings may be on a daily basis or on a per job basis.

One good technique to gain employee involvement in safety meetings is to have the employees organize and present the safety topics. It also provides a de-

Each employee should have a "one person" safety meeting before beginning any work to determine all the hazards involved in the work and what procedures or practices will be used to ensure that the work is performed safely.

velopment opportunity for employees to organize and make a presentation to their peers. Care should be taken, however, that this does not become a "spare time job" or low priority work that appears to be a real burden to employees relative to their regular work. Bringing in outside experts or interesting speakers can also improve safety meetings. Safety meetings should be interesting and useful to the employees. Varying the subject material, format, length, and employee participation can keep the meetings interesting. Too often safety meetings become a matter of rote, and employees become cynical concerning the supervision and company commitment to safety, which defeats the purpose of the safety meeting.

Process Considerations

Anyone who works on any instrument should know what process is involved. The instrument does not have to be directly connected to the process but may only control or influence the process. This knowledge is required not only to troubleshoot the problem but to troubleshoot the problem safely. Some of the considerations are: (1) what safety problems could arise when the loop is manipulated to troubleshoot the system, (2) what happens if the loop is shut down inadvertently, (3) what happens if some of the process is released, (4) what temperature and pressure are being dealt with, (5) what health hazards are involved, and (6) what other risks are involved? Physical hazards such as hot surface temperatures should not be overlooked. Discussion with the operator or plant personnel prior to the maintenance can provide valuable information. This discussion should also cover emergency procedures.

Communication

Communication is the life blood of safety. Seldom is it that only one person is involved in an activity. Good communication between operations, maintenance, and engineering is critical to a safe job. Communication must be continuous from the start of the job to the end of the job. This does not mean just the people directly involved in the work but also adjacent people or other people who might be affected if something unusual happens during the work.

Some communications must be formal in nature, i.e., hot and cold work permits, excavation permits, ready to work permits, lockouts, etc. Others are more informal, more common sense in nature, but not necessarily less important. All maintenance personnel should be encouraged to communicate with the appropriate people while doing any work.

> **A work permit was given to perform some instrument work in an area. What was not communicated was that there would be welders several stories above the instrument techs. When the welders began work, sparks and hot material rained down upon the instrument techs, burning one of them. Discussion with the permit issuer about who else was working in the area could have eliminated this accident. What we have here is a failure to communicate!**

Cryogenic Considerations

The issue of safety in itself is of paramount importance in any industry. It takes on a special meaning and urgency in dealing with cryogenic liquids because of the physical properties of such liquids. An accident due to mishandling of a cryogenic liquid can cause property damage and personal injury and could even lead to a fatality. A good understanding of the properties of a cryogenic liquid will thus allow everyone to handle and use these liquids safely.

Properties That Relate to Safety

There are two main properties of a cryogenic liquid that relate directly to its safe handling. First and foremost, cryogenic liquids are extremely *cold*. By definition, cryogenic liquids have a normal boiling point below –238°F (–150°C). Some liquids such as nitrogen (LN_2) and liquid hydrogen (LH_2) or liquid helium (LHe) are much, much colder at 0 psi saturation.

The second property of cryogenic liquids is that they have different boiling points depending on the saturation pressures. In other words, the liquid state is a function of specific combination of both temperature and pressure. If any heat is allowed to enter into the liquid, it will cause a proportionate amount of the liquid to turn into gas. This, in turn, will raise the pressure of the closed container in which the liquid is kept.

Two other properties of cryogenic gases have special significance in regard to safety. The gases are colorless and odorless; they are thus impossible to detect by ordinary human senses.

Finally, two gases deserve special mention because of their unique properties. Oxygen does not burn by itself but is a supporter of combustion. The rate combustion is thus accelerated in an oxygen-rich environment. Hydrogen, on the other hand, can mix with air or oxygen and form a very volatile mixture. This hydrogen oxygen/air mixture can explode very vigorously if it comes in contact with a source of ignition.

Asphyxiation

Asphyxiation may occur when high concentrations of all other gases reduce the available amount of breathable oxygen in the air. Since cryogenic gases are odorless and colorless, their pressure or concentration cannot be detected by ordinary human senses. As the oxygen content of air decreases, asphyxia develops slowly and the victim generally will not be aware of the condition or the symptoms.

The following are precautions:

(1) Keep the working area (around cryogenic gas storage and handling) well ventilated.

(2) When entering oxygen-deficient closed areas, use portable breathing air packs or a hose mask connected to a breathing air source. Also, use the buddy system (i.e., having someone outside to keep watch).

CAUTION: *The use of an absorptive gas mask will not prevent asphyxiation in an oxygen deficient environment.*

(3) For enclosed areas or closed vessels or tanks, the atmosphere should be monitored with instrumentation.

(4) Oxygen analyzers and monitors with alarms should be used to measure the oxygen content of areas where low oxygen content may be possible.

EFFECTS OF LOW TEMPERATURE OF CRYOGENIC LIQUIDS

Because of the extreme low temperature of cryogenic liquids, they present a special hazard of frostbite on skin and exposed eye tissue. Even very brief contact with the delicate eye tissue can cause permanent damage.

Cryogenic liquids tend to splash and boil when a warm object is inserted into them or when a warm container is filled with the liquid. Thus, always use protective clothing and equipment while handling cryogenic liquids, such as:

(1) insulated gloves (loose fitting for quick removal in case of spillage);

(2) trousers (no cuffs that can trip spilled liquid; put trousers over boots or shoes);

(3) footwear (work boots or shoes; no sandals, etc.);

(4) safety glasses (with slide shields; chemical goggles or face shield are better for added protection); and

(5) long sleeved shirt.

Two other points:

(1) When bare skin touches uninsulated pipes or vessels containing cryogenic liquids, the extremely cold metal may cause the skin to stick to the cold surface; removing the skin may cause flesh tearing.

CAUTION: **DO NOT** touch cold metals with bare skin; use gloves and protective clothing.

(2) The extreme cold temperature of cryogenic liquids cause embrittlement of many materials, including carbon steel and plastics.

CAUTION: Make sure all the materials used in storing, piping, and handling of cryogenic liquids are made of materials that can withstand cryogenic temperatures without fractures induced by embrittlement.

Pressure Relief: Prevention of Rupture

As mentioned earlier, cryogenic liquids will boil off into the gaseous state whenever it comes in contact with any heat source. This gas, in turn, will cause a buildup of pressure in the vessel or piping and can ultimately lead to rupture of the vessel or piping if the pressure is not relieved properly. This can be prevented by the use of appropriate devices that release the pressure at set levels. Relief valves and rupture disks are two such common devices.

The pressure rating of these devices must be lower than the safe working pressures of the vessel or piping being protected.

Never install two shutoff valves in a pipe that will transport cryogenic liquids without installing a relief valve between them. If both the shutoff valves are inadvertently closed, any liquid trapped between them will build up pressure and ultimately cause the pipe to rupture.

Combustion

Oxygen itself is nonflammable, but it vigorously accelerates and supports combustion. Hydrogen, on the other hand, when mixed with air forms a very explosive gas mixture.

Items that are combustible burn more vigorously in an oxygen-rich environment. Similarly, some materials that are not flammable by themselves in air may burn easily in pure oxygen.

The following should be observed:

(1) Keep area well ventilated.

(2) Flammable substances should not come in contact with pure oxygen or oxygen-rich air.

(3) Do not allow organic materials to come in contact with oxygen, as they might react and ignite from a spark; example of organic materials are oil, grease, asphalt, cloth and tar.

*In sealing pipe thread, use Teflon™ tape; **DO NOT** use any organic based sealant.*

***DO NOT** use any organic-based lubricant in any equipment (used to handle cryogenic liquids); use inert lubricants.*

(4) Keep work areas and handling equipment clean. Even dirt that contains oil or grease can burn in an oxygen-rich atmosphere.

(5) Keep all work areas well ventilated.

(6) Have a fire plan and adequately trained personnel for fire fighting. Fires involving cryogenic liquids require special procedures, tools, and training.

Cleanliness

In the course or performing maintenance tasks on an instrument to be used in a cryogenic application, care must be taken to ensure that the instrument is clean before the installation. This is especially critical in the case of liquid oxygen. The instrument must be free from any oil-based product before it is put back in service. This caution applies to all components, including O-ring, sealant, and fill fluid.

Nuclear Plants

Troubleshooting is the systematic problem-solving process for identifying and resolving problems. The process starts from reports of ill-defined symptoms and progressively refines an analysis of the problem until the source of the problem can be precisely located and resolved. The process is extremely challenging and potentially risky. In nuclear power plants, the costs of both problems and errors in their resolution can be extremely large. Consequently, the troubleshooting process becomes an effort to minimize and manage the overall risks.

The initial steps of the process involve defining exactly what has happened versus what should have been happening. The deviation is then specified as to its identity, location, timing, and magnitude. The information is essential to determining where the problem occurs versus where it does not occur. A list of potential causes or sources of the problem can then be developed. The list will not be definitive, but it can be repeatedly revisited for refinement as the troubleshooting process continues. By considering what specific changes have occurred and what failures have occurred in the past, the most probable causes can be selected for initial attack. The process then proceeds to the more difficult task of localization and verification of the true causes.

A basic tenet of the process is to minimize the impact of both the troubleshoot ing and the problem in the plant. The troubleshooter is trying to avoid:

(1) plant trips,

(2) component/engineered safety feature actuations, and

(3) component damage.

This effort must be pursued in a safe manner by:

(1) avoiding electrical shocks,

(2) avoiding entanglements in equipment starting up, and

(3) proceeding slowly, surely, and carefully.

Because of the difficulty of the troubleshooting task, the troubleshooter should be continuously discussing the matter with other knowledgeable personnel to obtain their collective wisdom and to provide oversight checks to avoid errors.

To efficiently attack the problem with the minimum risk, the troubleshooter must be knowledgeable about the system and components involved. Based on this knowledge, problem localization and verification efforts should be started by developing a plan for systematically pursuing those potential causes that are:

(1) most likely,

(2) easiest to test,

(3) most accessible, and

(4) least risky to the plant and safety

These four criteria are somewhat contradictory, so determining the order in which to pursue the potential causes becomes a matter of judgment based on experience, management direction, personal troubleshooting philosophy, and quantitative analysis. As the process proceeds, assessment of the most likely cause will change, so the list of potential causes and the plan for pursuing them must be periodically revised. The degree of involvement by supervisors in changes to the plan will depend on the potential impact on the plant and the capabilities of the troubleshooting team. Troubleshooters normally proceed by progressively partitioning the problem while being alert for the possibility of multiple problems instead of a single cause. Information to do so will come from a variety of sources including:

(1) system drawings,

(2) written maintenance records,

(3) operator logs,

(4) oral comments from operations personnel,

(5) oral comments from maintenance personnel,

(6) atmospheric conditions,

(7) steam or water leaks, and

(8) discolored components from overheating.

Even with the wisest, most careful personnel, troubleshooting, can result in errors, actuations, and trips. To mitigate the consequences, the following are advised:

(1) KEEP THE OPERATORS INFORMED.

(2) Bypass logic and isolate components being worked on whenever feasible and allowed by plant administrative limits.

(3) Consider the impact of the troubleshooting on the plant systems by:

(a) introduced signals,

(b) lifted leads,

(c) changing impedances,

(d) introducing circuit noise,

(e) overloading alternate power supplies, and

(f) exceeding physical capacity of test points.

Before proceeding with the troubleshooting, consider whether to delay some efforts to a more opportune time (e.g., weekend or upcoming outage). Management should make this decision based on input from maintenance, operations, and engineering for minimizing the operational risks and impact on the plant.

Since troubleshooting is a skill, it must be developed and practiced before proficiency can be obtained. Some theory can be obtained from training courses and books or similar sources. However, case study training and experience from accompanying experienced experts (on-the-job training) is essential to rapid learning. When personnel are picked for troubleshooting, a combination of experienced and "student" personnel should be selected so that the proper tactics are passed on to the next generation of troubleshooters. The amount of student involvement versus observation will depend on the students' capabilities and the potential risks in using students.

As experience is gained at troubleshooting, the tactics and insights should be preserved by developing and revising written guidelines and procedures. This might include annotated block diagrams and schematics that supplement the design drawings, especially the control wiring diagrams for troubleshooting.

Several people working together in a coordinated manner can achieve a collective insight that is greater than any single person's capability.

Multiple problems can be much more difficult to resolve than single ones because of the complications from feedback between the problems. Multiple problems can be avoided by promptly resolving them while they are still simple and by preemptive attacks on aging problems. Examples of periodic preventive maintenance to avoid some aging problems include the following:

(1) Periodic connector tightening (e.g., at terminal strips and bus bar connections).

(2) Periodic thermography for developing hot spots.

(3) Periodic surveys of cabling such as with time domain reflectometry to measure the AC and DC insulation dielectric characteristics. This technique can detect cable problems before they have degraded into operational problems.

(4) Periodic verification of the environment for electronics (e.g., humidity and temperature).

(5) Periodic cleaning of electronics, buses, and breakers. This task requires care and caution to avoid creating more problems than are avoided.

(6) Periodic capacitor replacement or testing on logic cards, power supplies, etc., as recommended by the vendor.

(7) Periodic replacement of semiconductor components, such as output diodes that are known to have relatively short useful lives or increasing probabilities of failure with age.

Once the problem has been identified and resolved for the specific application, the troubleshooter must consider whether the basic or root problem might be so widespread that additional corrective action is needed. If the problem resulted from a basic design deficiency, all components or systems that have the design characteristic of concern and are used in a similar manner might have the same problem. If the problem resulted from a maintenance error, all components or systems that received maintenance in the same manner for a similar application might have the same problem. Keen insight based on an understanding of the problem and of the various potentials for duplicating the problem is needed to identify additional applications for further troubleshooting and to determine how to pursue these additional applications. In deciding on the extent, thoroughness, and urgency of follow-up action, the following should be considered:

(1) Cost of a similar failure in the other applications.

(2) Difficulty, costs, and risks in verifying whether or not the problem really exists in the other applications.

(3) Whether precursor symptoms will occur that can be recognized in time to avoid serious costs of a failure.

(4) Timing of the next plant outage or other period during which plant operations will reasonably permit a troubleshooting check for the problem at the other applications.

(5) Whether surveillances are planned that would permit timely detection of the problem at the other applications.

(6) Availability of personnel, equipment, and supplies to investigate the other applications for the problem and to fix any problems that might be found.

Problems with generic applicability, particularly those for which a cure has been developed, should be reported to other plants with similar applications so that the costs of the problem can be avoided by more people. Design problems or problems that can be fixed through a design change should be reported to the equipment vendor so that the product can be changed and improved and advice can be issued to the vendor's other customers.

This distribution of operating experience benefits the first plant by encouraging other plants and the vendor to provide their own insights into solving this and other operating problems.

Most component design problems result from knowingly or unknowingly using a component in an application or environment for which the component was not intended or not informing the vendor of a problem so that an effective fix could be developed.

Ergonomics

Ergonomics refers to the engineering discipline that deals with how humans interact with a system. The intent is to provide ease of and proper operation and maintenance of the system in question. From a maintenance perspective this means that the system is designed with maintainability and testability in mind. The following are some of the design techniques used:

(1) Adequate working space and access to components provided.

(2) Adequate lighting.

(3) Built-in test points and access to test points.

(4) Consistent tagging of all components and wires.

(5) Location of components in relation to function — left to right, top to bottom.

(6) Consistent color coding.

(7) Readout, lights, digital displays, CRTs providing operational and maintenance information.

(8) Good drawings.

- Comprehensive — all tagging, calibrations, specs, etc.
- Reflect field installation, i.e., layout and wiring drawings are drawn to look like physical arrangement and terminal drawings drawn as the terminals would appear to the technician in the field.
- References to all appropriate drawings or other documents.

(9) No built-in hazards — trip hazards, head knockers, sharp edges, sharp points, exposed hot or cold surfaces, awkward access, exposure to moving surfaces, exposed energized surfaces, and so on.

(10) Appropriate voltage level separation. Do not mix 120 V AC with 24 V DC or mount a 120 V AC terminal next to a thermocouple terminal. People tend to think in a narrow range when working on a specific component or area, so when working on thermocouples they are not thinking 120V AC.

A technician was working in a thermocouple cabinet where there was a set of exposed terminals using nonthermocouple wire located on the side of the cabinet. The job required that the technician's elbow be in close proximity to the side terminals. Before the job was done, however, the technician got curious and checked out the terminals and found out that they were 120 V AC. A little foresight by the technician perhaps prevented an accident. A plexiglass guard and warning sign were installed for these terminals.

In another case, a technician was working in a relay control cabinet that had 120 V AC and low voltage DC relays. Site rules stated that higher voltages were not allowed in these types of control cabinets. It was noted, however, that some large contactors were located on the side of the cabinet, close to where the work was to be done. A check of the contactors showed that they were 480 V AC contactors. Again, observant technicians prevented a potential accident. The contactors were removed from the cabinet. Properly trained and observant technicians are the first line of defense in the battle for a safe work environment.

Acknowledgment

Minnesota Valley Engineering, New Prague, MN contributed information which was quite helpful in the preparation of the chapter.

Standards and Recommended Practices

ANSI/SA-12.00.01-2002 (IEC 60079-0 Mod) Electrical Apparatus for Use in Class I, Zones 0, 1,& 2 Hazardous (Classified) Locations: General Requirements is the most common ISA and UL Standard for Electrical apparatus for explosive gas atmospheres – Part 0: General Requirements.

Electrical Apparatus items applied as a whole or in part for the utilization of electrical energy. These include, among others, items for the generation, transmission, distribution, storage, measurement, regulation, conversion, and consumption of electrical energy and items for telecommunications. This standard refers mainly to explosion proof fittings and enclosures. The enclosure must contain the explosion.

ANSI/ISA-12.01.01-1999 Definitions and Information Pertaining to Electrical INSTRUMENTS in Hazardous (Classified) Locations. This standard provides definitions and information pertaining to protection techniques, terminology, and the installation of electrical apparatus in hazardous (classified) locations and provides an introduction and basic background to the ISA-SP12, Electrical Safety, series of publications and committee activities. This document provides a general review of applicable codes and standards, and it should not be used in lieu of those codes and standards for equipment design, manufacture, installation, maintenance and test criteria.

This standard provides general guidance for safe design, installation, and maintenance of electrical apparatus in hazardous (classified) locations using appropriate means to prevent ignition of flammable gases and vapors, flammable liquids, combustible dusts, or ignitable fibers or flyings. The Standard is not necessarily relevant to the hazards posed by pyrophoric materials such as explosives or propellants containing their own oxidizers.

ANSI/ISA-RP12.06.01-2003 Recommended Practice for Wiring Methods for Hazardous (Classified) Locations Instrumentation Part 1: *Intrinsic Safety* is intended to promote the uniform installation of intrinsically safe systems for hazardous (classified) locations. Information is provided to clarigy and explain the requirements of Article 504 of the National Electric Code (NEC). This document applies to the installation of intrinsically safe systems for use in hazardous (classified) locations.

This document provides guidance to those who design, install, and maintain intrinsically safe systems for hazardous locations.

Intrinsic safety is a type of protection in which a portion of the electrical system contains only intrinsically safe equipment (apparatus, circuits, and wiring) that is incapable of causing ignition in the surrounding atmosphere. No single device or wiring is intrinsically safe by itself (except for battery-operated self-contained apparatus such as portable pagers, transceivers, gas detectors, etc., which are specifically designed as intrinsically safe self-contained devices), but is intrinsically safe only when employed in a properly designed intrinsically safe system. The type of protection is referred to by the International Electrotechnical Commission (IEC) as "EX1."

ANSI/ISA-RP12.06.02 (Currently in draft) Wiring Practices for Hazardous (Classified) Locations Instrumentation Part 2: Installation in a Class I, Division 2/Zone 2 Hazardous Location provides guidance for wiring of instrumentation systems in Class I Division 1/Zone 2 hazardous (classified) locations. This recommended practice is primarily intended to provide guidance for installing electrical and electronic test, measuring, and controlling equipment (for example, equipment complying with ANSI/ISA-S82.02.01 and ANSI/ISA-12.12.01) This recommended practice should be used in conjunction with nationally recognized codes that cover wiring practices—such as ANSI/NFPA 70, National Electrical Code (NEC) and CSA C22.1 Canadian Electrical Code (CEC) Part 1.

ISA-TR12.2-1995 Intrinsically Safe System Assessment Using the Entity Concept. This Technical Report gives direction when electrical equipment is used in locations where fire or explosion hazardous may exist due to flammable gases or vapors, flammable liquids, combustible dust, or ignitable fibers of flyings, the National Electrical Code (NEC), NFPA 70, requires special precautions to be taken in equipment construction and installation to ensure safe performance. There are several protection techniques in common use, each of which has its own set of advantages and disadvanges.

For low-power applications in Division 1 hazardous (classified) locations, the safest and most easily implemented protection technique is intrinsic safety. Intrinsic safety is the use of equipment and wiring that is incapable of releasing sufficient electrical or thermal energy under normal or abnormal conditions to cause ignition of a specific hazardous atmospheric mixture in it most easily ignited concentration. This is achieved by limiting the energy available to, and stored or generated by, the electrical equipment in the hazardous (classified) location to a level below that which can ignite the hazardous atmosphere.

ISA-RP12.2.02-1996 Recommendations for the Preparation, Content, and Organization of Intrinsic Safety Control Drawings. This recommended practice provides guidance in the preparation of control drawing for intrinsically safe apparatus, associated apparatus, and intrinsically safe systems.

ISA-RP12.4-1996 Pressurized Enclosures applies to equipment made suitable for use in hazardous (classified) location by the use of pressuring systems. This Recommended practice applies both to equipment with and without an internal release of a flammable gas or vapor. This recommended practice provides guidance to those who design and install pressurized systems for hazardous (classified) locations, and should be used in conjunction with ANSI/NFPA496, *Purged and Pressurized Enclosures for Electrical Equipment.*

ISA-12.10-1998 Area Classification in Hazardous (Classified) Dust Locations. This standard recommends procedures for classifying locations made hazardous by the presence of a cloud or blanket of combustible dust. It conforms with the National Electric Code (NEC) and the Canadian Electrical Code (CEC), and is intended to expand and clarify both.

This standard refers only to combustible mixtures created by dusts – such as agricultural, carbonaceous, plastic, chemical, and metal dusts. Such materials classified in NEC Article 500 as Class II, Groups E, F and G.

ISA-RP12.12.03-2002 Recommended Practice for Portable Electronic Products Suitable for Use in Class I and II Division 2, Class 1 Zone 2 and Class III, Division 1 and 2 Hazardous (Classified) Locations. The purpose of this recommended practice is to provide guidance for the use of general-purpose electronic products in certain hazardous (classified) locations.

This recommended practice does not address other considerations involving the use of portable electronic products which may result in these devices being considered unsafe due to creation of a distraction from important work tasks or radio frequency interference with measurement and control equipment.

ANSI/ISA-91.00.01-2001 Identification of Emergency Shutdown Systems and Controls that are critical to maintaining safety in process industries establishes a procedure to identify the emergency shutdown systems and safety critical controls that are key to maintaining safety in the process industries as defined in the Mechanical Integrity and Maintenance sections of Process Safety Management (PSM) regulations such as Occupational Safety and Health Administration (OSHA) Federal Regulation 29 CFR 1910 (reference 5.1), and Risk Management Program (RMP) regulations such as U.S. Environmental Protection Agency Regulation 40 CFR Part 68 (reference 2).

Mechanical integrity and maintenance are major elements in PSM and RMP programs. In this standard, provisions for mechanical integrity and maintenance apply to the following equipment: Emergency shutdown systems and Safety critical controls (including monitoring devices and sensors, alarms, and interlocks). This standard addresses the instruments that are classified as emergency shutdown systems and safety critical controls and establishes requirements for testing and documenting the test results of these systems.

ISA-92.0.01, Part 1-1998 Performance Requirements for Toxic Gas-Detection Instruments: Hydrogen Sulfide.

ISA-RP92.0.02, Part II-1998 Installation, Operation, and Maintenance of Toxic Gas-Detection Instruments: Hydrogen Sulfide establishes user criteria for the installation, operation, and maintenance of toxic gas detection instruments.

ANSI/ISA S92.02.01, Part 1-1998 Performance Requirements for Carbon Monoxide Detection Instruments (50-1000 ppm Full Scale)

ISA-RP92.02.02, Part II-1998 Installation, Operation, and Maintenance of Carbon Monoxide Detection Instruments (50-1000 ppm Full Scale) applies to all carbon monoxide gas detection instruments that satisfy the performance requirements in ISA-S92.01, Part 1.

ISA-92.03.01-1998 Performance Requirements for Ammonia Detection Instruments (25-500 ppm)

ISA-RP92.03.02-1999 Installation, Operation, and Maintenance of Ammonia Detection Instruments (25-500 ppm Full Scale) applies to all ammonia gas-detection instruments that satisfy the performance requirements in ISA-292.03.01.

ANSI/ISA 92.04.01-1996, Part 1 - Performance Requirements for Instruments used to detect Oxygen-Deficient/Oxygen-Enriched atmospheres.

ISA-RP92.04.02, Part II-1996 Installation, Operation, and Maintenance of Instruments Used to Detect Oxygen-Deficient/Oxygen-Enriched Atmospheres applies to all oxygen detection instruments that satisfy the performance requirements in ANSI/ISA-92.04.01, Part 1, Performance requirements for Instruments used to Detect Oxygen-Deficient/Oxygen-Enriched Atmospheres.

ISA-96.06.01-1998 Performance Requirements for Chlorine Detection Instruments (0.5-30ppm Full Scale)

ISA-RP92.06.02-1999 Installation, Operation, and Maintenance of Chlorine Detection Instruments (0.5-30 ppm Full Scale) applies to all chlorine gas-detection instruments that satisfy the performance requirements in ISA-92.06.01.

18

Fiber Optics

Overview

Fiber optics offers many advantages and few disadvantages for use in the process control industry.

Optical fibers can reliably transmit high volumes of data over long distances without repeaters. Their unique combination of immunity, isolation, security, safety and environmental properties makes fiber optics exceptionally well suited for chemical and petroleum industries and other explosive industries. Fiber optic data highways and other communication networks are currently used in the process control and automation industries.

The telephone industry started using fiber optic communications links about 1872 because of its greater information carrying capacity. About 1980, a major instrumentation and control company used fiber optic cable as their data highway for their distributed control systems. Fiber optic switches and sensors are now used in process control applications.

Fiber optic communication is a technology in which light is transmitted along the inside of a thin, flexible glass or plastic fiber (Figure 18-1). An electrical signal is translated into a light signal, usually using light emitting diodes (LEDs) or laser diodes. This is generally called the emitter or source. The light signal may be turned on and off (modulated) at higher frequencies than electrical signals over conventional wires due to the greater bandwidth and lower attenuation. The optical fiber works on the principle of total internal reflection. Once light begins to reflect down the fiber, it will continue to do so. This light signal is received by a photodiode (detector or receiver), which converts the light signal into an electrical signal.

As the technology has evolved, other optical devices and components have been developed. Fiber optic systems are now used in a variety of industrial applications, such as data highways, sensors, and other signal lines.

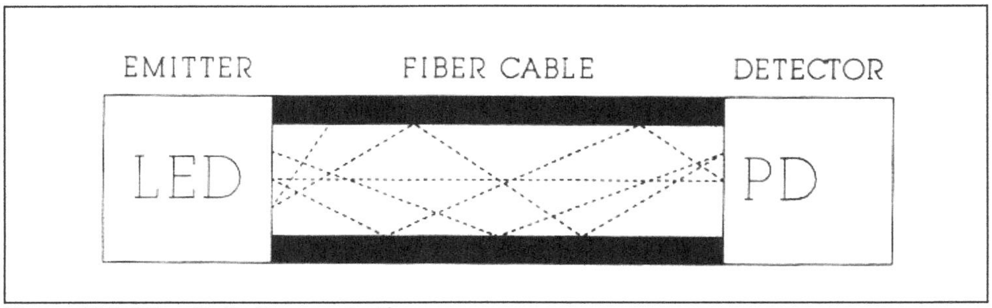

Figure 18-1. Fiber Optic Communication.

Construction

The fiber is constructed of two layers of glass or plastic, one layer surrounding the other. The inner layer, or core, has a higher refractive index than the outer layer, the cladding (Figure 18-2).

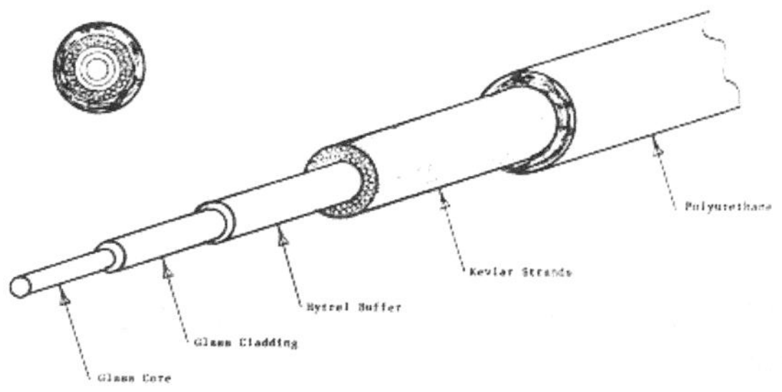

Figure 18-2. Fiber Optic Cable.

Classification

Fibers are classified by their refractive index profiles, core size, which is the basis of the number of modes, and material composition (Figure 18-3).

The index profile is the profile of the refractive index across the face of the fiber. There are two types of index profiles: Step and Graded.

The step has one change of the refractive index between core and cladding. Graded, on the other hand, has several different layers of different indexes covering the core.

Number of modes relates to how many light rays travel through the core at the same time. In some fibers many light rays travel through the core, each taking a different path. These are referred to as multimode fibers. If only one light ray moves through the core, these are referred to as single mode fibers. Notice the input pulse and output pulse waveforms from each classification.

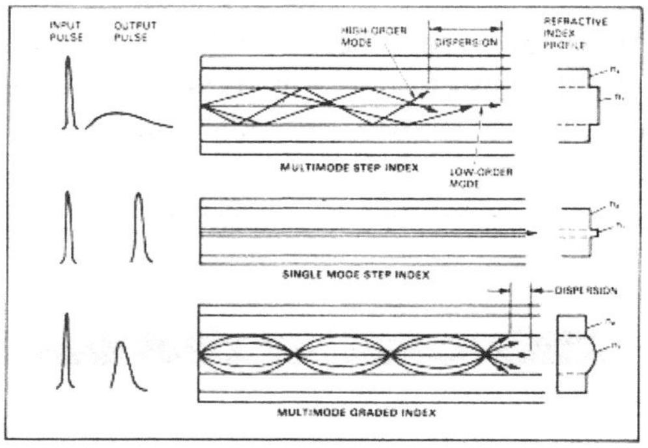

Figure 18-3. Types of Fibers.

Sensing Modes

Light waves coming from a fiber cable can be modulated internally or externally.

Intrinsic – Modulation of the light from within the fiber itself. Measured variable distorts the fiber core, causing light to be modulated. Intrinsic sensors use the chemistry of core glass to achieve changes in modulation. Some methods are absorption, scattering, fluorescence, changes in refractive index, or polarization.

Extrinsic – Modulation of the light from outside the fiber. Micro-bending, interruption or light path, and reflection of the light beams.

Advantages

Noise Immunity – Fiber optics are not affected by noise caused by electromagnetic interference (EMI), radio frequency Interference (RFI), electromagnetic pulses (EMP), lightning or cross-talk. Fiber optic cable can run next to high voltage lines and electrical machinery. No ground shielding is required; therefore all ground current and voltage problems are eliminated.

Safety – When using LEDs, light cannot initiate explosions in hazardous areas, shock people or start fires. Interferences will not cause unwanted pulses to transmit errors. Data communication is clean and safe. Isolation-optical isolation properties ensure ground-fault and fault current isolation.

Low Attenuation – Optical fibers offer lower power losses than do coaxial cable. More important, attenuation in a fiber does not increase with signaling frequency as it does with wires. An important advantage of fiber optics is that attenuation is constant at all modulation frequencies.

Security – It is virtually impossible to tap undetected into a fiber optic cable. Since fiber does not radiate energy, eavesdropping techniques are useless.

Small Size – Optical fibers are considerably smaller than copper cables. One fiber optic cable can transmit more information than many multiple wire cables. This one quality can save many dollars in installation costs.

Environment – Glass optical fiber is not affected by moisture, temperature, caustics, or acids, oxidation and corrosion. However, plastics used in some cores, cladding and jackets could be affected.

Accuracy – Because of noise immunity and low attenuation, signaling errors are low.

Long Life – With no electrical current the mechanical life of the system components becomes the useful life of the control system. Environment conditions affect the life of other mediums.

Fast Response – Transmission can be the speed of light. Temperature sensors are capable of responding to temperature changes up to 50,000 times/per second.

Disadvantages

Connections—Putting connectors on fiber optic cable requires time and expertise. With experience, a connector can be installed on a fiber optic cable in about the same time as a coaxial connector. Tapping a fiber optic line requires connectors, terminal expanders and couplers.

Bending – A sharp bend could break the glass fiber. A bend radius versus the pull strength should be a concern with fiber optics.

Unidirectional – Normally, data is transferred in only one direction due to the optical-electrical transmitter. Therefore, a return fiber optic cable is sometimes necessary.

Testing – Special attenuation test equipment is necessary. Continuity can be checked by moving a light past one end and seeing the light at the other end.

Analog Signals – Signals are on-off and normally cannot transmit voltage levels without analog to digital converters.

Experience/Education – Because fiber optics are fairly new, not many plant personnel have had experience installing fiber optic cables and sensors. There are very few educational and user publications written on fiber optic technology.

User Resistance – The main reason that fiber optic systems are not being used today is the resistance to change by design and operating people.

Standardization – There are no standards relative to core, cladding and connector sizes. With four multimode fibers in use today, problems exist when specifying cabling for generic use.

Hazards – There is little if any danger to the eye in any LED or laser-based fiber optic system. The bare fiber can be painful if stuck in the skin, and it is virtually impossible to find and remove. Various solvents and epoxies are used which require adequate ventilation.

Applications

Fiber optics can be used in a variety of process control applications.

Data Highways – The first use in process control applications. Used in distributed control systems between the operators console and the controller file and between controllers.

Signal Transmission between transmitter and receiver – Fiber optic transmission may be used to transmit the signal to or from the control system to the field devices.

Presence/Absence sensors – Photoelectric sensing detecting the presence or absence of an object. Generally used as counters or tachometers.

On-Off Switches – An optical switch used in place of an electrical switch can be used in many applications such as pressure, temperature, vibration or any other type of switch used with electrical switches.

Temperature Sensors – Several techniques are used to measure temperature. Bimetallic reflective, reflective, differential thermal expansion, active sensing materials placed in the optical path, and micro-bending.

Pressure Sensors – Several techniques are used to measure pressure. Displacement, transmissive, reflective, micro-bending, and active sensing materials placed in the optical path.

Level Sensors – High/low discrete level and continuous level monitoring. Pressure, reflective, refractive index, transmissive.

Refractive Index Sensors – Speed of light in a given medium. Used to sense levels by changes in the reflection due to different mediums.

Gap Sensor – Pulses of light in proportional to position.

Angular Position – Optical encoders can transmit light and receive light pulses proportional to angular displacement.

Digital Input Module – The digital input module shown in Figure 18-4 uses a LED to shine a light into a fiber optic cable; the switch interrupts the light received by a photo diode. On-off manual switches, temperature switches, pressure switches are some applications. The micro-switch in the pressure or temperature switch is replaced by an optical switch.

Digital Output Module – Figure 18-5 shows a digital output module to light a light or operate a relay. The DCS system turns on or off an LED and this light is transmitted down the fiber optic cable. If you use a digital output module to operate a relay. Additional power is normally needed. The photo diode receives a light or not light signal from the DCS. The photo diode turns on and off a relay through an amplifier.

Motor Control – Starting and stopping motors, pumps, fans, etc. from a hazardous area has been very expensive before fiber optic switches were available. Explosion proof switches and rigid conduit were required. Figure 18-6 shows a

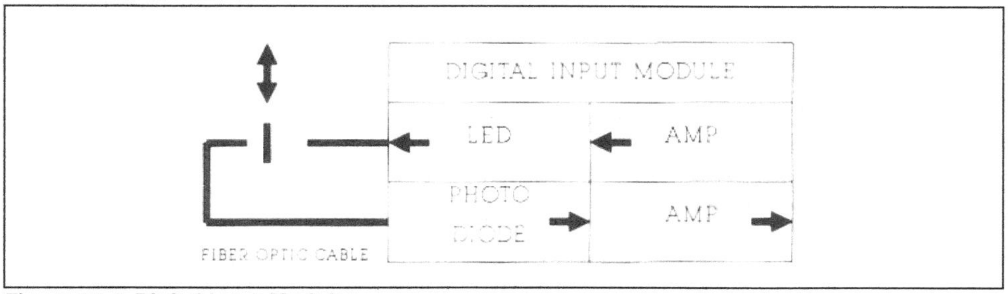

Figure 18-4. Digital Input Modules.

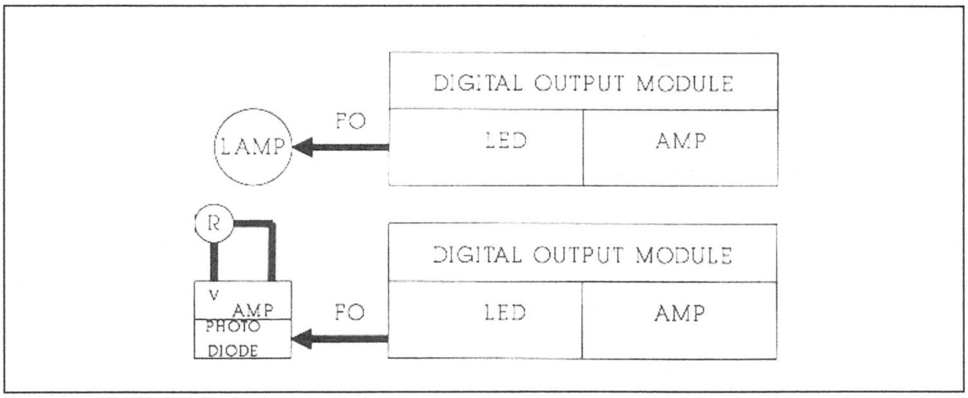

Figure 18-5. Digital Output Modules.

circuit to turn on or off a motor. Two fiber optic switches, one normally open and the other normally closed, are placed in the hazardous area and connected with fiber optic cable to an optical-to-electrical switch in a safe area.

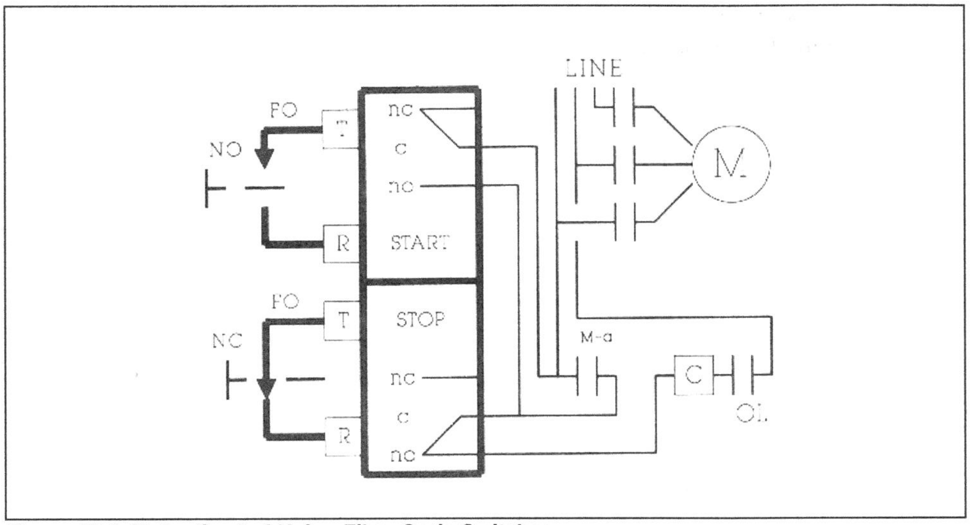

Figure 18-6. Motor Control Using Fiber Optic Switches.

Optical encoder – The standard optical encoders could not transmit signals more than 50 feet. Noise was a factor because low level voltage pulses were transmitted over copper wires at changing frequencies. Moving the LEDs and photo diodes into control I/O modules and transmit and receive light signals provides all the advantages of using fiber optic technology. See Figure 18-7.

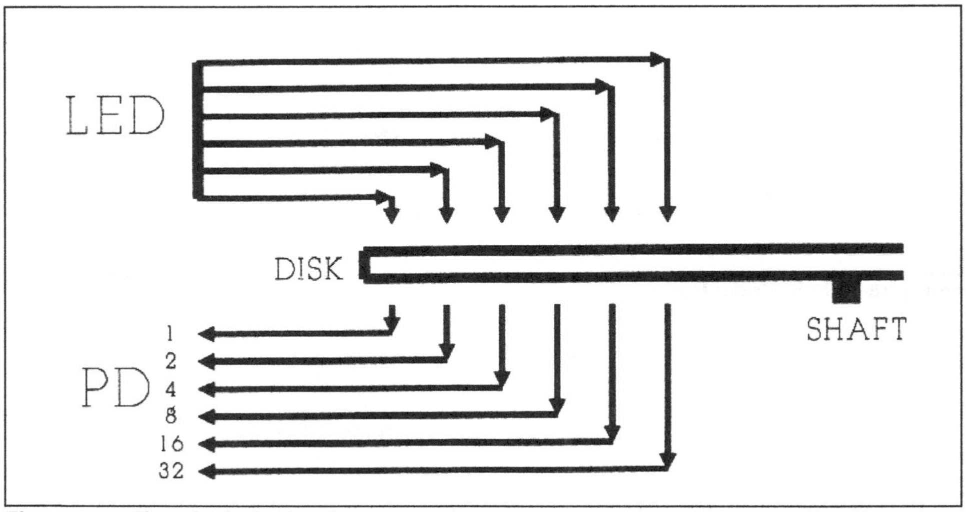

Figure 18-7. Fiber Optic Optical Encoder.

Analog Type Devices – To transmit analog signals from the field device to the control system input module, the analog signal is normally digitized and transmitted over the fiber optic cable, then demodulated and formed into an analog signal. For fiber optic sensor applications, intrinsic and extrinsic modulation and attenuation provides an analog signal that can be analog in nature. This distance form the sensor to the sensor receiver is generally quite close and the signal is changed to a digital signal for long distance transmission.

Analog Transmission – There are several ways that analog information can be transmitted over the fiber optic cables. Normally the analog signal is digitized and transmitted over the fiber optic cable, then demodulated and formed into an analog signal. Time proportioning and varying the frequency are digitizing methods. True analog light signals may be transmitted without digitizing.

Time Proportioning/Duty Cycle – Varying the on and off time can be an indication of an analog signal (See Figure 18-8).

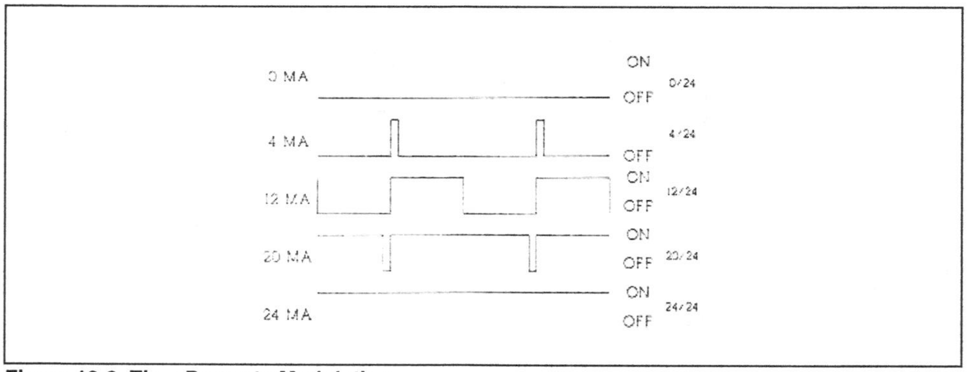

Figure 18-8. Time Property Modulation.

Frequency – Varying the frequency can be an indication of an analog signal (Figure 18-9).

- To indicate 4 mA we transmit light pulses at 4 KHz
- To indicate 12 mA we transmit light pulses at 12 KHz
- To indicate 20 mA we transmit light pulses at 20 KHz
- Less than 4 KHz and above 20 KHz indicates an error signal.

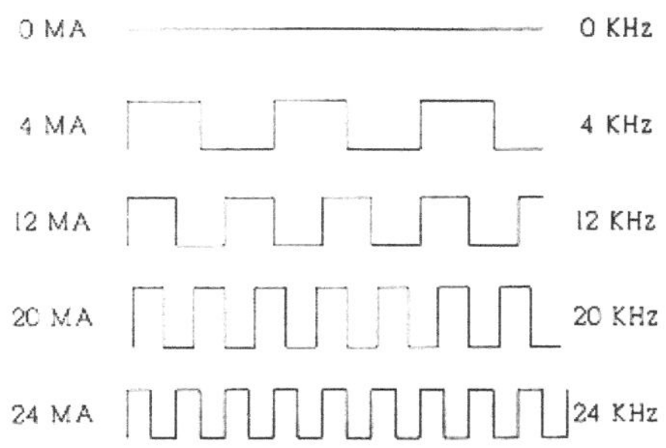

Figure 18-9. Frequency Modulation.

True Analog Light Signals – Light intensity or light phase shift can be transmitted short distances. Connections and distance attenuate the intensity of the light.

Analog Input/Output Modules

Figure 18-10 shows an analog input circuit, the signal from the transmitter modulates an LED. The light is transmitted over the fiber optic cable and received by a photo diode, it is amplified and demodulated and sent to the Distributed Control System as an analog input.

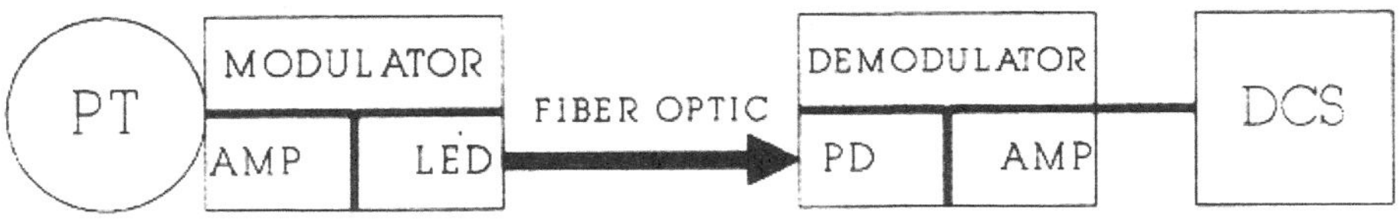

Figure 8-10. Analog Input to DCS Input Module.

In an analog output (Figure 18-11), the DCS system sends an analog signal to the modulator, the modulator digitizes the signal and the LED is modulated. This modulated light is received by the photo diode and demodulated. This signal operates the SCR or the I/P.

Multiplexed Signals – Multiplexers scan several inputs (analog and discrete, digitized and true analog intensities) transmits all these inputs over one fiber optic cable and is received by the receiver, which separates the signals. Some multiplexers use fiber optic sensors as input and others use conventional sensors.

Sensors

Fiber Optic sensors provide a means by which the measured variable can affect light traveling down an optical fiber.

Intensity Modulated Sensors – Sensors that detect the variation of the intensity of the light. Analog in nature, but have significant usage in digital (on-off) applications for switches and counters. Excellent sensitivity and are commonly used in industrial applications.

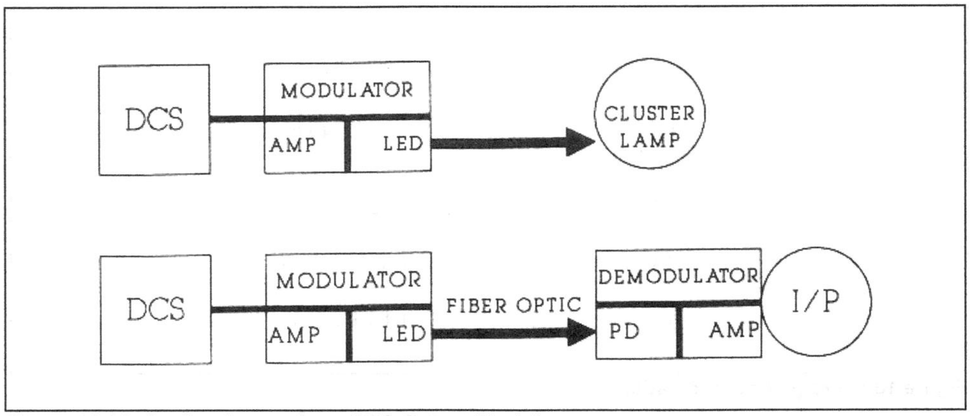

Figure 18-11. Analog Output from DCS Output Module.

- Transmission or Transmissive (Figure 18-12)
  a. Interruption of a light beam in a switch configuration
  b. Axial displacement
  c. Radial displacement
  d. Frustrated total reflection

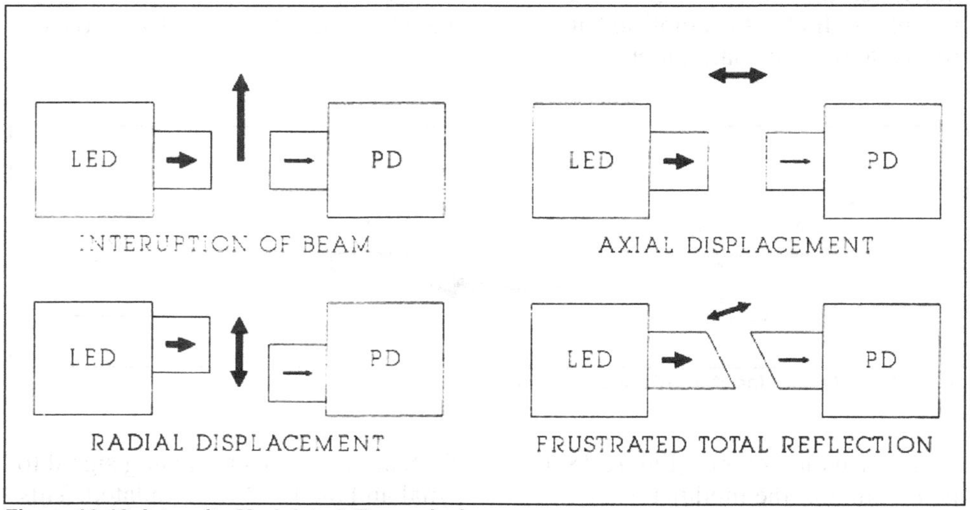

Figure 18-12. Intensity Modulated, Transmissive.

- Reflection (Figure 18-13)
  a. Used for broad sensor use due to accuracy, simplicity, and potential low cost.
  b. The sensor is comprised of two bundles of fiber or a pair of single fibers.
  c. One bundle of fibers transmits light to a reflecting target, the other bundle traps reflected light and transmits it to a detector.
  d. The intensity of the light depends on how far the reflecting target is from the fiber probe.
- Microbending Concept
  a. If a fiber is bent past the critical angle, small amounts of light are lost through the wall (into the cladding) of the fiber.

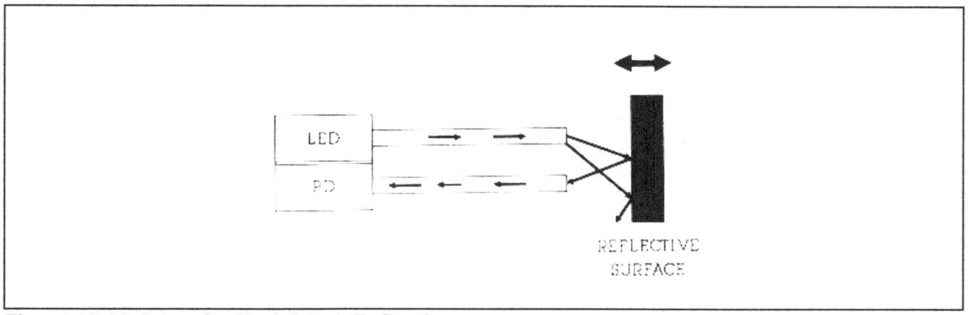

Figure 18-13. Intensity Modulated, Reflection.

b. If a transducer bend the fiber due to a change in some physical property as shown in Figure 18-14, then the amount of light is related to the value of the physical property.
c. Pressure caused the transducer to squeeze together and bend the fiber.
d. Low cost, accurate and immune to dirty environments.

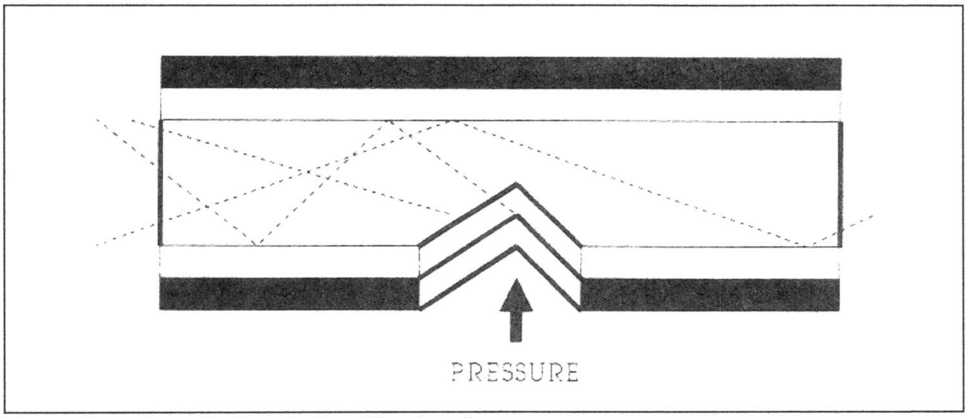

Figure 18-14. Intensity Modulating, Microbending.

Phase-Modulated Sensors – Phase modulated sensors detect the phase shift of the lightwave. The phase of the light wave represents the process variable value. Stretching one fiber of a pair and not the other will show a phase shift. Offers extreme sensitivity, needs laser light and temperature compensation. Seldom used in process control applications.

Color Change Sensors – Color change sensors, sometimes called spectral change, detect changes of the input light's spectrum. Light is sent to the sensor over a single glass fiber. The input light's spectrum is modified by the sensor in proportion to the parameter being sensed and reflected back to an instrument over the same fiber. These sensors are sometimes classified with the phase-modulated sensors because they basically use an interferometric cavity resonator "microshift" approach.

Intrinsic Concept – Intrinsic sensors change the intensity of the returning light form the sensor, but, unlike the transmissive, reflective and microbending concepts, no movement is required. Intrinsic sensors use the chemistry of the core glass to achieve the sensing activity. The prime mechanisms are:

a. Absorption—Doping the core glass (Nd doped). Used in temperature sensors 0-800 degree C range. Uses the ratio of two wave lengths.
b. Scattering—Same concepts as absorption

c. Fluorescence—Doping the glass with various additives. A light source can be used to stimulate fluorescence, which is affected by temperature or the fiber can be stimulated by outside radiation and fluorescence detected.
d. Refractive index—Varies the amount of received light by effectively changing the numerical aperture of the fiber. Many polymeric coating materials can be made to have index changes with temperature, thus providing a temperature sensor.
e. Polarization—Doping glass with earth oxides make fiber sensitive to magnetic fields.

Fiber optic communication, switches, analog instruments and sensors will take a greater share of the market because of all the advantages with only a few disadvantages. The biggest disadvantage is the resistance to change.

Appendix A

Glossary of Terms

The *Automation, Systems, and Instrumentation Dictionary, 4th Edition,* should be used for most definitions. Here are some definitions that pertain directly to the maintenance of instrumentation and control systems.

accessible area An area routinely or periodically entered by plant personnel in the performance of routine functions during normal plant operations and in accordance with applicable health physics procedures.

administrative controls These are controls placed on activities through the use of permits, standard operating procedures, standard maintenance procedures, supervision, etc., to ensure safe operation and maintenance of the facility as well as the normal operations.

analog control system Classically, a system that consists of electronic or pneumatic single-loop analog controllers, in which each loop is controlled by a single, manually adjusted device.

analytical Having a known, reliable, reproducible relationship between the component of interest and the equipment output signal.

analytical instrument An item of equipment that provides analytical data as its output. Liquid analysis and gas analysis are examples.

analyzer A specialized instrument that provides a known, reliable, reproducible relationship between a component of interest and the equipment's output signal.

annunciator A device or group of devices that call attention to changes in process conditions that have occurred. An annunciator usually points to abnormal process conditions but may be used also to show normal process status. Usually included are sequence logic circuits, labeled visual displays, audible devices, and manually operated push buttons.

ANSI Acronym for American National Standards Institute.

API Acronym for American Petroleum Institute.

Approved 1. Acceptable to the authority having jurisdiction. 2. Tested and certified by a national testing laboratory such as Underwriter Laboratories (UL) or Factory Mutual (FM), UL "lists" while FM "approves" equipment.

area classification Hazardous classifications for explosive gases, liquids, or dust. See *hazardous (classified) location.*

aspirator A device powered by pressurized gas or liquid that creates a positive pressure as its output. Typically used to provide flow to a sample, based on the venturi principle.

Authority Having Jurisdiction 1. Organization, office or individual responsible for approving equipment, an installation, or procedure (NFPA), 2. Acceptable to the Assistant Secretary of Labor for Occupational Safety and Health (OSHA).

AIT Autoignition temperature.

Autoignition Temperature The minimum temperature required to initiate or cause a self-sustaining combustion of a solid, liquid, or gas independently of the heating or heating element.

automate 1. To apply the principles of automation. 2. To operate or control by automation. 3. To install automatic procedures, as for manufacturing, servicing, etc.

automation 1. The implementation of processes by automatic means. 2. The theory, art, or technique of making a process more automatic. 3. The investigation, design, development, and application of methods of rendering processes automatic, self-moving, or self-controlling. 4. The conversion of a procedure, a process, or equipment to automatic operation. 5. Application of devices to a process or a machine to enable it to respond in a predetermined manner to a disturbance in the process or to a machine.

block diagram 1. Used to provide a simple pictorial representation of a control system. Block diagrams have two basic symbols: the circle and the function block. The arrows entering and leaving the circle represent the flow of information, and the head of each arrow has an algebraic sign associated with it, either plus or minus. 2. Block diagrams show the graphical representation of the hardware in a system. The primary purpose of a block diagram is to indicate the paths along which information or control flows between various parts of the system.

buffer A liquid used in field analytical instrument calibration that maintains the pH value stable even when subjected to dilution or contamination.

calibrate 1. To ascertain outputs of a device corresponding to a series of values of the quantity that the device is to measure, receive, or transmit. 2. Data so obtained are used to (a) determine the locations at which scale graduations are to be placed; (b) adjust the output, to bring it to the desired value, within a specified tolerance; and (c) ascertain the error by comparing the device output reading to a standard.

calibration 1. Determination of the experimental relationship between the quantity being measured and the output of the device that measures it, where the quantity measured is obtained through a recognized standard of measurement. 2. A test during which known values of measurement are applied to a device, and corresponding output readings are recorded under specified conditions. 3. The adjustment of a device or a series of devices in order to bring the output to a desired value, within a specified tolerance, for a particular value of input.

calibration cycle The frequency with which a device is due for calibration. This cycle could be dependent on calendar, cycles, or hours.

calibration record A record of the measured relationship of the transducer output to the applied measurand over the transducer range.

calibration report A table or graph of the measured relationship of an instrument as compared over its range to a standard.

calibration traceability The relationship of the calibration of an instrument through a step-by-step process to an instrument or group of instruments calibrated and certified by a national standardizing laboratory.

centralized To bring under one control.

centralized maintenance shop One maintenance shop that has responsibility to maintain all equipment in the facility. Usually, several crafts work out of this one centralized maintenance shop.

certify 1. To confirm formally as true, accurate, or genuine. 2. To guarantee as meeting a standard. 3. To issue a license or certificate.

certification 1. The act of certifying. 2. The state of being certified. 3. Certification of instrumentation and control technicians addresses qualifications based on education, experience, training, and job performance. 4. The attainment of certification is a means for individuals to indicate to the general public, co-workers, employers, and others that an impartial, nationally recognized organization has determined that they are qualified to perform specific technical tasks by virtue of their technical knowledge and experiences.

checkout Determination of the working condition of a system.

coalescing filter A filter that acts to bring small particles of liquid together as larger droplets, which can then be physically removed from the sample conditioning system.

computer control system A system in which all control action takes place within the control computer. Single or redundant computers may be used.

computer-integrated manufacturing (CIM) A central computer gathers all types of data, provides information stored in the database for decisions, and controls production input and output.

condensate The typical term for the product of a coalescing filter. Moisture in air lines caused by pressure and temperature changes.

configurable 1. A term applied to a device or system whose functional characteristics can be selected or rearranged through programming or other methods. The concept excludes rewiring as a means of altering the configuration. 2. A system feature that permits selection through entry of keyboard commands of the basic structure and characteristics of a device or system, such as control algorithms, display formats, or input/output terminations.

configuration 1. The arrangement of elements of a whole. 2. A low-level, fill-in-the-blank form of programming a process control device. 3. A particular selection of hardware devices or software routines and/or programs that function together. 4. A device or system whose functional characteristics can be selected or rearranged through programming or other methods is configurable.

conformance Action in accordance with a specified standard.

contamination The effect of undesirable components on the desired component (species of interest)

contract maintenance 1. Maintenance not normally done by plant personnel. 2. Specific maintenance performed by maintenance service organization from outside the plant.

control system A system in which deliberate guidance or manipulation is used to achieve a prescribed value of a variable.

Control Systems Engineer 1. One who designs, assembles, builds, and operates instrumentation and control systems. 2. Applies engineering standards and practices to the specification, sizing, and functional design of instrumentation hardware and control systems. Involves a clear understanding of the manufacturing or scientific process to be controlled. Serves as the key person on the instrumentation design and operation team, often supervising and reviewing the team's efforts. See *Instrument and Control Engineer.*

corrective maintenance 1. A repair activity that is not normal to the operation of the equipment, requires access to the interior, and is expected to be performed by qualified personnel who are aware of the hazards involved. 2. Such activities typically include locating causes of faulty performance, replacement of defective components, adjustment of service controls, or the like. 3. Maintenance is reduced or eliminated by analyzing prior maintenance work and making changes in design, installation materials, or procedures.

critical equipment Equipment failure that adversely affects safety, environment, system availability, yield, quality, or costs.

customer One who purchases or is furnished with a commodity or service.

decentralized Distribution of the functions among several authorities; decentralized maintenance distributes maintenance functions among areas of responsibilities or areas of the physical plant.

design equivalent replacement A replacement of material, component, consumable, or spare part that is similarly procured and is at least equivalent in performance to the item being replaced. It must also meet the requirements of form, function, and fit.

detector A device that converts the desired component (species of interest) into a know output signal.

digital computer A computing device that uses numerical digits to represent discretely all variables.

direct digital control (DDC) 1. A computer control technique that sets the final control element's position directly by the computer output. 2. A control system in which the computer carries out the functions normally performed by conventional controllers, for example, three-term control. 3. A term used to imply that a digital controller is connected directly to a final control element or actuator in a manufacturing process.

distributed control system (DCS) 1. A system that, while being functionally integrated, consists of subsystems that may be physically separated and remotely located from one another. 2. A system comprised of operator consoles, a communication system, and remote or local processor units that perform control, logic, calculations, and measurement functions. 3. Two meanings of distributed shall apply: (a) processors and consoles distributed physically in different areas of the plant or building and (b) data processing such as several processors running in parallel (concurrent), each with a different function. 4. The class of instrument (input/output devices, control devices, and operator interface devices) that, in addition to executing the stated control functions, permits transmission of control, measurement, and operating information to and from a single or a plurality or a user specifiable locations and is connected by a communication link. 5. A system of dividing plant or process control into several areas of responsibility, each managed by its own controller (processor), with the whole interconnected to form a single entity, usually by communication buses of various kinds.

dust-ignition proof A type of enclosure construction for Class II areas which means enclosed in such a manner that will exclude ignitable amounts of dusts or amounts that might affect performance or rating and that where installed properly will not permit arcs, sparks, or heat generated or liberated inside the enclosure to cause the ignition of exterior accumulations or airborne suspensions of a specified dust on or around the enclosure.

electromagnetic spectrum The total range of radiation present as energy ($E = mc^2$). This range is from relatively long wavelength (10 wavenumbers or 1×10^{-3} m or far IR) to very short wavelength (1 angstrom or 1×10^{-10} m or X-ray).

emission A term usually referring to the escape of an undesirable species (or components) from a process system. Usually has a negative connotation.

Engineering Controls Controls that are engineered into the installation to ensure safe operation and maintenance of the installation. These include interlocks, guards, signs, shutdowns, etc.

Engineering Technician One who applies principles, methods, and techniques especially prescribed by engineers and appropriate to a field of technology (instrumentation and control systems), combined with practical knowledge of the construction, application, properties, operation, and limitations of engineering systems, processes, structures, machinery, devices or materials, and (as required) related manual crafts, instrumental, mathematical, or graphic skills. Under profession direction, an engineering technician analyzes and solves technological problems, prepares formal reports on experiments, tests, and other projects, or carries out functions such as drafting, surveying, designing, technical sales, advising consumers, technical writing, teaching, or training.

experience Prior applicable work in design, construction, pre-operational and start-up testing activities, operation, maintenance, on-site activities, or technical services. Observation of others performing work in the above areas is not experience.

explosion-proof A type of enclosure construction for Class I areas capable of withstanding an explosion of the specified gases or vapors which may occur inside the enclosure and to prevent the ignition of the specified gases or vapors surrounding the enclosure by sparks, flashes, venting of gases, or the explosion within and which operates at an enclosure external temperature which will not ignite the specified gases or vapors surrounding the enclosure.

field analytical instrument See *analyzer*.

final control element 1. The element in a control system that directly changes the value of the manipulated variable. 2. The device that exerts a direct influence on the process.

function The purpose of, or an action performed by, a device.

functional specification A document that tells exactly what the system should do, what will be supplied to the system, and what is expected to come out of it.

gaseous sample A species in which the components of interest are not liquid, not solid, and can be compressed and expanded greatly upon heating.

hazardous (classified) location A location in which fire or explosion hazards may exist due to flammable gases or vapors, flammable liquids, combustible dust, or ignitable fibers or flyings.

HVAC Acronym for heating, ventilating, and air conditioning.

improvement maintenance Efforts to reduce or eliminate the need for maintenance. Reliability engineering efforts should emphasize elimination of failures that require maintenance. Includes modification, retrofit, redesign, or change-order.

inaccessible area An area considered inaccessible if entry may be dangerous without special controls.

in-house maintenance 1. Maintenance performed by plant maintenance personnel. 2. Not contract maintenance.

instrument A device used directly or indirectly to measure and/or control a variable. The term includes primary elements, final control elements, computing devices, and electrical devices such as annunciators, switches, and push buttons. The term does not apply to parts (e.g., a receiver, bellows, or resistor) that are internal components of an instrument.

instrument block valve The initial valve used to isolate the process material from a field instrument (field analytical) sample conditioning system.

Instrument Engineer 1. One who applies engineering standards and practices to the specification, sizing, and functional design of instrumentation hardware or control systems. Involves a clear understanding of the manufacturing or scientific process to be controlled. Serves as the key person on the instrumentation design and operation team, often supervising and reviewing the team's efforts. 2. Under supervision, participates in the design and planning of control and instrument systems as required by the project assignment, including: (a) collecting background information, (b) preparing drawings and calculations, (c) designing or modifying systems, (d) assisting in selection and procurement of equipment, (e) ensuring compliance with applicable standards and codes, (f) completing assigned tasks on schedule, (g) assisting technicians and designers as needed, and possible specialization in a specific engineering discipline. See *Instrumentation and Control Engineer.*

Instrument Engineering Technician One who helps engineers in the design of control and instrumentation systems by providing semi-professional technical assistance, including: (a) collecting background information, (b) performing calculations, (c) transmitting information to project team members, (d) preparing design specifications, (e) checking design documents to ensure compliance with applicable standards and codes, (f) preparing diagrams, (g) preparing requisitions, (h) executing necessary tests, collecting data, and making analyses, (i) maintaining equipment, (j) performing miscellaneous administrative work, and (k) assisting in testing, field start-ups, and training.

Instrument Field Engineer One who provides field engineering services at installation and start-up sites, ensuring specification fulfillment and operating ability. Also defines any potential problems that may arise, including: (a) managing installation and initial servicing, (b) maintaining own technical abilities and awareness of new methodologies, (c) checking out operation of panels and instruments, (d) providing instruction in the maintenance and repair of equipment to company service personnel, and (e) consulting on service problems in the field.

Instrument Field Service Representative One who fulfills customer service requirements on company-produced instruments and equipment, including: (a) analyzing and correcting instrument operating problems, (b) contacting supervisor or field engineer for assistance when necessary, (c) promoting scheduled maintenance agreements and sales of replacement or spare parts, (d) training new service personnel, (e) maintaining tools and equipment in good operating order, (f) reporting time, activities and expenses in accordance with company requirements, and (g) keeping up to date on new products and product applications.

instrument loop diagram Contains information useful to engineering, construction, commissioning and start-up, and maintenance work; extensions of P&IDs (see ISA-5.4); generally contains only one loop. General layout of a loop diagram is divided into sections for relative location of devices. Symbols used are from ISA-5.1, Instrument Symbols and Identification, and ISA-5.3, Graphic Symbols for Distributed Control/Shared Display Instrumentation. 2. A loop diagram contains the information needed to understand the operation of the loop and also shows all connections to facilitate instrument start-up and maintenance. The loop diagram must show the components and accessories of the instrument loop, highlighting special safety and other requirements.

Instrument Maintenance Technician One who assembles, installs, maintains, troubleshoots, and repairs various components of measurement and control systems. A high school diploma or the equivalent is required, but, with rapidly changing technology, vocational/technical school or community college training is strongly recommended. Most employers also provide on-the-job training.

Instrument Mechanic One who installs, calibrates, inspects, tests, and repairs instruments and control system devices. Mechanics must be able to work well with their hands, be willing to improve their skills through training, and be interested in learning new technologies.

Instrument Service Specialist One who maintains and/or troubleshoots control and instrumentation equipment and calibrates instrument hardware, including: (a) maintenance of thorough knowledge of equipment, (b) possible specialization in specific technology or complex equipment, (c) performance of analyses to ensure proper functioning of instruments, (d) surveys of available equipment, (e) installation and repair of equipment.

Instrument Supervisor Usually, an experienced instrument technician who supervises the work of a team of instrument specialists. In addition to mechanical ability, the supervisor must understand the entire system or process operations for which the team is responsible. Leadership ability is also important.

instrument specification A detailed and exact statement of particulars; especially, a statement that prescribes performance, dimensions, construction, tolerances, bill of materials, features, and operating conditions.

Instrument Technician 1. Usually requires certification or graduation from a technical college program. The technician works with theoretical or analytical problems, helping engineers find ways to improve the performance of an instrument or a system, as well as helping mechanics troubleshoot system components. 2. One who, in support of engineers or scientists, can carry out in a responsible manner either proven techniques, known to those who are technically expert in a particular technology (instrumentation and control systems), or those techniques especially prescribed by engineers and supervisors.

Instrument Technologist One who is qualified to practice engineering technology by reason of having the knowledge and the ability to apply well-established mathematical, physical science, and engineering principles and methods of technological problem solving that are acquired by engineering technology education and engineering technology experience; will usually have earned a baccalaureate degree in engineering technology or gained considerable technical experience on the job; is a member of the engineering team, which will normally include technicians and engineers and, for special projects, may include scientists, craftsmen, and other specialists; expected to have a thorough knowledge of the equipment, applications, and established state-of-the-art design and problem-solving methods in a particular field (instrumentation and control systems). The configuration of technical personnel possessing complementary capabilities that facilitate the engineering process is, by necessity, peculiar to each situation.

Instrumentation and Control Engineer One who applies engineering standards and practices to the specification, sizing, and functional design of instrumentation hardware or control systems. Involves a clear understanding of the manufacturing or scientific process to be controlled. Serves as the key person on the instrumentation design and operation team, often supervising and reviewing the team's efforts. 2. Under supervision, participates in the design and planning of instrument and control systems as required by the project assignment, including: (a) collecting background information, (b) preparing drawings and calculations, (c) designing or modifying systems, (d) assisting in selection and procurement of equipment, (e) ensuring compliance with applicable standards and codes, (f) completing assigned tasks on schedule, (g) assisting technicians and designers as needed, and (h) possible specializing in specific engineering discipline. See *Instrument Engineer.*

IR Acronym for infrared, a short wavelength (shorter than visible) portion of the electromagnetic spectrum. Units are usually cm^{-1} (wavenumbers).

I&C Technician See *Instrument Technician.*

Intrinsic Safe Electrical system designed such that under normal or abnormal conditions, sufficient energy cannot be released in the hazardous area so as to serve as an ignition source.

knowledge The fact of knowing, with familiarity with and detailed understanding of theory and concepts gained through experience or association.

ladder diagram Symbolic representation of a control scheme. The power lines form the two sides of a ladder-like structure, with the program elements arranged to form the rungs. The basic program elements are contacts and coils as in electromechanical logic systems.

liquid sample A species where the components of interest are not gaseous, are not solid, can be compressed only immeasurably, and expand only very slightly upon heating.

lockout The act of de-energizing equipment and circuits and ensuring that they remain de-energized. This normally includes the use of a locking device such as a padlock and a red tag indicating who, when, and why the equipment or circuits locked out.

logic diagram 1. In data processing, a diagram that represents a logic design and, sometimes, the hardware implementation. 2. Graphic method of representing a logic operation or set of operations.

loop 1. A sequence of instructions that is executed repeatedly until a terminal condition prevails. 2. A complete hydraulic, electric, magnetic, or pneumatic circuit. 3. In data processing, a closed sequence of instructions that are repeated.

maintenance management system (MMS) A part of the management information system (MIS) that is useful for maintaining the company's equipment. It accesses equipment information, spare parts availability and location, maintenance work order systems, preventive maintenance systems, maintenance personnel qualifications, equipment maintenance history, and any other information that will help the maintenance engineer, supervisors, technicians, or mechanics to be more proficient on the job.

maintain 1. To keep in continuance or in a certain state, as of repair. 2. To preserve or keep in a given existing condition, as of efficiency or repair.

maintainability 1. The relative ability of a device or system to remain in operation, requiring only scheduled routine maintenance, without extensive periods of downtime. 2. The inherent characteristic of a design or installation that determines the ease, economy, safety, and accuracy with which maintenance actions can be performed. 3. The ability to restore a product to service or to perform preventive maintenance within required limits.

maintenance Any activity intended to keep equipment in satisfactory working condition, including tests, measurements, replacements, adjustments, and repair. Routine operations required to keep the equipment functional.

maintenance conditions Conditions under which maintenance is performed.

Maintenance Engineer One who assists maintenance supervisors, technicians, and mechanics in maintaining the facilities/systems by providing in-depth engineering assistance on high-maintenance equipment; assists design and project engineers in plant start-up, retro-fits, and modifications to ensure that the design is engineered for maintenance; ensures standardization of equipment and systems; provides assistance in determining the frequency of preventive maintenance; determines service life of equipment by cost to repair and frequency of repair; ensures that the equipment continues to conform to the technical specifications; ensures that equipment is kept in calibration; works with management to ensure customer/production/maintenance cooperation by providing systems and equipment so that the end item meets all specifications and production schedules; ensures that new systems are designed to be maintainable; ensures proper documentation is provided; suggests and provides training when required to ensure that the maintenance personnel have the knowledge to maintain the equipment or system; helps determine if in-house maintenance or contract maintenance is more feasible; ensures operational maintenance is being performed; ensures maintenance mechanics, technicians, technologists, and engineers are involved in a project from the very beginning, looking at instrument/wiring/equipment accessibility and elevations, spare parts inventory, standardization, new equipment training, test equipment, and equipment calibration.

maintenance engineering Developing concepts, criteria, and technical requirements for maintenance during the conceptual and acquisition phases of a project; providing policy guidance for maintenance activities and exercising technical and management direction and review of maintenance programs.

management information system (MIS) A computerized system that uses a large database containing information on (a) customers, (b) equipment, (c) supplies, (d) spare parts, (e) personnel, (f) process, (g) sales forecast, (h) history, (i) costs, (j) profits, and so on. Selected information is available to those persons who make decisions.

Manager, Instrumentation and Control Systems One who is responsible for all departmental/sectional activities, including: (a) defining and reporting activities and needs to management; (b) maintaining liaisons with other departments/sections to coordinate work assignments; (c) providing administrative and technical support to other department/sections; (d) assigning of projects to personnel and manpower scheduling, recruitment, evaluation, and salary review; (e) managing development of training program; (f) making decisions on crucial or complex project activities; (g) supervising all departmental/sectional personnel; (h) preparing departmental/sectional budgets, forecasts and goals; (i) promoting safety.

means The method or approach.

multicraft Applied to maintenance personnel who are proficient in more than one craft, such as Instrument Technician, Electrician, and Instrument Mechanic and who are responsible for maintaining a variety of equipment used in control systems.

multiskilled Applied to maintenance personnel who are skilled in more than one craft. See *multicraft.*

NICET 1. Acronym for National Institute for Certification in Engineering Technologies. 2. NICET provides certification for "Industrial Instrumentation Engineering Technology Technician."

NIST Acronym for National Institute of Standards and Technology. Formerly NBS (National Bureau of Standards).

noncritical equipment Equipment failure that can be identified and repaired within a reasonable period of time so that adverse effects, if any, are within acceptable limits.

nonincendive Electrical system designed such that under normal conditions, sufficient energy cannot be released in the hazardous area so as to serve as an ignition source.

nonscheduled maintenance See unscheduled maintenance.

normal operational condition Equipment is in a normal operational condition when it conforms electrically and mechanically with its design specification and is used within the limits specified by the manufacturer.

OSHA Acronym for Occupational Safety and Health Administration.

OEM Acronym for original equipment manufacturer.

off-line diagnostics 1. Describing the state of a control system, subsystem, or piece of computer equipment that is operable but currently not actively monitoring or controlling the process. 2. A program to check out systems and subsystems, providing error codes if an error is detected. This diagnostic program is run while the system if off-line.

on-condition maintenance 1. Performed when equipment needs it. 2. Inspection of critical components, regarding safety as paramount, and repairing defects.

on-line diagnostics 1. Describing the state of a control system, subsystem, or piece of computer equipment that is operable and actively monitoring or controlling the process. 2. A program to check out systems and subsystems, providing error codes and alarms if errors are detected. This diagnostic program runs in the background while the control system is in the operating mode.

operating conditions Conditions to which a device is subjected during operation, not including the variable measured by the device.

operational Describing a state of readiness for immediate use, as may be said of equipment or systems.

operational maintenance Any maintenance activity, other than corrective maintenance, intended to be performed by the operator and required in order for the equipment to serve its intended purpose. Such activities typically include the correcting of "zero" on a panel instrument, changing charts, making records, adding ink, and the like.

operator The person who initiates and monitors the operation of a computer or process.

particulate Small solid species existing in gaseous or liquid sample streams.

personal computer (PC) Generally used at an office desk for word processing, databases, and spreadsheets. A personal computer may be used to aid in the configuration of programmable logic controllers (PLC) and distributed control systems (DCS) or may be used for data acquisition and control of small processes. Ruggedized PCs have been used on the process floor for control and data acquisition.

piping and instrumentation diagram (P&ID) Shows the interconnection of process equipment and the instrumentation used to control the process. In the process industry, a standard set of symbols is used to prepare drawings of processes. The instrument symbols used in these drawings are generally based on ISA-5.1. See *process and instrumentation diagram.*

planned maintenance Each equipment piece is identified and its maintenance defined.

ppB parts per billion; 1 part in 1,000,000,000 or 1×10^9 units.

ppM parts per million; 1 part in 1,000,000, or 1×10^6 units.

ppT parts per trillion; 1 part in 1,000,000,000,000 or 1×10^{12} units.

predictive control A type of automatic control in which the current state of a process is evaluated in terms of a model of the process and controller actions modified to anticipate and avoid undesired excursions. It is self-tuning and sometimes is accomplished using artificial intelligence.

predictive maintenance 1. A preventive maintenance program that anticipates failures that can be corrected before total failure. A variation from normal can indicate a system or equipment is approaching nonconformance. Vibration, eccentricity, and noise monitoring are measurements that can predict failure. Also, an increase in diagnostic errors and retries can indicate a failure about to happen. 2. Maintenance is performed when the need is indicated based on regular or continuous monitoring of the equipment.

preventive maintenance 1. Maintenance specifically intended to prevent faults from occurring during subsequent operation. (Contrast with corrective maintenance.) 2. Equipment maintenance is scheduled prior to failure on an estimate of the life of the equipment.

procedure A precise step-by-step method for effecting a solution to a problem.

process An assembly of equipment and material that relates to some manufacturing sequence.

process block valve See *instrument block valve.*

process and instrumentation diagram (P&ID) Shows the interconnection of process equipment and the instrumentation used to control the process. In the process industry, a standard set of symbols is used to prepare drawings of processes. The instrument symbols used in these drawings are generally based on ISA 5.1. See *piping and instrumentation diagram.*

process control 1. Descriptive of systems in which computers or controllers are used for automatic regulation of operations or processes. Typical are operations wherein the operation control is applied continuously, and adjustments to regulate the operation are directed by a computer to keep the value of a controlled variable constant. (Contrasted with numerical control.) 2. The regulation or manipulation of variables that influence the conduct of a process in such a way as to obtain a product of desired quality and quantity in an efficient manner. 3. An operation that regulates parameters by observation of the parameters, comparison with some desired value, and action to bring the parameters as close as possible to the desired value. 4. Adapting automatic regulatory procedures to the more efficient manufacture of products or the processing of material.

process control engineering A branch of engineering that deals with ways and means of keeping process variables as close as possible to the desired values or keeping them within specified ranges.

project engineering 1. Engineering activities associated with designing and constructing a manufacturing or processing facility. 2. Engineering activities related to a specific objective such as solving a problem or developing a product.

programmable logic controller (PLC) 1. A controller, usually with multiple inputs and outputs, that contains an alterable program. 2. A solid-state control system that has a user-programmable memory for storage of instructions to implement specific functions such as I/O control, logic, timing, counting, three-mode (PID) control, communication, arithmetic, and data and file manipulation. 3. Consisting of a central processor, input/output interface, and memory, a PLC is designed as an industrial control system. 4. A PLC may form part of a distributed control system (DCS).

purge 1. To ventilate an enclosure, 2. A common term to refer to using pressurization and ventilation to reduce the area classification of enclosure or room.

qualified Competent, suited, capable, or having met the requirements for a specific position or task.

reliability The probability that a component, piece of equipment, or system will perform its intended function for a specified period of time, usually operating for hours, without requiring corrective maintenance.

remedial maintenance The maintenance performed following equipment failure, as required, on an unscheduled basis. (Contrasted with preventive maintenance.)

repair 1. The act of restoring an item to serviceable condition following a failure or malfunction. 2. Narrowly, restoring an item to serviceable condition but not exactly to original design specifications.

repairs maintenance Equipment fails and the required maintenance is performed. See *corrective maintenance.*

resolution Separation for purposes of analysis.

retrofit 1. A word derived from "retroactive," which describes modification of a piece of equipment to incorporate design changes made in later models of the same equipment. 2. Modification and upgrading of older control systems. 3. Parts, assembly, or kit that will replace similar components originally installed on equipment. Retrofits are generally performed on a machine to correct a deficiency or to improve performance.

rework Restoring an item to a condition exactly conforming to original design specifications; usually applied to corrective action taken when an item has failed an inspection but requires a relatively simple operation, such as replacing a part or component, to enable the item to pass an identical inspection.

safe area 1. Nonhazardous area. 2. An area in which explosive gas/air mixtures are not expected to be present so that special precautions for the construction and use of electrical apparatus are not required.

SAMA Acronym for Scientific Apparatus Makers Association, which no longer supports its standards nor writes standards.

sample A part of a whole containing the component(s) of interest.

sample conditioning system A system of hardware items that act together as a whole to prepare process material (sample) for introduction into an analyzer.

sample probe A simple or complex item which removes sample from process and delivers it to the sample transfer line or the sample conditioning system. (sample probe diagram goes here.)

sample transfer line A means to get sample from the sample probe to the sample conditioning system. This can be simple tubing, pipe, or complex heated/refrigerated/insulated transfer line.

scheduled maintenance Maintenance carried out in accordance with an established plan; related to preventive maintenance. Scheduled maintenance may be scheduled on hours used, sequences, or the calendar.

service Operations requested to either optimize the equipment's operations or to return the failed equipment to operational status.

service life The length of time a mechanism or piece of equipment can be used before it becomes either unreliable or economically impractical to maintain in good working order.

set point 1. A predetermined level at which a bi-stable device changes state to indicate that the quantity under surveillance has reached the selected value. 2. A desired value of process variable.

setup 1. An arrangement of data or devices to solve a particular problem. 2. Preliminary operations—such as control adjustments, installation of tooling, or filling of process fluid reservoirs—that prepare a manufacturing facility or piece of equipment to perform specific work.

SMI Standard maintenance instruction.

SMP Standard maintenance procedure.

SOP Standard operating procedures.

specification 1. A list of requirements that must be met when making material, part, component or assembly; installing the system; or testing its attributes or functions. 2. A set of standard requirements applicable to any product or process within the jurisdiction of a given standards-making organization; an industry consensus standard.

spectral quality The resolution of the spectral data.

spectrometer An analytical device that identifies a particular energy region as its analytical measurement principle.

spectrophotometer An analytical device that identifies light energy as its analytical measurement principle.

standard 1. An acknowledged measure of comparison for quantitative or qualitative value. 2. Commonly used and accepted by an authority. 3. A degree or level of requirement, excellence, or attainment.

supervisory control and data acquisition (SCADA) A particular field of equipment control and data handling.

supervisory control system (SCS) Provides remote set point information to single-loop analog controllers by a digital computer.

symbol Graphic representations used on drawings to indicate devices, instruments, types of communication lines, connection points, valves, actuators, primary elements, and others.

system 1. An assembly of procedures, processes, methods, routines, or techniques united by some form of regulated interaction to form an organized whole. 2. The complex of hardware and software utilized to effect the control of a process. 3. An assemblage of equipment, machines, control devices, or a combination thereof, interconnected mechanically, hydraulically, pneumatically, or electrically, and intended to act together to perform a predetermined function. 4. Hardware that is connected to operate as a unit.

system concept The viewpoint that all components that make up a system (functional entity) interact with each other to produce a desired result.

technical specification A description of the technical characteristics of an item or system in sufficient detail to form the basis for design, development, production, and, in some cases, operation.

technician 1. Specialist in a technical process. 2. A person whose occupation requires training in a specific technical process.

test A standard procedure for determining an attribute or performance characteristic of a material, part, component, assembly or system; a test may be used to determine basic properties, verify a function or condition, establish a response characteristic or calibration, or provide information about operating behavior.

troubleshoot To search for the cause of a malfunction or erroneous problem behavior in order to remove the malfunction.

Type "X" Purge A purge that reduces the area classification in the purged enclosure from Division 1 to non-hazardous.

Type "Y" Purge A purge that reduces the area classification in the purged enclosure from Division 1 to Division 2.

Type "Z" Purge A purge that reduces the area classification in the purged enclosure from Division 2 to non-hazardous.

unscheduled maintenance An urgent need for repair or upkeep that was unpredicted or not previously planned and must be added to or substituted for previously planned work.

UV Acronym for ultraviolet, a long wavelength (longer than visible) portion of the electromagnetic spectrum. Units are usually nanometers (nm).

wavelength The distance from one wave's intersection with the x-axis to the next wave's intersection with the x-axis. Usually measured in nanometers or decimal fractions of a centimeter. Usually expressed by the lowercase Greek alphabet letter lambda (λ).

wavenumber The reciprocal of wavelength (1/cm or cm^{-1}). Usually measured in whole numbers of magnitude 400–13,000.

workaround An action required to complete the process run, even though all equipment is not working satisfactory. One may have to run part of the process on manual, or one might jump out an interlock until the maintenance can be scheduled.

Appendix B
Bibliography

Ambrose, Jon. "Connected Maintenance," Proceedings of *AUTOWEST,* 2004.

ANSI/ISA-7.0.01. *Quality Standard for Instrument Air.* American National Standards Institute/ISA, The Instrumentation, Systems, and Automation Society, 1996.

ANSI/ISA-RP12.06.01. *Recommended Practice for Wiring Methods for Hazardous (Classified) Locations Instrumentation Part 1: Intrinsic Safety.* American National Standards Institute/ISA, The Instrumentation, Systems, and Automation Society, 2003.

ANSI/ISA-51.1. *Process Instrumentation Terminology.* American National Standards Institute/ISA, The Instrumentation, Systems, and Automation Society, 1979 (R1993).

API RP 540. *Electrical Installations in Petroleum Processing Plants* - Fourth Edition. American Petroleum Institute, 1999.

ASME INTERNATIONAL BOILER AND PRESSURE VESSEL CODE - 2004 Edition. American Society of Mechanical Engineers, 2004.

Barron, Randall F. *Cryogenic Systems.* Second Edition. Monographs on cryogenics / Oxford science publications. Clarenden Press, 1985.

Bell, J. H. *Cryogenic Engineering.* Prentice Hall, 1963.

Beranek, Leo L., editor. *Noise and Vibration Control Engineering: Principles and Applications.* Wiley-Interscience, 1992.

Bever, Ken. "Utilizing Advanced Asset Management Technologies To Improve Productivity and Lower Maintenance Costs," Proceedings of *ISA 2002,* 2002.

Bock, Eileen, "Advanced Options for Asset Management Software Enhance Maintenance Practices Further than Ever," Proceedings of *ISA EXPO*, 2003.

Brown, G.R. "Replacing an Aging Distribution System," Paper #92-0379. *Advances in Instrumentation and Control, Vol.47- Part 2.* ISA, The Instrumentation, Systems, and Automation Society, 1992.

BSR/API RP-500. *Recommended Practice for_Classification of Locations for Electrical Installations at Petroleum Facilities Classified as Class I, Division 1, or Division 2* - Second Edition. American Petroleum Institute, 1997.

Callahan, F.J. *Swagelok® Tube Fitting and Instrument Manual.* Swagelock Co., 1985.

Carroll, Grady C. *Industrial Instrument Servicing Handbook.* McGraw-Hill, 1960.

CGA P-12. *Safe Handling of Cryogenic Liquids.* Compressed Gas Association, 1993.

Code of Federal Regulations, Title 49 (Transportation). US Government Printing Office.

DeLisio, Ralph. "Redefining Maintenance in Today's Changing Manufacturing Environment," Proceedings of *ISA EXPO*, 2003

Fribance, Austin E. *Industrial Instrumentation Fundamentals.* McGraw-Hill, 1962.

Gillum, Donald R. *Industrial Level Measurement.* ISA, the Instrumentation, Systems, and Automation Society, 1984.

Gillum, Donald R. *Industrial Pressure, Level, and Density Measurement.* ISA, the Instrumentation, Systems, and Automation Society, 1995.

Gillum, Donald R. *Industrial Pressure Measurement.* ISA, the Instrumentation, Systems, and Automation Society, 1982.

Goettsche, Lawrence D., "Maintenance in the Changing World," *Advances in Instrumentation and Control.* Vol. 45 - Part 4. ISA, the Instrumentation, Systems, and Automation Society, 1990.

Gotti, Franco. "e-Maintenance: Closing the Gap Between Automation and Information Technology," Proceedings of *IMS 2002*, 2002.

Haselden, GG. *Cryogenic Fundamentals.* Academic Press (London), 1971.

Hasselbaum, Robert. "21st Century Maintenance Technology to Meet Critical Industrial Requirements," *ISA 2001, Technology Update*, Vol. LVI - Part 1 (ISA Vol. 413). ISA, the Instrumentation, Systems, and Automation Society, 2001.

IEEE 488.1. *IEEE Standard for Higher Performance Protocol for the Standard Digital Interface for Programmable Instrumentation.* Institute of Electrical and Electronics Engineers, 2003.

ISA-5.1. *Instrumentation Symbols and Identification.* ISA, 1984 (R1992).

ISA-12.04.01(IEC 60079-2 Mod). *Electrical Apparatus for Explosive Gas Atmospheres - Part 2 Pressurized Enclosures.* ISA, the Society for Instrumentation, Systems, and Automation, 2003.

ISA-RP12.4. *Pressurized Enclosures.* ISA, 1996.

ISA-12.10. *Area Classification in Hazardous (Classified) Dust Locations.* ISA, 1988.

ISA. *The Automation, Systems, and Instrumentation Dictionary, 4th Edition.* ISA, The Instrumentation, Systems, and Automation Society, 2003.

ISA Standards Library CD, with Updates. ISA, The Instrumentation, Systems, and Automation Society.

Jenkins, Neil and Stanley Schatt. *Understanding Local Area Networks.* Fifth Edition. Sams Pub., 1995.

Jordan, O.C. and Colclazier, J.D. "Parts Are Not Just Parts for DCS Systems," *Advances in Instrumentation and Control,* Vol. 47, Part 2. ISA, the Instrumentation, Systems, and Automation Society, 1992.

Kallen, Howard P. *Handbook of Instrumentation and Controls*, McGraw-Hill, 1961.

Klerer, Melvin. *User-Oriented Computer Languages: Analysis and Design.* Macmillan, 1988.

Lukas, Michael P. *Distributed Control Systems: Their Evaluation and Design.* Van Nostrand Reinhold, 1986.

Magison, Ernest C. *Electrical Instruments in Hazardous Locations.* Fourth Edition. ISA, 1998.

Magison, Ernest C. *Intrinsic Safety*. ISA, 1984.

Magison, Ernest C. *Temperature Measurement in Industry*. ISA, 1990.

Maguire, Brian. "Doing More With Less: Effectively managing maintenance work with technology that supports an underlying business process, " Proceedings of *AUTOWEST*, 2004.

Mardiguian, Michel, *Grounding and Bonding*. Interference Control Technologies, Inc., 1988.

McCabe, Warren L., Julian C. Smith, and Peter Harriott. *Unit operations of chemical engineering.* Seventh Edition. McGraw-Hill chemical engineering series. McGraw-Hill, 2004.

McMillan, Gregory K., editor-in-chief, and Douglas M Considine, late editor-in-chief. *Process/Industrial Instruments and Controls Handbook*. Fifth Edition. McGraw-Hill, 1999.

Morrison, Ralph. *Grounding and Shielding Techniques.* Fourth Edition. Wiley-Interscience, 1998.

Morrison, Ralph. *Grounding and Shielding Techniques in Instrum*entation. Third Edition. Wiley-Interscience, 1986.

Morrison, Ralph and Warren. H. Lewis. *Grounding and Shielding Techniques in Facilities.* Wiley-Interscience, 1990.

NACE MR0175/ISO 15156. *Petroleum and natural gas industries Materials for use in H_2S-containing Environments in oil and gas production - Part 1: General principles for selection of cracking-resistant materials - Part 2: Cracking-resistant carbon and low alloy steels, and the use of cast irons - Part 3: Cracking-resistant CRAs (corrosion-resistant alloys) and other alloys.* NACE International, 2003.

NEMA ICS 1.1. *Safety Guidelines for the Application, Installation, and Maintenance of Solid State Control R(2003)* National Electrical Manufacturers Association, 2003.

NFPA 70. *National Electric Code*. National Fire Protection Association, 2005.

NFPA 496. *Standard for Purged and Pressurized Enclosures for Electrical Equipment*. National Fire Protection Association, 2003.

NFPA 497. *Classification of Flammable Liquids, Gases, or Vapors and of Hazardous (Classified) Locations for Electrical Installations in Chemical Process Areas*. National Fire Protection Association, 2004.

NFPA 499. *Classification of Combustible Dusts and of Hazardous (Classified) Locations for Electrical Installations in Chemical Process Areas*. National Fire Protection Association, 2004.

NIST Handbook Number 44. *Specifications, Tolerances and Other Technical Requirements for Weighing and Measuring Devices: As Adopted by the 88th National Conference on Weights and Measures*. Government Printing Office, 2003.

O'Connor, Patrick D.T., David Newton, and Richard Bromley. *Practical Reliability Engineering*. Fourth Edition. Wiley, 2002.

O'Reilly, Tom. "Collaborative Maintenance Strategies Help Manufacturers Become Lean, Mean and Agile," Proceedings of *AUTOWEST, 2004.*

Patton, Jr. Joseph D., *Service Parts Management.* ISA, 1984.

Patton, Jr. Joseph D., *Maintainability and Maintenance Management*. Third Edition. ISA, 1994.

Patton, Jr. Joseph D. and Herbert C. Feldmann. *Service Parts Handbook.* Solomon Press, 1997.

Patton, Jr. Joseph D., *Preventive Maintenance*. Third Edition. ISA, The Instrumentation, Systems, and Automation Society, 2004.

Petefish, Michael D. "Make Employee Training Your Competitive Advantage, Part 1," *InTech*, April 1991, p.63.

Peterson, Arnold P.G. *Handbook of Noise Measurement*. GenRad, 1980.

Sloan, Martha E. *Computer Hardware and Organization*. Second Edition. Science Research Associates, Inc., 1983.

Spitzer, David W. *Flow Measurement*. Second Edition. ISA, The Instrumentation, Systems, and Automation Society, 2001.

Thompson, Lawrence M. *Electrical Measurement and Calibration: Fundamentals and Applications.* Second Edition. ISA, 1994.

Thompson, Lawrence M. *Electronic Controllers.* Third Edition. ISA, 1989.

Thompson, Lawrence M. *Industrial Data Communications* Third Edition. ISA, the Society for Instrumentation, Systems, and Automation, 2002.

UL 913. Standard for Safety for Intrinsically Safe Apparatus and Associated Apparatus for Use in Class I, II, III, Division 1, Hazardous (Classified) Locations. Underwriters Laboratories, 2003.

Wang, Jim. "A Better Maintenance Decision Using the Internet," *48th International Instrumentation Symposium*, 2002.

Index

Printed in the USA/Agawam, MA
September 17, 2026

910661.231